Lecture Notes in Economics and Mathematical Systems 487

Springer
Berlin
Heidelberg
New York
Barcelona
Hong Kong
London
Milan
Paris
Singapore
Tokyo

Yacov Y. Haimes   Ralph E. Steuer (Eds.)

Research and Practice in Multiple Criteria Decision Making

Proceedings of the XIVth International Conference on Multiple Criteria Decision Making (MCDM) Charlottesville, Virginia, USA, June 8–12, 1998

Editors
Prof. Yacov Y. Haimes
Director of the Center for Risk Management
of Engineering Systems
103 Albert Small Building
University of Virginia
Charlottesville, VA 22903, USA

Prof. Ralph E. Steuer
Terry College of Business
University of Georgia
Athens, GA 30602-6253, USA

Library of Congress Cataloging-in-Publication Data

International Conference on Multiple Criteria Decision Making (14th : 1998 :
Charlottesville, VA)
Research and practice in multiple criteria decision making : proceedings of the XIV-th
International Conference on Multiple Criteria Decision Making (MCDM),
Charlottesville, Virginia, USA, June 8-12, 1998 / Yakov Y. Haimes, Ralph E. Steuer, eds.
p. cm. -- (Lecture notes in economics and mathematical systems ; 487)
Includes bibliographical references.
ISBN 3540672664 (softcover : alk. paper)
1. Decision making--Congresses. 2. Multiple criteria decision making--Congresses. 3.
Risk assessment--Congresses. 4. Risk management--Congresses. I. Haimes, Yacov Y. II.
Steuer, Ralph E. III. Title. IV. Series.

HD30.23 .I57 1998
658.4'03--dc21

00-026592

ISSN 0075-8442
ISBN 3-540-67266-4 Springer-Verlag Berlin Heidelberg New York

Springer-Verlag is a company in the BertelsmannSpringer publishing group.

Printed in Germany

Typesetting: Camera ready by author
Printed on acid-free paper SPIN: 10734520 42/3143/du-543210

Preface

During the past two decades, the consideration of multiple objectives in modeling and decision making has grown by leaps and bounds. The nineties in particular have seen the emphasis shift from the dominance of single-objective modeling and optimization toward an emphasis on multiple objectives. The proceedings of this Conference epitomize these evolutionary changes and contribute to the important role that the field of multiple criteria decision making (MCDM) now plays in planning, design, operational, management, and policy decisions. Of special interest are the contributions of MCDM to manufacturing engineering. For example, it has recently been recognized that optimal, single-objective solutions have often been pursued at the expense of the much broader applicability of designs and solutions that satisfy multiple objectives. In particular, the theme (MCDM and Its Worldwide Role in Risk-Based Decision Making) of the XIVth International Conference on Multiple Criteria Decision Making (Charlottesville, Virginia, USA, June 8-12, 1998) represents the growing importance of risk-cost-benefit analysis in decision making and in engineering design and manufacturing. In such systems, minimizing the risk of rare and extreme events emerges as an essential objective that complements the minimization of the traditional expected value of risk, along with the objectives attached to cost and performance. These proceedings include forty-five papers that were presented at the Conference.

A variety of techniques have been proposed for solving multiple criteria decision-making problems. The emphasis and style of the different techniques largely reflect the fields of expertise of their developers. Social scientists and economists, for example, tend to look at an MCDM problem from a "human factor" viewpoint and concentrate mainly on the role of the decision maker (DM). Here all major activities revolve around the DM's subjective value judgments and the task of translating those judgments into some form of preference function (e.g., multiattribute utility function or indifference function). On the other hand, those who are well-equipped with mathematical tools and more familiar with physical systems (e.g., in engineering design and manufacturing) tend to look at an MCDM problem from a physical (structural) viewpoint and try to attack it with mathematical rigor. While both groups have their merits, the former is often criticized for being less focused on physical elements whereas the latter is sometimes criticized for being too involved with only the physical aspects. While the latter group seemingly chooses to ignore other significant elements of the problem, it is mainly because such elements tend to defy quantification and hence rigorous mathematical treatment.

The XIVth International Conference was aimed at bridging the gap between the two groups. The conference program, themes, and sessions were organized around the fundamental premise that MCDM has emerged as a philosophy that integrates common sense with empirical, quantitative, normative, descriptive and value-judgment-based analyses. It is a philosophy supported by advanced systems concepts (e.g., data management procedures, modeling methodologies, optimization and simulation techniques, and decision-making approaches) that are grounded in both the arts and the sciences for the ultimate purpose of improving the decision-making process.

Furthermore, risk assessment should be an integral part of the multiple criteria modeling effort, and risk management should be an imperative part of the MCDM process -- not an after-the-fact vacuous exercise. The planning, design, and operations of most, if not all, publicly and privately owned systems are characterized by elements of risk and uncertainty -- the theme of this Conference and Workshop. Risk assessment is defined here as a process that encompasses all of the following four elements or steps: risk identification, risk quantification, risk evaluation, and risk acceptance. Risk management is defined as the formulation of policies and the development of risk-control options (i.e., measures to reduce or prevent risk). The obvious and inevitable overlapping of risk assessment and risk management has led many to consider the latter as a part of the former.

The Conference served as an excellent medium for the exchange of knowledge among scientists and engineers from the U.S. and other countries who are interested in the planning, design, and operations of technology and non-technology-based systems in which decisions are often made under conditions of risk and uncertainty.

For their leadership and guidance in carrying out and implementing the XIVth International Conference on Multiple Criteria Decision Making, we would like to recognize the members of the Local Organizing Committee

Yacov Y. Haimes (Chair), University of Virginia
Donald E. Brown, University of Virginia
Jared L. Cohon, Carnegie-Mellon University
Lorraine R. Gardiner, Auburn University
Ambrose Goicoechea, Lockheed-Martin Corporation
Keith W. Hipel, University of Waterloo (Canada)
Roman Krzysztofowicz, University of Virginia
James H. Lambert, University of Virginia
Elliot R. Lieberman, U.S. Environmental Protection Agency
Craig Piercy, Towson University
Andrew P. Sage, George Mason University
Richard M. Soland, George Washington University
Ralph E. Steuer, University of Georgia
Po-Lung Yu, University of Kansas
Stanley Zionts, State University of New York at Buffalo

and the members of the Executive Committee of the International Society on Multiple Criteria Decision Making

Pekka J. Korhonen, President of Society, Helsinki School of Economics (Finland)
Carlos A. Bana e Costa, IST/SAEG (Portugal)
Valerie Belton, President-Elect of Society, University of Strathclyde (Scotland)
Denis Bouyssou, Group ESSEC (France)
Rafael Caballero, University of Malaga (Spain)
Lorraine R. Gardiner, Auburn University (USA)
Yacov Y. Haimes, Meeting Ex-Officio, University of Virginia (USA)
Murat Köksalan, Future Meeting Ex-Officio, Middle East Technical University (Turkey)
Oleg I. Larichev, Institute for Systems Analysis (Russia)
Kaisa M. Miettinen, Secretary of Society, University of Jyväskylä (Finland)
Hirotaka Nakayama, Konan University (Japan)
Richard M. Soland, Vice President of Finance of Society, George Washington University (USA)

Theodor J. Stewart, Past Meeting Ex-Officio, University of Cape Town (South Africa)
Ralph E. Steuer, Immediate Past President of Society, University of Georgia (USA)
Jyrki Wallenius, Helsinki School of Economics (Finland)
Stanley Zionts, Past President of Society and Chairman of the Awards Committee, State University of New York at Buffalo (USA).

Also, we would like to thank Virginia A. McCowan, Office Manager of the Center for Risk Management of Engineering Systems, University of Virginia for all her work in assembling this volume.

Yacov Y. Haimes
Ralph E. Steuer

August 9, 1999
Charlottesville, Virginia

MCDM and Its Worldwide Role in Risk-Based Decision Making: Opening Remarks to the XIVth International Conference on MCDM

Yacov Y. Haimes
University of Virginia

I am pleased with the response of the MCDM community to the XIVth International Conference on Multiple Criteria Decision Making and to its theme: MCDM and Its Worldwide Role in Risk-Based Decision Making. During the last two decades, MCDM has emerged as a philosophy, theory, and methodology that integrates common sense with empirical, quantitative, normative, descriptive, and value-judgment-based analysis. The sessions of this Conference reflect this evolving nature of MCDM-- a field supported by the cognitive, social, and behavioral sciences, and advanced by holistic systems concepts such as data management procedures, modeling methodologies, optimization and simulation techniques, and decision-making approaches. These are grounded on both the arts and the sciences for the ultimate purpose of improving the decision-making process. Although the Conference received over 160 abstracts of papers submitted by MCDM experts from over 40 countries, not all presenters were able to attend. Many of these papers address the elements of risk and uncertainty that are inherent in the theory, methodology, and practice of MCDM.

Public interest in the field of risk analysis has expanded by leaps and bounds during the last decade. Furthermore, during the last two decades, risk analysis has emerged as an effective and comprehensive procedure that supplements and complements the management of almost all aspects of our lives. Managers of health care, the environment, and physical infrastructure systems such as water resources, transportation, and electric power -- to cite a few -- all incorporate risk analysis in their decision-making processes. The omnipresent adaptation of risk analysis by many disciplines and its deployment by industry and government agencies in decision making have led to an unprecedented development of theory, methodology, and practical tools. As a member of eight diverse professional societies, I find technical articles on risk analysis published in all of their journals. These articles address concepts, tools, and technologies that have been developed and practiced in such areas as design, development, system integration, prototyping, and construction of physical infrastructure; in reliability, quality control, and maintenance; and in the estimation of cost and schedule and in project management.

Risk managers at all levels of decision making must have the ability, and sometimes the courage, to make the necessary trade-offs among different kinds of risks and their associated costs and benefits. Furthermore, the multifarious characteristics of risks of all kinds and the ultimate need to determine the acceptability of the various risk levels and their associated costs and benefits, are at the heart of the MCDM field. Risk assessment should be an integral part of the multiple-objective modeling effort, and risk management should be an integral part of the multiple-objective decision-making process -- not an after-the-fact vacuous exercise. Risk assessment is defined here as a process that encompasses the following four elements or steps: risk identification, risk quantification, risk evaluation, and risk acceptance. Risk management is defined as the formulation of policies and the development of risk-control options (i.e., measures to reduce or prevent risk). Because of obvious and inevitable overlapping, many people consider risk assessment part of risk management.

There are at least three challenges facing theoreticians and practitioners in the MCDM community whose interests lie within this year's theme. The first is to identify all important sources of risk and uncertainty, to understand their interconnectedness and dependency on one another, and to distinguish among the multiple factors that characterize the nature and extent of what we generally term "risk" and "reliability." In a distributed system, for example, an initial failure of one or two components commonly may trigger and cause the failure of the system as a whole. Most available models assume independence in the failure of these components; fault-tree analysis is a classic example. Yet, it is clear that the failure of a subsystem or a system component may impose an unacceptable load on the remaining part of the system that ultimately may lead to overall failure.

The second challenge is to quantify the appropriate costs, benefits, and risks and to determine the extent of such quantification, which in turn influences the applicability of appropriate MCDM methodologies. Indeed, a systemic evaluation of these non-commensurate objectives and their associated trade-offs, the subsequent formulation of meaningful and responsive policy options, and the ultimate decisions made by the appropriate managers and decision makers, are all very much dependent on the successful quantification of the risks, costs, and benefits.

The third challenge is to be cognizant of and accept the following premise: To the extent that risk assessment and management is precise, it is not real. To the extent that risk assessment and management is real, it is not precise.

My best wishes to all of you -- contributors, participants, and guests. I hope you find the XIVth International Conference on Multiple Criteria Decision Making productive and beneficial, and your visit to Charlottesville rewarding and enjoyable.

Table of Contents

Leadoff Papers

Part I: MCDM Theory

Part II: MCDM Methodologies

Part III: MCDM in Applications

The Even- Swap Method for Multiple Objective Decisions

by

John S. Hammond, Ralph L. Keeney, and Howard Raiffa

Abstract

Many important decision problems require value tradeoffs among conflicting objectives. Typically, some alternatives are better on some objectives and some alternatives are better on other alternatives. In these situations, the even-swap method allows you to explicitly focus on the relevant value tradeoffs to make a smart choice. The method guides you to make value tradeoffs that allow you to sequentially eliminate objectives by rendering them irrelevant to the choice and then eliminate dominated alternatives. The procedure repeats until one dominant alternative – the smart choice – is left.

The even-swap method, which is a generalization of Benjamin Franklin's procedure of comparing two alternatives by weighing the pros and cons, has several desirable features. It allows one to address easier value tradeoffs first and perhaps avoid the very difficult ones, it logically addresses how much of each objective is being traded off, and it requires only the minimum number of value tradeoffs to select the smart choice. Perhaps more important, the even-swap method is intuitive because the tradeoffs are made in terms of the consequences relevant to the decision rather than requiring any mathematical calculations.

Introduction

When your decision has only one objective, it is often straightforward. If you want to fly from New York to San Francisco as cheaply as possible, for example, you'd simply find the airline offering the lowest fare and buy a ticket. But having only one objective is a rare luxury. Usually, you're pursuing many different objectives simultaneously. Yes, you want a low fare, but you also want a convenient departure time, a direct flight, and an airline with an outstanding safety record. And you'd also like to have an aisle seat and to earn frequent flyer miles in one of your existing accounts.

Now, the decision is considerably more complicated. Since you can't simultaneously fulfill all your objectives, you're forced to seek a balance among them. You have to make tradeoffs.

Important decisions usually have conflicting objectives--you can't have your cake and eat it, too--and therefore you have to make tradeoffs. You need to give up something on one objective to achieve more in terms of another.

In the early 1980s, for example, the United States enacted a national speed limit of 55 miles per hour to reduce gasoline consumption. The limit also led to a reduction in highway fatalities. Ten years later, however, a fresh debate broke out over the limit. Proponents pointed to the thousands of lives that had been saved. Opponents argued that with the oil crisis long past and today's cars more fuel-efficient, the national limit should be raised to allow drivers to get to their destinations more quickly. Some participants in the debate held that states should be free to set their own speed limits.

Each of these viewpoints stresses a different objective: lives saved, convenience, and states' rights. Finding an appropriate balance among them is difficult, but not trying to balance them misses the point. Suppose we all agreed that the 55 mile-per-hour limit was justified by the number of lives saved. Inevitably, a proposal for a 45 mile-per-hour limit, clearly preferable given an exclusive focus on saving lives, would quickly follow. Why not 35 miles per hour, then, or 20? Each reduction in the speed limit would, after all, save many additional lives. At some point, however, other objectives would come into play. The vast majority of people would not accept a speed limit of 20 miles per hour. They would, in fact, object strenuously, using such reasons as convenience or states' rights, or both. There's the rub. *Decisions with multiple objectives cannot be resolved by focusing on any one objective.*

Eliminating Dominated Alternatives

Making wise tradeoffs is one of the most important and most difficult challenges in decision making. The more alternatives you're considering and the more objectives you're pursuing, the more tradeoffs you'll need to make. Hence, if possible, it is helpful to rule out some of your remaining alternatives before having to make tough tradeoffs. The fewer the alternatives, the fewer the tradeoffs you'll need to make and the easier your decision will be. To identify alternatives that can be eliminated, follow this simple rule: If alternative A is better than alternative B on some objectives and no worse than B on all other objectives, B can be eliminated from consideration. In such cases, B is said to be *dominated* by A—it has disadvantages without any advantages.

Say you would like to take a relaxing weekend getaway. You have five places in mind and three objectives: low cost, good weather, and short travel time. In looking at your options, you notice that alternative C costs more, has worse weather, and requires the same travel time as alternative D. Alternative C is dominated and can therefore be eliminated.

You need not be rigid in thinking about dominance. In making further comparisons among your options, you may find, for example, that alternative E also has higher costs and worse weather than alternative D but has a slight advantage in travel time—it would take a half hour less to get to E. You may easily conclude that the relatively small time advantage doesn't outweigh the weather and cost disadvantages. For practical purposes, alternative E is dominated by D—we call this "practical dominance"—and you can eliminate it as well. By looking for dominance, you have just made your decision much simpler—you only have to choose among three alternatives, not five.

A consequences table can be a great aid in identifying dominated alternatives because it provides a framework that facilitates comparisons. To construct a consequences table. use pencil and paper or a computer spreadsheet. List your objectives down the left side of a page and your alternatives along the top. This will give you an empty matrix. In each box of the matrix, write a concise description of the consequence that the given alternative (indicated by the column) will have for the given objective (indicated by the row). You'll likely describe some consequences quantitatively, using numbers, while expressing others in qualitative terms, using words. The important thing is to use consistent terminology in describing all the consequences for a given objective—in other words, use consistent terms across each row

A consequences table puts a lot of information into a concise and orderly format that allows you to easily compare your alternatives, objective by objective. It gives you a clear framework for making comparisons and, if necessary, tradeoffs. Moreover, it imposes an important discipline, forcing you to bring together all your thinking about your alternatives, your objectives, and your consequences into a single, concise framework. Although this kind of table is not too hard to create, we're always surprised at how rarely decision makers take the time to put down on paper all the elements of a complex decision. Without a consequences table, important information can be overlooked and comparisons can be made haphazardly, leading to wrong-headed decisions.

.Now, compare pairs of alternatives, and eliminate any that are dominated. If only one alternative remains, your choice is now obvious. However, when this occurs, you should try to create some new nondominated alternatives. In any case, if you still have more than one alternative in contention, you're going to have to make tradeoffs, a task that is greatly facilitated by a consequences table.

Making Tradeoffs Using Even Swaps

At this point, it will be useful to recall what American sage Ben Franklin had to say about decision tradeoffs. More than two hundred years ago, Franklin's friend Joseph Priestly, a noted scientist, faced a tough decision and he wrote to Franklin to ask which of two alternatives he should choose. Franklin recognized that the choice would depend on Priestly's objectives and on his evaluation of the two alternatives with respect to those objectives. Rather than suggest a specific choice, therefore, Franklin outlined a reasonable *process* to help Priestly choose (Franklin, 1956). Here is Franklin's letter, sent from London on September 19, 1772.

Dear Sir,

In the affair of so much importance to you, wherein you ask my advice, I cannot, for want of sufficient premises advise you what to determine, but if you please I will tell you how.

When those difficult cases occur, they are difficult, chiefly because while we have them under consideration, all the reasons pro and con are not present to the mind at the same time; but sometimes some set present themselves, and at other times another, the first being out of sight. Hence the various purposes or inclinations that alternately prevail, and the uncertainty that perplexes us.

To get over this, my way is to divide half a sheet of paper by a line into two columns; writing over the one pro, and over the other con. Then during three or four days consideration, I put down under the different heads short hints of the different motives, that at different times occur to me, for or against the measure.

When I have thus got them all together in one view, I endeavor to estimate their respective weights; and where I find two, one on each side, that seem equal, I strike them both out. If I find a reason pro equal to two reasons con, I strike out the three. If I judge some two reasons con, equal to some three reasons pro, I strike out the five; and thus proceeding I find at length where the balance lies; and if, after a day or two of further consideration, nothing new that is of importance occurs on either side, I come to a determination accordingly.

And, though the weight of reasons cannot be taken with the precision of algebraic quantities, yet when each is thus considered, separately and comparatively, and the whole lies before me, I think I can judge better, and am less liable to make a rash step, and in fact I have found great advantage from this kind of equation, in what may be called moral or prudential algebra.

Wishing sincerely that you may determine for the best, I am ever, my dear friend, yours most affectionately.

B. Franklin

Ben Franklin proposed a wonderful way to simplify a complex problem. Each time he eliminated an item from his list of pros and cons, he replaced his original problem with an equivalent but simpler one. Ultimately, by honing his list, he revealed a clear choice. Although Franklin did not explicitly use a list of objectives, his caution in advising his friend "for want of sufficient premises," together with his focused approach to his lists of pros and cons, shows that he relied on them implicitly.

In this paper, we extend Franklin's ideas about a "moral or prudential algebra" to a choice among any number of alternatives, not just two. Using a consequences table, we show how to make tough tradeoffs and use them to replace a complex decision problem with a simpler one, just as Franklin did. We call this technique the *even swap method.* First we'll describe how the even swap method works, illustrating the process using a simple example with only two alternatives and two objectives, and later we'll apply it to a more complex situation with many objectives and alternatives

The Essence of the Even Swap Method

What do we mean by even swaps? To explain the concept, we need to first state an obvious but fundamental tenet of decision making: If all alternatives are rated equally for a given objective—for example, all cost the same—then you can ignore that objective in choosing among those alternatives. If all airlines charge the same fare for the New York-San Francisco flight, then cost doesn't matter. You decision will hinge only on the remaining objectives.

The even swap method provides a way to adjust the consequences of different alternatives in order to render them equivalent and thus irrelevant. As its name implies, an even swap increases the value of an alternative in terms of one objective while decreasing its value *by an equivalent amount* in terms of another objective. In essence, the even swap method is a form of bartering—it forces you to think about the value of one objective in terms of another. If, for example, American Airlines charged $100 more for a New York-San Francisco flight than did Continental, you might swap a $100 reduction in the American fare for 2000 fewer American frequent flyer miles. In other words, you'd "pay" 2000 frequent flyer miles for the fare cut. Now, American would score the same as Continental on the cost objective, so cost would have no bearing in deciding between them. Whereas the assessment of dominance enables you to eliminate alternatives, the even swap method allows you to eliminate objectives. As more objectives are eliminated, fewer comparisons need to be made and the decision becomes easier.

The Even Swap Method in Action

Let's apply the even swap method to a fairly simple problem to illustrate how it works. Imagine you're running a Brazilian cola company and a number of other companies have expressed interest in buying franchises to bottle and sell your product. Your company currently has a 20 percent share of its market, and it earned $20 million in the fiscal year that's just ended. You have two key objectives for the coming year: increasing profits and expanding market share. You estimate that franchising would reduce your profits to $10 million, due to startup costs, but it would increase your share to 26 percent. If you don't franchise, your profits would rise to $25 million, but your share would increase to only 21 percent. You put this all down in a consequences table (Table 1).

Table 1. Consequences Table for a Cola Company's Possible Marketing Strategies

	Alternatives	
Objectives	Franchising	Not Franchising
Profit (in millions)	$10	$25
Market Share	26%	21%

Which is the smart choice? As the table indicates, the decision boils down to whether the additional $15 million profit from not franchising is worth more or less than the additional 5 percent market share from franchising. To resolve this question, you can apply the even swap method following a straightforward process.

First, determine the change necessary to cancel out an objective. If you could cancel out the $15 million profit advantage of not franchising, the decision would depend only on market share.

Second, assess what change in another objective would compensate for the needed change. You must determine what increase in market share would compensate for the profit decrease of $15 million. After a careful analysis of the long-term benefits of increased share, you settle on a 3 percent increase.

Third, make the even swap. Reduce the profit of the not-franchising alternative by $15 million while increasing its market share by 3 percent, to 24 percent. In combination, the restated consequences (a $10 million profit and a 24 percent market share) shown in Table 2, are equivalent in value to the original consequences (a $25 million profit and a 21 percent market share).

Table 2. A Cola Company's Even Swap

	Alternatives	
Objectives	Franchising	Not Franchising
Profit (in millions)	$10	~~$25~~ $10
Market Share	26%	~~21%~~ 24%

Fourth, cancel out the now-irrelevant objective. Now that the profits for the two alternatives are equivalent, profit can be eliminated as a consideration in the decision. It all boils down to market share.

Finally, select the dominant alternative. The new decision, while equivalent to the original one, is now easy. The franchising alternative, better on market share, is the obvious choice.

For the cola company, only one even swap revealed the superior alternative. Usually, it takes more—often many more. The beauty of the even swap approach is that no matter how many alternatives and objectives you're weighing, you can methodically reduce the number of objectives you need to consider until a clear choice emerges. The method, in other words, is iterative. You keep switching between making even swaps (to eliminate objectives) and identifying dominance (to eliminate alternatives) until only one alternative remains.

Simplifying a Complex Decision with Even Swaps

Now that we've discussed each step of the process, let's apply the whole thing to a more complex problem. Alan Miller is a computer scientist who started a technical consulting practice three years ago. For the first year, he worked out of his home, but with his business growing he decided to sign a two-year lease on some space in the Pierpoint Office Park. Now that lease is about to expire. He needs to decide whether to renew it or move to a new location.

After considerable thought about his business and its prospects, Alan defines five fundamental objectives for an office: short commuting time, good access to his clients, good office services (clerical assistance, copy machines, faxes, mail service), sufficient space, and low costs. He surveys more than a dozen possible locations and, dismissing those that clearly fall short of his needs, he settles on five viable alternatives: Parkway, Lombard, Baranov, Montana, and his current building, the Pierpoint.

He then develops a consequences table (Table 3), laying out the consequences of each alternative for each objective. He uses a different measurement system for each objective. He describes commuting time as the average time needed to travel to work in minutes during rush hour. To measure access to clients, he determines the percentage of his clients whose business is within an hour's lunch-time drive of the office. He uses a simple three-point scale to describe the office services provided: "A" means full service, including copy and fax machines, telephone answering, and for-fee secretarial assistance; "B" indicates fax machines and telephone answering only; and "C" means that no services are available. Office size is measured in square feet, and cost is measured by monthly rent.

Table 3. Office Selection Consequences Table

	Alternatives				
Objectives	Parkway	Lombard	Baranov	Montana	Pierpoint
Alan's Commute (min.)	45	25	20	25	30
Client Access (%)	50	80	70	85	75
Office Services (constructed scale)	A	B	C	A	C
Office size (sq. ft.)	800	700	500	950	700
Monthly Cost (dollars)	1,850	1,700	1,500	1,900	1,750

With so many alternatives to compare, Alan immediately seeks to eliminate some by using dominance or practical dominance. To make this easier, he uses the descriptions in the consequences table to create a ranking table (Table 4).

Table 4. Ranking Alternatives on Each Objective for Office Selection

	Alternatives				
Objectives	Parkway	Lombard	Baranov	Montana	Pierpoint
Alan's Commute (min.)	5	2(tie)	1	2(tie)	4
Client Access (%)	5	2	4	1	3
Office Services (constructed scale)	1(tie)	3	4(tie)	1(tie)	4(tie)
Office size (sq. ft.)	2	3(tie)	5	1	3(tie)
Monthly Cost (dollars)	4	2	1	5	3

Scanning the columns, he quickly sees that the Lombard office dominates the current Pierpoint site, outranking it on four objectives and tying it on the fifth (office size). He eliminates Pierpoint from further consideration. He also sees that Montana almost dominates Parkway, falling behind in cost only. Can he eliminate Parkway, too? He flips back to his original consequences table and notices that for the small cost disadvantage of Montana—only $50 per month—he would gain an additional 150 square feet, a much shorter commute, and much better client access. He eliminates Parkway using practical dominance.

Alan has reduced his choice to three alternatives—Lombard, Baranov, and Montana—none of which dominates any other. He redraws his consequences table (Table 5a).

To further clarify his choice, Alan needs to make a series of even swaps. In scanning the table, he sees considerable similarity among the commuting times for the three remaining alternatives. If the Baranov's twenty-minute commute were increased to twenty-five minutes using an even swap, all three alternatives would have an equivalent commute time and that objective could then be dropped from further consideration. Alan decides that this five-minute increase in Baranov's commute time can be compensated for by an 8 percent increase in Baranov's customer access, from 70 to 78 percent. He makes the swap, rendering commute time irrelevant in his deliberations (Table 5b). Alan then checks this table for dominated alternatives, but finds none.

Alan then eliminates the office services objective by making two even

Table 5. Making a Series of Even Swaps to Select the Right Office

	a. Starting Table			**b. Eliminate Commute**		
	Alternatives			**Alternatives**		
Objectives	**Lombard**	**Baranov**	**Montana**	**Lombard**	**Baranov**	**Montana**
Alan's Commute (min.)	25	20	25	~~25~~	~~20~~ ~~25~~	~~25~~
Client Access (%)	80	70	85	80	~~70~~ 78	85
Office Services (constructed)	B	C	A	B	C	A
Office size (sq. ft.)	700	500	950	700	500	950
Monthly Cost (dollars)	1700	1500	1900	1700	1500	1900

swaps with monthly cost. Using the Lombard service level (B) as a standard, he equates an increase in service level from C to B for Baranov with a $250 increase in monthly costs. He also equates a decrease in service level from A to B for Montana with a savings of $100 per month (Table 5c).

Each time Alan makes an even swap, he changes the way the alternatives match up. With the office services objective eliminated, he finds that the Baranov alternative is now dominated by the Lombard alternative and can be eliminated. This highlights an important process consideration. In making even swaps, you should always seek to create dominance where it didn't exist before, thus enabling you to eliminate an alternative. In your decision process, you will want to keep switching back and forth between examining your columns (alternatives) and your rows (objectives), between assessing dominance and making even swaps.

With Baranov out of the picture, only Lombard and Montana remain. They have equivalent scores in commuting time and services,

Table 5 (continued).

	c. Eliminate Office Services and Baranov			**d. Eliminate Office Size; Select Montana**	
	Alternatives			**Alternatives**	
Objectives	**Lombard**	**Baranov**	**Montana**	**Lombard**	**Montana**
Alan's Commute (min.)	~~25~~	~~25~~	~~25~~	~~25~~	~~25~~
Client Access (%)	80	78	85	80	85
Office Services (constructed)	~~B~~	~~C B~~	~~A B~~	~~B~~	~~B~~
Office size (sq. ft.)	700	500	950	~~700 950~~	~~950~~
Monthly Cost (dollars)	1700	~~1500~~ 1750	~~1900~~ 1800	~~1700~~ 1950	1900

leaving only three objectives to consider. Alan next makes an even swap between office and monthly cost. Deciding that the 700-square-foot Lombard office will be cramped, he equates Montana's additional 250 square feet with a substantial cost increase--$250 per month. This swap cancels the office-size objective, revealing Montana to be the clearly preferable alternative, with advantages in both of the remaining objectives, cost and customer access. Montana now dominates Lombard (Table 5d).

Alan signs the lease for space at Montana, confident that he has thought through the decision carefully, considered every alternative and objective, and made the smart choice.

Practical Advice for Making Even Swaps

Once you get the hang of it, the mechanical part of the even swap method becomes easy, almost a game. Determining the relative value of different consequences—the essence of any tradeoff process—is the hard part. By design, the even swap method allows you to concentrate on the value determinations one at a time, giving each careful thought. While there's no easy recipe for deciding how much of one consequence to swap for some amount of another—every swap requires subjective judgment—you can help ensure your tradeoffs are sound by keeping the following suggestions in mind.

Make the easier swaps first. Determining the value of some swaps will be more difficult for you then determining the value of others. In choosing among airlines, for example, you may be able to calculate, in fairly precise terms, the monetary value of a certain number of frequent flyer miles. After all, you know how many miles it would take to earn a free flight and what a flight would cost. Swapping between fares and miles will therefore be a straightforward process. On the other hand, swapping between airline safety records and flight departure times will be much less clear-cut. In this case, you should make the fare-mile swap—the easier swap—first. Often, you will be able to reach a decision (or at least eliminate a number of alternatives) by just making the easier swaps, saving you from having to wrestle with the harder ones at all.

Concentrate on the amount of the swap, not on the perceived importance of the objective. It doesn't make sense to say that one objective is more important than another without considering the actual degree of variation among the consequences for the alternatives under consideration. Is salary more important than vacation? It depends. If the salaries of all the alternative jobs are similar but their vacation times vary widely, then the vacation objective may be more important than the salary objective.

Concentrating on an objective's perceived importance can get in the way of making wise tradeoffs. Consider the debate that might go in a town trying to decide whether public library hours should be cut to save money. The library advocate declares, "Preserving current library hours is much more important than cutting costs!" The fiscal watchdog counters, "No, we absolutely have to cut our budget deficit! Saving money is more important." Were the two sides to focus on the actual amounts of time and money in question, they might find it easy to reach agreement. If cutting branch hours by just two hours one morning a week saves $250,000 annually, the library advocate might agree that the harm to the library would be small compared to the amount saved, especially considering other possible uses for the money. If, instead, the savings were a mere $25,000 annually, even the fiscal watchdog might agree that harm to the library wouldn't be worth the savings. The point is this: When you make even swaps, concentrate not on the importance of the objectives but on the importance of the amounts in question.

Remember that the value of an incremental change depends on what you start with. When you swap a piece of a larger whole—for example, a portion of an office's overall square footage—you need to think of its value in terms of the whole. For example, adding 300 square feet to a 700 square-foot office may make the difference between being cramped and being comfortable, whereas adding 300 square feet to a spacious 1,000 square-foot-office may not be nearly as valuable to you. The value of the 300 square feet, like the value of anything being swapped, is relative to what you start with. It's not enough to look just at the size of the slice; you also need to look at the size of the pie.

Make consistent swaps. While the value of what you swap will be relative, the swaps themselves should be logically consistent. If you would swap A for B and B for C, you should be willing to swap A for C. Let's say you're in charge of an environmental protection program charged with preserving wilderness quality and expanding salmon spawning habitats for as low a cost as possible. In a cost-benefit analysis, you might calculate that one square mile of wilderness and two miles of river spawning habitat both have values equivalent to $400,000. In making your swaps, you should therefore equate one square mile of wilderness with two miles of river spawning habitat. From time to time, check your swaps for consistency.

Seek out solid information. Swaps among consequences require subjective judgments, but these judgments can be buttressed by solid information and analysis. In making your environmental tradeoffs, for example, you might ask a fish biologist to provide information about how many salmon would use a mile of newly-created spawning habitat, how many eggs might eventually hatch, how many fish would survive to swim downstream, and how many would return to spawn in the river years later.

Whether a mile of spawning habitat would result in an increase in the annual salmon run of twenty or two thousand adult salmon will likely make a big difference in the relative value you establish for that habitat.

For some decisions, you yourself will be the source of much of the relevant information. If you are trading off vacation time and salary in choosing among job offers, only you know how you would spend a two-week versus a four-week vacation and the value of the difference to you. You should be as rigorous in thinking through your own subjective inputs as you are in assessing objective data from outside sources. No matter how subjective a tradeoff, you never want to be guided by whim—think carefully about the value of each consequence *to you*.

Our final and maybe most important bit of advice is an old adage: *practice makes perfect*. Like any new approach to an old problem, the even swap method will take some getting used to. The first few times you make swaps, you may struggle with the overall process as well as with each assessment of relative value. Fortunately, the process itself is relatively simple, and it always works the same way. Once you get the hang of it, you'll never have to think about it again. Deciding on appropriate swaps, on the other hand, will never be easy—each swap will require careful judgment. As you gain experience, though, you'll also gain understanding. You'll become more and more skilled at zeroing in on the real sources of value. You'll know what's important and what's not. Perhaps the greatest benefit of the even swaps method is that it forces you to think through the value of every tradeoff in a rational, measured way. In the end, that's the secret of making smart choices.

Acknowledgement: This article was adapted from the book *Smart Choices: A Practical Guide to Making Better Decisions.* Parts of this paper were previously published by the *Harvard Business Review* in Hammond et al. (1998).

References

Franklin, Benjamin (1956). Letter to Joseph Priestly. *The Benjamin Franklin Sampler*. Fawsett, New York.

Hammond, John S., Keeney, Ralph L., Raiffa, Howard (1998). "Even Swaps: A Rational Method for Making Trade-offs", *Harvard Business Review*, March – April, 137 – 150.

Hammond, John S., Keeney, Ralph, L., Raiffa, Howard (1998). *Smart Choices: A Practical Guide to Making Better Decisions,* Harvard Business School Press, Boston.

A Methodology For Risk Management Of Compound Failure Modes

Lori R. Johnson-Payton, Yacov Y. Haimes, and James H. Lambert
Center for Risk Management of Engineering Systems, University of Virginia, Charlottesville

Abstract

Most real-world systems exhibit multiple paths to failure. Yet mainly the likelihoods of failure associated with these failure paths have been analyzed. In order to effectively manage the risk associated with systems, both the likelihoods of failure and the consequences of multiple failure paths should be considered. Therefore, this paper develops a methodology for evaluating risk from the perspective of compound failure modes. Compound failure modes are defined as two or more paths to failure with consequences that depend on the occurrence of combinations of failure paths. In such a situation (of compound failure modes), it is important to have knowledge of the joint probability distribution of the failure path occurrences and their consequences. Through using extreme events analysis the likelihoods of failure along with the consequences are evaluated to quantify the risk associated with compound failure modes. Subsequently, we present a multiobjective tradeoff analysis of cost vs. risk for failure modes where a decision maker is able to see the incremental risk of each failure mode of the system. Examples illustrate the usefulness of the methodology.

Keywords: Multiobjective tradeoff analysis; partitioned multiobjective risk method (PMRM); extreme events; failure modes; risk assessment; risk management

1 Introduction

Engineers and planners are always concerned with designing systems for the extreme conditions they may encounter. The failure of systems constitutes an occurrence of failure modes which is based upon the likelihoods of failure and the consequences. Consider a nuclear system where there is a radioactive fission product. The release of this product is considered to be a failure of the system. The different ways in which this release might occur constitute different failure modes. Associated with each of these failure paths are different types of consequences, such as air crashes, fires, explosions, and toxic contaminations. The consequences associated with any of these failure paths occurring separately could be detrimental to society because of the lives lost and subsequently more catastrophic when the consequences occur together.

There is a need to evaluate the risk of many systems from the perspective of compound failure modes. A failure mode is a path to a failure event that leads to adverse consequences. Subsequently, compound failure modes are defined as two or more paths to failure with consequences that depend on combinations of failure paths occurring. This is where knowledge of the joint probability distribution of the occurrences of the failure paths and their consequences is important.

2 Literature Review

Failure Modes of Engineering Systems

The concept of failure modes has been described in various ways by a host of disciplines of which some include electrical engineering, nuclear engineering, and civil engineering. Failure modes have been analyzed in terms of: (1) multiple/multi-failure modes, (2) joint/dual failure modes, (3) correlated/dependent failures, (4) common cause failures, (5) cascading failures, and (6) negative dependencies. The following sections provide background the various ways in which failure modes are classified and the criticality of failure modes.

Classification of Failure Modes

There are various ways in which failure modes have been classified. Each of these identifications is addressed in this section. Easa (1994) discusses the reliability of open drainage channels under three different failure modes. The three modes of failure which are used to find the system reliability of open drainage channels are: (1) Exceeding channel capacity, (2) Violating maximum velocity for erosion control, and (3) Violating minimum velocity for deposition control. The advanced first-order-second-moment (AFOSM) reliability method is used to estimate the failure probability of each mode. Consequently, the system failure probability is found which accounts for correlations between the failure modes.

Ang and Tang (1984) discuss multiple failure modes for engineering problems where these failure modes may be dependent. Fault trees are studied as a tool for modeling multiple failure modes. However, this discussion addresses a single top-level failure event. The authors refer to multiple failure modes as the underlying basic events of a single fault tree.

Armstrong (1995) identifies two components as joint failure modes. He discusses the joint-reliability importance which is a measure of how two interacting components contribute to the system reliability. The authors discuss how dependency among the components affects the overall system reliability.

Modarres (1993) identifies failure modes by the term dependent failures. These failure modes are identified as events in which the probability of each failure is dependent on the occurrence of other failures. This dependence among failure may be due to the internal or external systems environment.

Hoyland et al., (1994) describe common mode failures as multiple failures that are a direct result of a common or shared root cause. Three different models are discussed for analyzing common mode failures which include: (1) square root method. (2) the binomial failure rate model, and (3) multiple parameter model. Each of these models has drawbacks of which a common negative aspect is their inability to analyze more than two failures along with consequences associated with these failures. Common-mode failures are described as faults which occur when similar modules fail in the same way at the same time.

Hoyland et al., (1994) describe cascading failures a failures which are initiated by the failure of one component in the system that results in a chain reaction. An example of this type of failure involves the sharing of a load by many components. The failure of one component may lead to an increased load on the remaining components which ultimately leads to an increased likelihood of failure.

Hoyland et al., (1994) define negative dependencies as single failures that reduce the likelihood of failures of other components. An illustrated example involves the downtime of a system. Suppose the system is down for the repair of a specific component. The load on the other components is removed and thus the likelihood of failure is reduced during the system downtime.

Although, there have been many contributions to the analysis of reliability in both the scientific and professional communities, these accomplishments have been based mainly on analyzing systems only in terms of their failure likelihoods. Systems need to be analyzed based upon failure likelihoods and consequences to be able to effectively manage the risks in a system. This need is further illustrated in the next section which discusses the severity classes of failure modes.

Severity of Failure Modes

As engineering systems have become more and complex, the different ways in which systems may fail (failure modes) has become increasingly important. The severity of failure modes in systems is identified using four classifications: (1) catastrophic, (2) critical, (3) marginal, and (4) minor (Modarres, 1993). Catastrophic refers to a failure that may cause death. Criticality means a mode that may cause severe injury or major system degradation. Marginal alludes to a path to failure that may cause minor injury or degradation in system performance. Finally, minor refers to mode of failure that does not cause injury or system degradation but may result in system failure and unscheduled maintenance or repair.

Compound Failure Modes

Failure modes have been studied previously in reliability engineering using mainly failure likelihoods. Yet these failure modes have different levels of severity, catastrophic, critical, marginal, and minor. These different severity levels essentially allude to the consequences for failure modes in a general way. However, there is no mention of how to analyze both the consequences and likelihood of failures to obtain the risk associated with compound failure modes. Therefore, a compound failure modes perspective accounts for the inherent uncertainties associated with failure likelihood and consequences to obtain the risk. Failure modes in this research are defined as compound failure modes. Compound failure modes are defined as two or more failure paths with consequences that depend on the occurrence of different combinations of failure paths. In such a situation (of compound failure modes), knowledge of the joint probability distribution of the occurrences of the failure paths and their consequences is important to a meaningful tradeoff analysis in risk management.

3 Review of Compound Failure Modes

Compound failure modes consist of both the likelihoods of failure and the consequences (Johnson-Payton, 1997; Lambert et al., 1996). Therefore, consider that a system can fail in n different ways (failure modes). Failure mode i is described by the random variable X_i, which takes on the value

$x_i = 1$ if failure mode i occurs and $x_i = 0$ if failure mode i does not occur. In general X_i is a binary random variable defined by the following:

$$X_i = \begin{cases} 1, & \text{if failure mode } i \text{ occurs} \\ 0, & \text{if failure mode } i \text{ does not occur} \end{cases} \tag{1}$$

Two failure modes, each of which is represented by a binary random variable, are considered together throughout this paper. The joint distribution associated with these two failure modes is defined whereby X_1 and X_2 are a pair of binary random variables where their joint probability distribution is given by

$$P(X_1 = k, X_2 = l) = p_{kl}, \qquad k, l = 0, 1 \tag{2}$$

and p_{k_l} (point probability) represents the probabilities of failure for failure modes 1 and 2 where:

(1) failure mode 1 occurs and failure mode 2 does not occur if $k=1$ and $l=0$
(2) failure mode 2 occurs and failure mode 1 does not occur if $k=0$ and $l=1$
(3) both failure modes 1 and 2 occur if $k=1$ and $l=1$
(4)both failure modes do not occur if $k=0$ and $l=0$

Note that $\sum_{k=0}^{1} \sum_{l=0}^{1} p_{kl} = 1.$

Table 1 below illustrates the matrix of probabilities for two failure modes, *1* and *2*.

Table 1. Probabilities of Failure for Failure Modes *1* and *2*

	$X_1=0$	$X_1=1$
$X_2=0$	p_{00}	p_{10}
$X_2=1$	p_{01}	p_{11}

The occurrences of failure modes alone or in combination lead to consequences. Which could be minor, major, or up to catastrophic. Examples include fires, floods, and crashes, all of which may lead to lives lost, property damage, and other consequences. The consequences associated with the occurrence of a failure mode can take on two forms, either deterministic (known) or probabilistic (unknown). When consequences are uncertain, and are measured in terms of continuous random variables, they are represented using probability density functions. Table 2 illustrates the different consequences associated with failure modes. The consequences associated with two failure modes can be defined in the following way:

Let c_{kl} or $f_{kl}(\cdot)$ represent the consequences for failure modes *1* and *2* where

(1) failure mode *1* occurs and failure mode *2* does not if $k=1$ and $l=0$
(2) failure mode *2* occurs and failure mode *1* does not if $k=0$ and $l=1$
(3) both failure modes *1* and *2* occur if $k=1$ and $l=1$
(4) both failure modes do not occur if $k=0$ and $l=0$

Table 2. Consequences for Failure Modes *1* and *2*

	$X_1=0$	$X_1=1$
$X_2=0$	c_{00} or $f_{00}(\cdot)$	c_{10} or $f_{10}(\cdot)$
$X_2=1$	c_{01} or $f_{01}(\cdot)$	c_{11} or $f_{11}(\cdot)$

The evaluation of risk for compound failure modes can be demonstrated by using and expanding upon the PMRM (partitioned multiobjective risk method) as explained in the next section.

4 Review of the PMRM

Evaluating the risk associated with the occurrences of different failure modes is vital in the effective management of systems. In most cases, the traditional expected value has been used as the sole measure of risk for a system. To assess and ultimately prevent the risk associated with compound failure modes, it is desirable to have complete knowledge of all risks involved with the consequences of failure modes.

The traditional expected value is one measure that can be used to analyze the average risk associated with the consequences of a failure mode. Used alone, this measure of risk distorts the relative importance of low probability/high consequence events. Therefore, to accurately measure the risk associated with compound failure modes, it is necessary to consider another measure of risk, the conditional expected value.

Risk of extreme events is concerned with the tail of the distribution where low probability/high consequence events occur. The partitioned multiobjective risk method (PMRM) is one method that allows observing more than the traditional expected value (Asbeck and Haimes, 1984). The concept of the PMRM involves considering a discrete or continuous random variable C of damage (e.g., lives lost). The continuous random variable C has a cumulative distribution function (cdf), $F(c)$ and a probability density function (pdf), $f(c)$ which are defined for the following relationships:

$$F(c) = P[C \leq c], \qquad c \geq 0 \tag{3}$$

$$f(c) = \frac{dF(c)}{dc}, \quad c \geq 0 \tag{4}$$

The cdf represented by the continuous random variable C represents the nonexceedance probability of C, the probability that C is observed to be less than or equal to some value c. The exceedance probability of C is defined as the probability that C is observed .to be greater than or equal to c, and is equal to one (1) minus the cdf evaluated at C. The PMRM concept involves dividing a probability axis into three different damage domains: (i) high probability/low consequence events, (ii) moderate probability/moderate consequence events, and (iii) low probability/high consequence events. In each of these damage domains a conditional-expected-value risk function is generated. The risk function which represents the low probability/high consequence event damage domain is denoted by f_4.

Let $(1-\alpha_2)$, where $0 \le \alpha_2 \le 1$, denote the exceedance probability that partitions the domain of C into a range of extreme events. In Figure 1, which represents the exceedance probability, there is a unique harm β_2 on the damage axis that corresponds to $(1-\alpha_2)$ on the probability axis. Consequences greater than β_2 represent low probability/high consequence events. The low probability/high risk function, f_4, represents the expected value of C, given that c is greater than β_2 and is defined by

$$f_4 = E[C \mid c \ge \beta_2] = \frac{\int_{\beta_2}^{\infty} cf(c)dc}{\int_{\beta_2}^{\infty} f(c)dc} \qquad (5)$$

Thus, for a particular design option, the risk measure f_4 represents an additional measure of risk in addition to the unconditional expected value E[C]. This is denoted by f_5 which is defined as

$$f_5 = E[C] = \frac{\int_{0}^{\infty} cf(c)dc}{\int_{0}^{\infty} f(c)dc} = \int_{0}^{\infty} cf(c)dc \qquad (6)$$

where the denominator goes to 1 based on the definition of a probability density function.

Extreme-event analysis based on the f_4 and f_5 risk functions has been applied to a wide variety of problems where catastrophic circumstances have been involved (Karlsson and Haimes, 1988b; Haimes et al., 1990; Mitsiopoulos et al., 1989, and others.).

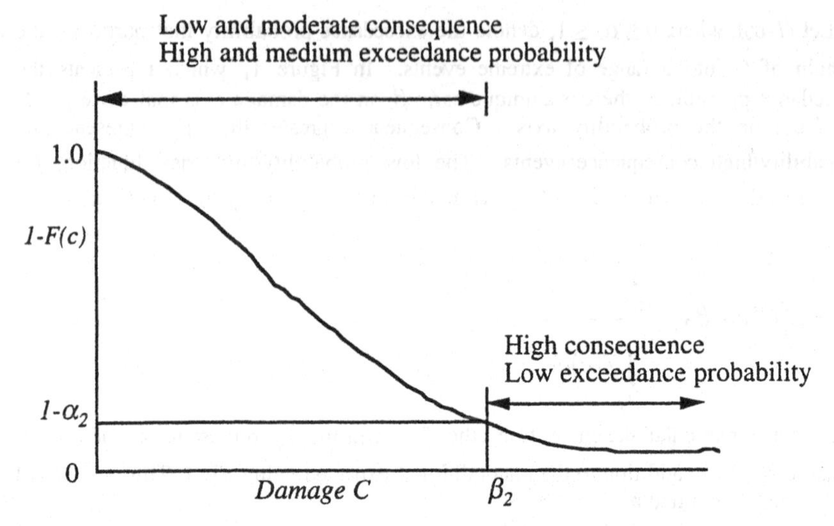

Figure 1. Conditional Expected Value for Probabilistic Consequences

Therefore, examining both the unconditional and conditional expected values will help decision makers recognize and eliminate all the risks associated with the consequences of different failure modes.

Another measure from the PMRM that is of interest is the cost function, f_1, associated with implementing a management policy or risk mitigation effort. This particular function is analyzed vs. both risk functions using multiobjective tradeoff analysis. The analysis allows decision makers to examine the incremental cost of mitigating risks associated with all failure modes. Decision makers will ultimately be able to ascertain and decide upon alternatives for reducing the risk associated with failure modes. In essence, through a risk-management approach involving the assessment of risks and a tradeoff analysis of cost vs. risk, all failure modes which occur in a system can be managed.

5 Risk of Extreme Events for Compound Failure Modes

Risk has previously been defined by Lowrance (1976) as the probability and severity of adverse effects. An adverse effect, in this case, is simply a consequence or outcome associated with the occurrence of a failure mode. The culmination of probability and severe effects is used to assess the risks associated with failure modes. There are two cases of compound failure modes as follows:

Deterministic risk assessment:	Probabilities of failure are represented point; probabilities and the consequences are known.
Probabilistic risk assessment:	Probabilities of failure are represented by point; probabilities and the consequences are random, represented by probability distributions.

Let C be a random variable which denotes the damage or consequence for a given system. Within a system, different modes of failure exist. Therefore, if a particular failure mode occurs, it may have a deterministic or a probabilistic consequence. In this case the following are four scenarios in which failure modes may or may not occur.

(1) X_1=0 and X_2=0 so both failure modes do not occur
(2) X_1=1 and X_2=0 so failure mode *1* occurs but *2* does not
(3) X_1=0 and X_2=1 so failure mode *2* occurs but *1* does not
(4) X_1=1 and X_2=1 so both failure modes occur

Ultimately, the risk probabilities are needed in order to examine the effects of the consequences on the system and for individual compound failure modes. In the following sections, the risk equations are formulated for cases I and II.

Case I:*Deterministic Risk Assessment*

For this case, the probabilities of failure are represented as point probabilities and the consequences for the system are known. System risk is defined as the risk of all failure modes occurring. For this case, system risk deals with joint probability mass functions (pmf) representing the failure likelihoods and deterministic consequences. An examination of the risk requires three elements: (1) the joint probability mass function (pmf), (2) the joint cumulative distribution function (cdf), and (3) the exceedance function which is one (1) minus the cdf. Recall that the pmf for the probabilities of failure is defined as

$$P(X_1 = k, X_2 = l) = p_{kl}, \qquad k,l = 0,1 \tag{7}$$

The joint cdf for the probabilities of failure is similarly defined as

$$F(x_1,x_2) = P(X_1 \le x_1, X_2 \le x_2) \tag{8}$$

and can be represented by

$$F(x_1,x_2) = \begin{cases} p_{00}, & x_1 = 0, x_2 = 0 \\ p_{00} + p_{10}, & x_1 = 0,1, x_2 = 0 \\ p_{00} + p_{01}, & x_1 = 0, x_2 = 0,1 \\ 1 & x_1 = 0,1, x_2 = 0,1 \end{cases} \tag{9}$$

Similarly, the exceedance function is represented by

$$1 - F(x_1,x_2) = \begin{cases} 1 - p_{00}, & x_1 = 0, x_2 = 0 \\ 1 - p_{00} - p_{10}, & x_1 = 0,1, x_2 = 0 \\ 1 - p_{00} - p_{01}, & x_1 = 0, x_2 = 0,1 \\ 0 & x_1 = 0,1, x_2 = 0,1 \end{cases} \tag{10}$$

Since the joint pmf, cdf, and exceedance functions have been provided for the probabilities of failure, it is now necessary to map the bivariate pmf of (X_1, X_2) into a univariate pmf for the consequences because the interest is in determining the risk associated with failure modes occurring. Recall that we have previously defined a random variable C, which denotes the damage or consequence for a given system. There are four deterministic consequences which the random variable C can take on; these are defined by c_{00}, c_{10}, c_{01}, and c_{11}. Each of these consequences is based on whether both failure modes do not occur, either failure mode occurs, or both occur. Furthermore, there are four probabilities of failure associated with the occurrence or nonoccurrence of failure modes; these are given by p_{00}, p_{10}, p_{01}, and p_{11}. Therefore, mapping the bivariate pmf of (X_1, X_2) into the univariate pmf of C can be done by first defining the random variable for the consequences C which, as we have seen, can take on four deterministic values. This is indicated below in equation (11).

$$C = \begin{cases} c_{00}, & if\,(X_1 = 0, X_2 = 0) \\ c_{10}, & if\,(X_1 = 1, X_2 = 0) \\ c_{01}, & if\,(X_1 = 0, X_2 = 1) \\ c_{11}, & if\,(X_1 = 1, X_2 = 1) \end{cases} \tag{11}$$

Since the random variable C has been defined, the pmf for C now can be defined as follows:

$$C(X_1, X_2) = \begin{cases} c_{00}, & \textit{with probability } p_{00} \\ c_{10}, & \textit{with probability } p_{10} \\ c_{01}, & \textit{with probability } p_{01} \\ c_{11}, & \textit{with probability } p_{11} \end{cases} \tag{12}$$

The exceedance function for the consequences and probabilities of failure is represented graphically in Figure 2, where it is assumed that $c_{11} > c_{01} > c_{10} > c_{00}$. The deterministic values for the consequences are on the horizontal axis, while the joint exceedance function for the probabilities of failure is on the vertical axis. Both the probabilities of failure and the consequences are based on the occurrence or nonoccurrence of failure modes. The expected value of the system f or all failure modes occurring is given by

$$f_5 = E[C(X_1, X_2)] = \sum_{k=0}^{1} \sum_{l=0}^{1} P(X_1 = k, X_2 = l) C(k, l) = \sum_{k=0}^{1} \sum_{l=0}^{1} p_{kl} c_{kl} \tag{13}$$

The conditional expected value of consequences in the low-probability/high-damage region, f_4, (extreme event) is based upon the location of the damage partitioning point, β_2, which essentially amounts to a hypothesis about the occurrence of events. For example, in Figure 2 the conditional expected value is the expected damage, conditional on the hypothesis that either failure mode 1 occurs (c_{01}) or both modes occur (c_{11}). The conditional expected value for this case is given by the following expression:

$$f_4 = E[C(X_1, X_2) \mid C(X_1, X_2) > \beta_2] = \frac{\sum_{k=0}^{1} \sum_{l=1}^{1} p_{kl} c_{kl}}{\sum_{k=0}^{1} \sum_{l=1}^{1} p_{kl}} = \frac{p_{01} c_{01} + p_{11} c_{11}}{p_{01} + p_{11}} \tag{14}$$

Similar hypotheses can be made for the conditional expected value depending upon the location of the damage partitioning point. Figure 2 shows how the various equations for f_4 are derived.

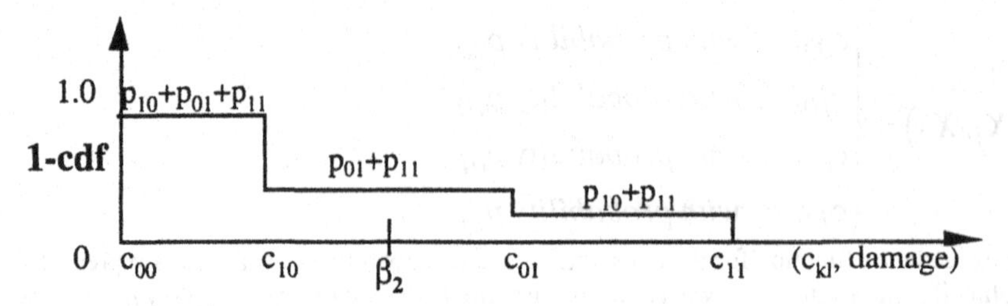

Figure 2. Exceedance Function

Therefore, (based on hypotheses regarding the occurrences or nonoccurrences of failure modes), the following rules are provided for determining the conditional expected value based on the location of the damage partitioning point. These rules can be generalized for any case depending upon where the different damages lie on the axis.

System Conditional Expected-Value Rules

(1) *If* $c_{00}, c_{10},$ *and* $c_{01} < \beta_2 < c_{11},$ *then* $f_4 = \dfrac{p_{10}c_{10} + p_{11}c_{11}}{p_{10} + p_{11}}$

(2) *If* c_{00} *and* $c_{10} < \beta_2 < c_{01}$ *and* $c_{11},$ *then* $f_4 = \dfrac{p_{01}c_{01} + p_{11}c_{11}}{p_{01} + p_{11}}$

(3) *If* $c_{00} < \beta_2 < c_{10},\ c_{01},$ *and* $c_{11},$

then $f_4 = \dfrac{p_{10}c_{10} + p_{01}c_{01} + p_{11}c_{11}}{p_{10} + p_{01} + p_{11}}$

(4) *If* $\beta_2 < c_{00}, c_{10},\ c_{01},$ *and* $c_{11},$

then $f_4 = \dfrac{p_{00}c_{00} + p_{10}c_{10} + p_{01}c_{01} + p_{11}c_{11}}{p_{00} + p_{10} + p_{01} + p_{11}} = f_5$

(5) *If* $\beta_2 > c_{00}, c_{10},\ c_{01},$ *and* c_{11} then $f_4 = 0$

The risks associated with individual failure modes are not discussed in this paper because in these cases there is only one event in the extreme-event region for one failure mode. Therefore, the conditional expected value reduces either to that event or to just the unconditional expected value. It is not really practical to consider failure modes individually when there are only two, but this analysis would be very practical if there are more than two failure modes.

Case II: *Probabilistic Risk Assessment*

For this situation, the failure likelihoods are denoted by point probabilities and the consequences are described by probability density functions. Two different methods are used to solve for the failure modes, risks for this case. The first method, *called separate failure mode risk assessment* allows for examining the risk for failure modes based on only the consequences. Here the probabilities of failure are used to indicate the probability of risk that occurs for a given failure mode. The second method, *called mixed density failure mode risk assessment*, allows for evaluating the risk for failure modes using both the probabilities of failure and the consequences. This approach is called a mixed-density approach because the failure likelihoods are point probabilities and the consequences are represented by density functions.

Method I, Separate Failure-Modes Risk Assessment

Uncertain consequences are measured in terms of four continuous random variables. These are shown as follows along with their associated density functions: $C_{00} \sim f_{00}(\cdot)$, $C_{10} \sim f_{10}(\cdot)$, $C_{01} \sim f_{01}(\cdot)$, and $C_{11} \sim f_{11}(\cdot)$. The general equations for the unconditional and conditional expected values, for the occurrences and nonoccurrences of failure modes, are given as follows:

$$E[C_{kl}] = \int_0^\infty c f_{kl}(c)dc, \qquad k,l = 0,1 \tag{15}$$

$$f_4 = E[C_{kl} \mid c \geq \beta_2] = \frac{\int_{\beta_2}^{\infty} c f_{kl}(c)dc}{\int_{\beta_2}^{\infty} f_{kl}(c)dc} = \frac{\int_{F^{-1}(\alpha_2)}^{\infty} c f_{kl}(c)dc}{1-\alpha_2},$$

$$\text{where } 1-\alpha_2 = \int_{\beta_2}^{\infty} f_{kl}(c)dc \tag{16}$$

For each random variable, the associated conditional and unconditional expected values can be found based on the equations given above. After finding these expected values, it is desirable to relate the probabilities of failure to the risk. Therefore, it can be said that the probability of the expected value (unconditional and conditional) of risk for failure mode *1* and not failure mode *2* is p_{10} and for failure mode *2* and not failure mode *1* is p_{01}. Likewise, the probability of the expected value (unconditional and conditional) of risk is p_{11} for both failure modes occurring and p_{00} for both failure modes not occurring. The next section discusses the use of mixed probability density functions for finding the risk associated with individual failure modes and the system.

Method II, Mixed-Density Failure-Mode Risk Assessment

The mixed-density failure-mode risk assessment allows for determining the expected values of risk for the system (all failure modes) or for individual paths to failure. The expected values are derived by finding a density function for the consequences by using point probabilities and density functions for the consequences. The density function for the consequences is
represented by

$$f(c) = \sum_{k=0}^{1} \sum_{l=0}^{1} p_{kl} f_{kl}(c) \tag{17}$$

In essence, this is a mixed density function, where there are different types of failure-mode scenarios and with each type there is a density function representing the consequences. The proof of Equation (17) is shown by Johnson-Payton (1997).

System Risk Assessment

Since it has been established that there is a single random variable which represents the density function of the consequences, the unconditional and conditional expected values for the system can be found using the traditional risk equations. However, first it is necessary to illustrate the mapping of the bivariate distribution into a single univariate pdf for the consequences as illustrated below:

$$C(X_1, X_2) = \begin{cases} C_{00}, & \text{with probability } p_{00} \\ C_{10}, & \text{with probability } p_{10} \\ C_{01}, & \text{with probability } p_{01} \\ C_{11}, & \text{with probability } p_{11} \end{cases} \tag{18}$$

This is done by using the random variable C for the consequences, in which there are four random variables that can result based on occurrences or nonoccurrences of the failure modes with associated probabilities of failure. Then there are associated density functions for the se random variables. Therefore, pdf of $C(X_1,X_2)$ is given by $p_{00}f_{00}(c)+p_{10}f_{10}(c)+p_{01}f_{01}(c)+p_{11}f_{11}(c)$ which is the mixed-density function $f(c)$.

The consequences are assumed to be independent of each other. Therefore, the unconditional expected value of the consequences can be found by

$$\begin{aligned} f_5 = E[C(X_1,X_2)] &= \int_0^{\infty} cf(c)dc \\ &= p_{00}E[C_{00}] + p_{10}E[C_{10}] + p_{01}E[C_{01}] + \\ &\quad p_{11}E[C_{11}] \end{aligned} \tag{19}$$

Similarly, the conditional expected value of the consequences is defined by

$$f_4 = E[C(X_1,X_2) \mid c \geq \beta_2] = \frac{\int_{\beta_2}^{\infty} cf(c)dc}{\int_{\beta_2}^{\infty} f(c)dc}$$

$$= \frac{\int_{\beta_2}^{\infty} p_{00}cf_{00}(c)dc + \int_{\beta_2}^{\infty} p_{10}cf_{10}(c)dc + \int_{\beta_2}^{\infty} p_{01}cf_{01}(c)dc + \int_{\beta_2}^{\infty} p_{11}cf_{11}(c)dc}{\int_{\beta_2}^{\infty} p_{00}f_{00}(c)dc + \int_{\beta_2}^{\infty} p_{10}f_{10}(c)dc + \int_{\beta_2}^{\infty} p_{00}f_{00}(c)dc + \int_{\beta_2}^{\infty} p_{10}f_{10}(c)dc} \tag{20}$$

Individual Failure Mode Risk Assessment

The risk associated with individual failure modes occurring can be found using mixed-probability density functions. The risks can be found for failure mode *1* occurring, given that failure mode *2* does not occur and vice-versa. These risks can be determined by essentially decomposing the original problem into two subproblems.

Subproblem 1 basically deals with the probabilities of failure associated with only failure mode *1* occurring (probabilities which have been normalized). Let C_1 represent the random variable associated with mixed-density function for failure mode *1*. The mixed-probability density function for the consequences of failure mode 1 is shown by

$$f(c_1) = \sum_{k=0}^{1}\sum_{l=0}^{0} p_{kl}' f_{kl}(c_1) = p_{00}' f_{00}(c_1) + p_{10}' f_{10}(c_1) \tag{21}$$

In order to do any further analysis, it is necessary to map the conditional distribution of (X_1, X_2) into a conditional pdf for the consequences. This is achieved by saying there are two random outcomes that can result for the random variable C, based upon whether failure mode 1 occurs or both failure modes do not occur. Therefore, the mapping of the conditional distribution of (X_1, X_2) into the conditional pdf of C can be seen as follows:

$$[C(X_1, X_2) \mid X_2 = 0] = \begin{cases} C_{00} & \text{with probability } \dfrac{p_{00}}{p_{00} + p_{10}} = p_{00}' \\ C_{10} & \text{with probability } \dfrac{p_{10}}{p_{00} + p_{10}} = p_{10}' \end{cases} \tag{22}$$

Therefore, the unconditional expected value is given as

$$f_5 = E[C(X_1, X_2) \mid X_2 = 0] = \int_0^{\infty} c_1 f(c_1) dc_1 = p_{00}' E[C_{00}] + p_{10}' E[C_{10}] \tag{23}$$

and the conditional expected value is given by

$$f_4 = E[C(X_1, X_2) \mid X_1 = 0, c_1 \geq \beta_2] = \frac{\int_{\beta_2}^{\infty} p_{00}' c_1 f_{00}(c_1) dc_1 + \int_{\beta_2}^{\infty} p_{10}' c_1 f_{10}(c_1) dc_1}{\int_{\beta_2}^{\infty} p_{00}' f_{00}(c_1) dc_1 + \int_{\beta_2}^{\infty} p_{10}' f_{10}(c_1) dc_1} \tag{24}$$

$$f(c_2) = \sum_{k=0}^{0}\sum_{l=0}^{1} p_{kl}' f_{kl}(c_2) = p_{00}' f_{00}(c_2) + p_{01}' f_{01}(c_2)$$

(25)

(26)

$$\begin{cases} C_{00} & \textit{with probability } \dfrac{p_{00}}{p_{00}+p_{01}} = p_{00}' \end{cases}$$

Therefore, the unconditional expected value can be found by

$$f_5 = E[C(X_1, X_2) \mid X_1 = 0] = \int_0^{\infty} c_2 f(c_2) dc_2 = p_{00}' E[C_{00}] + p_{01}' E[C_{01}] \tag{27}$$

$$f_4 = E[C(X_1, X_2) \mid X_1 = 0, c_2 \geq \beta_2] = \frac{\int_{\beta_2}^{\infty} p_{00}' c_2 f_{00}(c_2) dc_2 + \int_{\beta_2}^{\infty} p_{01}' c_2 f_{01}(c_2) dc_2}{\int_{\beta_2}^{\infty} p_{00}' f_{00}(c_2) dc_2 + \int_{\beta_2}^{\infty} p_{01}' f_{01}(c_2) dc_2} \tag{28}$$

In summary, this section on mixed-probability density functions is very useful for determining the risk associated with the failure modes and the system when there are point probabilities of failure and random consequences. The following section discusses how to manage compound failure modes.

6 Risk Management of Compound Failure Modes

Effective risk management of compound failure modes requires using a multiobjective tradeoff analysis such as the surrogate worth tradeoff (SWT) method used here(Chankong and Haimes, 1983). This method allows for better decisions in planning, design, and operation. In an SWT, one primary objective is selected to be optimized while the other objectives are used as constraints. The SWT can be performed for the system and for individual failure modes. The basic problem methodology for either individual or all failure modes occurring is as follows:

$$\text{Minimize}\begin{pmatrix} f_1 \\ f_4 \end{pmatrix} \text{ and minimize}\begin{pmatrix} f_1 \\ f_5 \end{pmatrix} \tag{29}$$

where tradeoffs among the cost and risk functions are represented by

$$\lambda_{1k} = -\frac{\Delta f_1}{\Delta f_k}, \qquad k = 4, 5 \tag{30}$$

Consequently, the SWT method is used to determine the tradeoffs for cost vs. risk for each design option for failure modes. Since there are a discrete number of options, the tradeoff functions are not continuous. The tradeoffs are performed between f_1 and f_5 and f_1 and f_4, for the ith and ith+1 design options as is indicated by (30). After obtaining the tradeoff analysis results, the decision maker will be able to examine the amount of risk that each failure mode contributes to the overall system. Furthermore, the decision maker will be able to determine which failure modes are the most risky and ultimately choose alternatives for decreasing the risks associated with them. Finally, this tradeoff analysis provides the decision maker with insights regarding the importance of considering different failure modes within a system vs. one failure mode when there are different outcomes.

In order to conduct the tradeoff analysis effectively, different design options are constructed based on changing the probabilities of failure or shifting the distributions for the consequences. The next section provides examples for cases I and II using risk of extreme events and multiobjective tradeoff analyses.

7 Risk Management Examples

This section discusses various examples for risk management of compound failure modes. It presents examples using fault trees and time-to-failure models to determine the probabilities of failure. Examples are solved using cases I and II for risk of extreme events with compound failure modes.

Case I: *Deterministic Risk Assessment Example*

This case is based on point probabilities and known consequences. The risk can be found for the system. However, it is necessary to determine different design options based on reducing the probabilities of failure and the consequences. Table 3 summarizes the data for the different design options.

Table 3. Design Option Data (System)

Design Option	Cost ($)	p_{00}	p_{10}	p_{01}	p_{11}	c_{00}	c_{10}	c_{01}	c_{11}
D1	0	0.9841	0.0027	0.0129	0.0003	0	3000	5000	10000
D2	1000	0.9866	0.0028	0.0104	0.0002	0	2700	4500	9000
D3	3000	0.9898	0.0029	0.0073	0.0001	0	2025	3375	6750
D4	5000	0.9924	0.0029	0.0046	0.0001	0	1215	2025	4050

Based on the different design options, the unconditional and conditional expected values can be found for the system. These cost-vs.-risk analysis results are provided in Table 4a. Likewise, the tradeoff values are contained in Table 4b for the *i*th and *i*th+1 design options where $i = 1,\dots,4$, for the system. The partitioning point on the damage axis for the system has been determined by the decision maker to be 2500 lives. Figure 4 is a graphical depiction of the risk for the system.

Table 4a. Cost vs. Risk Analysis Results (System)

	f_1	f_5	f_4
Design Options	Cost ($)	System	System
D1	0	76	4755
D2	1000	56	4191
D3	3000	31	3028
D4	5000	13	1743

Table 4b. Tradeoff Values (System)

Design Options	λ_{15}	λ_{14}
D1 vs. D2	50	1.77
D2 vs. D3	80	1.72
D3 vs. D4	111	1.57

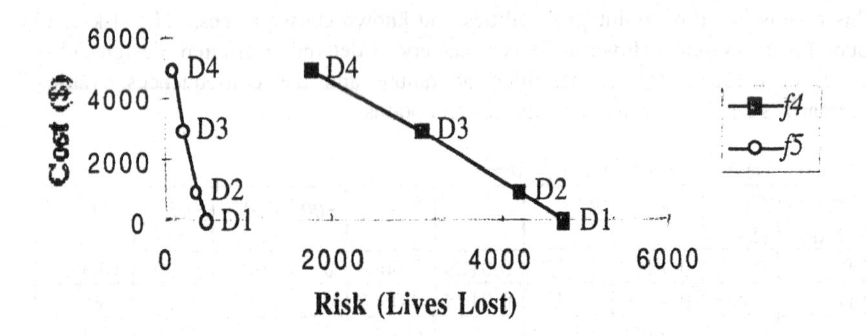

Figure 4. Cost vs. Risk (All Failure Modes)

From Table 4b and Figure 4, it can be seen that all design options are optimal, because all the tradeoff values are positive. The decision maker selects the most preferred of the four pareto optimal policies for the system according to personal preferences in the tradeoffs between the cost function and the risk functions.

This example illustrates how deterministic risk assessment is used for analyzing the risk associated with compound failure modes. Often engineers tend to look only at the unreliabilities as a measure for the system failing. In this case, the unreliabilities do not provide a good indication. It may appear that a reservoir system is in good shape, yet if it fails, a lot of lives could be lost. Therefore, it is necessary to look at the probabilities of failure and consequences as in Table 4a, which shows lives lost for all design options for a system. Under design option 4, there are 1743 lives lost with a reliability of 0.9924 and an unreliability of 0.0076 for the system. There is still room for improvement in determining design options to reduce the risk. However, looking only at the probabilities of failure, does not give the decision maker a true picture of the actual consequences associated with a system and the catastrophic effect on society if it should fail.

Case II: *Separate Failure Mode Risk Assessment Example*

Recall that separate failure mode risk assessment means the risk is analyzed based solely on the consequences and the probabilities of failure are related to the risk afterwards. This approach is very useful in industry because managers are able to relate the risk and its of occurrence.

Consider that within a system there are two subsystems which can fail. Suppose that there is a lack of data; then the triangular distribution can be used to represent the consequences. The triangular distribution is defined as follows:

$$f(c)=\begin{cases}\dfrac{2(c-a)}{(b-a)(d-a)} & a\le c\le d\\[2ex] \dfrac{2(b-c)}{(b-a)(b-d)} & d\le c\le b\\[2ex] 0 & otherwise\end{cases}\qquad(31)$$

where a is the minimum value, b is the maximum value, and d is the most likely value. The triangular distribution is represented by TR(a,d,b) for the consequences, which are measured in the number of lives lost. The design option data for this example is given in Tables 5a and 5b.

Table 5a. Design Option Data (System)

Design Option	Cost ($)	$P00$	$P10$	$P01$	$P11$
D1	0	0.9350	0.0429	0.0212	0.0010
D2	1000	0.9413	0.0387	0.0192	0.0008
D3	5000	0.9508	0.0325	0.0161	0.0006
D4	12000	0.9605	0.0262	0.0130	0.0004

Table 5b. Design Option Data (System)

Design Option	Cost ($)	$f_{00}(c)$	$f_{10}(c)$	$f_{01}(c)$	$f_{11}(c)$
D1	0	TR(0,2,5)	TR(0,450,650)	TR(0,150,300)	TR(0,600,1500)
D2	1000	TR(0,1,4)	TR(0,405,585)	TR(0,135,270)	TR(0,540,1350)
D3	5000	TR(0,1,3)	TR(0,338,488)	TR(0,113,225)	TR(0,450,1125)
D4	12000	TR(0,1,3)	TR(0,270,390)	TR(0,90,180)	TR(0,360,900)

Based on the different design options, the unconditional and conditional expected values can be found for the system and individual failure modes. The cost-vs.-risk analysis results are provided in Tables 6a and 6b. The partitioning point on the probability axis is α_2 = 0.99. The risks (unconditional and conditional expected values) for the consequences associated with each failure mode are determined using the following equations as derived by Schoof (1996) for the triangular distribution:

$$f_5 = \frac{a+b+d}{3} \tag{32}$$

$$f_4 = b - \frac{2}{3}\left[(1-\alpha_2)(b-a)(b-d)\right]^{1/2} \tag{33}$$

Based on the risk values in Tables 6a and 6b, the probability of failure can be related to the risks in the following ways. It can be said that the probability of the expected value (unconditional, 367 lives lost and conditional, 626 lives lost) of risk for failure mode *1*, for design option 1, is 0.0429. The probability of the expected value of risk for failure mode *2*, for design option 1, is 0.0212. Additionally, the probability of the expected value of risk for both failure modes occurring is 0.0010. These same statements can be made for design options 2-4 from observing Tables 5a, 6a, and 6b.

Table 6a. Cost-vs.-Risk Analysis Results (Individual Failure Modes)

		f_5	f_4	f_5	f_4
Design Options	Cost ($)	Failure Mode *1*	Failure Mode *1*	Failure Mode *2*	Failure Mode *2*
D1	0	367	626	150	286
D2	1000	330	563	135	257
D3	5000	275	470	113	214
D4	12000	220	376	90	172

Table 6b. Cost-vs.-Risk Analysis Results (Both Failure Modes)

		f_5	f_4
Design Options	Cost ($)	Both Failure Modes *1 & 2*	Both Failure Modes *1 & 2*
D1	0	700	1423
D2	1000	630	1280
D3	5000	525	1067
D4	12000	420	854

Figures 5 (a, b, and c) illustrate that all design options are optimal for individual failure modes and for both failure modes. This is also true if one considers the tradeoff values. These Figures when both failure modes occur, there is more risk associated with the system. Furthermore, the risk is worse when failure mode *1* occurs than it is for failure mode *2*. In this example, the unreliabilities are fairly low for design option 4 and the reliability is 0.9605, so one would tend to believe that the system is fairly safe. Yet, based on the calculation of the conditional expected value for this design option in Tables 6a and 6b, it is shown that 376 lives are lost when failure mode *1* occurs, 172 when failure mode *2* occurs, and 854 when failure modes *1* and *2* occur together. Obviously, when both failure modes occur together more lives are lost. The fact that the probability of failure is very low would tend to indicate to a decision maker that the system is fairly safe, at least for design options 2-4. However, by considering both the likelihoods of failure and the consequences, the decision maker can see the actual effect on the system in terms of potential lives lost either by individual failure modes or both occurring together.

Figures 5 a, b, and c depict the risk for failure modes. In each case the conditional expected value is worse, as expected. Finally, as in the previous case, it is ultimately up to the decision maker to decide which alternative is the most useful based on his or her preferences.

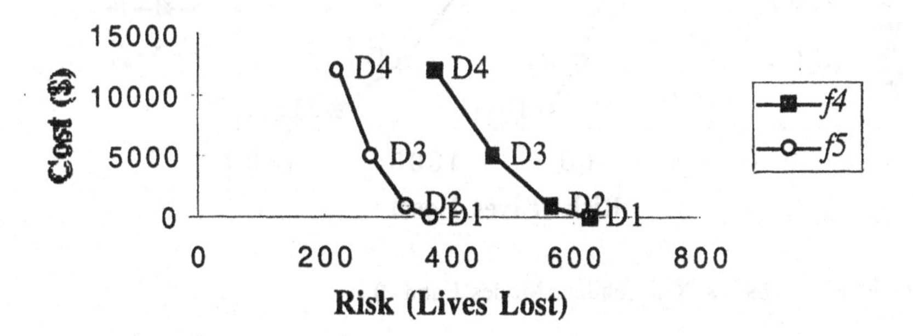

Figure 7a. Cost vs. Risk (Failure Mode *1*)

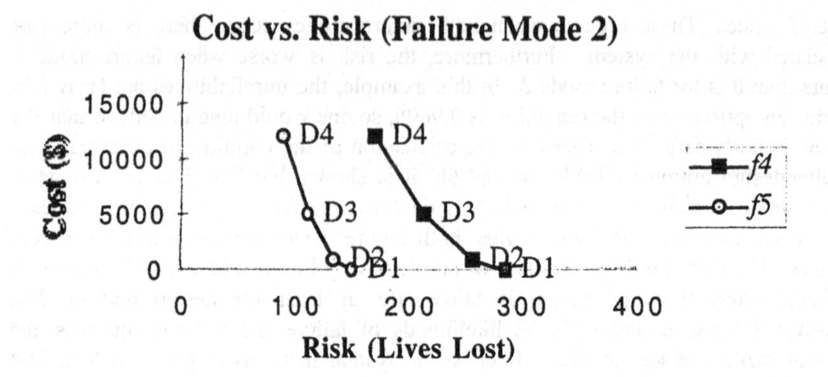

Figure 7b. Cost vs. Risk (Failure Mode 2)

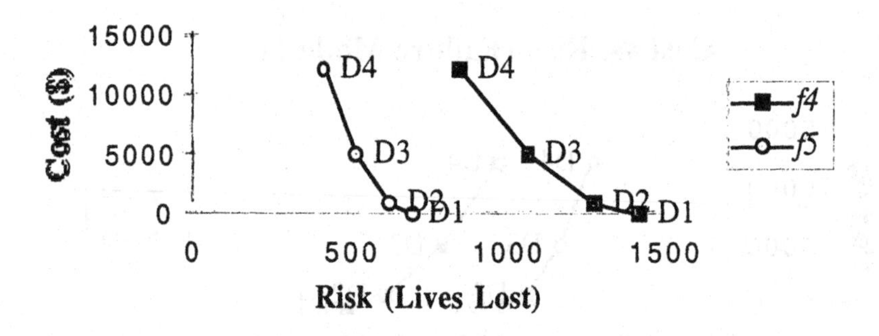

Figure 7c. Cost vs. Risk (Failure Modes 1 and 2)

Mixed Density Functions Risk Assessment Example

Recall that mixed density failure mode risk assessment analyses the risk using point probabilities and mixed density functions representing the consequences. For a normal distribution, the mean of a sum of normal random variables also happens to be the unconditional expected value for this case. The normal distribution is defined as follows:

$$f(c_i) = \frac{1}{\sigma\sqrt{2\pi}} e^{-\frac{1}{2}\left(\frac{c_i - \mu}{\sigma}\right)^2} \quad (34)$$

where μ is the mean and σ the standard deviation. The normal distribution is represented by $N(\mu, \sigma^2)$ for the consequences, which are measured in the number of lives lost. The design option data for this example is given in Tables 7a, b, c, and d. Table 7a summarizes of the data for failure mode *1*, Table 7b for failure mode *2*, and Tables 7c and d summarize the data for the system.

Table 7a. Design Option Data (Failure Mode *1*)

Design Option	Cost ($)	p_{00}	p_{10}	$f_{00}(c_1)$	$f_{10}(c_1)$
D1	0	.9901	.0099	N(0,1)	N(300,50)
D2	1000	.9910	.0090	N(0,1)	N(270,45)
D3	5000	.9926	.0074	N(0,1)	N(225,38)
D4	12000	.9940	.0060	N(0,1)	N(360,30)

Table 7b. Design Option Data (Failure Mode *2*)

Design Option	Cost ($)	p_{00}	p_{10}	$f_{00}(c_2)$	$f_{01}(c_2)$
D1	0	.9951	.0049	N(0,1)	N(800,90)
D2	1000	.9955	.0045	N(0,1)	N(720,81)
D3	5000	.9962	.0038	N(0,1)	N(600,68)
D4	12000	.9970	.0030	N(0,1)	N(480,54)

Table 7c. Design Option Data (System)

Design Option	Cost ($)	p_{00}	p_{10}	p_{01}	p_{11}
D1	0	0.9656	0.0097	0.0048	0.0198
D2	1000	0.9690	0.0088	0.0044	0.0179
D3	5000	0.9741	0.0073	0.0037	0.0149
D4	12000	0.9792	0.0059	0.0029	0.0119

Table 7d. Design Option Data (System)

Design Option	Cost ($)	$f_{00}(c)$	$f_{10}(c)$	$f_{01}(c)$	$f_{11}(c)$
D1	0	N(0,1)	N(300,50)	N(800,90)	N(1500,250)
D2	1000	N(0,1)	N(270,45)	N(720,81)	N(1350,225)
D3	5000	N(0,1)	N(225,38)	N(600,68)	N(1125,188)
D4	12000	N(0,1)	N(360,30)	N(480,54)	N(900,150)

Based on the different design options, the unconditional and conditional expected values can be found for the system and individual failure modes using the mixed probability density functions. The probability density functions for failure mode *1*, failure mode *2*, and the system can be found using the following equations:

$$f(c_1) = \sum_{k=0}^{1}\sum_{l=0}^{0} p_{kl}' f_{kl}(c_1) \tag{35}$$

$$f(c_2) = \sum_{k=0}^{0}\sum_{l=0}^{1} p_{kl}' f_{kl}(c_2) \tag{36}$$

$$f(c) = \sum_{k=0}^{1}\sum_{l=0}^{1} p_{kl} f_{kl}(c) \tag{37}$$

The damage partitioning point (β_2) for this case is 100 lives for failure mode *1*, 300 for failure mode *2*, and 500 for the system. The risk (unconditional and conditional expected values) are determined based on equations (19), (20), (23), (24), (27), and (28). The conditional expected values have been solved for using Maple. The cost-vs. -risk analysis results for this particular example are shown in Table 8.

Table 8. Cost Versus Risk Analysis Results

		f_5	f_4	f_5	f_4	f_5	f_4
Design Options	Cost ($)	Failure Mode *1*	Failure Mode *1*	Failure Mode *2*	Failure Mode *2*	System	System
D1	0	3	300	4	800	37	1212
D2	1000	2	270	3	720	30	1087
D3	5000	2	225	2	600	21	904
D4	12000	1	180	1	480	15	900

Figure 6 shows that all of the design options lie on the pareto-optimal frontier for the system and for individual failure modes. However, as the data in Table 8 indicates, the risk (unconditional expected value) associated with failure modes *1* and *2 is* nearly the same. This is also suggested by Figure 6 because the risk curves associated with failure modes *1* and *2* are overlapping for the unconditional and conditional expected values. Therefore, it is hard to tell which risk curve is for failure mode *1* and which one is for *2*. Note that $f_4(1)$ and $f_5(1)$ are the risk curves for the occurrence of failure mode *1*, $f_4(2)$ and $f_5(2)$ for the occurrence of failure mode *2*, and, $f_4(system)$ and $f_5(system)$ are the risk curves for the occurrence of all failure modes. Although the unconditional expected value for the system is larger than the expected values for failure modes *1* and *2*, it is still relatively small in comparison to the conditional expected values.

The conditional expected value for failure mode *2* is much larger than that for failure mode *1*. It should be noted that the conditional expected values for failure modes *1* and *2* are based on only one event in the extreme event region. Therefore, the conditional expected values for failure modes *1* and *2* reduce only to the expected values for the normal distribution for each of the design options. In Figure 6 we see that although the conditional expected value is large for failure modes *1* and *2, when* all the failure modes are considered together (system), the conditional expected value is much higher. Therefore, it is always necessary to consider the risk for individual failure modes as well as for the system to show the decision maker the amount of risk each failure mode contributes. The incremental risk of each failure mode can be calculated using the following equation:

$$Incremental\ Risk\ Contribution = \frac{Individual\ Failure\ Mode\ Risk}{System\ Risk} \tag{38}$$

In this case each failure mode *1* only contributes approximately 66% to the overall system risk for design option 1, while failure mode *2* contributes approximately 25. However, when both failure modes occur together the contribution is approximately 9%. Similar statements can be made about the incremental risk contribution for other design options. It is up to the decision maker to come up other alternatives to reduce the risk associated with the individual failure modes and the system.

Once again, merely observing the probabilities of failure is not enough to give an overall picture of the effects on the system in terms of lives lost. Therefore, it is necessary to analyze the risk for both the probabilities of failure and the consequences. The risk (conditional expected value) associated with the system for design option 4 is a little less, with 900 lives lost as opposed to 1212 for design option 1. However, the final decision is left up to the decision maker based on his/her preferences.

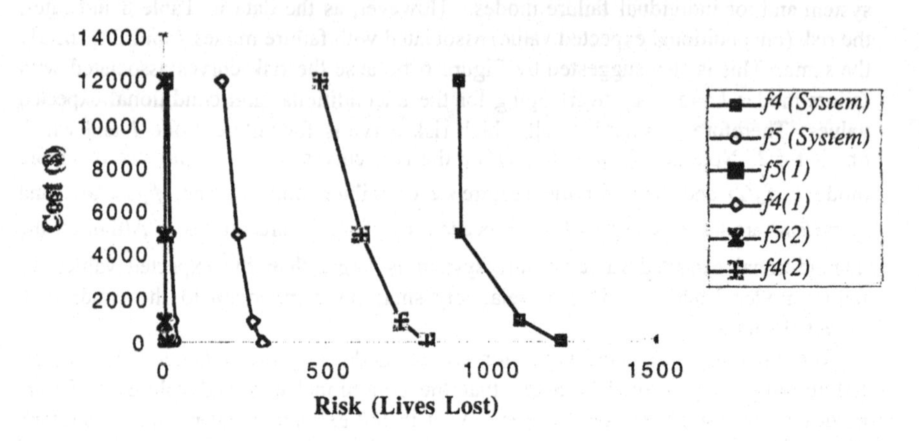

Figure 6. Cost vs. Risk (All Failure Modes)

The previous three examples provide a tutorial using the different methods derived for compound failure modes along with managing the risk. This section demonstrates the usefulness of considering all failure modes. If only individual failure modes are considered until the risk has been examined for the system the decision maker is not able to understand the true picture of compound failure modes. Only then will the decision maker be able to understand the impact on the system probabilities of failure and the consequences associated with failure modes.

8 Conclusion

The methodology presented in this paper provides insights to improve decision making for both simple and complex systems where it is not practical to evaluate the unreliability in terms of a single measure; the consequences must be considered also. Two cases have been explored extensively for the probabilities of failure and the consequences, where the failure likelihoods are represented by point probabilities and the consequences are either random or deterministic. Examples are presented which illustrate the usefulness of a risk-management framework in engineering, This methodology can be used across several disciplines to analyze compound failure modes and ultimately help manage systems where failure modes exist.

9 Acknowledgments

This research is supported by the Center for Risk Management of Engineering Systems. Special thanks to Ph.D. student Ganghuai Wang for weekly discussions and brainstorming sessions which have influenced and contributed to the work reported here.

References

[1] Asbeck, E. L., and Y. Y. Haimes (1984). "The Partitioned Multiobjective Risk Method (PMRM)," *Large Scale Systems*, 6(1), 13-38.

[2] Chankong, V. and Y. Y. Haimes (1983). *Multiobjective Decision Making Theory and Methodology*, Elsevier Science Publishing, New York.

[3] Haimes, Y.Y. (1993). "Risk of Extreme Events and Fallacy of the Expected Value," Control and Cybernetics.

[4] Haimes, Y.Y., J. H. Lambert, and D. Li (1992). "Risk of Extreme Events in a Multiobjective Framework," Water Resources Bulletin, Vol. 28, 201-209.

[5] Haimes, Y.Y., D. Li, P. Karlsson, and J. Mitsiopoulos (1990). "Extreme Events: Risk Management," *In Systems and Control Encyclopedia, Supplementary Volume 1*, M.G. Singh (Editor-in-Chief), Pergamon Press, Oxford .

[6] Hines, William and Douglas C. Montgomery (1990). Probability and Statistics in Engineering and Management Science, Third Edition.

[7] Johnson-Payton, L.R. (1997). *Risk Management of Compound Failure Modes in Engineering Systems*, Ph.D. Dissertation, Department of Systems Engineering, University of Virginia, Charlottesville, VA.

[8] Karlsson, P-O., and Y.Y. Haimes (1988). "Risk-based Analysis of Extreme Events," *Water Resources Research*, 24(1), 9-20.

[9] Lambert, J.H., L.R. Johnson, and Y.Y. Haimes (1996). "The Impact of Multiple Failure Modes in Risk Analysis for Civil Infrastructure Management," *Risk-Based Decision Making in Water Resources VI*, Y.Y. Haimes, D.A. Moser, and E.Z. Stakhiv, Eds., American Society of Civil Engineers, New York, 80-105.

[10] Lowrance, W. (1976). *Of Acceptable Risk*, William Kaufmann, Inc., Los Altos, CA.

[11] Mitsiopoulos, J., and Y.Y. Haimes (1989). "Generalized Quantification of Risk Associated with Extreme Events," *Risk Analysis*, 9(2), 243-254.

[12] Schooff, R. (1996). *HHM for Software Acquisition Risk*, Ph.D. Dissertation, Department of Systems Engineering, University of Virginia, Charlottesville, VA.

Y.Y. Haimes, Department of Systems Engineering and Center for Risk Management of Engineering Systems, University of Virginia Charlottesville, 22901.

J.H. Lambert, Department of Systems Engineering and Center for Risk Management of Engineering Systems, University of Virginia Charlottesville, 22901.

L.R. Johnson-Payton, Department of Systems Engineering and Center for Risk Management of Engineering Systems, University of Virginia Charlottesville, 22901.

Forming Win-win Strategy
------ A New Way to Study Game Problems

P. L. Yu* and J. M. Li**

* C. A. Scupin, Distinguished Professor, School of Business, University of Kansas, Lawrence, KS 66045, USA
** Visiting Scholar, School of Business, University of Kansas, Lawrence, KS 66045, USA (1997, 8 ~ 1998, 8). From Institute of Applied Mathematics, Guizhou University of Technology, Guiyang 550003, P. R. China

Abstract

In this paper, we briefly introduce the concept of habitual domains (HD) and its applications to conflict resolution. Relevant concepts of win-win strategy and mathematical methods for verifying if a joint strategy is a win-win strategy within a range of the relative weights (among the multiple criteria of each player) are derived. We also study the possibility of forming a win-win strategy by changing (or shifting) the habitual domains.

Keywords: Win-win strategy, weight, games with vector payoffs, habitual domains, second-order game, conflict resolution.

1 Introduction

Everything is changing with time. The only thing that is not changing is change. In fact, for a game problem, the number of criteria, the number of strategies, the preferences of the players, the players, and the payoffs perceived by each player etc., are subject to change. As such, if all the players are willing to cooperate to pursue potentially better results, then we should not doubt that there are many ways to acquire a good solution (or win-win strategy) for all the players involved by restructuring the above elements.

In order to deal with more realistic conflict problems, we present ideas of cooperation, non-cooperation, dynamics, and multicriteria payoffs in a more integrated and applicable way of habitual domains [1-3]. Our work is influenced by the concepts of second-order games [2 or 4], which is different from traditional game theory (for instance, see [5]) by allowing players' decision elements (strategy set, criteria, outcome, and preference) to vary with time. We will study the second-order game problems and conflict resolution from the viewpoint of the habitual domains, and formally introducing the concepts of forming win-win strategies

Recently, games with vector payoffs (or multicriteria games) get much attention (for instance, [6-9]). For games with vector payoffs, if we know each player utility function (or value function) or the relative weights of the criteria, then games with vector payoffs can be converted into games with scalar payoffs. Unfortunately, the players usually have hard time to identify precisely the relative weights of the criteria or utility functions. However, it is not so hard for them to estimate a range of the relative weights for the criteria. Therefore, it is important to study the existence of solutions for games with vector payoffs whose relative weights are within certain ranges. If we could restructure the weights of criteria for the players, we might find a win-win strategy for all the players, with each player claiming a victory. We shall report some important mathematical results in this approach.

The paper is organized as follows. In section 2, we introduce the concept of habitual domains and its applications to conflict resolution. In section 3, we introduce relevant concepts about win-win strategies and study the evolution of the game of single criterion. In section 4, we study the game with vector payoffs and derive methods to identify win-win strategies. In section 5, we study the possibility of forming a win-win strategy by changing (or shifting) the actual domain of each player. In the conclusion of section 5, we point out some further research problems.

2 Habitual domains and conflict resolution

Habitual Domains is the collection of all memory and thinking methods and judgment, together with their structures and dynamics. There are four basic concepts in habitual domains: (1) the actual domain (AD)--the set of ideas and actions that are actually activated at the present moment; (2) the potential domain (PD)--the collection of ideas and actions that can potentially be activated; (3) the activation probabilities (AP)--the probabilities that ideas and actions in PD also belong to AD; and (4) the reachable domain (RD)--the set of ideas and actions that can be attained from a given set in AD.

In order to fulfill our life goal, people need to interact with other people. People action or reaction is based on his actual domain and observable actual domains of the other parties. This may cause misunderstanding or misconception which may make the players' actual domains entangled. This entanglement usually causes conflict of the players.

Fortunately, as actual domain is only a small part of the potential domain, by expanding our habitual domains or by changing (or shifting) our actual domains we could disentangle our conflicts, and win-win situation could be obtained.

Let us give an example to illustrate the dynamics of conflict and the idea of expanding the habitual domains to find a solution to solve the conflict.

Example 2.1. Alinsky Strategy (adapted from [10])

During the days of the Johnson-Goldwater campaign (1960), commitments that were made by city authorities to the Woodlawn ghetto organization of Chicago were not being met. The organization could organize a demonstration or do nothing but wait patiently for the city authorities to keep the promise. After studying the possible outcome, they realized that these strategies could not change their unfair treatment of the city. Alinsky, a great social movement leader being asked to give a help, came up with a unique solvable situation. He would mobilize a large number of supporters to legally line up and occupy all the restroom facilities of the busy O are Airport. Imagine the chaotic situation of disruption and frustration that occurred when thousands of passengers who were hydraulically loaded (very high level of charge) rushed for restrooms but could not find the facility to relieve the charge.

How embarrassing when the newspapers and media around the world headlined and dramatized the situation. The supporters were extremely enthusiastic about the project, sensing the sweetness of revenge against the city. The threat of this tactic was leaked deliberately to the administration. The city image (not the budget) became the most important criterion of the administration . Within forty-eight hours the Woodlawn Organization was meeting with the city authorities, and the problem was, of course, solved graciously with each player releasing a charge and claiming a victory.

Consider the conflict situation in above story. First, the option of the Woodlawn organization was to organize a demonstration or do nothing but waiting passively. These options seem unable to change their unfair state. When Alinsky proposed a new strategy of legally lining up and occupying all the restroom facilities of the busy O are Airport, the conflict situation had been changed. The option of the organization became whether to execute the new strategy or not. The final result turned out to be satisfactory for both the Woodlawn organization and the city authorities. Think that only proposing a new strategy led to the resolution of the conflict! We can restructure the above conflict situation with matrix games in two stages. Interested readers see the unabridged version of this paper [11].

3 Win-win concepts and evolution of games

Let us consider the following game.

Let X_t^i be the *strategy set of the i-th player at time t*, i=1, 2, ..., n, $X_t=\prod_{i=1}^{n} X_t^i$ be the *set of joint strategies* (or simply call it a *set of points*). $\forall x \in X_t$, $x=(x_1, ..., x_n)$, when i is emphasized, we simply denote $x=(x_i, \hat{x}_i)$, where $x_i \in X_t^i$ and $\hat{x}_i \in \hat{X}_t^i = \prod_{j \neq i} X_t^j$. Let $F_t^i(x)$ be the value (or utility) payoff function of the i-th player.

Definition 3.1. A joint strategy $x^*=(x_i^*, \hat{x}_i^*)$, $i=1, ..., n$, is a *Nash-equilibrium point* (or simply call it *Nash point*) in X_t if

$$F_t^i(x_i^*, \hat{x}_i^*) \geq F_t^i(x_i, \hat{x}_i^*), \qquad \text{for all } i \text{ and } x_i \in X_t^i.$$

Definition 3.2. A joint strategy $x^*=(x_i^*, \hat{x}_i^*)$, $i=1, ..., n$, is a *self-interest point* in X_t if $F_t^i(x_i^*, \hat{x}_i)\geq F_t^i(x_i, \hat{x}_i)$, for all i and $x_i\in X_t^i$ and $\hat{x}_i \in \hat{X}_t^i$.

Definition 3.3. A joint strategy $x^*=(x_i^*, \hat{x}_i^*)$, $i=1, ..., n$, is an *ideal point* in X_t if $F_t^i(x_i^*, \hat{x}_i^*)\geq F_t^i(x_i, \hat{x}_i)$, for all i and $x_i\in X_t^i$ and $\hat{x}_i \in \hat{X}_t^i$.

Remark 3.1. Nash point is stable in the sense that no player can be better off if he unilaterally deviates from this point. The self-interest point corresponds to each player dominant strategy, which is more stable than Nash point (for further discussion, see [12]). At the ideal point, each player obtains his maximum payoff. In fact, an ideal point or a self-interest point must be a Nash point, but a Nash point is not necessary an ideal point or a self-interest point. An ideal point is not necessary a self-interest point and vice versa. We can see this from the forthcoming Example 3.1.

Definition 3.4. A joint strategy $x\in X_t$ is a *Pareto optimal point* (or simply call it *Pareto point*) if there is no $z\in X_t$ such that for each $i\in\{1, 2, ..., n\}$, $F_t^i(x)\leq F_t^i(z)$; and $\exists\ k\in\{1, 2, ..., n\}$, $F_t^k(x)\neq F_t^k(z)$.

Notice that an ideal point is also a Pareto optimal point, but the converse is not true.

Definition 3.5. Suppose that each player has a *minimal acceptable goal level* for satisfaction. For simplicity, we call it the *minimal acceptable level.* Let l_i denote the i-th player minimal acceptable level. A joint strategy which yields an outcome exceeding l_i is called a *satisfactory point* for the i-th player.

We assume that *each player is willing to cooperate to pursue a potentially better benefit.* Thus *a good solution (or a win-win strategy)* to the game *should at least be stable* (in the sense of self-interest or Nash equilibrium), *Pareto optimal* (with respect to the group payoffs), and *satisfactory* (for all the players). Therefore we give following definition.

Definition 3.6. Given that a joint strategy is a satisfactory point for all the players. We call it a *Nash win-win strategy*, if it is also a Pareto point and a Nash point; a *self-interest win-win strategy*, if it is also a Pareto point and a self-interest point; an *ideal win-win strategy*, if it is also an ideal point; a *perfect win-win strategy*, if it is also an ideal point and a self-interest point.

From Remark 3.1, we have:

Proposition 3.1. A perfect win-win strategy is also an ideal win-win strategy, a self-interest win-win strategy, and a Nash win-win strategy. Both self-interest win-win strategy and ideal win-win strategy are Nash win-win strategy.

Example 3.1. Suppose that A and B are the two players. Let N denote oncooperation", C denote ooperation". Let the payoff matrix be as follows:

$$\begin{array}{cc} & \text{B} \\ \text{A} & \begin{array}{c c} & \begin{array}{cc} N & C \end{array} \\ \begin{array}{c} N \\ C \end{array} & \begin{bmatrix} (4,4) & (10,2) \\ (2,10) & (8,8) \end{bmatrix} \end{array} \end{array}$$

This game is similar to the well-known game of he Prisoner Dilemma". Obviously, (N, N) is the unique Nash point, and it is also a self-interest point. But it is not a Pareto point because (N, N) is dominated by (C, C). Thus (N, N) is not a win-win strategy.

If every player is willing to cooperate to pursue a potentially better benefit, then unless a win-win strategy is formed, the game cannot have a stable solution. Suppose that both A and B minimal acceptable level are 8. Let us analyze the possible evolution of game in Example 3.1 from following stages.

Stage 1: (restructuring the game by changing the rule)

The joint strategies (N, C) and (C, C) are A satisfactory points. While (C, N) and (C, C) are B satisfactory points.

In this situation, (C, C) is a satisfactory point for both players. It is also a Pareto optimal point, but it is neither a self-interest point, nor a Nash point. If we can restructure the game (For more detailed discussion of this kind of restructuring, see [1, 12]) by adding a new rule that whoever deviates form " will have to pay 3 to the other side, then the payoff matrix becomes as follows:

$$\begin{array}{cc} & \text{B} \\ \text{A} & \begin{array}{c c} & \begin{array}{cc} N & C \end{array} \\ \begin{array}{c} N \\ C \end{array} & \begin{bmatrix} (4,4) & (7,5) \\ (5,7) & (8,8) \end{bmatrix} \end{array} \end{array}$$

In this situation, (C, C) is a common satisfactory point, a self-interest point, and an ideal point. Thus it is a perfect win-win strategy.

Stage 2: (restructuring the game by introducing new strategies)

If A or B or a third party proposes a new alternative (OC, OC), which means mutual overall cooperation, where OC denotes overall cooperation". Suppose the new payoff matrix is as follows:

$$\begin{array}{cc} & \text{B} \\ & \begin{array}{ccc} N & C & OC \end{array} \\ \text{A} \begin{array}{c} N \\ C \\ OC \end{array} & \begin{bmatrix} (4,4) & (10,2) & (10,1) \\ (2,10) & (8,8) & (9,8) \\ (1,10) & (8,9) & (10,10) \end{bmatrix} \end{array}$$

In this situation, (OC, OC) is both a common satisfactory point and an ideal point, thus it is an ideal win-win strategy. But it is not a perfect win-win strategy because it is not a self-interest point.

Stage : (restructuring the game by introducing new criteria)

Suppose A or B or a third party suggests that peaceful mind is very important. Assume that peaceful mind as a new criterion has caught the attention of A and B, and that the perceived payoff matrix is now as follows:

$$\begin{array}{cc} & \text{B} \\ & \begin{array}{cc} N & C \end{array} \\ \text{A} \begin{array}{c} N \\ C \end{array} & \begin{bmatrix} \left(\begin{pmatrix}4\\1\end{pmatrix}, \begin{pmatrix}4\\1\end{pmatrix}\right) & \left(\begin{pmatrix}10\\4\end{pmatrix}, \begin{pmatrix}2\\2\end{pmatrix}\right) \\ \left(\begin{pmatrix}2\\2\end{pmatrix}, \begin{pmatrix}10\\4\end{pmatrix}\right) & \left(\begin{pmatrix}8\\9.6\end{pmatrix}, \begin{pmatrix}8\\9.6\end{pmatrix}\right) \end{bmatrix} \end{array}$$

In this situation, how can we find a win-win strategy? Under what conditions (C, C) is a win-win strategy. We shall discuss this kind of game problems in the next section.

4 Matrix games with vector payoffs

For ease of presentation, let us focus on matrix games of two players. Its extension to n players is almost straightforward. We shall not stop to do so.

Let $X_1=\{x_1, x_2, ..., x_m\}$ and $X_2=\{y_1, y_2, ..., y_n\}$ be the strategy sets of player I and player II respectively, $X=X_1\times X_2=\{(x_i, y_j) \mid x_i\in X_1, y_j\in X_2\}$. Let $f(z)\in R^p$ and $g(z)\in R^q$, $\forall z\in X$, be player I and player II vector payoffs respectively. Let $f_{ij}=f(x_i, y_j)$, $g_{ij}=g(x_i, y_j)$, then the payoff matrix is as follows.

$$\begin{bmatrix} (f_{11}, g_{11}) & (f_{12}, g_{12}) & \cdots & (f_{1n}, g_{1n}) \\ (f_{21}, g_{21}) & (f_{22}, g_{22}) & \cdots & (f_{2n}, g_{2n}) \\ \cdots & \cdots & \cdots & \cdots \\ (f_{m1}, f_{m1}) & (f_{m2}, g_{m2}) & \cdots & (f_{mn}, g_{mn}) \end{bmatrix} \quad \text{(C)}$$

Denote the space of weight vector, which reflects the relative importance of criteria for player I by

$$\Delta_1=\{\lambda=(\lambda_1, \lambda_2, ..., \lambda_p)^T \mid \Sigma\lambda_i=1, \lambda_i\geq 0, i=1, 2, ..., p\}.$$

Denote the space of weight vector, which reflects the relative importance of criteria for player II by

$$\Delta_2=\{\mu=(\mu_1, \mu_2, ..., \mu_q)^T \mid \Sigma\mu_j=1, \mu_j\geq 0, j=1, 2, ..., q\}.$$

Suppose that player I estimates his relative weight vector to be in W_1, with $W_1=\{\lambda=(\lambda_1, \lambda_2, ..., \lambda_p)^T \mid \lambda_i\in[\lambda_i^0, \lambda_i^1]\subset[0, 1], i=1, 2, ..., p\}$;

and player II estimates his relative weight vector to be in W_2, with

$$W_2=\{\mu=(\mu_1, \mu_2, ..., \mu_q)^T \mid \mu_j\in[\mu_j^0, \mu_j^1]\subset[0, 1], j=1, 2, ..., q\}.$$

From the concept of relative weight and matrix (C), we have the following matrix game with payoffs as a function of weight variables:

$$\begin{bmatrix} (\lambda^T f_{11}, \mu^T g_{11}) & (\lambda^T f_{12}, \mu^T g_{12}) & \cdots & (\lambda^T f_{1n}, \mu^T g_{1n}) \\ (\lambda^T f_{21}, \mu^T g_{21}) & (\lambda^T f_{22}, \mu^T g_{22}) & \cdots & (\lambda^T f_{2n}, \mu^T g_{2n}) \\ \cdots & \cdots & \cdots & \cdots \\ (\lambda^T f_{m1}, \mu^T f_{m1}) & (\lambda^T f_{m2}, \mu^T g_{m2}) & \cdots & (\lambda^T f_{mn}, \mu^T g_{mn}) \end{bmatrix} \quad (D)$$

Definition 4.1. *A potential solution of the game with vector payoffs* (here, it means a potential Nash point, self-interest point, ideal point, Pareto point, or win-win strategy) is a solution to the game of (D) with weight variables satisfying the constraints of Δ_1 and Δ_2. It is called *a solution of the game with vector payoffs* if the weights are also satisfying the constraints of W_1 and W_2.

Note that for **a**, **b**$\in R^P$, $\forall\lambda\in\Delta_1$, if **a**$\geq$**b**, then λ^T**a**$\geq\lambda^T$**b**; if **a**>**b**, then λ^T**a**>λ^T**b**. Similarly, for **c**, **d**$\in R^q$, $\forall\mu\in\Delta_2$, if **c**$\geq$**d**, then μ^T**c**$\geq\mu^T$**d**; if **c**>**d**, then μ^T**c**>μ^T**d**.

From Definition 3.1-3.5, we have:

Proposition 4.1. Let us consider a joint strategy $(x_i, y_j)\in X$.

(i) If there is a $(x, y)\in X$ such that $f(x, y)\geq f(x_i, y_j)$, $g(x, y)\geq g(x_i, y_j)$, and at least one of the inequalities holds strictly, then (x_i, y_j) is not a potential Pareto optimal point.

(ii) If there is $(x, y_j)\in X$ such that $f(x, y_j)>f(x_i, y_j)$, or $(x_i, y)\in X$ such that $g(x_i, y)>g(x_i, y_j)$, then (x_i, y_j) is not a potential Nash point.

(iii) For each $x\in X_1$, if there is $y\in X_2$ such that $g(x, y)>g(x, y_j)$; or for each $y\in X_2$, there is $x\in X_1$ such that $f(x, y)>f(x_i,y)$; then (x_i, y_j) is not a potential self-interest point.

(iv) For each $(x, y)\in X$ such that $f(x_i, y_j)\geq f(x, y)$ and $g(x_i, y_j)\geq g(x, y)$, then (x_i, y_j) is an ideal point.

Note that this proposition is important for verifying if a joint strategy is a solution.

Proposition 4.2. For a joint strategy $(x_i, y_j)\in X$, let

$PN_{ij}=\{(\lambda, \mu) \mid \lambda^T f_{ij}\geq\lambda^T f_{kj}, k=1, ..., m; \mu^T g_{ij}\geq\mu^T g_{ih}, h=1, ..., n; \lambda\in\Delta_1, \mu\in\Delta_2\}$.

Then (x_i, y_j) is a *potential Nash point* if and only if $PN_{ij}\neq\phi$.

Proposition 4.3. For a joint strategy $(x_i, y_j)\in X$, let

$PI_{ij}=\{(\lambda, \mu) \mid \lambda^T f_{ij}\geq\lambda^T f_{kh}, \mu^T g_{ij}\geq\mu^T g_{kh}, \forall k=1, ..., m, h=1, ..., n; \lambda\in\Delta_1, \mu\in\Delta_2\}$.

Then (x_i, y_j) is a *potential ideal point* if and only if $PI_{ij} \neq \phi$.

Proposition 4.4. For a joint strategy $(x_i, y_j) \in X$, let
$PSI_{ij}=\{(\lambda, \mu) \mid \lambda^T f_{ih} \geq \lambda^T f_{kh}, \mu^T g_{kj} \geq \mu^T g_{kh}, \forall k=1, ..., m, h=1, ..., n; \lambda \in \Delta_1, \mu \in \Delta_2\}$.
Then (x_i, y_j) is a *potential self-interest point* if and only if $PSI_{ij} \neq \phi$.

Assumption 4.1. Assume the weights w_1 ($w_1>0$) and w_2 ($w_2>0$) reflect the power and/or importance of player I and player II, respectively.

Note that this assumption is innocent, because if $w_i=0$, i=1, 2, then the i-th player payoff become totally unimportant. The situation would not be a game problem.

Proposition 4.5. For a joint strategy $(x_i, y_j) \in X$, let
$PP_{ij}(w_1, w_2)=\{(\lambda, \mu) \mid w_1\lambda^T f_{ij}+w_2\mu^T g_{ij} \geq w_1\lambda^T f_{kh}+w_2\mu^T g_{kh}$, k=1, ..., m, h=1, ..., n; $\lambda \in \Delta_1, \mu \in \Delta_2\}$. Then (x_i, y_j) is a *potential Pareto point* with weights w_1 and w_2 if $PP_{ij}(w_1, w_2) \neq \phi$.

Proposition 4.6. For a joint strategy $(x_i, y_j) \in X$, if player I has a minimal acceptable level l_1, and player II has a minimal acceptable level l_2, let

$$PS_{ij}=\{(\lambda, \mu) \mid \lambda^T f_{ij} \geq l_1, \mu^T g_{ij} \geq l_2, \lambda \in \Delta_1, \mu \in \Delta_2\}.$$

Then (x_i, y_j) is a *potential common satisfactory point* if and only if $PS_{ij} \neq \phi$.

Let $W=W_1 \times W_2=\{(\lambda, \mu) \mid \lambda=(\lambda_1, \lambda_2, ..., \lambda_p)^T, \mu=(\mu_1, \mu_2, ..., \mu_q)^T, \lambda_i \in [\lambda_i^0, \lambda_i^1] \subset [0, 1]$, i=1, 2, ..., p; $\mu_j \in [\mu_j^0, \mu_j^1] \subset [0, 1]$, j=1, 2, ..., q$\}$ be a set of *weight combinations* of player I and player II. From above propositions and definitions, we have following theorem.

Theorem 4.1. A joint strategy (x_i, y_j) is a Nash (respectively, self-interest, ideal, or perfect) win-win strategy if $PN_{ij} \cap PP_{ij} \cap PS_{ij} \cap W \neq \phi$ (respectively, $PSI_{ij} \cap PP_{ij} \cap PS_{ij} \cap W \neq \phi$, $PI_{ij} \cap PS_{ij} \cap W \neq \phi$, or $PI_{ij} \cap PSI_{ij} \cap PS_{ij} \cap W \neq \phi$).

Definition 4.2. Given (w_1, w_2), $(x_i, y_j) \in X$, and $(\lambda^*, \mu^*) \in W$, (λ^*, μ^*) is called an *optimal weight combination* for (x_i, y_j) to be a Nash (repectively self-interest, ideal, or perfect) win-win strategy, if (λ^*, μ^*) is a solution of the linear programming $\{\max w_1\lambda^T f_{ij}+w_2\mu^T g_{ij}, (\lambda, \mu) \in PN_{ij} \cap PP_{ij} \cap PS_{ij} \cap W\}$
(respectively, $\{\max w_1\lambda^T f_{ij}+w_2\mu^T g_{ij}, (\lambda, \mu) \in PSI_{ij} \cap PP_{ij} \cap PS_{ij} \cap W\}$,
$\{\max w_1\lambda^T f_{ij}+w_2\mu^T g_{ij}, (\lambda, \mu) \in PI_{ij} \cap PS_{ij} \cap W\}$, or
$\{\max w_1\lambda^T f_{ij}+w_2\mu^T g_{ij}, (\lambda, \mu) \in PI_{ij} \cap PSI_{ij} \cap PS_{ij} \cap W\}$).
We can get following theorem, we omit the proof here.

Theorem 4.2. A joint strategy (x_i, y_j) is a Nash (respectively self-interest, ideal, or perfect) win-win strategy if and only if there exists a corresponding optimal weight combination (λ^*, μ^*) defined in Definition 4.2.

Example 4.1. Let us consider following 2×3 matrix game with vector payoffs, each player has 3 criteria.

$$\begin{bmatrix} \left(\begin{pmatrix}1\\4\\7\end{pmatrix},\begin{pmatrix}3\\2\\5\end{pmatrix}\right) & \left(\begin{pmatrix}3\\3\\1\end{pmatrix},\begin{pmatrix}6\\5\\6\end{pmatrix}\right) & \left(\begin{pmatrix}4\\6\\7\end{pmatrix},\begin{pmatrix}5\\8\\7\end{pmatrix}\right) \\ \left(\begin{pmatrix}6\\9\\2\end{pmatrix},\begin{pmatrix}5\\7\\6\end{pmatrix}\right) & \left(\begin{pmatrix}5\\4\\2\end{pmatrix},\begin{pmatrix}2\\6\\5\end{pmatrix}\right) & \left(\begin{pmatrix}3\\2\\4\end{pmatrix},\begin{pmatrix}5\\3\\8\end{pmatrix}\right) \end{bmatrix}$$

Note, $X_1=\{x_1, x_2\}$ and $X_2=\{y_1, y_2, y_3\}$ are the strategy sets of player I and player II respectively. Assume that player I weighted minimal acceptable level $l_1=6$, player II wighted minimal acceptable level $l_2=6$; and that player I and player II have same weight of power (and/or importance). Thus, $w_1=w_2=1$. Assume player I relative weight vector to be in $W_1=\{\lambda=(\lambda_1, \lambda_2, \lambda_3)^T \mid \lambda_1\in[0.1, 0.3], \lambda_2\in[0.2, 0.4], \lambda_3\in[0.6, 0.7]\}$, and player II relative weight vector to be in $W_2=\{\mu=(\mu_1, \mu_2, \mu_3)^T \mid \mu_1\in[0.2, 0.4], \mu_2\in[0.3, 0.5], \mu_3\in[0.45, 0.6]\}$.

From Proposition 4.1, it is easy to check that the joint strategies (x_1, y_1), (x_1, y_2), and (x_2, y_3) are not potential Nash point, and (x_2, y_2) is not a potential Pareto point.

Therefore, we only need to consider joint strategies (x_1, y_3) and (x_2, y_1).

For (x_2, y_1), notice that for each $\mu\in\Delta_2$,

$\mu^T g_{21}=\mu^T(5, 7, 6)^T\geq\mu^T(5, 8, 7)^T=\mu^T g_{13}$ if and only if $\mu_2=\mu_3=0$. But when $\mu_2=\mu_3=0$, $\mu^T g_{21}=\mu^T(5, 7, 6)^T<\mu^T(6, 5, 6)^T=\mu^T g_{12}$.

Thus (x_2, y_1) is not an ideal win-win strategy.

Next, by using Theorem 4.2, we solve following linear programming to see if (x_2, y_1) is a Nash win-win strategy. (Note, in the following, we delete those inequalities trivially held from the constraints.)

max $6\lambda_1+9\lambda_2+2\lambda_3+5\mu_1+7\mu_2+6\mu_3$

s.t.

$5\lambda_1+5\lambda_2-5\lambda_3\geq 0$, $\quad 4\mu_2-2\mu_3\geq 0$,

$5\lambda_1+5\lambda_2-5\lambda_3+2\mu_1+5\mu_2+\mu_3\geq 0$,

$3\lambda_1+6\lambda_2+\lambda_3-\mu_1+2\mu_2\geq 0$,

$2\lambda_1+3\lambda_2-5\lambda_3-\mu_2-\mu_3\geq 0$,

$3\lambda_1+7\lambda_2-2\lambda_3+4\mu_2-2\mu_3\geq 0$;

$6\lambda_1+9\lambda_2+2\lambda_3\geq 6$, $\quad 5\mu_1+7\mu_2+6\mu_3\geq 6$;

$\lambda_1+\lambda_2+\lambda_3=1$, $\quad \mu_1+\mu_2+\mu_3=1$,

$0.1\leq\lambda_1\leq 0.3$, $\quad 0.2\leq\mu_1\leq 0.4$,

$0.2\leq\lambda_2\leq 04$, $\quad 0.3\leq\mu_2\leq 0.5$,

$0.6\leq\lambda_3\leq 0.7$; $\quad 0.45\leq\mu_3\leq 0.6$,

Using LINDO [13], we know that the programming has no feasible solution. Thus (x_2, y_1) is not a Nash win-win strategy. But it is a potential Nash win-win strategy, because when we discard inequalities (10), (11), (12), (14), (15), and (16) from the above constraints, and let $\lambda_i \geq 0$, $\mu_i \geq 0$, $i=1, 2, 3$, then the corresponding linear programming has a solution: $\lambda_1=0$, $\lambda_2=1$, $\lambda_3=0$, $\mu_1=0$, $\mu_2=1$, $\mu_3=0$. maximum=16.

Is it possible for (x_2, y_1) to become a Nash win-win strategy? We will discuss this problem in the next section.

For (x_1, y_3), similar to (x_2, y_1), by Theorem 4.2 and Proposition 4.1, we can show that it is an ideal win-win strategy, but it is not a perfect win-win strategy (see [11] for details).

5 Forming a win-win strategy

Let $HD_i=\{PD_i, AD_i, AP_i, RD_i\}$, i=1, 2, represents the habitual domains of player I and player II, respectively. Recall that PD, AD, AP and RD are potential domain, actual domain, activation probability and reachable domain respectively. Note that

$$\{X_1, \Delta_1 \cap W_1, l_1, f_{ij}, i=1, ..., m; j=1, ..., n\} \subset AD_1, \; \Delta_1 \cup AD_1 \subset PD_1; \text{ and}$$
$$\{X_2, \Delta_2 \cap W_2, l_2, g_{ij}, i=1, ..., m; j=1, ..., n\} \subset AD_2, \; \Delta_2 \cup AD_2 \subset PD_2.$$

Recall that l_1 and l_2 are the minimal acceptable level of player I and player II, respectively, and f_{ij} and g_{ij}, i=1, ..., m; j=1, ..., n, are the payoffs perceived by player I and player II respectively. $\Delta_1 \cap W_1$ and X_1 is the observable actual domain of player I; while $\Delta_2 \cap W_2$ and X_2 is the observable actual domain of player II.

From the viewpoint of HD, if there is no win-win strategy for the game, then by changing (or shifting) the players' actual domains, or by expanding the players' potential domains, it is possible to form a win-win strategy. However, how much are the players willing to change? What is the best way to proceed? These are not only art of persuasion and negotiation, but also technical problems. Different person may have different answers to these questions. In the following, we only focus on the technical aspects.

Let us consider the joint strategy (x_2, y_1) in Example 4.1. We already know that it is not a Nash win-win strategy, but it is a potential Nash win-win strategy. When the players agree to adjust their relative weights to some extent, will (x_2, y_1) become a Nash win-win strategy ?

First, let us determine the *maximum range* of the relative weights for (x_2, y_1) to be a potential Nash win-win strategy by solving the following linear programming problems.

$$\max\{\lambda_i \mid (\lambda, \mu) \in PN_{21} \cap PP_{21} \cap PS_{21}\}, \quad \min\{\lambda_i \mid (\lambda, \mu) \in PN_{21} \cap PP_{21} \cap PS_{21}\}; \text{ and}$$
$$\max\{\mu_i \mid (\lambda, \mu) \in PN_{21} \cap PP_{21} \cap PS_{21}\}, \quad \min\{\mu_i \mid (\lambda, \mu) \in PN_{21} \cap PP_{21} \cap PS_{21}\}, \; i=1,$$
2, 3.

Using LINDO, we get that $\lambda_1 \in [0, 1]$, $\lambda_2 \in [0, 1]$, $\lambda_3 \in [0, 0.3125]$; and $\mu_1 \in [0, 0.5]$, $\mu_2 \in [0.25, 1]$, $\mu_3 \in [0, 0.66]$.

Recall that

$$W_1=\{\lambda=(\lambda_1, \lambda_2, \lambda_3)^T \mid \lambda_1 \in [0.1, 0.3], \lambda_2 \in [0.2, 0.4], \lambda_3 \in [0.6, 0.7]\}, \text{ and}$$
$$W_2=\{\mu=(\mu_1, \mu_2, \mu_3)^T \mid \mu_1 \in [0.2, 0.4], \mu_2 \in [0.3, 0.5], \mu_3 \in [0.45, 0.6]\}.$$

Comparing W_1 and W_2 with the maximum range of the relative weights for (x_2, y_1) to be a potential Nash win-win strategy, we know the main reason for (x_2, y_1) not being a Nash win-win strategy is the improper constraint $\lambda_3 \in [0.6, 0.7]$ in W_1. Thus to make (x_2, y_1) a Nash win-win strategy, player I should lower his weight for the third criterion to a proper range. Which means that the actual domain of player I should be changed (or shifted) by self-suggestion or by inputs of external information.

Note that comparing $W=W_1 \times W_2$ with the maximum range of the relative weights for (x_2, y_1) to be a potential Nash win-win strategy is a good way to find a direction for adjustment, but it is not a precise way because W and the maximum range of the relative weights having a non-empty intersection is only a necessary condition for the joint strategy (x_2, y_1) to be a Nash win-win strategy.

To be more precise, theoretically, by using computer, we can find an optimal approximation of the shortest distance between W and $PN_{ij} \cap PP_{ij} \cap PS_{ij}$. Thus we can roughly estimate the level of difficulty for changing (or shifting) the players' actual domains, and find a possible starting point and a possible target point for conducting persuasion to change the related actual domains.

In short, if we can change the environment of the game or the perceptions of the players, then a win-win strategy is possible to be formed. If we can create a joint strategy which maximizes each player' weighted payoff, then it can possibly be an ideal win-win strategy; if it also satisfies the conditions of self-interest point, then it can possibly be a perfect win-win strategy.

How to expand the players' HDs, and effectively restructure games as to generate a win-win strategy are certainly very important in our daily lives. The interested reader is referred to [1, 4] (especially Chapters 11-12 of [1]). We will not repeat it here.

6 Conclusions

We have studied the game problems from the view of habitual domain theory and derived relevant mathematical results about win-win strategy. The main idea is that by changing (or shifting) the actual domains of the players or by expanding the potential domains of the players, it is possible for the players to form a win-win strategy. There are many topics deserving further research. For examples, how to apply the ideas and methods of this paper to solve real-world conflicts and negotiation problems? How to make the computation more efficient? How to use computer to illustrate graphically the process of forming a win-win strategy? Many practical methods described in Chapters 11-12 of [1] can be mathematically formulated and solved in a similar way suggested by this paper. However, this still needs to be carried out.

References

[1] Yu, P. L.(1990). *Forming winning strategies: An Integrated Theory of Habitual Domains*. Springer Verlag, Heidelberg, Germany.

[2] Yu, P. L. (1985). *Multiple-Criteria Decision making: Concepts, Techniques, and Extensions*. Plenum, New York, NY.

[3] Yu, P. L. (1995). *Habitual Domains: Freeing Yourself from the Limits on Your Life*. Highwater Editions, Kansas.

[4] Yu, P. L. (1979). "Second-Order Game Problem: Decision Dynamics in Gaming Phenomena," *Journal of Optimization Theory and Applications*, 27, 147-166.

[5] Owen, G. (1995). *Game Theory*, 3rd Edition. Academic Press, New York.

[6] Bergstresser, K. and P. L. Yu (1977). "Domonation Structures and Multicriteria Problems in N-Person Games," *Theory and Decision*, 8, 5-48.

[7] Wang, S. Y. (1993). "Existence of a Pareto Equilibrium," *Journal of Optimization Theory and Applications*, 79, 373-384.

[8] Nishizaki, I. and M. Sakawa (1995). "Equilibrium Solutions for Multi-objective Bimatrix Games Incorporating Fuzzy Goals," *Journal of Optimization Theory and Applications*, 86, 433-457.

[9] Fernandez, F. R. and J. Purerto (1996). "Vector Linear Programming in Zero-Sum Multicriteria Matrix Games," *Journal of Optimization Theory and Applications*, 89, 115-127.

[10] Alinsky, S. D. (1972). *Rules for Radicals*. Vintage Books, New York.

[11] Yu, P. L. and J. M. Li (1998). *Forming Win-win Strategy--A New Way to Study Game Problems*. Working paper, School of Business, Uni. of Kansas.

[12] Kwon, Y. K., and P. L. Yu (1983). "Conflict Dissolution by Reframing Game Payoffs Using Linear Perturbations," *Journal of Optimization Theory and Applications*, 39, 187-214.

[13] Schrage, L. (1984). *Linear, Integer, and Quadratic Programming with LINDO*. Scientific Press, Palo Alto, California.

Relations among Several Efficiency Concepts in Stochastic Multiple Objective Programming

R. Caballero[1], E. Cerdá[2], M.M. Muñoz[1] and L. Rey[1]

[1] Department of Applied Economics (Mathematics), University of Málaga. Campus El Ejido s/n, 29071 Málaga, Spain.

[2] Department of Foundations of Economic Analysis, University Complutense of Madrid. Campus de Somosaguas, 28223 Madrid, Spain.

Abstract. In this paper, the resolution of stochastic multiple objective programming problems is studied. The existence of random parameters in the objective functions has resulted in the definition of several efficient solution concepts for such problems in the literature. We will focus our attention in the study of some of these concepts, namely, minimum risk and β probability. Once these concepts are defined, the relations among the sets of efficient solutions obtained are studied.

Keywords. Multiobjective programming, stochastic programming, efficiency.

1 Introduction

Let us consider the following Stochastic Multiple Objective Programming Problem:

$$\begin{aligned} &\operatorname*{Min}_{\mathbf{x}}\ \tilde{\mathbf{z}}(\mathbf{x},\tilde{\xi})=\left(\tilde{z}_1(\mathbf{x},\tilde{\xi}),\tilde{z}_2(\mathbf{x},\tilde{\xi}),\ldots,\tilde{z}_q(\mathbf{x},\tilde{\xi})\right)^t \\ &\text{s.t}\quad \mathbf{x}\in D \end{aligned} \qquad \text{(SMP)}$$

where $\mathbf{x} \in \mathbb{R}^n$ is the vector of decision variables of the problem and $\tilde{\xi}$ is a random vector defined on a set $\mathbb{E} \subset \mathbb{R}^s$. We assume that the family of events $\mathbb{F}$ is given and that for every $A \in \mathbb{F}$ the probability of A, P(A), is known. We also assume that the distribution of probability P is independent of the decision variables $x_1, x_2, \ldots x_n$.

We assume that the functions $\tilde{z}_1(\mathbf{x},\tilde{\xi}), \tilde{z}_2(\mathbf{x},\tilde{\xi}),\ldots, \tilde{z}_q(\mathbf{x},\tilde{\xi})$ are defined in the space $\mathbb{R}^n \text{x } \mathbb{E}$. We also assume that the set $D \subset \mathbb{R}^n$ is compact, convex, and nonempty and that it is a deterministic set or has been transformed into its deterministic equivalent by the criterion of chance constraints.

As the objective functions depend on the random parameter vector $\tilde{\xi}$ in the SMP problem, this implies that there can exist two different events of vector $\tilde{\xi}$, ξ_1 and ξ_2, such that, given two feasible vectors **x**, **y** $\in D$, then for some $k \in \{1, 2, ..., q\}$,

$$z_k(\mathbf{x},\xi_1) < z_k(\mathbf{y},\xi_1) \text{ and } z_k(\mathbf{x},\xi_2) > z_k(\mathbf{y},\xi_2)$$

Taking this fact into account, the necessity to specify some solution concepts for these problems seems evident. In the literature, several such concepts can be found. The basic idea of such definitions is to apply some of the criteria that exist in the stochastic programming literature to the stochastic objectives, in order to transform them into equivalent deterministic objectives. Using these transformations, an equivalent deterministic multiple objective problem is obtained, with a corresponding associated set of efficient solutions. Such solutions are considered to be efficient solutions to the original stochastic problem.

The existence of several transformation criteria from stochastic to deterministic objectives, gives rise to different efficient solution concepts. Therefore, associated with the same stochastic multiple objective problem, there can exist different sets of efficient solutions, one for each efficient solution concept considered. This fact can cause a certain confusion. Thus, it seems interesting to consider the existence of some kinds of relations among these concepts.

Next, two different efficient solution concepts for SMP problems will be considered: the *minimum risk efficiency for levels* $u_1, u_2, ..., u_q$, and the *efficiency with probabilities* $\beta_1, \beta_2, ..., \beta_q$. After defining these two concepts, the relationship between them is analyzed.

2 Minimum Risk Efficiency for Levels $u_1, u_2, ..., u_q$

This solution concept, defined by Stancu-Minasian and Tigan (1984), compares efficient solutions for the SMP problem to the efficient solutions for the multiple objective deterministic problem that are obtained when we apply the minimum risk criterion to each of the objective functions of the problem. To apply this criterion it is necessary to fix a level of minimum satisfaction for each of the stochastic objectives, $u_1, u_2, ..., u_q$, $u_k \in \mathbb{R}$, $k = 1, 2..., q$. When these values are fixed, the minimum risk problem, the deterministic equivalent of the SMP problem, consists in maximising the probability that each of the stochastic objectives does not surpass the fixed satisfaction level, in such a way that the deterministic equivalent to the SMP problem is:

$$\begin{aligned} &\operatorname*{Max}_{\mathbf{x}} \left(P(\tilde{z}_1(\mathbf{x},\tilde{\xi}) \le u_1),..., P(\tilde{z}_q(\mathbf{x},\tilde{\xi}) \le u_q)\right) \qquad \mathbf{MR(u)} \\ &\text{s.t} \quad \mathbf{x} \in D \end{aligned}$$

For this problem, Stancu-Minasian and Tigan (1984) define the concept of vectorial solution minimum risk of level $u_1, u_2, \ldots, u_q$ for the SMP problem in the following way:

Definition 1: Vectorial solution minimum risk of level u

$\mathbf{x} \in D$ is a vectorial solution minimum risk of level $\mathbf{u}$ for the SMP problem if it is an efficient solution to Problem MR(**u**).

From now on, we shall call these solutions efficient minimum risk solutions of $u_1, u_2, \ldots, u_q$. We denote by $\mathcal{E}_{MR}(\mathbf{u})$ the set of efficient solutions to the Problem MR(**u**).

The multiple objective deterministic equivalent problem that is obtained by applying this criterion, problem MR(**u**), generally depends on the fixed vector of satisfaction levels **u**, in such a way that, in general, given $\mathbf{u}, \mathbf{u}' \in \mathbb{R}^q$, if $\mathbf{u} \neq \mathbf{u}'$ then the sets of efficient minimum risk solutions of levels **u** and **u'** will be different: $\mathcal{E}_{MR}(\mathbf{u}) \neq \mathcal{E}_{MR}(\mathbf{u}')$.

3 Efficient Solutions with Probabilities $\beta_1, \beta_2, \ldots, \beta_q$

The concept of efficiency with probabilities $\beta_1, \beta_2, \ldots, \beta_q$ is a generalisation of a concept previously defined by Goicoechea, Hansen and Duckstein (1982), i. e., the concept of stochastic nondominated solution of level β, which they define in the following way:

Definition 2: Stochastic nondominated solution of level β

Let $z_k(\mathbf{x})$ be a value belonging to the rank of the random variable $\tilde{z}_k(\mathbf{x}, \tilde{\xi})$, $k = 1, 2, \ldots, q$. $\mathbf{x} \in D$ is a stochastic nondominated solution of level $\beta \in (0,1)$ if:

(i) $P\{\tilde{z}_k(\mathbf{x}, \tilde{\xi}) \leq z_k(\mathbf{x})\} = \beta,\ \forall k \in \{1,2,\ldots,q\}$

(ii) There does not exist a vector $\mathbf{y} \in D$ such that:

* $P\{\tilde{z}_k(\mathbf{y}, \tilde{\xi}) \leq z_k(\mathbf{y})\} = \beta,\ \forall k \in \{1, 2, \ldots, q\}$
* $\exists l \in \{1, 2, \ldots, q\}$ such that $z_l(\mathbf{y}) < z_l(\mathbf{x})$
* $z_k(\mathbf{y}) \leq z_k(\mathbf{x})\ \forall k \in \{1, 2, \ldots, q\},\ k \neq l$

From this definition, given the Stochastic Multiple Objective Programming Problem, if we apply the Kataoka criterion to each of the stochastic objective functions of the problem with a probability β, we obtain the following problem:

$$\begin{gathered}\underset{\mathbf{x},\mathbf{u}}{\text{Min}}\ \mathbf{u}=(u_1,\ldots,u_q)^t\\ \text{s.t}\ P\{\tilde{z}_k(\mathbf{x},\tilde{\xi})\le u_k\}=\beta,\ k=1,2,\ \ldots,q\\ \mathbf{x}\in D\end{gathered}$$

and we find that the set of efficient solutions to this problem is the set of nondominated solutions of level β previously defined, because for each $k \in \{1, 2, \ldots, q\}$ the variable u_k will be a function $z_k(\mathbf{x})$ that is obtained from the equality $P\{\tilde{z}_k(\mathbf{x},\tilde{\xi})\le u_k\}=\beta$. In this way, we find that the set of nondominated solutions of level β is obtained from the application of the Kataoka criterion to each of the objective functions of the stochastic multiple objective problem, fixing the same probability level for all the stochastic functions.

From this concept, it is possible to generalise the idea, considering different probability levels for the objective functions of the problem in the following way:

$$\begin{gathered}\underset{\mathbf{x},\mathbf{u}}{\text{Min}}\ \mathbf{u}=(u_1,\ldots,u_q)^t\\ \text{s.t}\ P\{\tilde{z}_k(\mathbf{x},\tilde{\xi})\le u_k\}=\beta_k,\ k=1,2,\ldots,q\\ \mathbf{x}\in D\end{gathered} \qquad \mathrm{K}(\beta)$$

Definition 3: Efficient solution with probabilities $\beta_1, \beta_2, \ldots, \beta_q$

Let $\mathbf{x} \in D$. We say that $\mathbf{x}$ is an efficient solution with probabilities $\beta_1, \beta_2, \ldots, \beta_q$ if there exists $\mathbf{u} \in \mathbb{R}^q$ such that $(\mathbf{x}^t, \mathbf{u}^t)^t$ is an efficient solution of the problem K(β).

We denote by $\mathcal{E}_K(\beta)$ the set of efficient solutions with probabilities $\beta = (\beta_1, \beta_2, \ldots, \beta_q)^t$. Note that the concept of efficient solution with probabilities $\beta_1, \beta_2, \ldots, \beta_q$ is defined for the vector $\mathbf{x}$, although the solutions of the problem to solve are vectors $(\mathbf{x}^t, \mathbf{u}^t)^t \in \mathbb{R}^{n+q}$.

As in the case of minimum risk, this concept of efficiency is associated with some previously fixed probability levels, and therefore the deterministic multiple objective problem in which efficient solutions with probabilities $\beta_1, \beta_2, \ldots, \beta_q$ are obtained (problem K(β)),generally depends on the fixed vector of probabilities, $\beta = (\beta_1, \beta_2, \ldots, \beta_q)^t$. Then, in general, given $\beta, \beta' \in \mathbb{R}^q$, if $\beta \neq \beta'$, then the set of efficient solutions for β is different to the one obtained for β': $\mathcal{E}_K(\beta) \neq \mathcal{E}_K(\beta')$.

4 Relations among the Efficient Minimum Risk Solutions of Levels $u_1, u_2, \ldots, u_q$ and the Efficient Solutions with Probabilities $\beta_1, \beta_2, \ldots, \beta_q$

With the SMP problem as our starting point, let us consider the following problems:

$$\begin{aligned} &\underset{\mathbf{x}}{\text{Max}} \ \left(P(\tilde{z}_1(\mathbf{x},\tilde{\xi}) \le u_1), \ldots, P(\tilde{z}_q(\mathbf{x},\tilde{\xi}) \le u_q)\right) \\ &\text{s.t} \quad \mathbf{x} \in D \end{aligned} \qquad \text{MR}(\mathbf{u})$$

$$\begin{aligned} &\underset{\mathbf{x},\mathbf{u}}{\text{Min}}\, \mathbf{u} = (u_1, \ldots, u_q) \\ &\text{s.t} \ P\left\{\tilde{z}_k(\mathbf{x},\tilde{\xi}) \le u_k\right\} = \beta_k, \ k = 1,2, \ldots, q \\ &\mathbf{x} \in D \end{aligned} \qquad \text{K}(\beta)$$

corresponding to the deterministic programs from which we obtain the efficient solutions minimum risk of levels $u_1, u_2, \ldots, u_q$ and to the efficient solutions with probabilities $\beta_1, \beta_2, \ldots, \beta_q$ for the SMP problem. Now we are going to analyse the relations among the sets of efficient points of these two problems.

We assume that the feasible sets of both problems, $D \subset \mathbb{R}^n$ and

$$\left\{(\mathbf{x}^t, \mathbf{u}^t)^t \in D \times \mathbb{R}^q \,/\, P\left\{\tilde{z}_k(\mathbf{x},\tilde{\xi}) \le u_k\right\} = \beta_k, k=1,2, \ldots, q\right\}$$

are closed, bounded and nonempty, and that therefore both problems have efficient solutions.

We also assume that for every $k \in \{1, 2, \ldots, q\}$, and for every $\mathbf{x} \in D$, the distribution function of the random variable $\tilde{z}_k(\mathbf{x},\tilde{\xi})$ is continuous and strictly increasing. These hypotheses imply that for every probability β_k there exists a unique real number u_k such that $P\left\{\tilde{z}_k(\mathbf{x},\tilde{\xi}) \le u_k\right\} = \beta_k$.

Let $\mathcal{E}_{MR}(\mathbf{u})$ be the set of efficient solutions to the problem MR(**u**), and $\mathcal{E}_K(\beta)$ the set of efficient solutions to the problem K(β). The following theorem relates both sets.

Theorem 1

Assume that the distribution function of the random variable $\tilde{z}_k(\mathbf{x},\tilde{\xi})$ is continuous and strictly increasing. Then **x** is an efficient solution to problem MR(**u**) if and only if $(\mathbf{x}^t, \mathbf{u}^t)$ is an efficient solution to problem K(β), with **u** and β such that:

$$P\{\tilde{z}_k(\mathbf{x},\tilde{\xi}) \le u_k\} = \beta_k, \quad \forall\, k \in \{1, 2, ..., q\}.$$

Proof

We demonstrate the theorem by *reductio ad absurdum*.

(a) If **x** is an efficient solution to problem MR(**u**), then $(\mathbf{x}^t, \mathbf{u}^t)$ is an efficient solution to problem K(β). It is clear that $(\mathbf{x}^t, \mathbf{u}^t)^t$ is a feasible solution to problem K(β).

Let us suppose that $(\mathbf{x}^t, \mathbf{u}^t)^t$ is not efficient for problem K(β). Then there exists a feasible vector $(\mathbf{x}'^t, \mathbf{u}'^t)^t$ that dominates $(\mathbf{x}^t, \mathbf{u}^t)^t$, and therefore it is verified that:

$$\begin{aligned}&\mathbf{x}' \in D\\ &P\{\tilde{z}_k(\mathbf{x}',\tilde{\xi}) \le u_k'\} = P\{\tilde{z}_k(\mathbf{x},\tilde{\xi}) \le u_k\} = \beta_k, \forall k \in \{1, 2, ..., q\}\\ &u_k' \le u_k \ \forall k \in \{1,2,...,q\} \text{ and } u_s' < u_s \text{ for some } s \in \{1, 2, ..., q\}\end{aligned}$$

In accordance with the properties of the distribution function of the random variable $\tilde{z}_k(\mathbf{x},\tilde{\xi})$, if $u_k' \le u_k$ and $u_s' < u_s$ then:

$$\begin{aligned}&P\{\tilde{z}_k(\mathbf{x}',\tilde{\xi}) \le u_k'\} \le P\{\tilde{z}_k(\mathbf{x}',\tilde{\xi}) \le u_k\}\\ &P\{\tilde{z}_s(\mathbf{x}',\tilde{\xi}) \le u_s'\} < P\{\tilde{z}_s(\mathbf{x}',\tilde{\xi}) \le u_s\}\end{aligned}$$

Therefore:

$$\begin{aligned}&P\{\tilde{z}_k(\mathbf{x},\tilde{\xi}) \le u_k\} = P\{\tilde{z}_k(\mathbf{x}',\tilde{\xi}) \le u_k'\} \le P\{\tilde{z}_k(\mathbf{x}',\tilde{\xi}) \le u_k\}\\ &P\{\tilde{z}_s(\mathbf{x},\tilde{\xi}) \le u_s\} = P\{\tilde{z}_s(\mathbf{x}',\tilde{\xi}) \le u_s'\} < P\{\tilde{z}_s(\mathbf{x}',\tilde{\xi}) \le u_s\}\end{aligned}$$

and **x** is not an efficient solution to problem MR(**u**),which contradicts the hypothesis.

(b) If $(\mathbf{x}^t, \mathbf{u}^t)^t$ is an efficient solution to problem K(β), then **x** is an efficient solution to problem MR(**u**). It is clear that $\mathbf{x} \in D$.

Suppose that **x** is not an efficient solution to problem MR(**u**), then there exists a feasible vector $\mathbf{x}' \in D$, and it is verified that:

$$\begin{aligned}&\beta_k = P\{\tilde{z}_k(\mathbf{x},\tilde{\xi}) \le u_k\} \le P\{\tilde{z}_k(\mathbf{x}',\tilde{\xi}) \le u_k\}, \forall k \in \{1,2,,...,q\}\\ &\beta_s = P\{\tilde{z}_s(\mathbf{x},\tilde{\xi}) \le u_s\} < P\{\tilde{z}_s(\mathbf{x}',\tilde{\xi}) \le u_s\}, \text{ for some } s \in \{1,2,,...,q\}\end{aligned}$$

For the properties of the distribution function we know that there exist $u_1', u_2', ..., u_q'$ with $u_k' \le u_k$, $\forall$ k $\in$ {1,2,...,q}, and that there exists at least one s $\in$ {1,2,...,q}, such that $u_s' < u_s$, verifying that:

$$\begin{aligned}&\beta_k = P\{\tilde{z}_k(\mathbf{x},\tilde{\xi}) \le u_k\} = P\{\tilde{z}_k(\mathbf{x}',\tilde{\xi}) \le u_k'\}, \forall k \in \{1, 2, ..., q\}\\ &\beta_s = P\{\tilde{z}_s(\mathbf{x},\tilde{\xi}) \le u_s\} = P\{\tilde{z}_s(\mathbf{x}',\tilde{\xi}) \le u_s'\}, \text{ for some } s \in \{1, 2, ..., q\}\end{aligned}$$

which is in contradiction to the hypothesis.

Corollary 1

$$\bigcup_{\mathbf{u}\in\mathbb{R}^q} \mathcal{E}_{MR}(\mathbf{u}) = \bigcup_{\beta\in B} \mathcal{E}_K(\beta)$$

with $B = \{\beta \in \mathbf{R}^q / \beta_k \in (0,1), k = 1, 2, ..., q\}$.

The proof of this corollary is immediate from the previous results.

From the results obtained we can see that the unions of the sets of efficient points of both problems coincide. Moreover, if $\mathbf{x} \in D$ is an efficient solution to problem $K(\beta)$, for some fixed probabilities $\beta = (\beta_1, \beta_2, ..., \beta_q)^t$ from theorem 1, we know that it is also an efficient minimum risk solution of levels $u_1, u_2, ..., u_q$, maintaining for the satisfaction levels and the probabilities the relation that appears in the theorem, and vice versa. This result permits us to perform the analysis of these efficient solutions by one of the two concepts and, from theorem 1, to obtain the level or the probability for which it is efficient in accordance with the other.

5 Application of the Cantelli Inequality to the Distribution Function of the Stochastic Objective

In the previous sections we have studied the concept of efficient minimum risk solution of levels $u_1, u_2, ..., u_q$ and the concept of efficient solution with probabilities $\beta_1, \beta_2, ..., \beta_q$. We also have obtained the relationship between the two concepts. In this section we are going to present an approach to try to study some cases in which it is very difficult, if not impossible, to obtain these solutions. Let us note that in order to obtain efficient solutions to problems $MR(\mathbf{u})$ and $K(\beta)$, it is necessary to know the probability distribution of the stochastic objectives of the SMP problem, which is not always possible. We propose using the Cantelli inequality (Rao (1973)) in order to obtain some insight into these cases.

Cantelli inequality

Let $\tilde{\xi}$ be a random variable, with the expected value $\bar{\xi}$ and finite variance σ_ξ^2. Then:

$$P\left(\tilde{\xi} - \bar{\xi} \le \lambda\right) \begin{cases} \le \dfrac{\sigma_\xi^2}{\sigma_\xi^2 + \lambda^2} & \text{if } \lambda < 0 \\ \ge 1 - \dfrac{\sigma_\xi^2}{\sigma_\xi^2 + \lambda^2} = \dfrac{\lambda^2}{\sigma_\xi^2 + \lambda^2} & \text{if } \lambda \ge 0 \end{cases}$$

Let us suppose that we know the expected value of the random variable $\tilde{z}_k(\mathbf{x},\tilde{\xi})$, $E\{\tilde{z}_k(\mathbf{x},\tilde{\xi})\}$ and its variance $\mathrm{Var}\{\tilde{z}_k(\mathbf{x},\tilde{\xi})\}$ Also suppose that the feasible set of the deterministic equivalent D is such that the variance is finite and its value is different from zero for all feasible $\mathbf{x}$.

In this case, if we apply the Cantelli inequality to the distribution function of the k objective, for $u_k \geq E\{\tilde{z}_k(\mathbf{x},\tilde{\xi})\}$ taking $\lambda = u_k - E\{\tilde{z}_k(\mathbf{x},\tilde{\xi})\} \geq 0$, we obtain:

$$P\{\tilde{z}_k(\mathbf{x},\tilde{\xi}) \leq u_k\} = P\left(\tilde{z}_k(\mathbf{x},\tilde{\xi}) - E\{\tilde{z}_k(\mathbf{x},\tilde{\xi})\} \leq u_k - E\{\tilde{z}_k(\mathbf{x},\tilde{\xi})\}\right) \geq$$
$$\geq \frac{\left(u_k - E\{\tilde{z}_k(\mathbf{x},\tilde{\xi})\}\right)^2}{\mathrm{Var}\{\tilde{z}_k(\mathbf{x},\tilde{\xi})\} + \left(u_k - E\{\tilde{z}_k(\mathbf{x},\tilde{\xi})\}\right)^2}$$

If we substitute the distribution functions of the objectives by these bounds in problem MR(**u**), we obtain the following new problem:

$$\begin{array}{ll} \underset{\mathbf{x}}{\mathrm{Max}} & \dfrac{\left(u_k - E\{\tilde{z}_k(\mathbf{x},\tilde{\xi})\}\right)^2}{\mathrm{Var}\{\tilde{z}_k(\mathbf{x},\tilde{\xi})\} + \left(u_k - E\{\tilde{z}_k(\mathbf{x},\tilde{\xi})\}\right)^2} \quad k = 1,2,\ldots,q \\ \text{s.t} & E\{\tilde{z}_k(\mathbf{x},\tilde{\xi})\} \leq u_k, \quad k = 1,2,\ldots,q \\ & \mathbf{x} \in D \end{array} \qquad (\mathrm{AMR}(\mathbf{u}))$$

It is clear that the set of efficient solutions to problem AMR(**u**), denoted by $\mathcal{E}_{AMR}(\mathbf{u})$, in general does not coincide with the set of efficient solutions to problem MR(**u**). That is, $\mathcal{E}_{AMR}(\mathbf{u}) \neq \mathcal{E}_{MR}(\mathbf{u})$, and the set $\mathcal{E}_{AMR}(\mathbf{u})$ can only be taken as an approximation of the set $\mathcal{E}_{MR}(\mathbf{u})$.
In order to obtain the efficient set for this problem, the transformation

$$g(y) = \sqrt{\frac{y}{1-y}} \quad \text{with } g:(0,1) \to \mathbf{R}$$

can be applied to each objective function. This way, the following problem is obtained:

$$\begin{array}{ll} \underset{\mathbf{x}}{\mathrm{Max}} & \left(\dfrac{u_1 - E\{\tilde{z}_1(\mathbf{x},\tilde{\xi})\}}{\sqrt{\mathrm{Var}\{\tilde{z}_1(\mathbf{x},\tilde{\xi})\}}}, \ldots, \dfrac{u_q - E\{\tilde{z}_q(\mathbf{x},\tilde{\xi})\}}{\sqrt{\mathrm{Var}\{\tilde{z}_q(\mathbf{x},\tilde{\xi})\}}} \right) \\ \text{s.t} & E\{\tilde{z}_k(\mathbf{x},\tilde{\xi})\} \leq u_k, k = 1,2,\ldots,q \\ & \mathbf{x} \in D \end{array} \qquad (1)$$

Given that g is a strictly increasing function in its domain, then the efficient set of problem (1) coincides with the efficient set of problem AMR(**u**) (see White (1982), p. 13).

On the other hand, if we define the set:

$$S=\left\{(\mathbf{x}^t,\mathbf{u}^t)^t \in D\times\mathbb{R}^q \,/\, \sqrt{\frac{\beta_k}{1-\beta_k}}\sqrt{\mathrm{Var}\{\tilde{z}_k(\mathbf{x},\tilde{\xi})\}}+E\{\tilde{z}_k(\mathbf{x},\tilde{\xi})\}\le u_k,\ k=1,2,\ldots,q\right\}$$

from the Cantelli inequality, it can be proved that:

$$S\subset\left\{(\mathbf{x}^t,\mathbf{u}^t)^t \in D\times\mathbb{R}^q \,/\, P\{\tilde{z}_k(\mathbf{x},\tilde{\xi})\le u_k\}\ge\beta_k,\ k=1,2,\ldots,q\right\}$$

and we state the problem:

$$\begin{aligned}&\operatorname*{Min}_{\mathbf{x},\mathbf{u}}\ \mathbf{u}=(u_1,\ldots,u_q)\\ &\text{s.t}\ \ E\{\tilde{z}_k(\mathbf{x},\tilde{\xi})\}+\sqrt{\frac{\beta_k}{1-\beta_k}}\sqrt{\mathrm{Var}\{\tilde{z}_k(\mathbf{x},\tilde{\xi})\}}\le u_k,\ k=1,2,\ldots q\\ &\qquad \mathbf{x}\in D\end{aligned} \qquad \mathrm{AK}(\beta)$$

In order to obtain the efficient set for problem AK(β), we can consider the following problem:

$$\begin{aligned}&\operatorname*{Min}_{\mathbf{x}}\ E\{\tilde{z}_k(\mathbf{x},\tilde{\xi})\}+\sqrt{\frac{\beta_k}{1-\beta_k}}\sqrt{\mathrm{Var}\{\tilde{z}_k(\mathbf{x},\tilde{\xi})\}},\ \ k=1,2,\ldots,q\\ &\text{s.t}\ \ \mathbf{x}\in D\end{aligned} \qquad (2)$$

As in the minimum risk case, the set of efficient solutions to problem AK(β), denoted by $\mathcal{E}_{AK}(\beta)$ will be different from the set of efficient solutions to problem K(β). This is $\mathcal{E}_{AK}(\beta)\neq\mathcal{E}_{K}(\beta)$, in general, but we can take the first one as an approximation of the second one.

The following theorem gives us the relation between the efficient solutions to problems AMR(**u**) and AK(β).

Theorem 2

x is an efficient solution to problem AMR(**u**) if and only if $(\mathbf{x}^t, \mathbf{u}^t)^t$ is an efficient solution to problem AK(β), with **u** and β such that:

$$u_k=\sqrt{\frac{\beta_k}{1-\beta_k}}\sqrt{\mathrm{Var}\{\tilde{z}_k(\mathbf{x},\tilde{\xi})\}}+E\{\tilde{z}_k(\mathbf{x},\tilde{\xi})\} \qquad (3)$$

or equivalently:

$$\beta_k=\frac{\left(u_k-E\{\tilde{z}_k(\mathbf{x},\tilde{\xi})\}\right)^2}{\mathrm{Var}\{\tilde{z}_k(\mathbf{x},\tilde{\xi})\}+\left(u_k-E\{\tilde{z}_k(\mathbf{x},\tilde{\xi})\}\right)^2} \qquad (4)$$

for $k = 1, 2, \ldots, q$.

Proof

We demonstrate the theorem by *reductio ad absurdum*.

(a) Let $\mathbf{x}$ be an efficient solution to problem AMR(**u**). Suppose that $(\mathbf{x}^t, \mathbf{u}^t)^t$ is not efficient for problem AK(β), and thus $\mathbf{x}$ is neither efficient for problem (2) with $\beta = (\beta_1, \beta_2, \dots, \beta_q)^t$, and β_k given by (4), $\forall\, k = 1,2,\dots,q$.

There exists a solution $\mathbf{x}'$ such that:

$$E\{\tilde{z}_k(\mathbf{x}',\tilde{\xi})\}+\sqrt{\frac{\beta_k}{1-\beta_k}}\sqrt{\mathrm{Var}\{\tilde{z}_k(\mathbf{x}',\tilde{\xi})\}}\leq E\{\tilde{z}_k(\mathbf{x},\tilde{\xi})\}+\sqrt{\frac{\beta_k}{1-\beta_k}}\sqrt{\mathrm{Var}\{\tilde{z}_k(\mathbf{x},\tilde{\xi})\}}$$

for each $k \in \{1, 2, \dots, q\}$ and:

$$E\{\tilde{z}_s(\mathbf{x}',\tilde{\xi})\}+\sqrt{\frac{\beta_s}{1-\beta_s}}\sqrt{\mathrm{Var}\{\tilde{z}_s(\mathbf{x}',\tilde{\xi})\}}< E\{\tilde{z}_s(\mathbf{x},\tilde{\xi})\}+\sqrt{\frac{\beta_s}{1-\beta_s}}\sqrt{\mathrm{Var}\{\tilde{z}_s(\mathbf{x},\tilde{\xi})\}}$$

for some $s \in \{1, 2,\dots, q\}$.

We know that:

$$u_k = \sqrt{\frac{\beta_k}{1-\beta_k}}\sqrt{\mathrm{Var}\{\tilde{z}_k(\mathbf{x},\tilde{\xi})\}}+E\{\tilde{z}_k(\mathbf{x},\tilde{\xi})\}$$

wich is equivalent to $\sqrt{\frac{\beta_k}{1-\beta_k}} = \frac{u_k - E\{\tilde{z}_k(\mathbf{x},\tilde{\xi})\}}{\sqrt{\mathrm{Var}\{\tilde{z}_k(\mathbf{x},\tilde{\xi})\}}}$, then:

$$E\{\tilde{z}_k(\mathbf{x}',\tilde{\xi})\}+\frac{u_k - E\{\tilde{z}_k(\mathbf{x},\tilde{\xi})\}}{\sqrt{\mathrm{Var}\{\tilde{z}_k(\mathbf{x},\tilde{\xi})\}}}\sqrt{\mathrm{Var}\{\tilde{z}_k(\mathbf{x}',\tilde{\xi})\}}\leq u_k$$

for each $k \in \{1, 2, \dots, q\}$ and:

$$E\{\tilde{z}_s(\mathbf{x}',\tilde{\xi})\}+\frac{u_s - E\{\tilde{z}_s(\mathbf{x},\tilde{\xi})\}}{\sqrt{\mathrm{Var}\{\tilde{z}_s(\mathbf{x},\tilde{\xi})\}}}\sqrt{\mathrm{Var}\{\tilde{z}_s(\mathbf{x}',\tilde{\xi})\}}< u_s$$

for $s \in \{1, 2,\dots, q\}$.

Therefore, we obtain for each $k \in \{1, 2, \dots, q\}$ and for $s \in \{1, 2, \dots, q\}$:

$$\frac{u_k - E\{\tilde{z}_k(\mathbf{x},\tilde{\xi})\}}{\sqrt{\mathrm{Var}\{\tilde{z}_k(\mathbf{x},\tilde{\xi})\}}}\leq\frac{u_k - E\{\tilde{z}_k(\mathbf{x}',\tilde{\xi})\}}{\sqrt{\mathrm{Var}\{\tilde{z}_k(\mathbf{x}',\tilde{\xi})\}}} \quad\text{and}\quad \frac{u_s - E\{\tilde{z}_s(\mathbf{x},\tilde{\xi})\}}{\sqrt{\mathrm{Var}\{\tilde{z}_s(\mathbf{x},\tilde{\xi})\}}}<\frac{u_s - E\{\tilde{z}_s(\mathbf{x}',\tilde{\xi})\}}{\sqrt{\mathrm{Var}\{\tilde{z}_s(\mathbf{x}',\tilde{\xi})\}}}$$

and so $\mathbf{x}$ is not an efficient solution to problem (1), and thus neither is it efficient for problem AMR(**u**), which is in contradiction to the hypothesis.

(b) Let $(\mathbf{x}^t, \mathbf{u}^t)^t$ be an efficient solution to problem AK(β) with β given, and with u_k given by (3) for all $k = 1, 2, \dots, q$. Then, $\mathbf{x}$ is an efficient solution to problem (2). It is clear than $\mathbf{x}$ is a feasible solution to problem AMR(**u**).

Suppose that **x** is not an efficient solution to problem AMR(**u**) and, thus, neither is it efficient for problem (1). Then there exists a vector **x'** $\in$ D, verifying that $E\{\tilde{z}_k(\mathbf{x}',\tilde{\xi})\} \leq u_k$ for each k = 1, 2,...,q such that **x'** dominates **x**, that is, for each $k \in \{1, 2, \ldots, q\}$ and for some $s \in \{1, 2, \ldots, q\}$:

$$\frac{u_k - E\{\tilde{z}_k(\mathbf{x},\tilde{\xi})\}}{\sqrt{\mathrm{Var}\{\tilde{z}_k(\mathbf{x},\tilde{\xi})\}}} \leq \frac{u_k - E\{\tilde{z}_k(\mathbf{x}',\tilde{\xi})\}}{\sqrt{\mathrm{Var}\{\tilde{z}_k(\mathbf{x}',\tilde{\xi})\}}} \quad \text{and} \quad \frac{u_s - E\{\tilde{z}_s(\mathbf{x},\tilde{\xi})\}}{\sqrt{\mathrm{Var}\{\tilde{z}_s(\mathbf{x},\tilde{\xi})\}}} < \frac{u_s - E\{\tilde{z}_s(\mathbf{x}',\tilde{\xi})\}}{\sqrt{\mathrm{Var}\{\tilde{z}_s(\mathbf{x}',\tilde{\xi})\}}}$$

We know that $\sqrt{\dfrac{\beta_k}{1-\beta_k}} = \dfrac{u_k - E\{\tilde{z}_k(\mathbf{x},\tilde{\xi})\}}{\sqrt{\mathrm{Var}\{\tilde{z}_k(\mathbf{x},\tilde{\xi})\}}} \quad \forall k \in \{1, 2, \ldots, q\}$, then:

$$\sqrt{\frac{\beta_k}{1-\beta_k}} \leq \frac{\sqrt{\frac{\beta_k}{1-\beta_k}}\sqrt{\mathrm{Var}\{\tilde{z}_k(\mathbf{x},\tilde{\xi})\}} + E\{\tilde{z}_k(\mathbf{x},\tilde{\xi})\} - E\{\tilde{z}_k(\mathbf{x}',\tilde{\xi})\}}{\sqrt{\mathrm{Var}\{\tilde{z}_k(\mathbf{x}',\tilde{\xi})\}}}$$

and

$$\sqrt{\frac{\beta_s}{1-\beta_s}} < \frac{\sqrt{\frac{\beta_s}{1-\beta_s}}\sqrt{\mathrm{Var}\{\tilde{z}_s(\mathbf{x},\tilde{\xi})\}} + E\{\tilde{z}_s(\mathbf{x},\tilde{\xi})\} - E\{\tilde{z}_s(\mathbf{x}',\tilde{\xi})\}}{\sqrt{\mathrm{Var}\{\tilde{z}_s(\mathbf{x}',\tilde{\xi})\}}}$$

which is equivalent to:

$$E\{\tilde{z}_k(\mathbf{x}',\tilde{\xi})\} + \sqrt{\frac{\beta_k}{1-\beta_k}}\sqrt{\mathrm{Var}\{\tilde{z}_k(\mathbf{x}',\tilde{\xi})\}} \leq E\{\tilde{z}_k(\mathbf{x},\tilde{\xi})\} + \sqrt{\frac{\beta_k}{1-\beta_k}}\sqrt{\mathrm{Var}\{\tilde{z}_k(\mathbf{x},\tilde{\xi})\}}$$

for every $k \in \{1, 2,\ldots, q\}$ and:

$$E\{\tilde{z}_s(\mathbf{x}',\tilde{\xi})\} + \sqrt{\frac{\beta_s}{1-\beta_s}}\sqrt{\mathrm{Var}\{\tilde{z}_s(\mathbf{x}',\tilde{\xi})\}} < E\{\tilde{z}_s(\mathbf{x},\tilde{\xi})\} + \sqrt{\frac{\beta_s}{1-\beta_s}}\sqrt{\mathrm{Var}\{\tilde{z}_s(\mathbf{x},\tilde{\xi})\}}$$

for $s \in \{1, 2,\ldots, q\}$, which contradicts the hypothesis that **x** is an efficient solution to problem (2) or to problem AK(β).

6 Conclusions

Taking into account the results obtained, it can be affirmed that, given the stochastic multiple objective problem, if the conditions established in theorem 1 are satisfied, then the sets of minimum risk efficient solutions for aspiration levels

$u_1, u_2, \ldots, u_q$, and efficient solutions with probabilities $\beta_1, \beta_2, \ldots, \beta_q$ are closely related. In fact, these two sets coincide if the levels and the probabilities satisfy the relation established in theorem 1.

Moreover, this reciprocity between the two efficient solution concepts also holds when the Cantelli inequality is applied to approximate the distribution function of each stochastic objective.

The importance of these results resides in the fact that they yield relations between two efficient solution concepts for the SMP problem, originally defined under different philosophies, and thus not apparently related. Besides this, they allow us to determine these efficient sets by calculating just one of them, with the corresponding theoretical and computational advantages.

Acknowledgements

The authors wish to express their gratitude to two referees for their valuable and helpful comments.

References

Ben Abdelaziz, F., *L'efficacité en Programmation Multiobjectifs Stochastique*. Ph.D. Thesis Dissertation., Université de Laval, Québec (1992).

Goicoechea, A., Hansen, D.R. and Duckstein, L., *Multiobjective Decision Analysis with Engineering and Business Applications*. John Wiley and Sons. New York (1982).

Rao, C.R., *Linear Statistical Inference and its Applications*. John Wiley and Sons. New York (1973).

Stancu-Minasian, I.M., *Stochastic Programming with Multiple Objective Functions*. D. Reidel Publishing Company. Dordrecht (1984).

Stancu-Minasian, I.M. and Tigan, S., The Vectorial Minimum Risk Problem. *Proceedings of the Colloquium on Approximation and Optimization*. Cluj-Napoca, 321-328 (1984).

Stancu-Minasian, I.M., Stochastic Programming with Multiple Fractile Criteria. *Rev. Roumaine Math. Pures Appl.* **37**, 10, 939-941 (1992).

Szidarovszky, F., Gershon, M.E. and Duckstein, L., *Techniques for Multiobjective Decision Making in Systems Management*. Elsevier. Amsterdam (1986).

White, D.J., *Optimality and Efficiency*. John Wiley and Sons. New York (1982).

Hard to Say It's Easy – Four Reasons Why Combinatorial Multiobjective Programmes Are Hard

Matthias Ehrgott[1]

[1] Department of Mathematics, University of Kaiserslautern, PO Box 3049, 67653 Kaiserslautern, Germany, e-mail: ehrgott@mathematik.uni-kl.de;

Abstract. In this paper we address the questions related to "hardness" of combinatorial multiobjective programmes. These are the difficulty of finding and counting Pareto optimal solutions ($\mathbb{NP}$- and $\#\mathbb{P}$-completeness), the size of the set of Pareto solutions and efficient objective vectors, and limits for the approximability of Pareto solutions. We demonstrate that combinatorial multiobjective programmes are hard in these four respects.

Keywords. Multicriteria Optimization, Combinatorial Problems, $\mathbb{NP}$-completeness, $\#\mathbb{P}$-completeness, Intractability, Approximation Algorithms

1 What's the Problem?

Multiobjective Programming has gained a lot of attention during the last few decades (see [12] for a survey). However, research in the field of combinatorial multiobjective programmes has only started at the end of the 80's. And even then, a major part of literature has been devoted to heuristic methods, or the solution of scalarized problems. See the survey [13]. Why is this the case? These approaches avoid certain inherent difficulties encountered in the determination of all Pareto optimal solutions of combinatorial multiobjective programmes. In this paper we want to emphasize these difficulties. We present a collection of results from literature, as well as some new conclusions on $\#\mathbb{P}$-completeness and new results on approximability.
The feasible set of a combinatorial problem is defined as a subset $\mathcal{F} \subseteq 2^E$ of the power set of a finite set $E = \{e_1, \ldots, e_m\}$. A combinatorial optimization problem is formulated as follows:

$$\min_{S \in \mathcal{F}} \sum_{e \in S} w(e) \tag{P}$$

In a multicriteria context several weight functions $w_q : E \to \mathbb{Z}$ are given, yielding several objective functions f_q, $q = 1, \ldots, Q$; $f_q(S) = \sum_{e \in S} w_q(e)$

The combinatorial multiobjective problem is then to solve

$$\text{``}\min_{S\in\mathcal{F}}\text{''}(f_1(S),\ldots,f_Q(S)) \qquad \text{(CMOP)}$$

in the sense of Pareto optimality (or efficiency). A subset $S \in \mathcal{F}$ is called Pareto optimal if there does not exist another feasible solution $S' \in \mathcal{F}$ such that $f_q(S') \leq f_q(S)$ for all $q = 1, \ldots, Q$ with strict inequality for at least one of the objectives. The corresponding vector $f(S) = (f_1(S), \ldots, f_Q(S))$ is called efficient or non-dominated. The set of Pareto optimal solutions of (CMOP) will be denoted by $\mathcal{P}$, the set of efficient values by $\mathcal{E}$ throughout the paper. For a specific (CMOP) we will use the notation

$$Q\text{-}\sum P$$

E.g. 3-$\sum TSP$ denotes a travelling salesman problem with three objectives.

Here we remark that quite often combinatorial problems with bottleneck objectives $f(S) = \max_{e\in S} w(e)$ are considered in the literature. This gives rise to (CMOP) with only bottleneck or mixed sum and bottleneck objectives. These have not been extensively studied, some first results can be found in [3] and [4].

2 Four Reasons Why (CMOP) Is Hard

In this section we will outline four reasons, why (CMOP) problems are particularly hard optimization problems. These include, of course their $\mathbb{NP}$-completeness (using an appropriate decision version of (CMOP)). However, they are commonly also $\#\mathbb{P}$-complete (i.e. the related counting problem is difficult). Thus, finding one solution, or counting the number of solutions of (CMOP) is difficult. But that's not the end of the story. Often, with the formulation (CMOP) as finding all Pareto optimal solutions we face just too big a problem: (CMOP) may be intractable. Even if we could find every single Pareto solution efficiently, the sheer number of them is prohibitively large (i.e. exponential in problem size). Such problems are called intractable. In this situation we may think of heuristics to find "reasonably" good solutions for (CMOP), such that we may be able to guarantee certain worst case deviations from Pareto optimal solutions. This is certainly a valuable effort, but its success is limited, as we will see.

2.1 $\mathbb{NP}$-completeness and $\#\mathbb{P}$-completeness

Computational complexity is generally related to decision problems[1]. So we will use the following decision version of (CMOP).

[1] For the theory of $\mathbb{NP}$-completeness we refer to [7], for $\#\mathbb{P}$-completeness see [16].

Given constants $k_1, \dots, k_Q \in \mathbb{Z}$, *does there exist a feasible solution* $S \in \mathcal{F}$ *such that* $f_q(S) \leq k_q$, $q = 1, \dots, Q$? D(CMOP)

D(CMOP) is basically the question of finding one Pareto optimal solution. Since we are dealing with vectors, efficient objective values are not unique. This generates a natural interest (even more than for single objective problems) in the question of the number of Pareto optimal, resp. efficient solutions. Thus we are dealing with the counting version of (CMOP).

How many feasible solutions $S \in \mathcal{F}$ *do satisfy* $f_q(S) \leq k_q$, $q = 1, \dots, Q$? #(CMOP)

Below we will summarize some complexity results for (CMOP). We start with the following trivial observation.

Remark 1 *If a combinatorial optimization problem* $1\text{-}\sum P$ *is* $\mathbb{NP}$*-hard, then the same is true for the multicriteria counterpart* $Q\text{-}\sum P$, *for all* $Q \geq 2$.

An important question concerning complexity is, of course, whether problems which are in class $\mathbb{P}$ (polynomially solvable) if a single objective is considered remain within that class in the presence of multiple objectives. The answer to this question will be negative, even for the "simple" cases we present here. We will refer to the following problem, which is well known to be $\mathbb{NP}$- and $\#\mathbb{P}$-complete (see [7] or [3]).

- 0-1-KNAPSACK: Given $(c_1, \dots, c_n, r)$ and $(p_1, \dots, p_n, d) \in \mathbb{Z}^{n+1}$, does there exist an $x \in \{0,1\}^n$ such that $cx \leq r$ and $px \geq d$?

To begin the investigation of computational complexity of (CMOP) let us consider a problem without any constraints, i.e. $E = \{e_1, \dots, e_m\}$, where the feasible set is the power set of $E, \mathcal{F} = 2^E$. Therefore any subset of E is a feasible solution and the problem is to find a subset of E of minimal weight. This problem is called the unconstrained combinatorial optimization problem (UCP). With only one criterion this problem is trivial. What about the bicriteria version?
The decision version of the bicriteria unconstrained combinatorial optimization problem $2\text{-}\sum UCP$

$$\min_{S \in 2^E} \left(\sum_{e \in S} w_1(e), \sum_{e \in S} w_2(e) \right)$$

is as follows: Given w_1 and $w_2 \in \mathbb{Z}^m$, two constants k_1 and $k_2 \in \mathbb{Z}$, does there exist a subset $S \subseteq \{e_1, \dots, e_m\}$ such that $\sum_{e \in S} w_1(e) \leq k_1$ and $\sum_{e \in S} w_2(e) \leq k_2$?
If we formulate $2\text{-}\sum UCP$ in terms of binary variables we have $\mathcal{F} = \{0,1\}^m$. Then the decision problem is as follows: does there exist an $x \in \{0,1\}^m$ such

that $w_1 x \leq k_1$ and $w_2 x \leq k_2$? We define a parsimonious transformation[2] from 0-1-KNAPSACK by letting $w_1 = c$, $k_1 = r$, $w_2 = -p$, and $k_2 = -d$. Thus

$$\text{0-1-KNAPSACK} \propto_p 2\text{-}\sum UCP$$

and hence we have proven:

Proposition 1 *$D(2\text{-}\sum UCP)$ is $\mathbb{NP}$-complete, $\#(2\text{-}\sum UCP)$ is $\#\mathbb{P}$-complete.*

From this result, we may argue that $\mathbb{NP}$- and $\#\mathbb{P}$-completeness are features of multicriteria optimization problems which occur very often. That this is indeed the case is illustrated by some examples from the literature.

The second example is the shortest path problem, S-T SHORTEST PATH. Given a graph $G = (V, E)$ or a digraph $D = (V, A)$ and two nodes s and t in the respective node set, the feasible set $\mathcal{F}$ is the set of paths from s to t in G or D. The bicriteria shortest path problem is

$$2\text{-}\sum \text{ S-T SHORTEST PATH.}$$

Proposition 2 ([11]) *The decision problem $D(2\text{-}\sum$S-T SHORTEST PATH$)$ is $\mathbb{NP}$-complete, the counting problem $\#(2\text{-}\sum$S-T SHORTEST PATH$)$ is $\#\mathbb{P}$-complete.*

Proof:
The decision problem clearly is in $\mathbb{NP}$. Therefore $\mathbb{NP}$-completeness is shown by the following transformation 0-1-KNAPSACK $\propto$ 2-$\sum$S-T SHORTEST PATH given in [11].
From $(c_1, \ldots, c_n), (p_1, \ldots, p_n), r$, and d we construct the following bicriteria shortest path problem on a digraph $D = (V, A)$ defined as follows:

$$\begin{aligned} V &= \{v_0, \ldots, v_n\} \\ A &= \{(v_{i-1}, v_i) : i = 1, \ldots, n\} \cup \{(v_{i-1}, v_i)' : i = 1, \ldots, n-1\} \\ w_1((v_{i-1}, v_i)) &= c_i \\ w_2((v_{i-1}, v_i)) &= 0 \\ w_1((v_{i-1}, v_i)') &= 0 \\ w_2((v_{i-1}, v_i)') &= p_i. \end{aligned}$$

Then there exists a path P from v_0 to v_n in D satisfying both $\sum_{e \in P} w_1(e) \leq k_1$ and $\sum_{e \in P} w_2(e) \leq \sum_{i=1}^n p_i - k_2$ if and only if there exists a subset $S \subseteq \{1, \ldots, n\}$ such that $\sum_{i \in S} c_i \leq k_1$ and $\sum_{i \in S} p_i \geq k_2$ ($i \in S$ corresponds to selecting (v_{i-1}, v_i) and $i \notin S$ to selecting $(v_{i-1}, v_i)'$ in D). Finally we remark

[2] A parsimonious transformation is a polynomial time transformation, which preserves the number of solutions. It is denoted by $\propto_p$.

that the transformation is parsimonious and therefore the $\#\mathbb{P}$-completeness is also proven. □

The $\#\mathbb{P}$-completeness conclusion above is an immediate observation, but has not been stated in the paper cited.

The third problem we consider is the assignment problem. Given a complete bipartite graph $K_{n,n}$ we have the set of all perfect matchings of $K_{n,n}$ as feasible set $\mathcal{F}$. Again we consider the bicriteria case, i.e. $w_1, w_2 \in \mathbb{R}^{n^2}$ are two costs on the edges of $K_{n,n}$, and we consider 2-$\sum$ ASSIGNMENT.

Proposition 3 ([11, 9]) *The decision problem D(2-$\sum$ ASSIGNMENT) is $\mathbb{NP}$-complete, the corresponding counting problem #(2-$\sum$ ASSIGNMENT) is $\#\mathbb{P}$-complete.*

Proof:
We present the reduction of a version of PARTITION which is also $\mathbb{NP}$-complete according to [7] to 2-$\sum$ ASSIGNMENT.

- EQUI-PARTITION: Let $(c_1, \ldots, c_{2n}) \in \mathbb{Z}^{2n}$ be such that $\sum_{i=1}^{2n} c_i = 2C$. Does there exist a subset $S \subseteq \{1, \ldots, 2n\}$, $|S| = n$ such that $\sum_{i \in S} c_i = \sum_{i \notin S} c_i = C$.

The reduction EQUI-PARTITION $\propto$ 2-$\sum$ ASSIGNMENT is taken from [11]. We use c_i, $i = 1, \ldots, 2n$ to construct the following bicriteria assignment problem. Let $\hat{c}$ be a real number such that $\hat{c} > \max_i c_i$. Let the node set V of $K_{2n,2n}$ be $V = U \cup R$. Define the weights on the edges of $K_{2n,2n}$ as follows:

$$w(u_i, r_i) = \begin{cases} (\hat{c} + c_i, \hat{c} - c_i) & r_i \text{ odd} \\ (\hat{c}, \hat{c}) & r_i \text{ even.} \end{cases}$$

Then there exists a subset $S \subset \{1, \ldots, 2n\}$, $|S| = n$ such that $\sum_{i \in S} c_i = C$ if and only if $K_{2n,2n}$ with the above weights contains a perfect matching M with $w_1(M) \leq 2n\hat{c} + C$ and $w_2(M) \leq 2n\hat{c} - C$.
The main result of [15] mentioned in the previous section has been used in [9] to show that #(2-$\sum$ ASSIGNMENT) is $\#\mathbb{P}$-complete. □

$\mathbb{NP}$-completeness can also be proved for the multicriteria matroid optimization problem of finding Pareto optimal bases of a matroid, denoted by Q-$\sum MB$. Note that single objective matroid problems are easily solved by the greedy algorithm. We cite from [2]:

Proposition 4 *The decision problem $D(Q\text{-}\sum MB)$ is $\mathbb{NP}$-complete.*

Here, $\#\mathbb{P}$-completeness is still an open problem.

As the single criterion counterparts of all problems presented here, 2-$\sum$ UCP, 2-$\sum$ s-t Shortest Path, 2-$\sum$ Assignment, and Q-$\sum MB$ are solvable in polynomial time, Propositions 1 - 3 are rather disappointing. They show that even bicriteria combinatorial problems are among the hardest problems. All results above have been proved for the case of 2 criteria. However, we remark that $\mathbb{NP}$-completeness results remain valid in the general case of Q criteria. Just let $w_q(e) := 0$ for all $e \in E$ and $q \geq 3$.

2.2 Intractability

So far we know that finding and counting Pareto optimal solutions of (CMOP) is hard. Let us now turn to the problem of how many Pareto solutions and efficient vectors we may have to find. It turns out that there may exist exponentially many, i.e. by considering a large enough problem, we may even be unable to just write down all such solutions.

The following is a general result from [3].

Proposition 5 *Let (CMOP) be a multiple criteria combinatorial optimization problem with the properties*

- $|S| = n$ *for all* $S \in \mathcal{F}$ *and*
- $w(e_i) = (2^{i-1}, 2^m - 2^{i-1})$.

Then all feasible solutions are Pareto optimal, i.e. $\mathcal{P} = \mathcal{F}$ *and* $|\mathcal{E}| = |\mathcal{P}| = |\mathcal{F}|$.

Proof:
Let $S \in \mathcal{F}$ be any feasible solution. Then

$$f_1(S) + f_2(S) = \sum_{l=1}^{n} (w_1(e_{i_l}) + w_2(e_{i_l})) = \sum_{l=1}^{n} (2^{i_l-1} + 2^m - 2^{i_l-1}) = n2^m.$$

Therefore, by the uniqueness of binary representations of numbers, all objective values $f(S_1)$ and $f(S_2)$ are pairwise non-comparable. □

In Proposition 5 all feasible solutions are also Pareto optimal and they have different objective value vectors. Other results in this respect are known in the literature. We cite two analogous results for the shortest path problem and the TSP, see [8] and [5], respectively.

Proposition 6 *Consider the problem 2-$\sum$s-t Shortest Path where* $s = v_1, t = v_n$ *in the digraph* $D = (V, A)$ *with* $|V| = n$ *being odd and the edges* *and their weights given by*

$$\begin{aligned} w(v_i, v_{i+2}) &= \left(2^{\frac{i-1}{2}}, 0\right) \; i = 1, 3, \ldots, n-2 \\ w(v_i, v_{i+1}) &= \left(0, 2^{\frac{i-1}{2}}\right) \; i = 1, 3, \ldots, n-2 \\ w(v_{i+1}, v_{i+2}) &= (0, 0) \; i = 1, 3, \ldots, n-2. \end{aligned}$$

Then $\mathcal{P} = \mathcal{F}$ and $|\mathcal{P}| = |\mathcal{E}| = |\mathcal{F}| = 2^{\frac{n-1}{2}}$.

Proposition 7 *For each number n of cities and each number $Q > 2$ of objectives there exist distance matrices $(w^q(i,j))$, $q = 1, \ldots, Q$ such that all possible TSP-tours are also Pareto optimal, i.e.*

$$|\mathcal{P}| = |\mathcal{E}| = \frac{(n-1)!}{2}.$$

2.3 Limits for Approximability

If, as indicated in the previous sections, solving (CMOP) is too difficult, we may think of heuristics. It is then interesting to investigate how good heuristic solutions can approximate Pareto optimal solutions. Below we will therefore discuss approximation ratios for (CMOP).
The concept of approximability and performance ratios of algorithms is thoroughly studied in combinatorial optimization, see e.g. [1] for a list of results. The performance ratio $R(S, S^*)$ of a feasible solution S of an instance of a (minimization) problem (P) with respect to an optimal solution S^* is defined as

$$R(S, S^*) := \frac{f(S)}{f(S^*)}.$$

Accordingly a (polynomial time) algorithm A for problem (P) is called an $r(n)$-approximate algorithm if $R(A(P), S^*) \leq r(|(P)|)$ for all instances of problem (P), where $A(P)$ is the solution found by algorithm A and $|(P)|$ denotes the size of a problem instance. Furthermore, $r : \mathbb{N} \to [1, \infty]$ is an arbitrary function. Note that $r(n) \equiv 1$ means that problem P can be solved (in polynomial time) by Algorithm A. We also remark that $R(S, S^*) = q$ can be equivalently stated as

$$\frac{f(S) - f(S^*)}{f(S^*)} = q - 1.$$

For multicriteria problems we cannot directly transfer this definition, since we have objective value vectors. So we have to decide how to compare these vectors, i.e. a norm is necessary. We will assume that $\mathbb{R}^Q$ is equipped with a monotone norm $||.||$. A norm is monotone if, whenever $|a_q| \leq |b_q|$, $q = 1, \ldots, Q$ holds for two elements $a, b \in \mathbb{R}^Q$, then $||a|| \leq ||b||$.

Furthermore, since efficient vectors are not unique, we have the options to define approximability with respect to one or to all Pareto optimal solutions. We will use the second approach. We will now provide two possible definitions of performance ratios in multiple criteria optimization. We shall consider algorithms which find **one** solution only.
In the first, we compare the norms of the vectors. Note that the norm of a heuristic solution may be larger or smaller than that of Pareto optimal solutions, wherefore the absolute value is needed.

Definition 1 *1. Let $S \in \mathcal{F}$ be a feasible solution of (CMOP) let $S^* \in \mathcal{P}$ be a Pareto optimal solution. The performance ratio R_1 of S with respect to S^* is defined as*

$$R_1(S, S^*) := \frac{|||f(S)|| - ||f(S^*)|||}{||f(S^*)||}.$$

2. An algorithm A for (CMOP) is an $r_1(n)$-approximate algorithm if the solution S found by the algorithm satisfies

$$R_1(A(CMOP), S^*) \leq r_1(|CMOP|)$$

for all Pareto optimal solutions $S^* \in \mathcal{P}$.

The second option of comparing the vectors directly, i.e. measuring the norm of the difference between heuristic and Pareto optimal solutions, is taken care of in the following definition.

Definition 2 *1. Let $S \in \mathcal{F}$ be a feasible solution of (CMOP), let $S^* \in \mathcal{P}$ be a Pareto optimal solution of (CMOP). The performance ratio R_2 of S with respect to S^* is defined as*

$$R_2(S, S^*) := \frac{||f(S) - f(S^*)||}{||f(S^*)||}.$$

2. An algorithm A for (CMOP) is an $r_2(n)$-approximate algorithm for (CMOP) if the solution S found by the algorithm satisfies

$$R_2(A(CMOP), S^*) \leq r_2(|CMOP|)$$

for all Pareto optimal solutions $S^* \in \mathcal{P}$.

Note that if any feasible solution S satisfies $R_2(S, S^*) \leq \rho$ for some Pareto solution S^* then also $R_1(S, S^*) \leq \rho$. Therefore we can observe:

Remark 2 *An $r(n)$-approximate algorithm for (CMOP) in the sense of Definition 2 is also an $r(n)$-approximate algorithm in the sense of Definition 1.*

Below we will discuss the first definition and provide a tight bound of $R_1 = 1$ for (CMOP). The second definition is discussed in [4], tight bounds are not known yet.
The first definition hints to the idea of using a solution with minimal norm as an approximate heuristic solution for the multicriteria problem. That this is indeed possible is shown in Theorem 1.

Theorem 1 *Let S_n^* be a feasible solution of (CMOP) with minimal norm, i.e.*

$$||S_n^*|| = \min_{S \in \mathcal{F}} ||S||.$$

Then the performance ratio $R_1(S_n^*, S^*) \leq 1$ for all $S^* \in \mathcal{P}$.

Proof:

$$\frac{|||f(S_n^*)|| - ||f(S^*)|||}{||f(S^*)||} = \frac{||f(S^*)|| - ||f(S_n^*)||}{||f(S^*)||} = 1 - \frac{||f(S_n^*||}{||f(S^*)||} \leq 1$$

□

Note that there always exists an S_n^* that is itself a Pareto optimal solution. So two questions arise: The first, whether the ratio 1 is tight, i.e. whether instances can be constructed which achieve this ratio. The second is whether S_n^* can be found.

Theorem 2 *For the general (CMOP), the approximation ratio $R_1 = 1$ is tight.*

Proof:
We consider a very simple example of a (CMOP). Let $E = \{a, b, c, d\}$ and let $\mathcal{F}$ consist of all two element subsets of E. Weights are given in the table below, M representing a large number.

e	a	b	c	d
$(w_1(e), w_2(e))$	$(M, 0)$	$(0, M)$	$(1, 1)$	$(1, 1)$

Then the five solutions $\{a, c\}, \{a, d\}, \{b, c\}, \{b, d\}$, and $\{c, d\}$ are Pareto optimal, with the efficient vectors $(M+1, 1)$, $(1, M+1)$, and $(2, 2)$. It is easy to see that both $R_1(\{c, d\}, \{a, c\})$ and $R_2(\{c, d\}, \{a, c\})$ approach 1 as M approaches infinity. This observation is true for all l_p norms, $||x||_p = \left(\sum_{q=1}^{Q} |x_q|^p\right)^{\frac{1}{p}}$ with $1 \leq p \leq \infty$. □

For most problems it will be possible to construct examples with a behaviour like this simple one. For specific problems, however, improved results may be possible.

Addressing the second question of whether S_n^* can be found the answer is: it depends on the specific (CMOP) and on the norm chosen. The problem to find S_n^* is

$$\min_{S \in \mathcal{F}} ||f(S)||. \tag{NMP}$$

When $||.||$ is the maximum norm l_∞ the problem is the so called max-ordering problem. We refer to [3] and references therein for results on this type of problems. Here, we will restrict ourselves to a result on a condition when the problem (NMP) is indeed solvable in polynomial time.

Proposition 8 *Problem (NMP) is solvable in polynomial time if 1-$\sum P$ is solvable in polynomial time and* $||x|| = \sum_{q=1}^{Q} |x_q|$.

Proof:
Problem (NMP) is equivalent to a single criterion problem, which by assumption can be solved in polynomial time:

$$\begin{aligned}
\min_{S \in \mathcal{F}} ||f(S)|| &= \min_{S \in \mathcal{F}} \left(\sum_{q=1}^{Q} f_q(S) \right) \\
&= \min_{S \in \mathcal{F}} \sum_{q=1}^{Q} \left(\sum_{e \in S} w_q(e) \right) \\
&= \min_{S \in \mathcal{F}} \sum_{e \in S} \hat{w}(e),
\end{aligned}$$

where $\hat{w}(e) = \sum_{q=1}^{Q} w_q(e)$. □

Proposition 8 has immediate consequences for such polynomially solvable problems as the shortest path problem, the spanning tree problem and the assignment problem. The well known algorithms that solve these (single objective) problems are 1-approximate algorithms for the multicriteria counterparts, where the l_1-norm is used.

3 Conclusions

In our paper we have investigated four sources of the hardness of multiobjective combinatorial programmes (CMOP). The results obtained point out several worthwhile research directions: the investigation of approximation algorithms for (CMOP), both for problems which are polynomially solvable or $\mathbb{NP}$-hard in the single objective case. Other definitions, e.g. component-wise, of approximation ratios is another area.

Due to the enormous difficulty encountered for (CMOP) recent efforts in applications of metaheuristics such as Simulated Annealing or Tabu Search should be a good approach for the solution of (CMOP). This area has gained considerable interest recently, see e.g. [10], [14], or [6].

References

[1] P. Crescenzi and V. Kann. A compendium of NP optimization problems. http://www.nada.kth.se/theory/problemlist.html, 1997.

[2] M. Ehrgott. On matroids with multiple objectives. *Optimization*, 38(1):73–84, 1996.

[3] M. Ehrgott. *Multiple Criteria Optimization – Classification and Methodology.* Shaker Verlag, Aachen, 1997.

[4] M. Ehrgott. Approximation algorithm for combinatorial multicriteria optimization problems. Technical report, University of Kaiserslautern, Department of Mathematics, 1998. Report in Wirtschaftsmathematik No. 39.

[5] V.A. Emelichev and V.A. Perepelitsa. On cardinality of the set of alternatives in discrete many-criterion problems. *Discrete Mathematics and Applications*, 2(5):461–471, 1992.

[6] X. Gandibleux, N. Mezdaoui, and A. Fréville. A multiobjective tabu search procedure to solve combinatorial optimisation problems. Technical report, Universiteé de Valenciennes, 1996.

[7] M.R. Garey and D.S. Johnson. *Computers and Intractability – A Guide to the Theory of NP-Completeness.* Freeman, San Francisco, 1979.

[8] P. Hansen. Bicriterion path problems. In G. Fandel and T. Gal, editors, *Multiple Criteria Decision Making Theory and Application*, number 177 in Lecture Notes in Economics and Mathematical Systems, pages 109–127, 1979.

[9] P. Neumayer. Complexity of optimization on vectorweighted graphs. In A. Bachem, U. Derigs, M. Jünger, and R. Schrader, editors, *Operations Research 93*, pages 359–361. Physica Verlag, Heidelberg, 1994.

[10] P. Searfini. Simulated annealing for multiple objective optimization problems. In *Proceedings of the Tenth International Conference on Multiple Criteria Decision Making*, pages 87–96, 1992.

[11] P. Serafini. Some considerations about computational complexity for multi objective combinatorial problems. In *Recent advances and historical development of vector optimization*, number 294 in Lecture Notes in Economics and Mathematical Systems. Springer-Verlag, Berlin, 1986.

[12] R.E. Steuer, L.R. Gardiner, and J. Gray. A bibliographic survey of the activities and international nature of multiple criteria decision making. *Journal of Multi-Criteria Decision Analysis*, 5:195–217, 1996.

[13] E.L. Ulungu and J. Teghem. Multi-objective combinatorial optimization problems: A survey. *Journal of Multi-Criteria Decision Analysis*, 3:83–104, 1994.

[14] E.L. Ulungu, J. Teghem, and P. Fortemps. Heuristics for multi-objective combinatorial optimization by simulated annealing. In *Multiple Criteria Decision Making: Theory and Applications, Proceedings of the 6th National Conference on Multiple Criteria Decision Making*, pages 228–238, 1995.

[15] L.G. Valiant. The complexity of computing the permanent. *Theoretical Computer Science*, 8:189–201, 1979.

[16] L.G. Valiant. The complexity of enumeration and reliability problems. *SIAM Journal of Computing*, 8(3):410 –421, 1979.

Multiattribute Utility Analysis in the IctNeo System

M. Gómez, S. Ríos-Insua, C. Bielza and J.A. Fernández del Pozo

Department of Artificial Intelligence, Madrid Technical University
Campus de Montegancedo, 28660 Boadilla, Madrid, Spain

Abstract

We have constructed IctNeo, a decision support system for neonatal jaundice management which integrates beliefs and preferences with an influence diagram. This paper considers the problem of modelling preferences by means of a multiattribute utility function. We describe the objectives hierarchy including a hierarchical structure of attributes, tests of independence which lead to a nested utility function and the assessment of imprecise weights and component utility functions.

Keywords: Multiattribute utility, utility and weight assessment, neonatal jaundice

1 Introduction

Jaundice in newborns has raised a challenge for hundreds of years, Gartner (1995), because nearly all neonates have a higher bilirubin level than healthy adults and over fifty per cent are visibly jaundiced during the first week of lifetime. An important aim is to differentiate the pathology from the variation levels within the standard range, since bilirubin may produce toxic effects in the central nervous system, Mollison and Cutbush (1951).

In the fifties, blood exchange was the only way of treatment, and it was applied when the infant had a bilirubin level over 20 mg/dl, since such level was considered a risk factor of neurologic abnormalities. However, later studies modified the guide line and proved that lower values of bilirubin were able to promote less showy neurological alterations, bilirubin encephalopathy being the most dangerous one. The introduction of phototherapy in jaundice management led to the treatment of lower bilirubin cases and produced a drastic reduction of blood exchanges.

Current jaundice protocols take into account the newborn weight and age, and are based on frequent controls of bilirubin and hemoglobin levels, suggesting several medical tests when some of them go over certain levels. The type of treatment at each moment, observe, phototherapy or exchange, depends on bilirubin levels and is independent of the cause of jaundice. Recently, Newman and Maisels (1992) have proposed a less aggressive approach based on the fact that the neurological risk associated with high concentrations of bilirubin in healthy full-term newborns is small compared to the risk due to treatments whose benefits are very limited or no longer existent. Thus, the morbidmortality derived from exchanges is more relevant than the benefits obtained from avoiding an each more infrequent pathology as nuclear jaundice is. The use of phototherapy has shown an important decrease in the number of exchanges and it does not seem to have any effect over the neurological evolution and development of the baby, although requires to keep away the baby from the mother, hence disrupting lactation.

The new recommendation implies reducing the number of laboratory tests, postponing the beginning of a phototherapy treatment in cases of (a) a jaundiced newborn without hemolysis until bilirubin levels are higher than the classic ones (23-29 mg/dl), and (b) having a greater control of newborns with hemolysis in an attempt to keep bilirubin levels at lower values (19-23 mg/dl).

We have constructed IctNeo, Ríos-Insua *et al.* (1998), a decision support system (DSS) for jaundice management, which integrates beliefs and preferences by means of an influence diagram, Shachter (1986), and aims at aiding doctors in neonatal jaundice management decisions. It yields a decrease in the costs of diagnostic and therapeutic phases, defines better the limits to change the treatment, and takes into account the preferences of parents and doctors. This system is based on Decision Analysis (DA), to integrate several uncertain factors and decisions to control neonatal jaundice, and thus, determine the expected utilities of different medical treatments.

The aim of this paper is to describe the multi-attribute utility analysis which the system uses. The paper includes four more sections. In Section 2 we provide the generation of the objectives hierarchy with the attributes and their respective measure scales. In Section 3 we postulate various sets of assumptions about the basic preference attitudes of the decision maker (DM) to derive the functional form of the multiattribute utility function consistent with these assumptions. Section 4 describes the assessment of imprecise utilities and weights and, Section 5, the application of an alternative test of additivity. Finally, we provide some conclusions.

2 Generation of objectives and attributes

Any strategy of treatment must specify objectives and attributes (i.e., measures of effectiveness) to measure the degree of achievement of objectives by each of the treatments. The distinct aspect of the DA approach is the degree of formality with which this specification is conducted.

The starting point for specifying objectives are the general concerns: direct and intangible effects. The questions we wished to answer in determining objectives and attributes were, for example, what is the intangible effect of concern in our particular problem? and how do we measure it? The process of answering such questions was essentially a creative task, although several aids, like information gathering, were of significant help in articulating objectives. With the aid of the doctors, we constructed an objectives hierarchy, Keeney and Raiffa (1976), with major objective "the well being of the newborn". We went down considering "direct effects" and "intangible effects" as general concerns of the major objective. Several subobjectives for direct effects, like costs in the plant and in the intensive care unit, number of days in both places,... could be considered, but we note that, to avoid redundancies and to have a simpler multiattribute model, all these might be included and condensed in only one objective named "minimize cost of stay at hospital". We took care to assure that all facets of the higher objective were accounted for in the subobjectives. Furthermore, *tests of importance* were repeatedly considered before any objective was included in the hierarchy, and the doctors were asked if they felt the best course of action could be altered if that objective was excluded. For intangible effects we included the subobjectives "baby-mother gap", "minimize risk of being admitted to hospital" and "injuries". Finally, baby-mother gap was broken into "minimize social cost", and "minimize emotional cost", and injuries was broken into two lower-level objectives, named "minimize injuries due to treatment" and "minimize injuries due to hyperbilirubinemia". Figure 2.1 shows the objectives hierarchy for this problem.

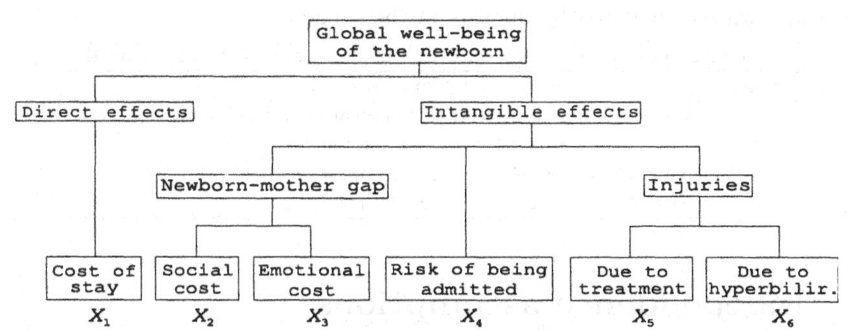

Fig. 2.1. An objectives hierarchy for the jaundice management problem.

Next, taking into account the desirable properties of a set of attributes, see e.g., Keeney and Raiffa (1976), we identified an attribute for each one of the six lowest-level objectives. We considered for the objective "minimize cost of stay at hospital" the natural-proxy attribute thousands of pesetas. However, for the remaining objectives it was much harder to determine an attribute and we had to construct one specially for each objective, thus considering scales developed *ad hoc*. We passed the *clairvoyance test*, Kirkwood (1997), trying to avoid ambiguous scales and looking at an analysis of tradeoff between the effort spent to develop the scale, and the ease of assessing alternatives and communicating the results of the analysis. Table 2.1 shows the attributes X_i, $i = 1, ..., 6$, with their measurement units. It is simple to check that for all attributes, less was preferred to more. Hence, we list also worst and best levels in the table.

As we have noted, for all attributes except for X_1, we introduced constructed scales. We describe now some of them. Let us first consider the social cost of the baby-mother gap (X_2). The scale of this attribute was constructed from interviews with parents and with the aid of doctors. We posed three questions: 1) To the mother: Is it a disturbance for you to come every day to hospital to nurse your baby?; 2) To the father: Can you come to visit the baby?; and 3) To both: Is there any repercussion in the care of other children or relatives?

Table 2.1. Attributes for evaluating medical treatments

Attribute	Measure (units)	Level	
		Worst x_{i*}	Best x_i^*
X_1 : cost of stay	thousands of ptas	1260	0
X_2 : social cost	constructed scale	2	0
X_3 : emotional cost	constructed scale	2	0
X_4 : risk of being admitted	constructed scale	3	0
X_5 : injuries due to treatment	constructed scale	4	0
X_6 : injuries due to hyperbilir.	constructed scale	5	0

For each one of those questions there was a particular scale and from them a global value was obtained for attribute X_2 in the range $0-2$. Another scale is shown in Table 2.2. This corresponds to the risk of being admitted to hospital (X_4) and was constructed by doctors in the range $0-3$.

Table 2.2. Constructed scale for risk of being admitted to hospital

Value	Level of impact
0	Newborn not admitted to hospital. Under observation.
1	Newborn admitted + receives some test.
2	Newborn admitted + receives medium care.
3	Newborn admitted + receives intensive care.

3 Independence assumptions

Once an objectives hierarchy and appropriate attributes X_i, $i = 1, ..., 6$, have been specified, next task is to assess a utility function $u(x_1, x_2, x_3, x_4, x_5, x_6)$ over the six attributes, where x_i designates a specific level of X_i. A direct assessment of u presents major practical shortcomings, so we investigated various sets of *independence assumptions* about the basic preferences attitudes of the DM, to derive a functional form of the *multiattribute utility function* consistent with them.

The first important step in selecting the form of the utility function involves investigating the reasonableness of *preferential independence* and *utility independence conditions.* To facilitate checking independence conditions and due to the homogeneity, on one hand, of attributes X_2 and X_3, and on the other, X_5 and X_6, we assumed that such attributes might be structured temporarily, substituting X_2 and X_3 by only one attribute Y_2, which represents "newborn-mother gap", and X_5 and X_6 by Y_4 meaning "injuries". Hence, we should have by the moment four attributes denoted $Y_1 = X_1$, $Y_2 = (X_2, X_3)$, $Y_3 = X_4$ and $Y_4 = (X_5, X_6)$. Then, we intended to determine a utility function of the form $u(y_1, y_2, y_3, y_4) = f[u_1(y_1), u_2(y_2), u_3(y_3), u_4(y_4)]$ where f is a scalar-valued function, and u_i a utility function over y_i.

To determine the functional form of f, the process began by examining whether an attribute was *utility independent* (u.i.) of its complement. After the motivation and familiarization with the terminology, we verified that attribute Y_4 was u.i. of its complement $\overline{Y}_4$. For that, we had to check whether the preference order for lotteries involving only changes in the level of Y_4 does not depend on the levels at which attributes Y_1, $Y_2 = (X_2, X_3)$ and Y_3 are fixed. We asked the doctors to determine a value $(x_5, x_6) = y_4$ for Y_4, such that they felt indifference between the sure consequence

$$(y_1^*, (x_2^*, x_3^*), y_3^*, (x_5, x_6))$$

and the lottery

$$\left(\begin{array}{cc} .5 & .5 \\ (y_1^*, (x_2^*, x_3^*), y_3^*, (x_5^*, x_6^*)) & (y_1^*, (x_2^*, x_3^*), y_3^*, (x_{5*}, x_{6*})) \end{array} \right).$$

After the dialogue process, they converged to the indifference point $(x_5, x_6) = (2, 2)$. Then, the process continued to determine if this indifference point was the same regardless of the levels of the other attributes. The tests were passed individually to three doctors and, afterwards, we conducted joint sessions to solve possible discrepancies. Table 3.1 shows some levels of different attributes presented to the doctors to verify the indifference (remember that $y_2 = (x_2, x_3)$).

Table 3.1. Attribute levels to check the u.i. of Y_4

y_1	y_2		y_3
400	0	0	0
700	0	0	1
1000	1	1	2
800	2	2	2
⋮	⋮	⋮	⋮

This implied that Y_4 was u.i. of $\overline{Y}_4$. In similar manner, we verified that Y_2 was u.i. of $\overline{Y}_2$.

Next, we decided to check whether $\{Y_4, Y_i\}$ was *preferential independent* (p.i.) of its complement, for $i = 1, 2, 3$ (note that if sure, we had a multiplicative or additive decomposition of the utility function). Hence:

1) First, we verified that $\{Y_4, Y_1\}$ was p.i. of $\overline{\{Y_4, Y_1\}}$, i.e., the preference order for consequences involving only changes in the levels of Y_4 and Y_1 does not depend on the levels at which attributes Y_2 and Y_3 are fixed. In this particular case, we fixed attributes Y_2 and Y_3 at their best levels, $Y_2 = (X_2^*, X_3^*) = (0, 0)$ and $Y_3 = X_4^* = 0$. We had, for attributes $(Y_1, Y_4) = (X_1, X_5, X_6)$, that

$$\begin{array}{cccccccccccccc} Y_1 & \multicolumn{2}{c}{\overbrace{\quad}^{Y_2}} & Y_3 & \multicolumn{2}{c}{\overbrace{\quad}^{Y_4}} & & Y_1 & \multicolumn{2}{c}{\overbrace{\quad}^{Y_2}} & Y_3 & \multicolumn{2}{c}{\overbrace{\quad}^{Y_4}} \\ X_1^* & X_2^* & X_3^* & X_4^* & X_5^* & X_6^* & & X_{1*} & X_2^* & X_3^* & X_4^* & X_{5*} & X_{6*} \\ (0, & 0, & 0, & 0, & 0, & 0) & \succ & (1260, & 0, & 0, & 0, & 4, & 5) \end{array}$$

and the doctors revealed the same preference order if we fixed attributes Y_2 and Y_3 at their worst levels, $Y_2 = (X_{2*}, X_{3*}) = (2,2)$ and $Y_3 = X_{4*} = 3$, i.e.,

$$\begin{array}{cccccc}
Y_1 & \multicolumn{2}{c}{\overbrace{\qquad\qquad}^{Y_2}} & Y_3 & \multicolumn{2}{c}{\overbrace{\qquad\qquad}^{Y_4}} \\
X_1^* & X_{2*} & X_{3*} & X_{4*} & X_5^* & X_6^* \\
(0, & 2, & 2, & 3, & 0, & 0)
\end{array} \succ \begin{array}{cccccc}
Y_1 & \multicolumn{2}{c}{\overbrace{\qquad\qquad}^{Y_2}} & Y_3 & \multicolumn{2}{c}{\overbrace{\qquad\qquad}^{Y_4}} \\
X_{1*} & X_{2*} & X_{3*} & X_{4*} & X_{5*} & X_{6*} \\
(1260, & 2, & 2, & 3, & 4, & 5)
\end{array}$$

or at any other level.

Similarly, we verified that: 2) $\{Y_4, Y_2\}$ is p.i. of $\{\overline{Y_4, Y_2}\}$; and 3) $\{Y_4, Y_3\}$ is p.i. of $\{\overline{Y_4, Y_3}\}$.

These assumptions, together with the u.i. for Y_4 of the other attributes, imply that the utility function u must be either *additive*

$$u\left(y_1, y_2, y_3, y_4\right) = \sum_{i=1}^{4} k_i u_i(y_i) \tag{1}$$

or *multiplicative*

$$\begin{aligned} u\left(y_1, y_2, y_3, y_4\right) &= \textstyle\sum_{i=1}^{4} k_i u_i(y_i) + k \sum_{\substack{i=1 \\ j>i}}^{4} k_i k_j u_i(y_i) u_j(y_j) \\ &+ k^2 \textstyle\sum_{\substack{i=1 \\ j>i \\ l>j}}^{4} k_i k_j k_l u_i(y_i) u_j(y_j) u_l(y_l) + k^3 \prod_{i=1}^{4} k_i u_i(y_i) \end{aligned} \tag{2}$$

where k, k_i, $i = 1,2,3,4$, are the scaling constants.

The logical next step in assessing u was to try to identify functions f_2 and f_3 such that $u_2\left(y_2\right) = f_2\left[u_2^x(x_2), u_3^x(x_3)\right]$ and $u_4\left(y_4\right) = f_3\left[u_5^x(x_5), u_6^x(x_6)\right]$, where the u_i^x's are utility functions over their respective domains. Note that, since Y_2 is u.i. of $\overline{Y}_2$ and Y_4 is also u.i. of $\overline{Y}_4$, we must just worry about whether X_2 and X_3 are *conditionally additive independent* (c.a.i.) given that $\overline{Y}_2$ is fixed at any level and whether X_5 and X_6 are also c.a.i. given that $\overline{Y}_4$ is fixed at any level.

We first examined the appropriateness of the c.a.i. assumption for X_5 and X_6. We tried to check the additive independence axiom, i.e., whether doctors were indifferent between lotteries

$$\begin{pmatrix} .5 & .5 \\ (4,5) & (0,0) \end{pmatrix} \quad \text{and} \quad \begin{pmatrix} .5 & .5 \\ (4,0) & (0,5) \end{pmatrix}$$

for any fixed level of attributes X_1, X_2, X_3 and X_4. Doctors found involved to provide an answer to the above comparison (it was hard to imagine X_5 and X_6 at opposite levels), so we decided to consider an alternative test of additivity. This was based on, see Keeney and Raiffa (1976): 1) *mutual conditional utility independence* (c.u.i.) between X_5 and X_6 and, 2) there are levels x_5^a, x_5^b, x_6^a and x_6^b, such that

$$\begin{pmatrix} .5 & .5 \\ (x_5^a, x_6^a) & (x_5^b, x_6^b) \end{pmatrix} \sim \begin{pmatrix} .5 & .5 \\ (x_5^a, x_6^b) & (x_5^b, x_6^a) \end{pmatrix} \tag{3}$$

both lotteries for any fixed level of attributes X_1, X_2, X_3 and X_4. To see 1), we asked the doctors, for fixed levels of the remaining attributes, to determine the certainty equivalent of lottery (for X_5 levels) $\begin{pmatrix} .5 & .5 \\ 0 & 4 \end{pmatrix}$, with attribute X_6 at its best level (0). The certainty equivalent was 3 and this was also the indifference point if we changed X_6 to any level. Similarly, the doctors determined the certainty equivalent of lottery (for X_6 levels) $\begin{pmatrix} .5 & .5 \\ 0 & 5 \end{pmatrix}$, with attribute X_5 at its best level (0). The certainty equivalent was 3 and this was also the indifference point if we changed X_5 to any level. Thus, X_5 and X_6 were mutual c.u.i. To test 2), doctors provided for (3), levels $x_5^a = 4$, $x_5^b = 1$, $x_6^a = 4$ and $x_6^b = 0$.

In analogous way, we found easy to test the c.a.i. between X_2 and X_3. Hence, we obtained additive utility functions for Y_2 and Y_4 given by

$$u_2(y_2) = k_2^x u_2^x(x_2) + k_3^x u_3^x(x_3) \quad \text{and} \quad u_4(y_4) = k_5^x u_5^x(x_5) + k_6^x u_6^x(x_6). \quad (4)$$

Moreover, to gain more confidence in the consequences, we decided to apply an alternative test of additivity not involving lotteries. That will be provided in Section 5, after the utility assessment.

4 Utility and weight assessment

We determined the component utility functions from the combination of two standard procedures: the probability equivalent method (PE) and the certainty equivalent method (CE), see e.g. Farquhar (1984), to mitigate the bias and inconsistencies of the elicitation process. Furthermore, instead of assessing only one number at each probability question as each method demands, we have assessed a class of utility functions for each attribute, von Nitzsch and Weber (1988) and Ríos *et al.* (1994). This is less demanding, since we ask doctors and parents to provide only incomplete preference statements by means of intervals, rather than unique numbers.

The system IctNeo uses the PE-method known as *extreme gambles*, where the DM has to specify probability intervals $[p_t^L, p_t^U]$ such that $\begin{pmatrix} p_i & 1-p_i \\ x_i^* & x_{i*} \end{pmatrix}$ $\sim x_i^t$ for all $p_i \in [p_t^L, p_t^U]$, where x_i^* and x_{i*} represent the best and worst outcomes for attribute X_i and we consider (for increasing preferences) some selected amounts $x_i^t \in [x_{i*}, x_i^*]$. In case of decreasing preferences, we should take the interval $[x_i^*, x_{i*}]$. We have taken three amounts, denoted $x_i^1 = \frac{1}{4}(x_{i*} + x_i^*)$, $x_i^2 = \frac{1}{2}(x_{i*} + x_i^*)$, $x_i^3 = \frac{3}{4}(x_{i*} + x_i^*)$, but some others could be chosen for comparison.

To obtain probability intervals with the PE-method, the system uses a graphical representation of probabilities for lotteries by means of colored sectors of a circle, to provide the probabilistic questions and guide the expert until an interval of indifference probabilities is obtained. A number of additional questions are included as consistency checks. Table 4.1 presents the results of the elicitation for attribute X_1. The intervals obtained in the elicitation are ranges of utilities for the same amount x_i^t.

Table 4.1. Assessed probability intervals for u_1

Attribute levels	x_1^*	x_1^1	x_1^2	x_1^3	x_{1*}
Amount	0	315	630	945	1260
Utility range	1	[.79, .89]	[.58, .70]	[.31, .40]	0

The CE-method used in IctNeo is the *fractile method.* The DM is asked to provide certainty equivalent intervals for lotteries whose results are the extreme values x_i^* and x_{i*} and probabilities p^t and $1-p^t$. We have taken $p^1 = .25$, $p^2 = .50$ and $p^3 = .75$. This means that the DM considers $\begin{pmatrix} p^t & 1-p^t \\ x_i^* & x_{i*} \end{pmatrix} \sim x_i^t$, for all amounts $x_i^t \in \left[x_{p^t}^L, x_{p^t}^U\right]$, for $t = 1, 2, 3$. Table 4.2 shows the results of the elicitation for attribute X_1. The intervals obtained in the elicitation are ranges of the attribute with the same utility p^t.

Table 4.2. Assessed attribute intervals for u_1

Probability levels p	0.0	p^1	p^2	p^3	1.0
Utility	1	.75	.50	.25	0
Attribute range	0.0	[354, 512]	[696, 821]	[927, 1068]	1260

Figure 4.1a shows the ranges of this utility function obtained with both methods, represented by the bounding utility functions u_1^L and u_1^U, where L (U) means lower (upper).

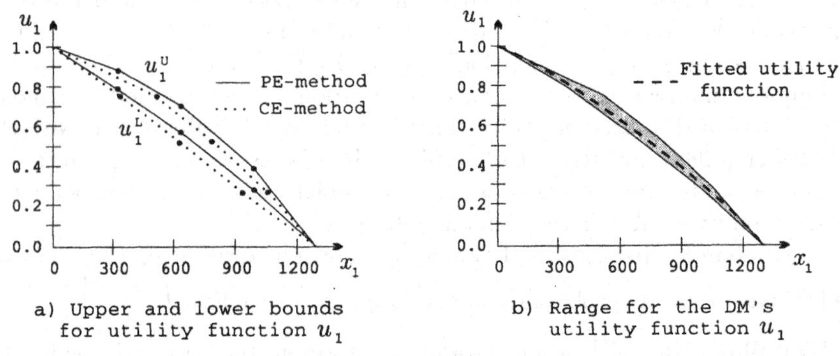

a) Upper and lower bounds for utility function u_1

b) Range for the DM's utility function u_1

Fig. 4.1.

We compare the responses given by both methods to detect inconsistencies. There will be inconsistencies if the intersection area obtained from both types of responses were empty in some range of the attribute and we should reassess the preferences until the DM provides a consistent range for the utility function. Thus, the intersection will be the range for the DM's utility functions. For attribute X_1 the utility intervals are shown in Table 4.3. Figure 4.1b shows that range. It will be used in sensitivity analysis to gain insight about the ranking of strategies.

Table 4.3. Utility intervals for u_1

Attribute	X_1		
Amount	315	630	945
Utility	$[.79, .85]$	$[.58, .67]$	$[.31, .39]$

Once with these assessments, since we need to rank the strategies, we build the component utility functions u_i. We fitted piecewise exponential functions $a + be^{-cx}$, using least squares with the midpoints of utility intervals, for the values $x_{i*}, x_i^1, x_i^2, x_i^3, x_i^*$, in X axis of each range. Table 4.4 shows the fitted component utility functions for all attributes.

Table 4.4. The single-attribute utility functions

Attribute	u_i	Range
X_1	$u_1(x_1) = 1.604 - .604\exp(.00077x_1)$	$[0, 1260]$
X_2	$u_2^x(x_2) = -.1108 + 1.111\exp(-1.153x_2)$	$[0, 2]$
X_3	$u_3^x(x_3) = -.225 + 1.225\exp(-.8473x_3)$	$[0, 2]$
X_4	$u_4(x_4) = 1.277 - .2766\exp(.5098x_4)$	$[0, 3]$
X_5	$u_5^x(x_5) = 1.361 - .361\exp(.3316x_5)$	$[0, 4]$
X_6	$u_6^x(x_6) = 1.408 - .4083\exp(.2476x_6)$	$[0, 5]$

The utility functions were all assessed by doctors, except for attribute "emotional cost" that was assessed with the aid of a doctor and parents.

Once the measures had been made comparable by defining a utility function for each attribute, the next step was to combine the individual utility functions into the overall function (1) or (2), with components (4). To establish the relative importance of each attribute, we first assessed the weights or scaling constants k_i. The key element to establish such relative importances is a *tradeoff*, which is the one used by the system. We proceeded as follows. Let us consider the case of attribute Y_1: if y_1^m represents the average value over its range, we consider comparisons of the form

$$\begin{pmatrix} p_1 & 1-p_1 \\ (y_1^*, y_2^*, y_3^*, y_4^*) & (y_{1*}, y_{2*}, y_{3*}, y_{4*}) \end{pmatrix} \sim (y_1^m, y_{2*}, y_{3*}, y_{4*}) . \qquad (5)$$

Doctor must provide p_1, such that he is indifferent between the lottery and the sure consequence in (5). As in the case of utility assessment, we allow for imprecision since it may be more demanding for the DM to provide a unique value p_1 instead of an interval $[p_1^L, p_1^U]$. Then, from the properties of the utility function, we have that $p_1^L = k_1^L u_1(y_1^m)$, $p_1^U = k_1^U u_1(y_1^m)$ and, hence, $k_1^L = p_1^L / u_1(y_1^m)$, $k_1^U = p_1^U / u_1(y_1^m)$. Note that to compute $u_2(y_2^m)$ and $u_4(y_4^m)$ we have used (4), with precise constants k_i^x, shown in Table 4.7, below. Table 4.5 shows the weight intervals for the Y_i's.

Table 4.5. Weight intervals for attributes Y_i

Attribute	Y_1	Y_2	Y_3	Y_4
k_i-weight interval	$[.073, .145]$	$[.019, .043]$	$[.091, .271]$	$[.065, .152]$

Now, since $\sum_{i=1}^4 k_i \in \left[\sum_{i=1}^4 k_i^L, \sum_{i=1}^4 k_i^U\right] = [.248, .611] \not\ni \{1\}$, the multiplicative utility function (2) is always appropriated and the additional constant k must be found. Moreover, since it is always $\sum_{i=1}^4 k_i < 1$, it follows that

$k \in (0, \infty)$ and we shall determine k as the solution to $1+k = \prod_{i=1}^{4}(1+kk_i)$, Keeney (1974, 1980). Since we have interval weights, we obtain a range $[2.514, 19.163]$ for k, where the extremes are obtained for the lower and upper values, respectively, from Table 4.5.

For assessing the scaling constants k_i^x in the additive utility functions (4), we used the same procedure based on tradeoffs but taking into account the consistency requirements $k_2^x + k_3^x = 1$ and $k_5^x + k_6^x = 1$. For that, consider the case of function u_2, where we show the procedure for assigning constant k_2^x. Suppose that the DM provides for probability comparisons analogous to (5), the extreme values p_2^{xL} and p_2^{xU}. Then, from properties of the utility function, $p_2^{xL} = k_2^{xL} u_2^x(x_2^m)$ and $p_2^{xU} = k_2^{xU} u_2^x(x_2^m)$ and, hence $k_2^{xL} = p_2^{xL}/u_2^x(x_2^m)$ and $k_2^{xU} = p_2^{xU}/u_2^x(x_2^m)$. Table 4.6 presents the results of the elicitations.

Table 4.6. Weight intervals for attributes X_i

Attribute	X_2	X_3	X_5	X_6
$[p_i^{xL}, p_i^{xU}]$	$[.262, .434]$	$[.247, .388]$	$[.374, .522]$	$[.231, .467]$

Since we have to rank the strategies with (2) we need precise values for the scaling constants. Hence, we provide the average values for constants k_i given by $k_i = (k_i^L + k_i^U)/2$, and from them, we find constant k. For constants k_i^x in the additive functions, we provide the normalized average values

$$k_i^x = \frac{k_i^{xL} + k_i^{xU}}{\sum_{j=2}^{3}\left(k_j^{xL} + k_j^{xU}\right)}, \; i = 2,3 \quad \text{and} \quad k_i^x = \frac{k_i^{xL} + k_i^{xU}}{\sum_{j=5}^{6}\left(k_j^{xL} + k_j^{xU}\right)}, \; i = 5,6.$$

Then, the precise values of the scaling constants for (2) and (4) are shown in Table 4.7.

Table 4.7. Weights for the utility functions

Multiplicative function		Additive functions	
k_i	Value	k_i^x	Value
k_1	.109	k_2^x	.578
k_2	.031	k_3^x	.422
k_3	.181	k_5^x	.558
k_4	.109	k_6^x	.442
k	6.329		

Note that constant k determines the type and degree of interaction between attributes. Because $k = 6.329 > 0$ and $\sum_{i=1}^{4} k_i < 1$, we have a destructive interaction: low utility on one attribute can result in a low overall utility.

5 Alternative checking of additivity

To gain more confidence on the additive decompositions (4), we applied an alternative *test of additivity based on tradeoffs*, not involving lotteries, Delquié and Luo (1997). To conduct such test on (X_5, X_6), we applied the *simple tradeoffs test* to attributes X_5 and X_6. We consider an option with intermediate levels given by (x_5^i, x_6^i) and we produce another option (x_5^u, x_6^u) such that $(x_5^i, x_6^i) \sim (x_5^u, x_6^u)$, where x_5^u is a change in x_5^i and we have to determine x_6^u that provides the above indifference. Similarly, we determine (x_5^l, x_6^l) such that x_6^l is a change in x_6^i and we have to determine x_5^l such that $(x_5^i, x_6^i) \sim (x_5^l, x_6^l)$. Then, if

$$\frac{u_6^x(x_6^i) - u_6^x(x_6^u)}{u_5^x(x_5^i) - u_5^x(x_5^u)} = \frac{u_6^x(x_6^i) - u_6^x(x_6^l)}{u_5^x(x_5^i) - u_5^x(x_5^l)} \tag{6}$$

the model is additive. We began with $(x_5^i, x_6^i) = (2,3)$ and the doctors provided the indifferences $(x_5^i, x_6^i) = (2,3) \sim (3,1) = (x_5^u, x_6^u)$ and $(x_5^i, x_6^i) = (2,3) \sim (1,4) = (x_5^l, x_6^l)$. We obtained -1.213 and -1.217 for the first and second quotient in (6), respectively, being very close. Thus, an additive decomposition was acceptable as we had proved above.

The same test was applied to attributes (X_2, X_3), with similar results.

6 Conclusions

IctNeo is a DSS which is currently being tested. It is supported by an influence diagram integrating beliefs and preferences. In this paper we have considered the problem of preference modelling by means of an imprecise nested multiattribute utility function and we have described in some detail the structuring and imprecise assessment process to reach the valid utility function implemented into the system. It will become an important aid for doctors, who will have deeper insight and understanding of the jaundice problem. The feature that will be incorporated to IctNeo in the future is the sensitivity analysis option that will permit to check the robustness to the structure, and to probabilities and utilities assignments to aid doctors in choosing a final treatment, see Bielza *et al.* (1996), Ríos Insua (1990) and Ríos Insua and French (1991).

Acknowledgments This paper has been supported by DGESIC project PB97-0856 and FIS project 97/ 0003-02. We are grateful to doctors M. Sánchez Luna, S. Caballero and D. Blanco, who were the experts in the elicitation processes and to the referees for helpful comments.

References

[1] Bielza, C., D. Ríos Insua and S. Ríos-Insua (1996). "Influence Diagrams under Partial Information", in *Bayesian Statistics 5,* J.M. Bernardo, J.O. Berger, A.P. Dawid and A.F.M. Smith (eds.), Oxford U. Press, 491-497.

[2] Ph. Delquié and M. Luo (1997). "A Simple Trade-off Condition for Additive Multiattribute Utility", *Journal of Multi-Criteria Decision Analysis* **6**, 248-252.

[3] Farquhar, P.H. (1984). "Utility Assessment Methods", *Management Science* **30**, 1283-1300.

[4] Gartner, L.M. (1995). "Neonatal Jaundice", *Pediatrics in Review* **16**, 22-31.

[5] Keeney, R.L. (1974). "Multiplicative Utility Functions" *Operations Research* **22**, 22-34.

[6] Keeney, R.L. (1980). *Siting Energy Facilities,* Academic Press, New York.

[7] Keeney, R.L. and H. Raiffa (1976). *Decisions with Multiple Objectives: Preferences and Value Tradeoffs,* Wiley, New York.

[8] Kirkwood, C.W. (1997). *Strategic Decision Making,* Duxbury Press, Belmont.

[9] Mollison, P.L. and M. Cutbush (1951). "A Method of Measuring the Severity of a Series of Cases of Hemolytic Disease of the Newborn", *Blood* **6**, 777-788.

[10] Newman, T.B. and M.J. Maisels (1992). "Evaluation and Treatment of Jaundice in the Term Infant: A Kinder, Gentler Approach", *Pediatrics,* **89**, 809-830.

[11] Ríos, S., S. Ríos-Insua, D. Ríos Insua and J.G. Pachón (1994). "Experiments in Robust Decision Making", in *Decision Theory and Analysis: Trends and Challenges,* S. Ríos (ed.), Kluwer, Boston, 233-242.

[12] Ríos Insua, D. (1990). *Sensitivity Analysis in Multiobjective Decision Making,* LNEMS 347, Springer, Berlin.

[13] Ríos Insua, D. and S. French (1991). "A Framework for Sensitivity Analysis in Discrete Multi-Objective Decision-Making", *European Journal of Operational Research* **54**, 176-190.

[14] Ríos-Insua, S., C., Bielza, M. Gómez, J.A. Fdez del Pozo, M. Sánchez Luna, and S. Caballero (1998). "An Intelligent Decision System for Jaundice Management in Newborn Babies", in *Applied Decision Analysis,* F.J. Girón (ed.), Kluwer, Boston, 133-144.

[15] Shachter, R.D. (1986). "Evaluating Influence Diagrams", *Operations Research* **34**, 871-882.

[16] Von Nitzsch, R. and M. Weber (1988). "Utility Function Assessment on a Micro-Computer: An Interactive Procedure", *Annals of Operations Research* **16**, 149-160.

Maximal Point Theorems in Product Spaces and Applications for Multicriteria Approximation Problems

A. Göpfert and Chr. Tammer *

Abstract

The paper contains new existence results for approximately efficient elements of a finite dimensional vector optimization problem. Then necessary conditions for approximately efficient solutions of an inverse Stefan problem are derived using a maximal point theorem in a product space.

Keywords: Maximal point theorem, variational principle, ϵ - Kolmogorov condition, inverse Stefan problem

AMS subject classification: 90C26, 49J40, 90C29

1 Introduction

Approximation problems have been studied by many authors from the theoretical as well as the computational point of view. Such problems play an important role in optimization theory and many practical problems can be described as approximation problem. Beside problems with one objective function several authors investigated even multicriteria approximation problems. The aim of our paper is to derive necessary conditions for approximate solutions of a multicriteria approximation problem, so-called ϵ - Kolmogorov conditions. We will explain this approach at the example of the approximative solution of partial differential equations, especially for the inverse Stefan problem. In order to derive these necessary conditions for approximate solutions we will show in Section 2 existence results for approximately efficient elements which are important for the proof of the existence of conical support points in product spaces in Section 3. We will use conical support points for sets in product space $X \times R^p$ (X for the variables of the approximation problem and R^p for the p goal functions), which imply a variational principle

*Martin-Luther-University Halle-Wittenberg, Dept. for Mathematics and Informatics, D-06099 Halle, Germany

for multicriteria optimization problems. An existence theorem for conical support points of a set $\mathcal{A} \subset X \times R^p$ with respect to a certain cone in $X \times R^p$ is called a maximal point theorem in a product space. Theorem 1 in Section 3 is such a maximal point theorem where we use the cone

$$\mathcal{K}_\epsilon = \{(x,y) \in X \times R^p \; : \; y + \epsilon k^0 ||x|| \in -R^p_+\},$$

with $\epsilon > 0$, $k^0 \in int\ R^p_+$ and R^p_+ is the usual ordering cone in R^p. Taking $p = 1$ and $k^0 = 1$ the set $\mathcal{K}_\epsilon$ coincides with the well-known cone introduced by Phelps [16]. If we put $\epsilon = 0$ in the definition of the cone $\mathcal{K}_\epsilon$ then elements $(x_0, y_0) \in \mathcal{A}$ which are maximal elements with respect to $\mathcal{K}_\epsilon$ have the property that the y-component of y_0 is maximal with respect to the cone R^p_+. This means that y_0 is a global maximal element.
The construction of the cone $\mathcal{K}_\epsilon$ shows that $\mathcal{K}_\epsilon$ as well as R^p_+ have a not too big width: They have a bounded base. Regarding this fact we have to study the following problems in the case $p > 1$:

- How to ensure the existence of sufficiently good elements of $\mathcal{A}$ (approximately maximal points)? Some new results follow in Section 2.
- How to ensure that certain (conical) sections in $\mathcal{A}$ become sufficiently small? Lemma 4 and assumption (A) in Section 3 give the answer.

In Section 4 we study some applications. As usual, our maximal point theorem implies a variational principle of Ekeland's type for multicriteria optimization problems (cf. [1], [7], [8], [9], [10], [13], [14], [15], [18]). In particular, we derive from the variational principle ϵ- Kolmogorov - conditions for suboptimal solutions of multicriteria approximation problems.
Finally, in Section 5 we discuss our method at the example of the inverse Stefan problem following Reemtsen [17] and Jahn [11]. We show necessary conditions for approximative solutions of the inverse Stefan problem, which are important for numerical algorithms. We consider the problem of melting ice, where the temperature distribution $u(x,t)$ in the water at the time t is described by the heat-flow equation $u_{xx}(x,t) - u_t(x,t) = 0$. We assume that the motion of the melting interface is known and some other boundary condition has to be determined, i.e., the ablating boundary $\delta(.)$ is a known function of t and the heat input $g(t)$ along $x = 0$ is to be determined. Physically, the boundary condition has to be determined such that the melting interface moves in the prescribed way $x = \delta(t)$, $t \geq 0$.
Suppose that $\delta(t) \in C^1[0,T]$, $T > 0$, is a given function, $0 \leq t \leq T$, $0 \leq x \leq \delta(t)$ and $\delta(0) = 0$. Put

$$D(\delta) := \{(x,t) \in R^2 \mid 0 < x < \delta(t),\ 0 < t \leq T\} \quad \text{for} \quad \delta \in C^1[0,T].$$

Now, consider the parabolic boundary value problem

$$u_{xx}(x,t) - u_t(x,t) = 0, \quad (x,t) \in D(\delta), \tag{1}$$

$$u_x(0,t) = g(t), \quad 0 < t \leq T, \tag{2}$$

where $g \in C([0,T])$, $g(0) < 0$ is to be determined,

$$u(\delta(t),t) = 0, \quad \dot{\delta}(t) = -u_x(\delta(t),t), \quad 0 < t \leq T. \tag{3}$$

2 On approximately efficient points

In this section we derive existence results for approximately efficient elements of a multicriteria optimization problem which we will use in the proof of our main result - a maximal point theorem (Theorem 1, Section 3). Well-known scalarization techniques and proper efficiency concepts use convex cones of the kind $B \subset R^p$ with $(R^p_+ \setminus \{0\}) \subset int\, B$ (cf. Kaliszewski [12], Dubov [4], Gerth, Weidner [6], Weidner [20] and Tammer [18]). At first we recall cones which are used in the scalarization technique in the sense of Dubov, where $\lambda \in int R^p_+$, $\alpha \in (0,1)$, $v \in R^p$ and $F \subset R^p$ is a given nonvoid set:

$$\min_{y \in F} \max_{i=1,\ldots,p} \lambda_i (y_i + \alpha \sum_{j=1 j \neq i}^{p} y_j - v_i), \tag{4}$$

The scalarization in (4) corresponds to the efficiency with respect to the cone

$$D^\alpha = \{y \in R^p \mid y_i + \alpha \sum_{j=1, j\neq i}^{p} y_j \geq 0 \quad \forall i = 1,...,p\},\ \alpha \in (0,1). \tag{5}$$

In the following we consider the cone $B^\alpha \subset R^p$ as a slight modification of Dubov's cone D^α

$$B^\alpha := \{y \in R^p \mid \alpha y_i + \sum_{j=1, j\neq i}^{p} y_j \geq 0 \quad \forall i = 1,...,p\} \text{ with } \alpha > 0. \tag{6}$$

It is very easy to see that the statements in the following Lemma 1 are true.

Lemma 1 *The cone $B^\alpha \subset R^p$, $\alpha > 0$, has the following properties*

(i) *$B^\alpha \subset R^p$ is a convex cone for all $\alpha > 0$, pointed if $\alpha \neq 1$.*

(ii) *$(R^p_+ \setminus \{0\}) \subset int\, B^\alpha$ for all $\alpha > 0$.*

In the next lemmata we derive a new assertion about the existence of approximately efficient elements of a certain set F with respect to the cone B^α in the form $\{z \in F \mid F \cap (z - \epsilon k^0 - (B^\alpha \setminus \{0\})) = \emptyset\} \neq \emptyset$ (compare Definition 1 in Section 4). Some other existence results for approximate solutions are shown by Loridan [14] and Tammer [19].

Lemma 2 *Let F be a proper subset of R^p. Let the convex cone $B^\alpha \subset R^p$, $\alpha > 0$, be given by (6), $k^0 \in int\ B^\alpha$. Suppose $\epsilon > 0$ and $\inf_{y\in F} z_{B^\alpha}(y) > -\infty$ for a continuous, subadditive, strictly int B^α - monotone functional z_{B^α} : $R^p \longrightarrow R$ with $z_{B^\alpha}(0) = 0$. Then it holds*

$$\{z \in F \mid F \cap (z - \epsilon k^0 - (B^\alpha \setminus \{0\})) = \emptyset\} \neq \emptyset.$$

Proof: For $k^0 \in int\ B^\alpha$ and $\epsilon > 0$ let be

$$\mathcal{A}_{B^\alpha,k^0,\epsilon} := \{\bar{y} \in F \mid z_{B^\alpha}(\bar{y}) < \inf_{y\in F} z_{B^\alpha}(y) - z_{B^\alpha}(-\epsilon k^0)\}.$$

From $z_{B^\alpha}(0) = 0$, $\epsilon k^0 \in int\ B^\alpha$ and the strict $int\ B^\alpha$ - monotonicity of z_{B^α} it follows $z_{B^\alpha}(-\epsilon k^0) < z_{B^\alpha}(0) = 0$. Now $\inf_{y\in F} z_{B^\alpha}(y) > -\infty$ and $0 < -z_{B^\alpha}(-\epsilon k^0)$ imply $\mathcal{A}_{B^\alpha,k^0,\epsilon} \neq \emptyset$.
Further we have

$$\mathcal{A}_{B^\alpha,k^0,\epsilon} \subset \{z \in F \mid F \cap (z - \epsilon k^0 - (B^\alpha \setminus \{0\})) = \emptyset\}.$$

If we assume that there exists an element $\bar{y}$ of $\mathcal{A}_{B^\alpha,k^0,\epsilon}$ with

$$\bar{y} \notin \{z \in F \mid F \cap (z - \epsilon k^0 - (B^\alpha \setminus \{0\})) = \emptyset\},$$

then there must be an element $y' \in F$ with $y' \in \bar{y} - \epsilon k^0 - (B^\alpha \setminus \{0\})$. Now we get from the strict $int\ B^\alpha$ -monotonicity, the subadditivity of z_{B^α} and the definition of $\mathcal{A}_{B^\alpha,k^0,\epsilon}$ that

$$z_{B^\alpha}(y') \leq z_{B^\alpha}(\bar{y} - \epsilon k^0) \leq z_{B^\alpha}(\bar{y}) + z_{B^\alpha}(-\epsilon k^0) < \inf_{y\in F} z_{B^\alpha}(y).$$

This is a contradiction to $y' \in F$. ∎

Lemma 3 *Suppose that F is a proper subset of R^p and $B^\alpha \subset R^p$, $\alpha > 0$, is a cone given by (6) with $k^0 \in int\ B^\alpha$ and $F \cap -intB^\alpha = \emptyset$.
Then, we have for all $\epsilon > 0$ that*

$$\{z \in F \mid F \cap (z - \epsilon k^0 - (B^\alpha \setminus \{0\})) = \emptyset\} \neq \emptyset.$$

Proof: Regarding $F \cap -int\ B^\alpha = \emptyset$ for the convex cone B^α given by (6) we can apply a separation theorem for nonconvex sets (Corollary 2.1 in [6]) which implies $z_{B^\alpha}(0) \leq z_{B^\alpha}(y)$ for all $y \in F$ and $-\infty < \inf_{y\in F} z_{B^\alpha}(y)$ for the continuous, sublinear, strictly $int\ B^\alpha$-monotone functional $z_{B^\alpha} : R^p \longrightarrow R$,

$$z_{B^\alpha}(y) := inf\{t \in R \mid y \in -cl\ B^\alpha + tk^0\}, \quad y \in R^p,$$

with $z_{B^\alpha}(0) = 0$.
So the assumptions of Lemma 3 are fulfilled and we get the existence of an element $y_\epsilon \in F$ with $y_\epsilon \in \{z \in F \mid F \cap (z - \epsilon k^0 - (B^\alpha \setminus \{0\})) = \emptyset\} \neq \emptyset$. ∎

3 Maximal point theorems

The main result in this section is Theorem 1. In order to prove this maximal point theorem we need the following assertion:

Lemma 4 *Suppose that the convex cone B^α, $\alpha > 0$, is given by (6) and $k^0 \in int\ R^p_+$. Furthermore, consider*

$$D_n := (y_n - R^p_+) \cap (R^p \setminus \{y_n - \frac{1}{n}k^0 - (B^\alpha \setminus \{0\})\})$$

and

$$d_n := sup\{||u_n - v_n|| \quad | \quad u_n,\ v_n \in D_n\}.$$

Then it holds $d_n \longrightarrow 0$ for $n \longrightarrow \infty$.

The proof of Lemma 4 is based on results about the parametrization of spaces (cf. [6], Lemma 2.1) regarding that the cone R^p_+ has a bounded base.

Remark 1: In particular, the point of Lemma 4 is, that *diam* D_n can be made arbitrary small if n is sufficiently big.

Phelps [16] proved a maximal point theorem in $X \times R$ which implies Ekeland's variational principle [5]. In order to get a variational principle also for multicriteria optimization problems we derive a maximal point theorem in a product space $X \times R^p$.
In the following theorem the assertion is shown without an additional requirement with respect to the cone as in Theorem 3, assumption (c), in [7]. We only assume that certain sets fulfil a weak boundedness condition in difference to the sharper assumptions in Theorem 3 in [7]. Furthermore, we prove the following theorem without making use of a scalarization in difference to the proof of Theorem 3 in [7]. Corresponding results as in the following Theorem 1 for more general spaces we have derived under an additional assumption with respect to the cone and under a sharper boundedness condition in [8], compare also [9].
In the proof of the following theorem we use the existence result for approximately efficient elements given in Lemma 3.
The main result of this section is the following maximal point theorem, where we prove, that any point of a closed subset A of the product space $X \times R^p$ which fulfils a certain boundedness condition, is dominated by a maximal point of A with respect to the cone $\mathcal{K}_\epsilon$. We use the cone

$$\mathcal{K}_\epsilon := \{(x, y) \in X \times R^p \mid y + \sqrt{\epsilon}k^o \parallel x \parallel \in -R^p_+\}$$

in the product space $X \times R^p$, where X is a Banach space, $\epsilon > 0$ and $k^0 \in$ *int* R^p_+ in this theorem.

Theorem 1 *Assume that A is a closed subset of $X \times R^p$, where X is a Banach space, $\epsilon > 0$ and moreover,*

(A) *Suppose that B^α, $\alpha > 0$, is given by (6) and*

$$\{y \in P_A\} \cap (\bar{y} - int\ B^\alpha) = \emptyset, \quad \bar{y} \in R^p,$$

where $P_A := \{y \in R^p \mid (x, y) \in A \quad for\ some \quad x \in X\}$.

Then for any point $(x, y) \in A$ there exists a point $(x_o, y_o) \in A$ such that

$$(x_o, y_o) \in A \cap (\mathcal{K}_\epsilon + (x, y)) \quad and \quad \{(x_o, y_o)\} = A \cap (\mathcal{K}_\epsilon + (x_o, y_o)).$$

Proof: Consider a sequence $\{A_n\}_{n \in N}$ of sets:

$$A_n := A \cap (\mathcal{K}_\epsilon + (x_n, y_n)). \tag{7}$$

Under the given assumptions the sets A_n are closed. We define the sequence $\{(x_n, y_n)\}$ inductively as follows: $(x_1, y_1) = (x, y)$, when we have obtained (x_n, y_n) then we choose $(x_{n+1}, y_{n+1}) \in A_n$ such that

$$\nexists (x, y) \in A_n \qquad \text{with} \qquad y \in y_{n+1} - \frac{1}{n+1} k^o - (B^\alpha \setminus \{0\}). \tag{8}$$

Such an element $y_{n+1} \in PA_n$ (with $PA_n := \{y \in R^p \ : \ (x, y) \in A_n\}$) is nothing else than an $(n+1)^{-1} k^0$-approximately efficient element of the of the set PA_n with respect to the ordering induced by B^α.
Regarding Lemma 3 y_{n+1} must exist under the given assumptions.
The inclusion $(x_{n+1}, y_{n+1}) \in A_n$ implies the inclusion $(x_{n+1}, y_{n+1}) \in \mathcal{K}_\epsilon + (x_n, y_n)$ and so we can conclude that

$$\mathcal{K}_\epsilon + (x_{n+1}, y_{n+1}) \subset \mathcal{K}_\epsilon + \mathcal{K}_\epsilon + (x_n, y_n) \subset \mathcal{K}_\epsilon + (x_n, y_n),$$

where the last inclusion follows from the fact that $\mathcal{K}_\epsilon$ is a cone. So we get from the definition of the sets A_n in (7) that $A_{n+1} = A \cap (\mathcal{K}_\epsilon + (x_{n+1}, y_{n+1})) \subset A \cap (\mathcal{K}_\epsilon + (x_n, y_n)) = A_n$, and hence

$$A_{n+1} \subset A_n. \tag{9}$$

Further, we have for an arbitrary element $(x, y) \in A_n$ the inclusion $(x, y) \in \mathcal{K}_\epsilon + (x_n, y_n)$ such that $(x, y) - (x_n, y_n) \in \mathcal{K}_\epsilon$ and $y - y_n + \sqrt{\epsilon} \parallel x - x_n \parallel k^o \in -R^p_+$ hold, and so

$$y \in y_n - \sqrt{\epsilon} \parallel x - x_n \parallel k^o - R^p_+, \tag{10}$$

which implies

$$y \in y_n - R^p_+. \tag{11}$$

Using Lemma 5, (8) and (11) yield

$$d_n \longrightarrow 0 \qquad \text{for} \quad n \longrightarrow \infty, \tag{12}$$

where $d_n := sup\{||u_n - v_n|| \mid u_n, v_n \in D_n\}$ and

$$D_n := (y_n - R^p_+) \cap (R^p \setminus \{y_n - \frac{1}{n}k^0 - (B^\alpha \setminus \{0\})\}).$$

Because of (8) we get $\not\exists(x,y) \in A_{n-1}$ with $y \in y_n - \frac{1}{n}k^o - (B^\alpha \setminus \{0\})$. Regarding $A_n \subset A_{n-1}$ we can conclude $\not\exists(x,y) \in A_n$ with $y \in y_n - \frac{1}{n}k^o - (B^\alpha \setminus \{0\})$. So we derive using (10) and regarding $R^p_+ \setminus \{0\} \subset int\ B^\alpha$ the inequality $\sqrt{\epsilon} \parallel x - x_n \parallel < \frac{1}{n}$ and so

$$\parallel x - x_n \parallel < \frac{1}{\sqrt{\epsilon}n}. \tag{13}$$

Using (12) and (13) we get $\parallel x - x_n \parallel + \parallel y_n - y \parallel \leq \frac{1}{\sqrt{\epsilon n}} + d_n$ and so $diam\ A_n \longrightarrow 0$. The completeness of R^p_+ and the closedness of the sets A_n yield that $\cap_n A_n$ contains only one element. Let (x_o, y_o) be this element. Of course we have $(x_o, y_o) \in A_1 = A \cap (\mathcal{K}_\epsilon + (x,y))$, which implies $(x_o, y_o) \in \mathcal{K}_\epsilon + (x,y)$.
Further, $(x_o, y_o) \in A_n = A \cap (\mathcal{K}_\epsilon + (x_n, y_n))$ $(n \in N)$ implies $(x_o, y_o) \in \mathcal{K}_\epsilon + (x_n, y_n)$ $(n \in N)$ and so $(x_o, y_o) + \mathcal{K}_\epsilon \subset \mathcal{K}_\epsilon + (x_n, y_n) + \mathcal{K}_\epsilon \subset \mathcal{K}_\epsilon + (x_n, y_n)$ $(n \in N)$. Finally, we get $A \cap (\mathcal{K}_\epsilon + (x_o, y_o)) \subset A \cap (\mathcal{K}_\epsilon + (x_n, y_n)) = A_n$ $(n \in N)$ and this implies $A \cap (\mathcal{K}_\epsilon + (x_o, y_o)) = \{(x_o, y_o)\}$. ■

4 Applications for approximation problems

Now, as said in the Introduction, a variational principle for optimization problems with an objective function which takes its values in a partially ordered space is a direct consequence of our Theorem 1. This variational principle (Theorem 2) is an assertion about the existence of an efficient solution of a slightly perturbed vector optimization problem being in a certain neighbourhood of an approximately efficient element of the original vector optimization problem.
In the sequel X is considered to be a real Banach space, $S \subset X$ and R^p_+ is the usual ordering cone in R^p, $k^0 \in int\ R^p_+$. We introduce a function $f : S \longrightarrow R^p$ and assume that f is bounded from below and *epi* f is closed.
First, we will introduce approximately efficient elements of vector optimization problems (see Tammer [19]).

Definition 1 *An element $f(x_\epsilon) \in f(S)$ is called an approximately efficient point of $f(S)$ with respect to R^p_+, $k^o \in int\ R^p_+$ and $\epsilon \geq 0$, if*

$$f(S) \cap (f(x_\epsilon) - \epsilon\, k^o - (R^p_+ \setminus \{0\})) = \emptyset.$$

The set of approximately efficient points of $f(S)$ with respect to R^p_+, k^o and ϵ is denoted by $\epsilon k^o - Eff(f(S), R^p_+)$. If we put $\epsilon = 0$ the set $\epsilon k^o -$

$Eff(f(S), R^p_+)$ coincides with the set of efficient points of $f(S)$ with respect to R^p_+, denoted by $Eff(f(S), R^p_+)$.
Following the lines of the proof of Theorem 4 in [7], we get the variational principle mentioned above:

Theorem 2 *(Variational principle) Assume that $f : S \longrightarrow R^p$ is a function which is bounded from below and epi f is closed. Then for any $\epsilon > 0$ and any $f(x^o) \in \epsilon k^o - Eff(f[S], R^p_+)$ there exists an element $x_\epsilon \in S$ with the following properties:*

1. $f(x_\epsilon) \in f(x^o) - \sqrt{\epsilon} \parallel x_\epsilon - x^o \parallel k^o - R^p_+$,

2. $\parallel x_\epsilon - x^o \parallel \leq \sqrt{\epsilon}$,

3. $f_{\epsilon k^o}(x_\epsilon) \in Eff(f_{\epsilon k^o}(S), R^p_+)$, *where* $f_{\epsilon k^o}(x) := f(x) + \sqrt{\epsilon} \parallel x - x_\epsilon \parallel k^o$.

Remark 2: Theorem 2 is a special case of Theorem 4 in [7]. As proved in Lemma 1 (ii), the cone B^α from Section 2 fulfils condition c) in Theorem 3 in [7], which implies the variational principle (Theorem 4 in [7]). Instead of the closedness of *epi f* it is possible to assume that S is closed and f has a lower semicontinuity property.

Now, we will show, taking a vectorial approximation problem as example, that it is possible to use the third assertion in Theorem 2 in order to derive ϵ-Kolmogorov conditions. Let us assume now that $x \in R^n$, $A_i \in \mathcal{L}(R^n, Y_i)$, Y_i are reflexive Banach spaces, $a^i \in Y_i$ and $\parallel . \parallel_i$ are norms in Y_i $\forall i = 1, ..., p$, and $S \subset R^n$ is a closed convex set. Suppose that $k^0 \in int\ R^p_+$. In order to formulate our vector optimization problem, we introduce the vector-valued objective function

$$f(x) := \begin{pmatrix} \parallel A_1(x) - a^1 \parallel_1 \\ \parallel A_2(x) - a^2 \parallel_2 \\ \cdots \\ \parallel A_p(x) - a^p \parallel_p \end{pmatrix}.$$

Under the assumptions given above we formulate the following multicriteria approximation problem (P):

$$(P): \qquad Eff(f[S], R^p_+).$$

The following corollary is a direct consequence of Theorem 2 if we use the directional derivative of the p norms $\parallel A_i x - a^i \parallel_i$. So we obtain the promised ϵ-Kolmogorov conditions as necessary conditions for approximately efficient solutions of the problem (P):

Corollary 1 *Under the assumptions given above for any $\epsilon > 0$ and any approximately efficient element $f(x^o) \in \epsilon k^o - Eff(f[S], R^p_+)$ there exists an element $x_\epsilon \in S$ with $f(x_\epsilon) \in \epsilon k^o - Eff(f[S], R^p_+)$,*
$\| x_\epsilon - x^o \| \leq \sqrt{\epsilon}$, such that for any feasible direction v at x_ϵ with respect to S having $\| v \| = 1$ there is a linear continuous mapping $l_\epsilon \in \mathcal{L}_1$, where

$$\mathcal{L}_1 := \{l = (l_1, ..., l_p),\ l_i \in L(Y_i, R) \mid \| l_i \|_{i^*} = 1,$$

$$l_i(A_i(x_\epsilon)) - a^i) = \| A_i(x_\epsilon) - a^i \|_i \ \forall i = 1, ..., p\},$$

with

$$\begin{pmatrix} l_{\epsilon 1} A_1(v) \\ \cdots \\ l_{\epsilon p} A_p(v) \end{pmatrix} \in \{ \begin{pmatrix} l_1 A_1(v) \\ \cdots \\ l_p A_p(v) \end{pmatrix} \} + R^p_+ \quad \text{for all} \quad l = (l_1, ..., l_p) \in \mathcal{L}_1$$

and

$$\begin{pmatrix} (A_1^* \, l_{\epsilon 1})(v) \\ (A_2^* \, l_{\epsilon 2})(v) \\ \cdots \\ (A_p^* \, l_{\epsilon p})(v) \end{pmatrix} \notin -\sqrt{\epsilon} k^o - int\ R^p_+, \tag{14}$$

$$\sum_{i=1}^{p} y_i^* ((A_i^* l_{\epsilon i})(v)) \geq -\sqrt{\epsilon} y^*(k^o) \text{ for a } y^* \in R^p_+ \setminus \{0\}, \text{ respectively.}$$

Moreover, in the case that $S = R^n$ and f is Gâteaux-differentiable in a neighbourhood of x^o we have

$$\| \sum_{i=1}^{p} y_i^* (A_i^* l_{\epsilon i}) \|_* \leq \sqrt{\epsilon} y^*(k^o).$$

Remark 3: Although the existence of x_ϵ and l_ϵ is proved only for $\epsilon > 0$ we would like to check Corollary 1 putting $\epsilon := \bar{\epsilon} = 0$. Then we had elements $x_{\bar{\epsilon}}$ and $l_{\bar{\epsilon}} \in \mathcal{L}_1$ such that in condition (14) it holds

$$\begin{pmatrix} (A_1^* \, l_{\bar{\epsilon} 1})(v) \\ (A_2^* \, l_{\bar{\epsilon} 2})(v) \\ \cdots \\ (A_p^* \, l_{\bar{\epsilon} p})(v) \end{pmatrix} \notin -int\ R^p_+,$$

for all directions v at $x_{\bar{\epsilon}}$ with $\| v \|_X = 1$. For the special case $A = I$ this is the well-known necessary Kolmogorov condition for weakly efficient elements of a certain class of vector-valued approximation problems (see [11]). But it has already been pointed out that the assumptions of the variational principle (Theorem 2) by themselves do not guarantee that such an element $x_{\bar{\epsilon}}$ exists.

5 The inverse Stefan - Problem

Finally, we come back to the inverse Stefan - problem (1), (2), (3), (cf. Crank [3]) which we have introduced in Section 1. In the following we apply the results of Section 4 for a characterization of approximate solutions of this problem in form of ϵ- Kolmogorov-conditions using the settings

$$\bar{u}(x,t,a) = \sum_{i=0}^{l} a_i w_i(x,t), \quad l > 0 \text{ integer, fixed,}$$

$$\text{with} \quad w_i(x,t) = \sum_{k=0}^{[\frac{i}{2}]} \frac{i!}{(i-2k)!k!} x^{i-2k} t^k, \quad i = 0,...,l$$

and as an ansatz $g(t) = c_0 + c_1 t + c_2 t^2$, $c_0 \le 0, c_1 \le 0, c_2 \le 0$.

Now, we get an objective function given by three error functions

$$\varphi_1(t,a,c) = \bar{u}(\delta(t),t,a) - 0, \quad \varphi_2(t,a,c) = \bar{u}_x(0,t,a) - g(t),$$

$$\varphi_3(t,a,c) = \bar{u}_x(\delta(t),t,a) - (-\dot{\delta}(t)),$$

$$\varphi(a,c) := \begin{pmatrix} \| \varphi_1(.,a,c) \|_1 \\ \| \varphi_2(.,a,c) \|_2 \\ \| \varphi_3(.,a,c) \|_3 \end{pmatrix}.$$

Moreover, assume $S \subset R^l \times R^3$ and $S := \{s \in R^l \times R^3 \mid s_i \in R \ \forall\, i = 1,...,l+3;\ s_i \le 0 \ \forall\, i = l+1,...,l+3\}$. Now, we study the problem to determine the set $Eff(\varphi[S], R^3_+)$ in order to compute approximate solutions of the inverse Stefan problem. This is a special case of problem (P) with

$$f(s) := \begin{pmatrix} \| A_1 s - a^1 \|_1 \\ \| A_2 s - a^2 \|_2 \\ \| A_3 s - a^3 \|_3 \end{pmatrix},$$

where $A_i \in \mathcal{L}(R^l \times R^3, Y_i)$, Y_i are reflexive L_q-spaces, especially

$$A_1(t) = (w_1(\delta(t),t), w_2(\delta(t),t), ..., w_l(\delta(t),t), 0, 0, 0),$$

$$A_2(t) = (w_{1x}(0,t), w_{2x}(0,t), ..., w_{lx}(0,t), -1, -t, -t^2),$$

$$A_3(t) = (w_{1x}(\delta(t),t), w_{2x}(\delta(t),t), ..., w_{lx}(\delta(t),t), 0, 0, 0),$$

$$s^T = (a_1, a_2, ..., a_l, c_0, c_1, c_2)^T,$$

$$a^1 = (0,...,0) \in Y_1, \quad a^2 = (0,...,0) \in Y_2, \quad a^3 = -\dot{\delta} \in Y_3 = L_q[0,T],$$

and $||.||_i$ $(i = 1,2,3)$ denote a norm in a reflexive L_q - space Y_i.
Now, it is possible to apply Corollary 1:

Under the assumptions given above for any $\epsilon > 0$ and any approximately efficient element $\varphi(s^o) \in \epsilon k^o - Eff(\varphi[S], R^3_+)$ there exists an element $s_\epsilon \in S$ with $\varphi(s_\epsilon) \in \epsilon k^o - Eff(\varphi[S], R^3_+)$, $\| s_\epsilon - s^o \|_{R^{l+3}} \leq \sqrt{\epsilon}$, such that for any feasible direction v at s_ϵ with respect to S having $\| v \| = 1$ there is a linear continuous mapping $l_\epsilon \in \mathcal{L}_1$, where $\mathcal{L}_1 := \{l = (l_1, l_2, l_3),\ l_i \in L(Y_i, R) :$ $\| l_i \|_{i^*} = 1, l_i(A_i(s_\epsilon) - a^i) = \| A_i(s_\epsilon) - a^i \|_i\ \forall i = 1,2,3\}$, with

$$\begin{pmatrix} l_{\epsilon 1}A_1(v) \\ l_{\epsilon 2}A_2(v) \\ l_{\epsilon 3}A_3(v) \end{pmatrix} \in \{ \begin{pmatrix} l_1A_1(v) \\ l_2A_2(v) \\ l_3A_3(v) \end{pmatrix} \} + R^3_+ \quad \forall\, l = (l_1, l_2, l_3) \in \mathcal{L}_1 \quad \text{and}$$

$$\begin{pmatrix} (A_1^* l_{\epsilon 1})(v) \\ (A_2^* l_{\epsilon 2})(v) \\ (A_3^* l_{\epsilon 3})(v) \end{pmatrix} \notin -\sqrt{\epsilon}k^o - int\ R^3_+,$$

$$\sum_{i=1}^{3} y_i^*((A_i^* l_{\epsilon i})(v)) \geq -\sqrt{\epsilon} y^*(k^o) \text{ for a } y^* \in R^3_+ \setminus \{0\}, \text{ respectively.} \quad (15)$$

Remark 4: For $\epsilon = 0$ the condition (15) coincides with the well-known Kolmogorov condition (cf. Jahn [11]), which means that the directional derivative at the optimal point is greater than or equal to zero.
Moreover, necessary conditions for approximate solutions of the typ (15) are important for numerical algorithms, especially for proximal point algorithms (cf. Benker, Hamel and Tammer [2]). In a coming paper Gergele, Göpfert, Tammer (1999) we derive a proximal point algorithm and corresponding computer programs written in C for the approximative solution of the inverse Stefan problem (1)-(3).

References

[1] Attouch, H.; Riahi, H. (1993): *Stability results for Ekeland ϵ - variational principle and cone extremal solutions.* Math. Oper. Res.18, 173-201.

[2] Benker, H.; Hamel, A.; Tammer, Chr. (1996) : A Proximal Point Algorithm for Control Approximation Problems. ZOR - Mathematical Methods of Operations Research 43 (3), 261-280.

[3] Crank, J. (1984): Free and moving boundary problems. Clarendon Press Oxford.

[4] Dubov, J.A. (1981): *Ustojcivost optimalnych po Pareto vektornych ocenok i ϵ- ravnomernye resenija.* Avtom. telem. 6, 139-146.

[5] Ekeland, I. (1974): *On the variational principle.* J. Math. Anal. Appl. 47, 324 - 353.

[6] Gerth (Tammer), Chr. and P. Weidner (1990): *Nonconvex separation theorems and some applications in vector optimization.* J. Optim. Theory Appl. 67, 297 - 320.

[7] Göpfert, A. and Chr. Tammer (1995): *A new maximal point theorem.* ZAA 14 No. 2, 379-390.

[8] Göpfert, A.; Tammer, Chr. and Zalinescu, C. (1997): *On the vectorial Ekeland's variational principle and minimal points in product spaces.* Submitted to Nonlin. Anal.

[9] Göpfert, A.; Tammer, Chr. and Zalinescu, C. (1998): *Maximal point theorems in product spaces.* In preparation.

[10] Isac, G. (1977): *Sur les points support coniques dans des espaces localement convexes.* Ann. Fac. Sci. Kinshasa; Sect. Math.-Phys. 3 (2).

[11] Jahn, J. (1986): *Mathematical vector optimization in partially ordered spaces.* Frankfurt am Main: Verlag P. Lang.

[12] Kaliszewski, I. (1994): *Quantitative Pareto analysis by cone separation technique.* Kluwer Academic Publishers, Boston.

[13] Khanh, P. O. (1986): *On Caristi-Kirk's theorem and Ekeland's variational principle for Pareto extrema.* Preprint. Warsaw: Polish Acad. Sci., Inst. Math.: Preprint 357, 1 - 7.

[14] Loridan, P. (1984): *ε-solutions in vector minimization problems.* J. Optim. Theory Appl. 43, 265 - 276.

[15] Nemeth, A. B. (1986): *A nonconvex vector minimization problem.* Nonlin. Anal.: Theory, Methods and Appl. 10, 669 - 678.

[16] Phelps, R. R. (1993): *Convex Functions, Monotone Operators and Differentiability* (2nd Ed.). Lect. Notes Math. 1364, 1 - 118.

[17] Reemtsen, R. (1981): *On level sets and an approximation problem for the numerical solution of a free boundary problem.* Computing 27, 27-35.

[18] Tammer, Chr. (1992): *A Generalization of Ekeland's principle.* Optimization 25, 129 - 141.

[19] Tammer, Chr. (1993): *Existence results and necessary conditions for ϵ-efficient elements.* In: Multicriteria Decision. (Eds.: B. Brosowski, J. Ester, S. Helbig and R. Nehse). Frankfurt: Verlag P. Lang, 97 - 109.

[20] Weidner, P. (1991): *Ein Trennungskonzept und seine Anwendung auf Vektoroptimierungsverfahren.* Habilitationsschrift, Martin-Luther-Universität Halle-Wittenberg.

A Robustness Index of Binary Preferences

Servio T. Guillén [1], Mayra S. Trejos [1,2] and Roberto Canales [3]

[1] Instituto de Ingeniería, UNAM, C.U., AP 70-472, Coyoacán 04510, México, D.F.
[2] Universidad de Panamá, Facultad de Ciencias Exactas, Panamá, Panamá
[3] Instituto de Investigaciones Eléctricas, AP 1-475, Cuernavaca 62000, Morelos, México

Abstract

An index of robustness of the preference between two alternatives is proposed. Given a finite number of alternatives, n conflicting criteria and weights $w_i \geq 0$, i=1, ...n representing the preferences of the decision maker, a robustness index $r(x,y) \in [-1,1]$ is defined. This index can be seen as a measure of the "robustness" of the preference order of two alternatives x and y with respect to the chosen weights w_i, i=1, ...n. If $r(x,y)$ is closed to zero, only minor changes of the weights will change the preference order of the alternatives x and y, whereas e.g. a value of $r(x,y)$ close to 1 implies a "strong" preference of x over y. It is shown that the index can also be defined for general additive preference models. A proof that the proposed index, for the additive case, is moderated stochastic transitive is given.

Key words: Preferences models, general additive models, robustness indexes

1. Introduction

A complete binary preference model is a rule that determines, for each pair of alternatives, if one is better than the other one or if there is indifference between them. This model has a mathematical structure associated to the axioms, that comply with the decision maker's preferences (French, 1988; Vincke, 1992), and a set of parameters. Additive models are widely used (Keeney and Raiffa 1976), either as an adequate representation of many preference structures, or as an approximate representation to facilitate their handling. One of the major difficulties is to assign the criteria weights to tune these models (Roy and Mousseau, 1995). There have been multiple contributions for the determination of the criteria weights (Fishburn, 1967; Saaty, 1980; Schoemaker and Carter, 1982; Von Winterfeldt and Edwards, 1986). Recently, Bana e Costa and Vansnick

(1995) introduced MACBETH, an interactive method, to help decision making, in which the criteria's weights are determined through a cardinal scale, considering not only preferences, but also the degree of preference over a set of alternatives.
In practical applications it is common that the decision maker does not accept *a posteriori* all the results of the preference model, and proposes to modify the weights. Bana e Costa and Vincke (1995), and Trejos (1991) dealt with this problem and proposed a method to consider weight intervals as imprecision of the weights values. One of the inconveniences of this method is that the decision maker must provide, a priori, the intervals of the weights, and occasionally, falls into the same problem of modifying the proposed intervals. In this work, an index that allows the decision maker to determine the robustness of the preference between two alternatives is proposed. It is the proportion in which he must modify the weights to change the preferences between the two alternatives.

A linear multicriteria preference model of a set of alternatives A, $A \subseteq R^n$, is given by a function v, v: $A \rightarrow R$, of the form:

$$v(x) = \sum_{i \in I} w_i x_i \qquad \forall\, x \in A,$$

where $(w_1,\ldots, w_n) \in R^n$ is a constant non-zero vector with non-negative components named *weights*; $I = \{1,\ldots, n\}$, $n > 1$ is the set of criteria, and $x_1,\ldots,x_n$ the evaluations of alternative x in the n criteria; v is called *value function* and $v(x)$ *the value of alternative* x. The preference model associates the difference $v(x) - v(y)$, for any pair of alternatives, $\forall\, x,y \in A$, to the preference between them, as follows:

x is preferred to y $\Leftrightarrow$ $v(x) - v(y) > 0$

x is indifferent to y $\Leftrightarrow$ $v(x) - v(y) = 0.$

Sometimes the difference $v(x) - v(y)$ is interpreted as an *intensity index* of the preference of x over y (Fishburn, 1986; Vansnick, 1984). Nevertheless, it doesn't provide any information about *robustness* of the preference, that is, if the relative weights are modified by a certain amount, does the preference remain unchanged?

In a multicriteria case, the fixed positive difference in the values between two alternatives x, y, and $v(x) - v(y) > 0$, can be distributed in different ways on the criteria. An extreme case occurs when $w_i(x_i - y_i) \geq 0$ for all i and $w_j(x_j - y_j) > 0$ for some j. This means that x *dominates* y. This corresponds to the maximum robustness of the preference of x over y, since this result does not depend on the values of the weights. In another situation, when the preference is reversed with small changes in the weights of the model, the robustness is weak. For example, when two criteria, 1 and 2, cancel each other, $w_1(x_1 - y_1) = - w_2(x_2 - y_2)$, but any of the two is much more important than all the other criteria together, that is:

$$|\sum_{i \neq 1,2} w_i (x_i - y_i)| << |w_1 (x_1 - y_1)|$$

then, with small changes of w_1 or w_2, x is preferred to y, or y is preferred to x.

Indices have been proposed with different purposes to the one presented here, though they can admit similar interpretations. Jacquet-Lasgrèze (1982), introduced indices which he called "credibility indices", to generalize the procedure of criteria aggregation. In a different context, because the author doesn't explicitly consider a multicriteria structure, Fishburn (1982) generalized the well-known theory of linear utility of von Neumann and Morgenstern (1967) through an index to model preference under risk. He suggests that the index could be interpreted as a measure of the intensity of preference.

The concept, and all properties of the robustness index, except the transitivity, are valid for the general additive preference model, which includes, as particular cases, the additive differences model (Tversky, 1969), and the usual additive model (Keeney and Raiffa, 1976).

This work is organized as follows: Section 2 describes the general additive model, in section 3, the robustness index is defined and some of its properties and their interpretation are presented, in section 4 a numerical example is given, and finally, in section 5, the main conclusions are presented.

2. Preference Models

Let A be a finite set of alternatives, and I as a set of criteria, $I = \{1, \ldots, n\}$, $n > 1$. The *general additive* model of the preferences over A is defined as : $\forall$ x, y$\in$A

$$\begin{aligned} &x \text{ is preferred to } y \iff \sum w_i p_i(x,y) > 0 \\ &x \text{ is indifferent to } y \iff \sum w_i p_i(x,y) = 0 \end{aligned} \qquad (1)$$

$w \neq 0$, where $w_i \geq 0$, $i \in I$, are the weight factors associated to the criteria i and p_i, p_i: $A \times A \rightarrow R$, $i \in I$, n antisymmetric functions (that is, $p_i(x,y) = - p_i(y,x)$ $\forall x,y \in A$, $i \in I$). The value of $p_i(x,y)$ is the evaluation of alternative x over y in the criteria i. When there are functions $v_i(\cdot)$ for all i, such that $p_i(x, y) = v_i(x) - v_i(y)$ then, the model is said to be *additive*. On the other hand, when $v_i(x) = x_i$ the model is said to be *linear*.

3. Robustness Index

The *robustness index of the preferences*, *r* (Guillén, 1993), is defined for two alternatives x, y, such that x is preferred to, or indifferent to, y, and that they differ in their evaluation, in at least one criteria, $w_i p_i(x, y) \neq 0$ for some i, as:

$$r(x, y) = \left(\sum w_i p_i(x, y) \Big/ \sum | w_i p_i(x, y) | \right)$$

Properties:

a) $r(x, y)$ includes the preferences model (Equations 1),

$$\text{x is preferred to y} \Leftrightarrow r(x,y) > 0$$

$$\text{x is indifferent to y} \Leftrightarrow r(x,y) = 0$$

b) $r(x,y)$ takes its values in the interval [-1,1].

c) If $\sum w_i p_i(x, y) > 0$, then $r(x,y) = \sup\{\theta \in R: \max_i |\delta_i| < \theta \Rightarrow \sum(1 + \delta_i) w_i p_i(x, y) > 0\}$,
that is to say , *r* is the supreme of the per unit absolute values of increments and decrements of the weights w_i that does not alter the preference expressed by the model.

d) For the additive case, the moderated stochastic transitivity property (Suppes *et al.* 1989) holds

$$r(x, y) \geq 0, r(y, z) \geq 0 \Rightarrow r(x, z) \geq \min \{r(x, y), r(y, z)\}$$

Proof:

Let two alternatives x, y be such that x is preferred or indifferent to y, and $w_i p_i(x, y) \neq 0$ for some i.

a) By definition $r(x, y) > 0$ is equivalent to $\sum w_i p_i(x, y) > 0$ and $r(x, y) = 0$ is equivalent to $\sum w_i p_i(x, y) = 0$. Therefore, the robustness index implies the preference model.

b) This statement is an immediate consequence of:

$$\sum w_i p_i(x, y) \leq \sum | w_i p_i(x, y) |$$

c) By hypothesis,

$$\sum w_i p_i(x, y) > 0$$

Let λ be the supreme of the per unit absolute values of increments and decrements of the weights w_i that does not alter the preference of x over y, or,

$$\lambda = \sup\{\theta \in R: \max_i |\delta_i| < \theta \Rightarrow \sum (1 + \delta_i) w_i p_i(x, y) > 0\}.$$

It can be shown that for the sup of θ is obtained when the absolute values of δ_i are the same for all i. In that case, $|\delta_i| = \lambda$, and if $\delta_i > 0$ when $w_i p_i(x, y) > 0$, and $\delta_i < 0$ when $w_i p_i(x, y) < 0$, for all $i \in I$, the indifference between alternatives x and y is obtained, which is equivalent to:

$$\sum_{i \in P(x,y)} (1 - \lambda) w_i p_i(x, y) + \sum_{i \in N(x,y)} (1 + \lambda) w_i p_i(x, y) = 0 \qquad (2)$$

where $P(x,y) = \{i: w_i p_i(x,y) > 0\}$ and $N(x,y) = \{i: w_i p_i(x,y) < 0\}$, $\forall$ x,y$\in$ A are the subsets of criteria such that "x is preferred to y" and "y is preferred to x", respectively. From the equation (2),

$$\sum_{i \in P(x,y)} w_i p_i(x, y) + \sum_{i \in N(x,y)} w_i p_i(x, y) = \lambda[\sum_{i \in P(x,y)} w_i p_i(x, y) - \sum_{i \in N(x,y)} w_i p_i(x, y)]$$

from which it is concluded that:

$$\lambda = (\sum w_i p_i(x, y) \Big/ \sum | w_i p_i(x, y) |)$$

therefore,

$r(x,y) = \lambda$

d) It will be shown that $r(x, y) \geq 0$, $r(y, z) \geq 0 \Rightarrow r(x, z) \geq \min \{r(x, y), r(y, z)\}$, using the fact that

$$A' > 0, B' > 0, A \geq 0, B \geq 0, B/B' \geq A/A' \Rightarrow (A + B)/(A' + B') \geq A/A'. \qquad (3)$$

If $r(y, z) \geq r(x, y)$, it must be proved that $r(x, z) \geq r(x, y)$. In this case let

$A = \sum w_i [v_i(x) - v_i(y)]$,

$B = \sum w_i [v_i(y) - v_i(z)]$,

$A' = \sum | w_i [v_i(x) - v_i(y)] |$ and

$B' = \sum | w_i [v_i(y) - v_i(z)] |$,

then, the left hand side conditions of the implication (3) are satisfied, because $r(x, y) = A/A'$ and $r(y, z) = B/B'$.

Under these conditions,

$$r(x,z) = \sum w_i [v_i(x) - v_i(z)] \Big/ \sum | w_i [v_i(x) - v_i(z)] | =$$

$$= \{\sum w_i [v_i(x) - v_i(y)] + \sum w_i [v_i(y) - v_i(z)]\} \Big/ \{\sum | w_i [v_i(x) - v_i(y)] + w_i [v_i(y) - v_i(z)] | \}$$

$$\geq \{\sum w_i [v_i(x) - v_i(y)] + \sum w_i [v_i(y) - v_i(z)]\} \Big/ \{\sum | w_i [v_i(x) - v_i(y)] | + \sum | w_i [v_i(y) - v_i(z)] | \}$$

$$= (A + B)/(A' + B') \geq A/A' \qquad (4)$$

and since $A/A' = r(x, y)$, it follows that

$r(x,z) \geq r(x, y)$.

On the other hand, when $r(x, y) \geq r(y, z)$ it must to proved that $r(x, z) \geq r(y, z)$. In this case let

$$A = \Sigma\, w_i[v_i(y) - v_i(z)],$$

$$B = \Sigma\, w_i[v_i(x) - v_i(y)],$$

$$A' = \Sigma \left| w_i[v(y) - v_i(z)] \right|,$$

$$B' = \Sigma \left| w_i[v_i(x) - v_i(y)] \right|,$$

and then all conditions on the left side of the implication (3) are satisfied, because $r(y, z) = A/A'$, $r(x, z) = B/B'$. The inequalities of expression (4) also hold under the new definitions of A, A', B and B', therefore

$$r(x,z) \geq (A + B)/(A' + B') \geq A/A' = r(y, z)$$

This finishes the proof.

Comments:

If x is preferred to y, the indifference is reached with an increment in the proportion $r(x, y)$ for the weights of the criteria in which y is better than x and a decrement in the same proportion for those in which x is preferred to y. In geometrical terms, $r(x, y)$ is the Tchebycheff's weighted distance (Tchebycheff's weighted distance, with weights $\mu_j > 0$, between points z and v, is defined as $\max_j\{\mu_j\ |z_j - v_j|\}$), with weights $1/w_i$, from the point $(w_1,\ldots,w_n) \in R^n$ to the hyperplane of weights for which the indifference between x and y is given (Figure 1).

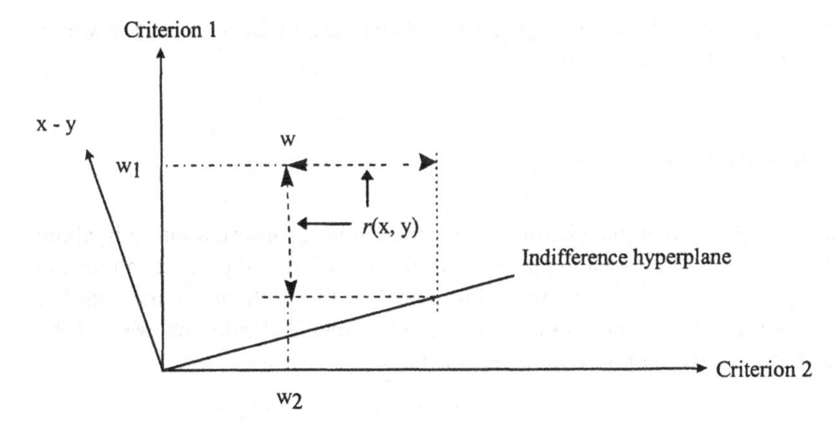

Figure 1. Geometrical representation of r(x, y) ($w_1 = w_2 = 1$)

Clearly $r(x, y) = 1$ only when $\sum w_i(x_i - y_i) = \sum |w_i(y_i - z_i)|$) and this happens only when y is not preferred to x in any of the criteria. In this situation, it is said that x *dominates* y.

4. Numerical Example

A three criteria linear model is considered. Each of the four alternatives **a**, **b**, **c** and **d** are evaluated as in Table 1.

Table 1. Decision table

		Alternatives			
Criteria	Weights	**a**	**b**	**c**	**d**
1	1	100	97	97	100
2	2	100	249	102	99
3	3	100	1	99	100
	v(.) =	600	598	598	598

The differences of value between alternatives do not discriminate between any of the preferences **a** to **b**, **a** to **c**, **a** to **d**, because v(**a**) = 600 > v(**b**) = v(**c**) = v(**d**) = 598. Nevertheless, the values of the robustness index are quite different r, r(**a,b**) = 100/299 = 0.0033, r(**a,c**) = 0.20, r(**a,d**) = 1. It is observed that the robustness of the preference of **a** over each of the alternatives **b**, **c** and **d**, goes from very small value (**a** to **b**), which almost corresponds to indifference, to the unit value in which there is the dominance (**a** over **d**).

5. Conclusions

For the general additive models, the value function gives information only about the preference between two alternatives. On the other hand, the index introduced in this paper, not only contains the information on the preferences but supplies additional data on its robustness of it, e.g. how much should the weights be changed before a given preference is reversed.

6. References

Bana e Costa, C. and J-C. Vansnick (1995). "A theoretical framework for measuring attractiveness by a categorical based evaluation technique (MACBETH)," in *Advances in Multicriteria Analysis*, Pardalos, P. M., Siskos, Y. and Zopounidis (eds.), 93-100, Kluwer Academic Publishers, Dordrech, Netherland.

Bana e Costa, C. and Vincke, Ph. (1995). "Measuring credibility of compensatory preference statements when trade-offs are interval determined," *Theory and Decision*, 39, 127-155.

Fishburn, P. C. (1967). "Methods of estimating additive utilities," *Management Science*, 13, 435-453.

Fishburn, P. C. (1982). "Nontransitive measurable utility," *Journal of Mathematical Psychology*, 26, 31-67.

Fishburn, P. C. (1986). "Ordered preference differences without ordered preferences," *Synthese*, 67, 361-368.

French, S. (1988). Decision Theory: *An Introduction to the Mathematics of Rationality*. Halsted Press, Wiley, New York.

Guillén, S. (1993). "Índices de preferencia asociados a modelos aditivos multicriterio," *Instituto de Ingeniería*, No. 556, Universidad Nacional Autónoma de México, México.

Jacquet-Lasgrèze, E. (1982). "Binary preference indices: a new look on multicriteria aggregation procedures," *European Journal of Operational Research*, 10, 26-32.

Keeney, R. L. and Raiffa, H. (1976). *Decisions with Multiple Objectives: Preferences and Value Trade-Offs*. John Wiley & Sons, New York.

Roy, B., Mousseau V. (1995). "Theoretical framework for analyzing the notion of relative importance of criteria. Journal of Multicriteria," *Decision Analysis*, 5, 145-159.

Saaty, T. L. (1980). *The Analytic Hierarchy Process*. McGraw-Hill.

Schoemaker, P. and Carter, W. (1982). "An experimental comparison of different approaches to determining weights in additive utility model," *Management Science*, 28, 182-196.

Suppes, P., Krantz, D.; Luce, R. and Tversky, A. (1989). *Foundations of Measurement*, Vol. 2, Academic Press Inc., New York.

Trejos, M. (1991), "Método de relaciones binarias de sobreclasificación que usa una familia de funciones de utilidad," Doctoral Thesis, Facultad de Ingeniería de la Universidad Nacional Autónoma de México, México

Tversky, A. (1969), "Intransitivity of preferences," *Psychological Review* 76, 31-48.

Vincke, Ph. (1992), *Multicriteria Decision Aid*, Chichester: Wiley, New York.

Vansnick, J-C. (1984). "Strength of preference, theoretical and practical aspects," *Operational Research*, Brans, J. P. (Ed.),Elsevier Science Publishers B.V. (North-Holland), 449-462.

Von Neumann, J. and Morgenstern, O. (1967). *Theory of Games and Economic Behavior*, Princeton University Press, 3nd edition, Princeton.

Von Winterfeldt, D., and Edwards, W. (1986), *Decision and Behavioral Research*, Cambridge University Press, Cambridge.

Servio T. Guillén sgb@pumas.iingen.unam.mx
Mayra S. Trejos mta@pumas.iingen.unam.mx
Roberto Canales rcanales@iie.org.mx

Tangent and Normal Cones in Nonconvex Multiobjective Optimization

Kaisa Miettinen and Marko M. Mäkelä

Department of Mathematical Information Technology, University of Jyväskylä, P.O. Box 35 (MaE), FIN-40351 Jyväskylä, Finland

Abstract. Trade-off information is important in multiobjective optimization. It describes the relationships of changes in objective function values. For example, in interactive methods we need information about the local behavior of solutions when looking for improved search directions.

Henig and Buchanan have generalized in Mathematical Programming 78(3), 1997 the concept of trade-offs in convex multiobjective optimization problems. With the help of tangent cones they define a cone of trade-off directions.

In this paper, we examine the possibility of extending the results of Henig and Buchanan for nonconvex multiobjective optimization problems. We carry out the generalization in the sense of Clarke's nonconvex analysis.

Keywords. Nonlinear multiobjective optimization, Pareto optimality, nonconvexity, trade-off directions

1 Introduction

The need of trading off plays an elementary role when searching for desirable solutions for multiobjective optimization problems, where it is impossible to meet all the objectives simultaneously. Trading off is of particular value in interactive methods (see, e.g. [2], [15]) since the knowledge of how changes in some objective function values affect the others is useful. So far, ways of generating trade-off information have been suggested based on either using certain scalarizing functions and under rather restricting assumptions (see, e.g. [2], [8], [18], [20]) or relying on special characteristics of the problems solved (like linearity, see, e.g. [9]). Another approach has been to create scalarizing functions that generate solutions satisfying pre-specified bounds on trade-off information (see, e.g. [12], [13]).

Recently, the trade-off ideas have been generalized into a cone of trade-off directions by Henig and Buchanan [11] for convex problems. This is a promising approach because of its independence of the scalarizing function used and its minor presumptions set to the problem treated. In [11], the cone of trade-off directions is defined as a Pareto optimal surface of a tangent cone located at the point considered. The calculation of trade-off directions is based on the characterizations of tangent cones and their polar cones, normal cones. Special attention is paid to proper Pareto optimality which is equivalent to the nonemptiness of the cone of trade-off directions.

The treatment in [11] has its basis in classical convex analysis in the sense of Rockafellar [17]. Clarke [3] has generalized this theory for a nonconvex case. Our intention here is to examine how the results in [11] can be generalized in the spirit of Clarke.

Giving up convexity brings along the need of dealing with local instead of global analysis. It is well-known that sufficiency conditions usually necessitate convexity. However, there is no guarantee that even convex necessary results would be valid for nonconvex problems.

In what follows, we begin by defining optimality concepts. For clarity of notations, we concentrate only on global analysis. However, the results are valid also for local optima. According to Clarke we define tangent and normal cones for nonconvex sets. The results given in [11] are then studied under the nonconvexity assumption.

2 Foundations

We consider a multiobjective optimization problem of the form

$$
\text{(2.1)} \qquad \begin{array}{ll} \text{minimize} & \{f_1(x), f_2(x), \ldots, f_k(x)\} \\ \text{subject to} & x \in S = \{x \in \mathbf{R}^n \mid (g_1(x), g_2(x), \ldots, g_m(x))^T \leq 0\}, \end{array}
$$

where we have k *objective functions* $f_i \colon \mathbf{R}^n \to \mathbf{R}$ and m *constraint functions* $g_i \colon \mathbf{R}^n \to \mathbf{R}$. The *decision vector* x belongs to the closed (nonempty) *feasible set* $S \subset \mathbf{R}^n$.

In the following, we denote the image of the feasible set by $Z \subset \mathbf{R}^k$. The set Z is called a *feasible criterion set* and its elements are termed *criterion vectors*, denoted by $z = f(x) = (f_1(x), f_2(x), \ldots, f_k(x))^T$. Thus, we have $Z = f(S)$.

All the functions are assumed to be locally Lipschitz continuous. A function $h \colon \mathbf{R}^n \to \mathbf{R}$ is *locally Lipschitz continuous* at a point $x \in \mathbf{R}^n$ if there exist scalars $K > 0$ and $\delta > 0$ such that $|h(x^1) - h(x^2)| \leq K\|x^1 - x^2\|$ for all $x^1, x^2 \in B(x;\delta)$, where $B(x;\delta) \subset \mathbf{R}^n$ is an open ball with centre x and radius δ.

Problem (2.1) is convex if all the objective functions and the feasible set are convex.

The sum of two sets A and E is defined by $A + E = \{a + e \mid a \in A,\ e \in E\}$. The interior, closure, boundary and convex hull of a set A are denoted by $\operatorname{int} A$, $\operatorname{cl} A$, $\operatorname{bo} A$ and $\operatorname{conv} A$, respectively.

We denote the negative orthant of $\mathbf{R}^k$ by $\mathbf{R}^k_- = \{d \in \mathbf{R}^k \mid d_i \leq 0 \text{ for } i = 1, \ldots, k\}$.

As the concepts of optimality we employ Pareto optimality and proper Pareto optimality.

Definition 2.2. A decision vector $x^* \in S$ and the corresponding criterion vector $z^* \in Z$ are *Pareto optimal* if there does not exist another decision vector $x \in S$ such that $f_i(x) \leq f_i(x^*)$ for all $i = 1, \dots, k$ and $f_j(x) < f_j(x^*)$ for at least one index j. The set of Pareto optimal criterion vectors is called a *Pareto optimal set* and denoted by

$$P(Z) = \{z^* \in Z \mid (z^* + \mathbf{R}^k_- \setminus \{0\}) \cap Z = \emptyset\}.$$

Definition 2.3. Vectors $x^* \in S$ and $z^* \in Z$ are *properly Pareto optimal* if there does not exist another decision vector x and corresponding criterion vector $z = f(x) \in Z$ such that $z^* \in z - C \setminus \{0\}$, in other words, $(z^* + C \setminus \{0\}) \cap Z = \emptyset$ for some convex cone C such that $\mathbf{R}^k_- \setminus \{0\} \subset \operatorname{int} C$. The *properly Pareto optimal set* is denoted by

$$PP(Z) = \{z^* \in Z \mid (z^* + C \setminus \{0\}) \cap Z = \emptyset\}.$$

Definition 2.3 of proper Pareto optimality was originally given by Henig [10].

Note that the necessary conditions to be given in Theorems 3.3, 3.5 and 4.1 are also sufficient in the convex case. The proofs can be found in [11].

3 Tangent and Normal Cones

Next we define two geometrical basic tools: tangent and normal cones for nonconvex problems. The cones are defined at a vector in the feasible criterion set. Then we can establish some connections to proper Pareto optimality.

Definition 3.1 (Clarke). The *tangent cone* of a set $Z \subset \mathbf{R}^k$ at $z \in Z$ is given by the formula

$$T_z(Z) = \{d \in \mathbf{R}^k \mid \text{for all } t_j \searrow 0 \text{ and } z_j \to z \text{ with } z_j \in Z, \\ \text{there exists } d_j \to d \text{ with } z_j + t_j d_j \in Z\}.$$

The *normal cone* of Z at $z \in Z$ is the polar cone of the tangent cone, that is,

$$N_z(Z) = T_z(Z)^\circ = \{y \in \mathbf{R}^k \mid y^T d \leq 0 \text{ for all } d \in T_z(Z)\}.$$

In convex problems, Definition 3.1 is equal to the convex notions of [11] as shown in Theorem 4.1.5 of [14]. It is important to note that tangent and normal cones are convex even in the case Z is nonconvex.

Lemma 3.2. $T_z(Z)$ and $N_z(Z)$ are closed and convex cones such that $0 \in T_z(Z) \cap N_z(Z)$.

Proof. See Theorems 4.1.2 and 4.1.4 of [14]. ∎

The next two theorems characterize the relations between properly Pareto optimal solutions and the corresponding tangent and normal cones.

Theorem 3.3. If $z \in PP(Z)$, then

$$T_z(Z) \cap \mathbf{R}^k_- \setminus \{0\} = \emptyset.$$

Proof. If $z = f(x) \in PP(Z)$, then

$$(z + C \setminus \{0\}) \cap Z = \emptyset. \tag{3.4}$$

Let us suppose that there exists $\hat{d} \in T_z(Z) \cap \mathbf{R}^k_- \setminus \{0\}$. If $t_j \searrow 0$, then it follows from the definition of the tangent cone that there exists $d_j \to \hat{d}$ with $z + t_j d_j \in Z$.

Let C be a convex cone such that $\mathbf{R}^k_- \setminus \{0\} \subset \operatorname{int} C$. Since $\hat{d} \in \mathbf{R}^k_- \setminus \{0\} \subset \operatorname{int} C$ and $d_j \to \hat{d}$, there exist j_0 such that $0 \neq d_j \in \operatorname{int} C \subset C \setminus \{0\}$ for all $j \geq j_0$. Because C is a cone and $t_j > 0$, we have $t_j d_j \in C \setminus \{0\}$ for all $j \geq j_0$. Then we have

$$z + t_{j_0} d_{j_0} \in (z + C \setminus \{0\}) \cap Z,$$

which is a contradiction with (3.4). ∎

Figure 1 depicts the empty intersection in Theorem 3.3.

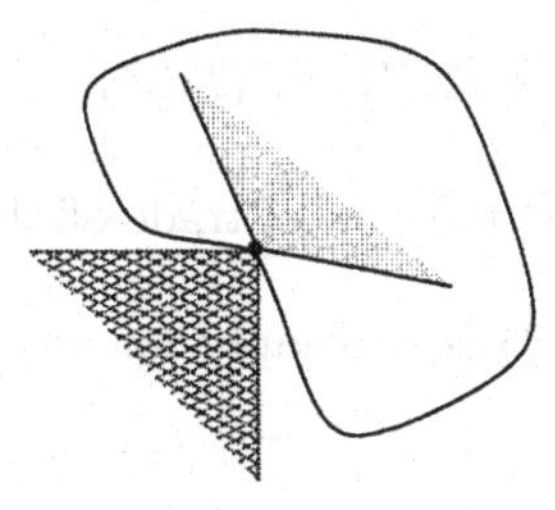

Figure 1. Empty intersection.

Theorem 3.5. If $z \in PP(Z)$, then

$$N_z(Z) \cap \operatorname{int} \mathbf{R}^k_- \neq \emptyset.$$

Proof. Let $z \in PP(Z)$, then by Theorem 3.3 we have $T_z(Z) \cap \mathbf{R}_-^k \setminus \{0\} = \emptyset$, which is equivalent to

$$T_z(Z) \cap \mathbf{R}_-^k = \{0\}.$$

Then

$$(T_z(Z) \cap \mathbf{R}_-^k)^\circ = \{0\}^\circ = \mathbf{R}^k.$$

On the other hand, by Corollary 16.4.2 of [17] and by the fact that the sum of closed convex cones containing zero is closed, we get

$$\begin{aligned}(T_z(Z) \cap \mathbf{R}_-^k)^\circ &= \text{cl}\,(T_z(Z)^\circ + (\mathbf{R}_-^k)^\circ) \\ &= \text{cl}\,(N_z(Z) + \mathbf{R}_+^k) \\ &= N_z(Z) + \mathbf{R}_+^k.\end{aligned}$$

Then for any $p \in \mathbf{R}^k$, there exist $y \in N_z(Z)$ and $d \in \mathbf{R}_+^k$ such that $p = y + d$. If in particular $p \in \text{int}\,\mathbf{R}_-^k$, then $p_i < 0$ for every $i = 1, \ldots, k$. Because $d_i \geq 0$ for all $i = 1, \ldots, k$, then $y_i < 0$ for all $i = 1, \ldots, k$ and thus

$$y \in N_z(Z) \cap \text{int}\,\mathbf{R}_-^k. \quad \blacksquare$$

4 Generalizing the Cone of Trade-off Directions

Henig and Buchanan [11] have generalized the notion of traditional trade-offs into a cone of trade-off directions for convex problems. This cone of trade-off directions is the Pareto optimal subset of a tangent cone. We generalize this cone for nonconvex problems.

Let us denote the Pareto optimal surface of a tangent cone by

$$PT_z(Z) = P(T_z(Z)).$$

Note that in a convex case the cone of trade-off directions is defined in [11] as $PT_z(Z)$.

We can show that $PT_z(Z)$ is nonempty for any properly Pareto optimal point.

Theorem 4.1. If $z \in PP(Z)$, then

$$PT_z(Z) \neq \emptyset.$$

Proof. Let $z \in PP(Z)$ and let us suppose that $PT_z(Z) = \emptyset$. Then by the definition of Pareto optimality

$$(d + \mathbf{R}^k_- \setminus \{0\}) \cap T_z(Z) \neq \emptyset$$

for all $d \in T_z(Z)$. Especially by choosing $d = 0$, we get

$$(\mathbf{R}^k_- \setminus \{0\}) \cap T_z(Z) \neq \emptyset,$$

which by Theorem 3.3 is a contradiction to the assumption that $z \in PP(Z)$. ∎

We close this section by formulating a result enabling the generation of vectors in $PT_z(Z)$. The proof of the corresponding Theorem 5 in [11] is here valid as such.

Theorem 4.2. If $d \in PT_z(Z)$, then $d^T y \leq 0$ for all $y \in N_z(Z)$ and $d^T y = 0$ for some $y \in N_z(Z) \cap \mathbf{R}^k_- \setminus \{0\}$.

The relations of the cones and perpendicular vectors in Theorem 4.2 are illustrated in Figure 2. The figure depicts the properly Pareto optimal point in both a locally convex and a nonconvex environment.

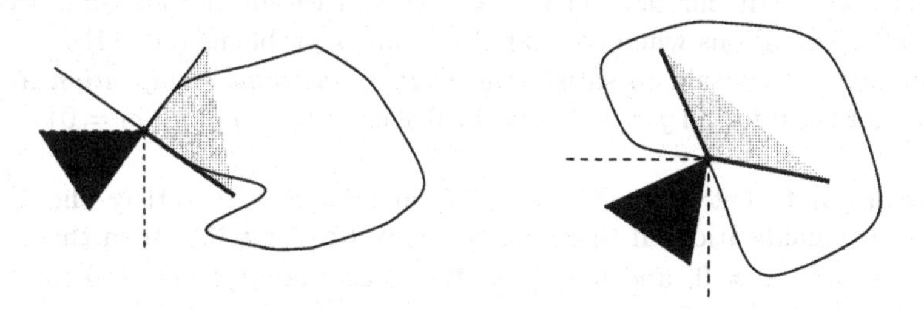

Figure 2. The results of Theorem 4.2 in two different cases.

5 Necessary Conditions for Normal Vectors

Let us continue with a collection of definitions and results from nonsmooth analysis.

Definition 5.1 (Clarke). The *subdifferential* of a locally Lipschitz continuous function $h\colon \mathbf{R}^n \to \mathbf{R}$ is a set

$$\partial h(x) = \{\xi \in \mathbf{R}^n \mid h^\circ(x; v) \geq \xi^T v \text{ for all } v \in \mathbf{R}^n\},$$

where

$$h^\circ(x; v) = \limsup_{\substack{u \to x \\ t \searrow 0}} \frac{h(u + tv) - h(u)}{t}$$

is the *generalized directional derivative* of h at x in the direction v.

For subdifferentials we have

$$\partial\Big(\sum_{i=1}^{k} w_i f_i(\hat{x})\Big) \subset \sum_{i=1}^{k} w_i \partial f_i(\hat{x}), \tag{5.2}$$

where $w_i \in \mathbf{R}$ for $i = 1, \ldots, k$. In addition,
(5.3)

$$\partial \left(\max_{i=1,\ldots,k} f_i(\hat{x}) \right) \subset \operatorname{conv} \left\{ \partial f_i(\hat{x}) \mid i \text{ such that } \max_{j=1,\ldots,k} f_j(\hat{x}) = f_i(\hat{x}) \right\}.$$

For details, see [3].

We bring this paper to an end with some optimality results concerning normal cones. In our nonconvex case we can establish somewhat weaker optimality conditions when compared to convex problems (cf. [11]).

Problem (2.1) is said to satisfy the *Cottle constraint qualification* at $\hat{x}$ if either $g_j(\hat{x}) < 0$ for all $j = 1, \ldots, m$, or $0 \notin \operatorname{conv}\{\partial g_j(\hat{x}) \mid g_j(\hat{x}) = 0\}$.

Theorem 5.4. Let $\hat{z} = f(\hat{x}) \in P(Z)$ and let $\hat{x} \in S$ satisfy the Cottle constraint qualification. If there exists $y \in N_{\hat{z}}(Z) \cap \operatorname{int} \mathbf{R}^k_-$, then there exist $0 \leq \lambda \in \mathbf{R}^k$, $\lambda \neq 0$, and $0 \leq \mu \in \mathbf{R}^m$ such that $\mu_j g_j(\hat{x}) = 0$ for every $j = 1, \ldots, m$, and

$$0 \in \sum_{i=1}^{k} -\frac{\lambda_i}{y_i} \partial f_i(\hat{x}) + \sum_{j=1}^{m} \mu_j \partial g_j(\hat{x}).$$

Proof. The assumption $y \in N_{\hat{z}}(Z) \cap \operatorname{int} \mathbf{R}^k_-$ means that $y_i < 0$ for every $i = 1, \ldots, k$.

Let us define for every $x \in S$ a function

$$F(x) = \max_{i=1,\ldots,k} \left[-\frac{1}{y_i} (f_i(x) - \hat{z}_i - y_i) \right].$$

This function attains its minimum at $\hat{x} \in S$. If this was not the case, there would exist $x^\circ \in S$ such that

$$\max_{i=1,\ldots,k} \left[-\frac{1}{y_i}(f_i(x^\circ) - \hat{z}_i - y_i) \right] < \max_{i=1,\ldots,k} \left[-\frac{1}{y_i}(f_i(\hat{x}) - \hat{z}_i - y_i) \right] = 1.$$

This means that $-\frac{1}{y_i}(f_i(x^\circ) - \hat{z}_i - y_i) < 1$ for every $i = 1, \ldots, k$. In other words, $f_i(x^\circ) < \hat{z}_i$ for every $i = 1, \ldots, k$, which contradicts the Pareto optimality of $\hat{z}$.

Now we know that $F(\hat{x}) \leq F(x)$ for every $x \in S$. Applying necessary Karush-Kuhn-Tucker optimality conditions (see, e.g. [14]) we know that there exists $0 \leq \mu \in \mathbf{R}^m$ such that $\mu_j g_j(\hat{x}) = 0$ for every $j = 1, \ldots, m$ and $0 \in \partial F(\hat{x}) + \sum_{j=1}^m \mu_j \partial g_j(\hat{x})$.

According to (5.2) and (5.3) we have

$$\begin{aligned} \partial F(\hat{x}) &\subset \operatorname{conv} \left\{ \partial\Big(-\frac{1}{y_i}(f_i(\hat{x}) - \hat{z}_i - y_i)\Big) \mid i = 1, \ldots, k \right\} \\ &\subset \operatorname{conv} \left\{ -\frac{1}{y_i} \partial (f_i(\hat{x}) - \hat{z}_i - y_i) \mid i = 1, \ldots, k \right\} \\ &\subset \operatorname{conv} \left\{ -\frac{1}{y_i} \partial f_i(\hat{x}) \mid i = 1, \ldots, k \right\}. \end{aligned}$$

From the definition of convex hulls we know that there exists $0 \leq \lambda \in \mathbf{R}^k$ such that $\sum_{i=1}^k \lambda_i = 1$ and $\partial F(\hat{x}) \subset \sum_{i=1}^k -\frac{\lambda_i}{y_i} \partial f_i(\hat{x})$. This completes the proof. ∎

Note that, as proved in Theorem 3.5, the assumptions of Theorem 5.4 are satisfied whenever $\hat{z} \in PP(Z)$.

Under the convexity assumption the Cottle constraint qualification is equivalent to the so-called Slater constraint qualification, that is, there exists at least one $x \in \operatorname{int} S$.

We state the next theorem from [11] in order to facilitate the comparison between convex and nonconvex cases.

Theorem 5.5. Let problem (2.1) be convex and satisfy the Slater constraint qualification. If $\hat{z} = f(\hat{x}) \in Z$, then $y \in N_{\hat{z}}(Z)$ if and only if there exists $0 \leq \mu \in \mathbf{R}^m$ such that $\mu_j g_j(\hat{x}) = 0$ for every $j = 1, \ldots, m$ and

$$0 \in \sum_{i=1}^k -y_i \partial f_i(\hat{x}) + \sum_{j=1}^m \mu_j \partial g_j(\hat{x}).$$

We write down the following general optimality results which can directly be derived from the theorems above. The convexity part is derived in [11].

Corollary 5.6. Let $\hat{x} \in S$ satisfy the Cottle constraint qualification. A necessary condition for $\hat{z} = f(\hat{x}) \in PP(Z)$ being valid is that there exist $0 \leq \lambda \in \mathbf{R}^k$, $\lambda \neq 0$, and $0 \leq \mu \in \mathbf{R}^m$ such that $\mu_j g_j(\hat{x}) = 0$ for every $j = 1, \ldots, m$ and

$$0 \in \sum_{i=1}^{k} \lambda_i \partial f_i(\hat{x}) + \sum_{j=1}^{m} \mu_j \partial g_j(\hat{x}).$$

If problem (2.1) is convex, then $\lambda > 0$. Further, the condition is also sufficient.

Note that the optimality conditions above are a byproduct of the actual treatment of this paper. More general conditions can be derived with other approaches. Fritz John and Karush-Kuhn-Tucker type optimality conditions for nonconvex multiobjective optimization problems have been given by different authors in different spaces. Fritz John type necessary optimality conditions for weak Pareto optimality in Euclidean spaces are studied in [5]. The corresponding results in Banach spaces for weak and/or Pareto optimality are given in [6], [16] and [19]. A more general formulation is given in [7]. Some special cases with proper Pareto optimality are considered in [4]. More recently, in [1] the authors have provided Fritz John and Karush-Kuhn-Tucker type optimality conditions for convex multifunctions in Banach spaces. More references related to optimality conditions can be found in [15].

6 Final Remarks

We have studied the possibility of generalizing into nonconvex cases the convex results originally given in [11] concerning the cone of trade-off directions, tangent and normal cones and proper Pareto optimality. Some of the results can be extended analogously. Nevertheless, the necessary optimality result concerning normal cones and Pareto optimal solutions differs from its convex counterpart in both assumptions and formulation.

The starting point of our treatment was to utilize the convexity property of Clarke's tangent cones in nonconvex cases. This enables the flexible generalization of the convex results. However, using tangent cones in defining trade-off directions in nonconvex cases may blur the original idea (see Figure 1). In this sense, starting from feasible directions would reflect the trade-off idea better. Unfortunately, the cone of feasible directions is nonconvex whenever the feasible criterion set is not locally convex. This means that handling trade-off directions on the basis of feasible directions requires other tools and further research.

Acknowledgements

This research was supported by the grants 22346 and 8583 of the Academy of Finland.

References

1. Amahroq, T., Taa, A. (1997), *On Lagrange-Kuhn-Tucker Multipliers for Multiobjective Optimization Problems*, Optimization **41**, No. 2, 159–172.
2. Chankong, V., Haimes, Y.Y. (1983), "Multiobjective Decision Making Theory and Methodology," North-Holland, New York.
3. Clarke, F.H. (1983), "Optimization and Nonsmooth Analysis," John Wiley & Sons, Inc., New York.
4. Coladas, L., Li, Z., Wang, S. (1994), *Optimality Conditions for Multiobjective and Nonsmooth Minimization in Abstract Spaces*, Bulletin of the Australian Mathematical Society **50**, No. 2, 203–218.
5. Craven, B.D. (1989), *Nonsmooth Multiobjective Programming*, Numerical Functional Analysis and Optimization **10**, No. 1&2, 49–64.
6. Doležal, J. (1985), *Necessary Conditions for Pareto Optimality in Nondifferentiable Problems*, Problems of Control and Information Theory **14**, No. 2, 131–141.
7. El Abdouni, B., Thibault, L. (1992), *Lagrange Multipliers for Pareto Nonsmooth Programming Problems in Banach Spaces*, Optimization **26**, No. 3–4, 277–285.
8. Haimes, Y.Y., Chankong, V. (1979), *Kuhn-Tucker Multipliers as Trade-Offs in Multiobjective Decision-Making Analysis*, Automatica **15**, No. 1, 59–72.
9. Halme, M., Korhonen, P. (1989), *Nondominated Tradeoffs and Termination in Interactive Multiple Objective Linear Programming*, in "Improving Decision Making in Organisations," Lecture Notes in Economics and Mathematical Systems 335, Edited by Lockett, A.G, Islei, G., Springer-Verlag, Berlin, Heidelberg, 410–423.
10. Henig, M.I. (1982), *Proper Efficiency with Respect to Cones*, Journal of Optimization Theory and Applications **36**, No. 3, 387–407.
11. Henig, M.I., Buchanan, J.T. (1997), *Tradeoff Directions in Multiobjective Optimization Problems*, Mathematical Programming **78**, No. 3, 357–374.
12. Kaliszewski, I., Michalowski, W. (1995), *Generation of Outcomes with Selectively Bounded Trade-Offs*, Foundations of Computing and Decision Sciences **20**, No. 2, 113–122.
13. Kaliszewski, I., Michalowski, W. (1997), *Efficient Solutions and Bounds on Tradeoffs*, Journal of Optimization Theory and Applications **94**, No. 2, 381–394.

14. Mäkelä, M.M., Neittaanmäki, P. (1992), "Nonsmooth Optimization: Analysis and Algorithms with Applications to Optimal Control," World Scientific, Singapore.
15. Miettinen, K. (1999), "Nonlinear Multiobjective Optimization," Kluwer Academic Publishers, Boston.
16. Minami, M. (1983), *Weak Pareto-Optimal Necessary Conditions in a Nondifferentiable Multiobjective Program on a Banach Space*, Journal of Optimization Theory and Applications **41**, No. 3, 451–461.
17. Rockafellar, R.T. (1970), "Convex Analysis," Princeton University Press, Princeton, New Jersey.
18. Sakawa, M., Yano, H. (1990), *Trade-Off Rates in the Hyperplane Method for Multiobjective Optimization Problems*, European Journal of Operational Research **44**, No. 1, 105–118.
19. Wang, S. (1984), *Lagrange Conditions in Nonsmooth and Multiobjective Mathematical Programming*, Mathematics in Economics **1**, 183–193.
20. Yano, H., Sakawa, M. (1987), *Trade-Off Rates in the Weighted Tchebycheff Norm Method*, Large Scale Systems **13**, No. 2, 167–177.

Multiobjective Linear Production Programming Games

Ichiro Nishizaki1 and Masatoshi Sakawa[1]

Department of Industrial and Systems Engineering, Faculty of Engineering, Hiroshima University, 1-4-1 Kagamiyama, Higashi-Hiroshima, Hiroshima, 739-8527 Japan

Abstract. In this paper we consider a production model in which multiple decision makers pool resources to produce finished goods. Such a production model, which is assumed to be linear, can be formulated as a multiobjective linear programming problem. It is shown that a multi-commodity game arises from the multiobjective linear production programming problem with multiple decision makers and such a game is referred to as a multiobjective linear production programming game. The characteristic sets in the game can be obtained by finding the set of all the Pareto extreme points of the multiobjective programming problem. It is proven that the core of the game is not empty, and points in the core are computed by using the duality theory of multiobjective linear programming problems.

Key Words Multiobjective linear production programming problem, multi-commodity game, the core, the least core, the nucleolus

1 Introduction

A conventional mathematical programming problem is supposed to have a single decision maker or to have multiple decision makers who have the same interests. In managerial and public decision making problems, however, there are multiple decision makers who have different interests from one another. In this paper, we deal with a production programming problem with multiple decision makers who have different interests. In such a problem, unless all the decision makers conclude an allocation of profit or cost among them to be fair, they do not always support the allocation scheme, even if the whole profit is maximized. Cooperative games in characteristic function form have often been used to analyze and resolve allocation problems of joint profit or cost.

By using the cooperative game theory, Owen considered linear production programming problems in which multiple decision makers pool resources to produce some goods [14]. An objective function of the linear production programming problem was represented as a total revenue from selling some kinds of goods, and the problem was formulated as a linear programming problem in which, subject to resource constraints, the revenue is maximized. He gave an allocation scheme of the total revenue by adopting a point in the core of the cooperative game arising from the production programming problem. Subsequently extensions of the production model and relationship between other optimization problems and the cooperative games have been studied in relation to Owen's work [2,5,7,9,13,15].

Kalai and Zemel investigated an optimization problem in a network whose arcs were owned by different individuals [13]. A similar network flow problem with multiple commodities was examined by Derks and Tijs [3,4] while a single commodity was flowed in the network dealt with by Kalai and Zemel. van den Nouweland, Aarts and Borm referred to such a game as a multi-commodity game, extended the concept of the core, and applied to other problems, such as production programming problems [17]. By regarding multiple commodities as multiple objectives, multi-commodity games can be identified with multiobjective games. Tanino, Muranaka and Tanaka [16] also studied the concept of the core in the multiobjective cooperative game independently of the above works. Bergstresser and Yu also treat multiobjective cooperative games in the study on multicriteria problems in n-person games [1]. They examine cooperative games with vector-valued characteristic function while, in this paper, we will deal with cooperative games with characteristic sets. Bergstresser and Yu mainly considered the core defined by domination structures and referred to a couple of solution concepts which yield a unique solution such as the nucleolus in n-person cooperative games.

In this paper, we consider the linear production programming problem with multiple decision makers in multiobjective environments. Consider a joint venture with multiple decision makers who produce some goods. During manufacturing the goods on a commercial basis, if some pollutant is discharged for a unit of production, it is required to formulate a programming problem not only with an objective function representing revenue but also with an objective function representing quantity of the pollutant as a by-product and then such a situation can be formulated as a multiobjective programming problem. The revenue yielded by the joint venture, that is, by cooperation among the multiple decision makers should be maximized and be allocated among them fairly. The pollutant must be treated with a suitable manner and the joint venture should also compensate neighbor residents who suffer from the pollutant. The joint venture must allocate costs required from the treatment and the compensation among the multiple decision makers in proportion to the quantity of the pollutant.

For the multi-commodity game arising from the multiobjective linear production programming problem, it should be noted that the joint venture formulates the multiobjective linear programming problem, but all the objectives dealt with in the problem may be not always allocated among the multiple decision makers, the members of the joint venture.

In Section 2, some concepts in multi-commodity games are briefly reviewed. We formulate a production programming problem as a multiobjective linear programming problem and show that a multi-commodity game arises from the problem. We illustrate the arising multi-commodity game with a numerical example. In Section 3, it is proven that the multi-commodity game has the nonempty core, and it is shown that points in the core can be computed by using dual Pareto optimal solutions to the multiobjective linear production programming problem. Especially, all the Pareto optimal extreme points are computed by using one of algorithms for obtaining Pareto optimal solutions to multiobjective linear programming problems

[20,18,8,10,6] and the points in the core are obtained by the duality theorem of multiobjective linear programming problems [11,12].

2 A multi-commodity game arising from a multiobjective programming problem

Let $\mathbb{R}^p_+$ denote the nonnegative orthant of the p-dimensional real space $\mathbb{R}^p$, *i.e.*,

$$\mathbb{R}^p_+ \triangleq \{x \in \mathbb{R}^p \mid x_j \geqq 0,\ j = 1,2,\ldots,p\}, \tag{1}$$

and let $\mathbb{R}^p_{++} \triangleq \mathbb{R}^p_+ \backslash \{0\}$. We conform to the following conventional notations in the vector optimization and use the notations consistently throughout the paper. For given two vectors $\boldsymbol{a},\boldsymbol{b} \in \mathbb{R}^p$, define $\boldsymbol{a} = \boldsymbol{b}$ if and only if $a_j = b_j$ for all $j = 1,2,\ldots,p$, $\boldsymbol{a} \geqq \boldsymbol{b}$ if and only if $a_j \geqq b_j$ for all $j = 1,2,\ldots,p$, $\boldsymbol{a} \geq \boldsymbol{b}$ if and only if $a_j \geqq b_j$ for all $j = 1,2,\ldots,p$ and $\boldsymbol{a} \neq \boldsymbol{b}$ (there is at least one j such that $a_j > b_j$), and $\boldsymbol{a} > \boldsymbol{b}$ if and only if $a_j > b_j$ for all $j = 1,2,\ldots,p$. For a set $A \subset \mathbb{R}^p$, define the set of all Pareto maximal points

$$\mathrm{Max}A \triangleq \{\boldsymbol{a} \in A \mid (A - \boldsymbol{a}) \cap (\mathbb{R}^p_+) = \{0\}\}. \tag{2}$$

If $\boldsymbol{a} \in \mathrm{Max}A$, there exists no $\boldsymbol{a}' \in A$ such that $\boldsymbol{a} \leq \boldsymbol{a}'$.

In this section, it is shown that a multi-commodity game arises from a multiobjective programming problem. First, we briefly describe multi-commodity games. Let n be a fixed positive integer, $N \triangleq \{1,2,\ldots,n\}$, and Λ denote a family of nonempty subsets of N. An element $i \in N$ are called a player and an element $S \in \Lambda$, a coalition. Let ℓ be a fixed positive integer, $K \triangleq \{1,2,\ldots,\ell\}$, and an element $k \in K$ denote a commodity or an objective.

For a coalition $S \in \Lambda$, consider a set $V(S)$ satisfying the following conditions:

(i) For a coalition $S \in \Lambda$, $V(S)$ is a nonempty closed subset of $\mathbb{R}^\ell_+$.
(ii) If $\boldsymbol{u} \leqq \boldsymbol{v}$ for $\boldsymbol{v} \in V(S)$ and $\boldsymbol{u} \in \mathbb{R}^\ell$, then $\boldsymbol{u} \in V(S)$.

The first condition (i) of $V(S)$ means upper boundedness, and the second condition (ii) means comprehensiveness. For a family of sets $V = \{V(S) \mid S \in \Lambda\}$, suppose that, for each commodity, a vector of multiple payoffs $\boldsymbol{v} = (v^1, v^2, \ldots, v^\ell) \in V(S)$ can be shared by members of a coalition S, and that $v^k \geqq \sum_{i \in S} u^k_i$, $k = 1,2,\ldots,\ell$ for a payoff variable $\boldsymbol{u} = (\boldsymbol{u}^1, \boldsymbol{u}^2, \ldots, \boldsymbol{u}^\ell) \in \mathbb{R}^{n \times \ell}$, $\boldsymbol{u}^k = (u^k_1, u^k_2, \ldots, u^k_n) \in \mathbb{R}^n$, $k = 1,2,\ldots,\ell$. Then, an ℓ-commodity game can be represented by (N, V). In our model, commodities are corresponding to objectives.

We can try to predict the outcome of bargaining among the players or to present an arbitration scheme to the players, once a representation in a multi-commodity game has been specified. Such analysis is based on the assumption that the players will form the grand coalition N and divide a bundle of multi-commodity in the set of all Pareto maximal points among themselves.

A multiobjective linear production programming problem is described as follows. Each of the n decision makers is in possession of a resource vector $\boldsymbol{b}^i =$

$(b_1^i, b_2^i, \ldots, b_m^i)$, $i = 1, 2, \ldots, n$, and p kinds of goods are produced by cooperation of the decision makers. A coalition $S \in \Lambda$ will have a total of

$$b_r(S) = \sum_{i \in S} b_r^i \tag{3}$$

units of the rth resource. A unit of the jth good , $j = 1, 2, \ldots, p$ requires a_{rj} units of the rth resource, $r = 1, 2, \ldots, m$. We formulate the production model as an ℓ objective linear programming problem. For a coalition $S \in \Lambda$, the ℓ-objective linear programming problem is represented as

$$\left.\begin{array}{l} \text{maximize } z_1(\boldsymbol{x}) = c_{11}x_1 + c_{12}x_2 + \cdots + c_{1p}x_p \\ \quad \cdots\cdots\cdots \\ \text{maximize } z_\ell(\boldsymbol{x}) = c_{\ell 1}x_1 + c_{\ell 2}x_2 + \cdots + c_{\ell p}x_p \\ \text{subject to } a_{11}x_1 + a_{12}x_2 + \cdots + a_{1p}x_p \leqq b_1(S) \\ \quad \cdots\cdots\cdots\cdots \\ \qquad a_{m1}x_1 + a_{m2}x_2 + \cdots + a_{mp}x_p \leqq b_m(S) \\ \qquad x_1, x_2, \ldots, x_p \geqq 0 \end{array}\right\}, \tag{4}$$

equivalently

$$\left.\begin{array}{l} \textbf{maximize } z(\boldsymbol{x}) = C\boldsymbol{x} \\ \text{subject to } \boldsymbol{x} \in T_S \triangleq \{\boldsymbol{x} \mid A\boldsymbol{x} \leqq \boldsymbol{b}(S),\ \boldsymbol{x} \in \mathbb{R}_+^p\} \end{array}\right\}, \tag{5}$$

where "**maximize**" means vector maximization, C is an $\ell \times p$ matrix, A is an $m \times p$ matrix, and $\boldsymbol{b}(S) = (b_1(S), b_2(S), \ldots, b_m(S))$ is an m-dimensional column vector. Let

$$\hat{T}_S \triangleq \{z \in \mathbb{R}^\ell \mid z = C\boldsymbol{x},\ \forall \boldsymbol{x} \in T_S\}, \tag{6}$$

and the set of all Pareto optimal values to the multiobjective linear production programming problem (5) can be represented by Max $\hat{T}_S$. Then a multi-commodity game (N, V) can be constructed by the set of players N and the characteristic sets

$$V(S) = (\text{Max}\, \hat{T}_S - \mathbb{R}_+^\ell) \cap \mathbb{R}_+^\ell. \tag{7}$$

We refer to the multi-commodity game as a multiobjective linear production programming game. We have defined $\hat{T}_S$ as a subset of $\mathbb{R}^\ell$. However, when all of the ℓ objectives do not always need to be allocated, the set $\hat{T}_S$ may be defined as a subset of a subspace of $\mathbb{R}^\ell$, *i.e.*,

$$\hat{T}_S \triangleq \left\{ z \in \mathbb{R}^{\ell'} \;\middle|\; z_k = \sum_{j=1}^{p} c_{kj}x_j,\ \forall k \in K' \subset K,\ \forall \boldsymbol{x} \in T_S \right\}, \tag{8}$$

where ℓ' is the number of objectives to be allocated, and K' is a set of the objectives to be allocated.

If a feasible solution area T_S to the multiobjective linear production programming problem (5) is a nonempty bounded set, the set T_S is a bounded convex polyhedron and the characteristic set $V(S)$ is a comprehensive and compact subset of $\mathbb{R}^{\ell}_{+}$.

[Example] Consider the following bi-objective linear production programming problem with three decision makers (players).

$$\left.\begin{array}{l} \text{maximize } z_1(\boldsymbol{x}) = 2.5x_1 + 5x_2 \\ \text{maximize } z_2(\boldsymbol{x}) = 3x_1 + 2x_2 \\ \text{subject to } 2x_1 + 9x_2 \leqq 430 \\ \qquad\qquad 6x_1 + 4x_2 \leqq 410 \\ \qquad\qquad 8x_1 + 9x_2 \leqq 570 \\ \qquad\qquad x_1,\ x_2 \geqq 0 \end{array}\right\} \tag{9}$$

Each of the players initially possesses three kinds of resources as shown in Table 1.

Table 1. Initial resources.

Resources	Players			The grand coalition
	1	2	3	
R_1	139	181	110	$b_1(N) = 430$
R_2	140	87	183	$b_2(N) = 410$
R_3	130	225	215	$b_3(N) = 570$

Then the characteristic sets of coalitions become the following.

$$\begin{aligned}
V(\{1,2,3\}) &= \{z \in \mathbb{R}^2_+ \mid z_1 \leqq 271.3,\ z_2 \leqq 205,\ 11z_1 + 17.5z_2 \leqq 5700\}, \\
V(\{1,2\}) &= \{z \in \mathbb{R}^2_+ \mid z_1 \leqq 197.2,\ z_2 \leqq 113.5,\ 11z_1 + 17.5z_2 \leqq 3550\}, \\
V(\{1,3\}) &= \{z \in \mathbb{R}^2_+ \mid z_1 \leqq 160.6,\ z_2 \leqq 129.4,\ 11z_1 + 17.5z_2 \leqq 3450\}, \\
V(\{2,3\}) &= \{z \in \mathbb{R}^2_+ \mid z_1 \leqq 196.2,\ z_2 \leqq 135,\ 11z_1 + 17.5z_2 \leqq 4400\}, \\
V(\{1\}) &= \{z \in \mathbb{R}^2_+ \mid z_1 \leqq 72.2,\ z_2 \leqq 48.8,\ 11z_1 + 17.5z_2 \leqq 1300\}, \\
V(\{2\}) &= \{z \in \mathbb{R}^2_+ \mid z_1 \leqq 102.3,\ z_2 \leqq 43.5\}, \\
V(\{3\}) &= \{z \in \mathbb{R}^2_+ \mid z_1 \leqq 85.4,\ z_2 \leqq 80.6,\ 11z_1 + 17.5z_2 \leqq 2150\}.
\end{aligned}$$

The characteristic sets of the coalitions arising from the problem are depicted in Figure 1. An area enclosed dashed lines in Figure 1 represents the feasible solution area of (9) in the objective space.

3 The core of the multiobjective linear production programming game

For conventional cooperative games in characteristic function form (N, v), the concept of the core is defined by the domination relation between two imputations.

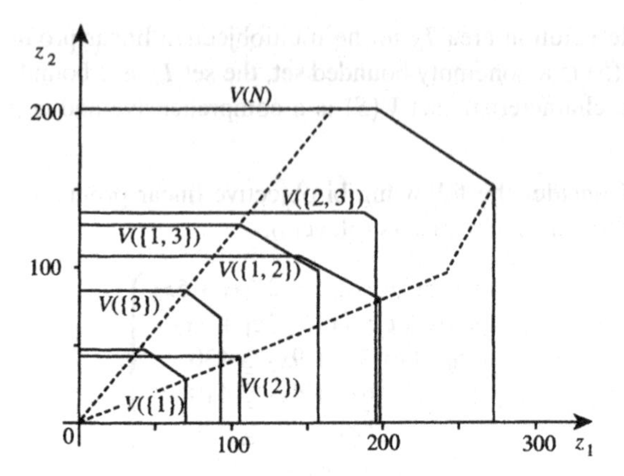

Fig. 1. The characteristic sets.

The core of the game is the set of all nondominated imputations. Especially, if the game has superadditivity property, the core of the game is the set of all imputations satisfying the condition of the coalitional rationality. For multi-commodity games, similar concepts have been studied. van den Nouweland, Aarts and Borm referred to a solution set defined by the domination relation as the dominance core and a solution set defined by the coalitional rationality as the stable set [17].

The set of payoff vectors satisfying the individual rationality is defined as the set $IR(N,V) \triangleq \{\boldsymbol{u} \in \mathbb{R}^{\ell \times n} \mid \boldsymbol{u}_i \notin V(\{i\}) \setminus \operatorname{Max} V(\{i\}),\ \forall i \in N\}$ and the set of payoff vectors satisfying the collective rationality is defined as $GR(N,V) \triangleq \{\boldsymbol{u} \in \mathbb{R}^{\ell \times n} \mid \boldsymbol{u}_N \in \operatorname{Max} V(N)\}$, where $\boldsymbol{u}_N = \sum_{i \in N} \boldsymbol{u}_i \in \mathbb{R}^{\ell}$. The set $I(N,V)$ of all imputations, which is a set of payoff vectors satisfying the conditions of the individual rationality and the collective rationality, is defined as

$$I(N,V) = \{\boldsymbol{u} = ((u_1^1, \dots, u_1^\ell), \dots, (u_n^1, \dots, u_n^\ell)) \in \mathbb{R}_+^{\ell \times n} \mid \boldsymbol{u}_N \in \operatorname{Max} V(N),$$
$$\boldsymbol{u}_i \notin V(\{i\}) \setminus \operatorname{Max} V(\{i\}),\ \forall i \in N\}. \quad (10)$$

For $S \in \Lambda$ and $\boldsymbol{u}, \boldsymbol{v} \in I(N,V)$, we say that $\boldsymbol{u}$ dominates $\boldsymbol{v}$ through S if $\boldsymbol{u}_i > \boldsymbol{v}_i$ for all $i \in S$ and $\boldsymbol{u}_S \in V(S)$, where $\boldsymbol{u}_S = \sum_{i \in S} \boldsymbol{u}_i$. Let $\boldsymbol{u}$ dom$_S$ $\boldsymbol{v}$ denote that $\boldsymbol{u}$ dominates $\boldsymbol{v}$ through S. We say that $\boldsymbol{u}$ dominates $\boldsymbol{v}$ if there is any coalition $S \in \Lambda$ such that $\boldsymbol{u}$ dom$_S$ $\boldsymbol{v}$. Let $\boldsymbol{u}$ dom $\boldsymbol{v}$ denote that $\boldsymbol{u}$ dominates $\boldsymbol{v}$. The dominance-core $DC(N,V)$ is defined as the set of all nondominated imputations, *i.e.*,

$$DC(N,V) = \{\boldsymbol{u} \in I(N,V) \mid \text{there is no } S \in \Lambda \text{ and no } \boldsymbol{v} \in I(N,V)$$
$$\text{such that } \boldsymbol{v} \text{ dom}_S\ \boldsymbol{u}\}. \quad (11)$$

We say that a payoff vector $\boldsymbol{u}$ is feasible if $\boldsymbol{u}_N \in V(N)$, and a coalition S can improve on a payoff vector $\boldsymbol{u}$ if $\boldsymbol{u}_S \in V(S)$. The stable set is defined as a set of all feasible payoff vectors which no coalition S can improve on, *i.e.*,

$$SO(N,V) = \{\boldsymbol{u} \in GR(N,V) \mid \boldsymbol{u}_S \notin V(S) \setminus \operatorname{Max} V(S),\ \forall S \subset N\}. \quad (12)$$

We say that a multi-commodity game has superadditivity property if, for $S,T \in \Lambda$, $S \cap T = \emptyset$,

$$V(S) + V(T) \subset V(S \cup T). \tag{13}$$

Tanino, Muranaka and Tanaka [16] defined a multiobjective cooperative game the definition of which is slightly different from that of the multi-commodity game, but both representations are essentially equivalent to each other. They showed that $DC(N,V) = SO(N,V)$ on the assumption that $V(\emptyset) = \{0\}$, $\sum_{i \in N} V(\{i\}) \subset V(N)$. We also assume the above conditions consistently throughout the paper.

We examine some properties of the multiobjective linear production programming game.

Theorem 1. *The multiobjective linear production programming game (N,V) defined by Problem (5) and the sets (7) has superadditivity property.*

Proof Consider $\boldsymbol{s} \in V(S)$ and $\boldsymbol{t} \in V(T)$ such that $\boldsymbol{x}_s$ and $\boldsymbol{x}_t$ are feasible to Problem (5), *i.e.*, $\boldsymbol{s} = C\boldsymbol{x}_s$ and $\boldsymbol{t} = C\boldsymbol{x}_t$, where C is an $\ell \times p$ matrix in (5). Then we have $\boldsymbol{x}_s \in T_S$ and $\boldsymbol{x}_t \in T_T$, where T_S and T_T are feasible solution areas in (5), and $\boldsymbol{x}_s$ and $\boldsymbol{x}_t$ satisfy the following inequalities, respectively:

$$A\boldsymbol{x}_s \leqq \boldsymbol{b}(S), \quad A\boldsymbol{x}_t \leqq \boldsymbol{b}(T),$$

where A and $\boldsymbol{b}(S)$ are an $m \times p$ matrix and an m-dimensional column vector in (5), respectively. Thus, we have $A\boldsymbol{x}_s + A\boldsymbol{x}_t \leqq \boldsymbol{b}(S) + \boldsymbol{b}(T)$, and from (3), $A(\boldsymbol{x}_s + \boldsymbol{x}_t) \leqq \boldsymbol{b}(S \cup T)$.

Therefore $\boldsymbol{x}_s + \boldsymbol{x}_t \in T_{S \cup T}$. Let $\boldsymbol{x}_s + \boldsymbol{x}_t = \boldsymbol{x}_{s+t}$ and then we have

$$C\boldsymbol{x}_s + C\boldsymbol{x}_t = C\boldsymbol{x}_{s+t}, \; \boldsymbol{s} + \boldsymbol{t} = C\boldsymbol{x}_{s+t}.$$

Because the following two statements are equivalent, we have $\boldsymbol{s} + \boldsymbol{t} \in V(S \cup T)$.

1) $\boldsymbol{s} + \boldsymbol{t} \in V(S \cup T)$.

2) There exists $\boldsymbol{x}_{s+t}$ such that $\boldsymbol{s} + \boldsymbol{t} = C\boldsymbol{x}_{s+t}$.

Consider $\boldsymbol{s}' \in V(S)$ and $\boldsymbol{t}' \in V(T)$ such that $\boldsymbol{x}_{s'}$ and $\boldsymbol{x}_{t'}$ are not feasible to Problem (5). We can find $\boldsymbol{s} \in V(S)$ and $\boldsymbol{t} \in V(T)$ such that $\boldsymbol{s}' \leq \boldsymbol{s}$ and $\boldsymbol{t}' \leq \boldsymbol{t}$, and $\boldsymbol{x}_s$ and $\boldsymbol{x}_t$ are feasible to the Problem (5). Therefore, if $\boldsymbol{s}' \in V(S)$ and $\boldsymbol{t}' \in V(T)$ then $\boldsymbol{s}' + \boldsymbol{t}' \in V(S \cup T)$.

For the other cases, the same property can be easily demonstrated from the above facts. ■

It follows from Theorem 1 that $DC(N,V) = SO(N,V)$ in the multiobjective linear production programming game (N,V). We will refer to the solution concepts as the core and use the symbol $C(N,V) = DC(N,V) = SO(N,V)$.

In multi-commodity games, the concept of balancedness is defined by van den Nouweland, Aarts and Borm [17] as follows. A multi-commodity game is said to be balanced if, for each balanced map $\lambda : \Lambda \to \mathbb{R}_+$ such that

$$\sum_{\substack{S \in \Lambda \\ S \ni i}} \lambda(S) = 1, \; \forall i \in N, \tag{14}$$

we have

$$\sum_{S\in\Lambda} \lambda(S)V(S) \subset V(N), \tag{15}$$

where $S \ni i$ means a coalition S including the player i. They showed the result that each balanced multi-commodity game has at least one stable outcome.

Theorem 2. *The multiobjective linear production programming game (N,V) defined by Problem (5) and the sets (7) is balanced.*

Proof Let $\lambda(S)$, $S \in \Lambda$ be the balanced map for Λ. For all $r = 1,2,\ldots,m$,

$$\sum_{S\in\Lambda} \lambda(S)b_r(S) = \sum_{S\in\Lambda}\sum_{i\in S} \lambda(S)b_r^i = \sum_{i\in N}\left\{\sum_{\substack{S\in\Lambda\\ S\ni i}} \lambda(S)\right\} b_r^i = \sum_{i\in N} b_r^i = b_r(N).$$

Let $\boldsymbol{x}^S = (x_1^S, x_2^S, \ldots, x_p^S)$ be a Pareto optimal solution to the multiobjective linear production programming (5), and then

$$z(\boldsymbol{x}^S) = (z_1(\boldsymbol{x}^S), z_2(\boldsymbol{x}^S), \ldots, z_\ell(\boldsymbol{x}^S)) \in V(S).$$

For two vectors $\boldsymbol{v}, \boldsymbol{w} \in \mathbb{R}^p$, let $\langle \boldsymbol{v} \cdot \boldsymbol{w}\rangle = \sum_{j=1}^p v_j w_j$. We have

$$\begin{aligned}
\sum_{S\in\Lambda} \lambda(S)z(\boldsymbol{x}^S) &= \sum_{S\in\Lambda} \{\lambda(S)(\langle \boldsymbol{c}_{1\cdot} \cdot \boldsymbol{x}^S\rangle, \langle \boldsymbol{c}_{2\cdot} \cdot \boldsymbol{x}^S\rangle, \ldots, \langle \boldsymbol{c}_{\ell\cdot} \cdot \boldsymbol{x}^S\rangle)\} \\
&= \left(\left\langle \boldsymbol{c}_{1\cdot} \cdot \sum_{S\in\Lambda} \lambda(S)\boldsymbol{x}^S \right\rangle, \left\langle \boldsymbol{c}_{2\cdot} \cdot \sum_{S\in\Lambda} \lambda(S)\boldsymbol{x}^S \right\rangle, \ldots, \left\langle \boldsymbol{c}_{\ell\cdot} \cdot \sum_{S\in\Lambda} \lambda(S)\boldsymbol{x}^S \right\rangle\right) \\
&= (\langle \boldsymbol{c}_{1\cdot} \cdot \hat{\boldsymbol{x}}\rangle, \langle \boldsymbol{c}_{2\cdot} \cdot \hat{\boldsymbol{x}}\rangle, \ldots, \langle \boldsymbol{c}_{\ell\cdot} \cdot \hat{\boldsymbol{x}}\rangle) = z(\hat{\boldsymbol{x}}),
\end{aligned}$$

where $\boldsymbol{c}_{k\cdot} = (c_{k1}, c_{k2}, \ldots, c_{kp})$, $k = 1,2,\ldots,\ell$ and $\hat{\boldsymbol{x}} = \sum_{S\in\Lambda} \lambda(S)\boldsymbol{x}^S \in \mathbb{R}^p$.

For all $r = 1,2,\ldots,m$,

$$\sum_{j=1}^p a_{rj}\left(\sum_{S\in\Lambda} \lambda(S)x_j^S\right) \leqq \sum_{S\subset N} \lambda(S)b_r(S) = \sum_{i\in N} b_r^i \left(\sum_{S\in\Lambda} \lambda(S)\right) = b_r(N),$$

and then

$$\sum_{j=1}^p a_{rj}\hat{x}_j \leqq b_r(N).$$

From $x_j^S \geqq 0$ and $\lambda(S) > 0$, we have $\hat{\boldsymbol{x}} \geqq 0$. Because $\hat{\boldsymbol{x}}$ satisfies the constraints of the multiobjective linear production programming problem (5) with $S = N$,

$$z(\hat{\boldsymbol{x}}) = \sum_{S\in\Lambda} \lambda(S)z(\boldsymbol{x}^S) \in V(N).$$

Thus, from $z(x^S) \in \operatorname{Max} V(S)$ and $V(S)$ is comprehensive, $\sum_{S\in\Lambda}\lambda(S)v \in V(N)$ for all $v \in V(S)$, and we have

$$\sum_{S\in\Lambda}\lambda(S)V(S) \subset V(N).$$

Therefore the multiobjective linear production programming game (N,V) is balanced. ■

It is shown from Theorem 2 that the multiobjective linear production programming game (N,V) has a nonempty core. It is, however, important to find points in the core $C(N,V)$ and we consider a dual problem to the multiobjective linear production programming problem in order to find them.

To prepare for doing so, we briefly review the results on the duality of multiobjective linear programming problems in accordance with the studies by Isermann [11,12]. In general, a primal problem of a multiobjective linear programming problem is expressed as

$$\left.\begin{array}{l} \textbf{maximize } z(x) = Cx \\ \text{subject to } x \in T_p \triangleq \{x \mid Ax = b,\ x \in \mathbb{R}^p_+\} \end{array}\right\}, \tag{16}$$

where "**maximize**" means vector maximization, $z(x) = (z_1(x), z_2(x), \ldots, z_\ell(x))$, C is an $\ell \times p$ matrix, A is an $m \times p$ matrix, and b is an m-dimensional column vector. The corresponding dual problem to Problem (16) is represented as

$$\left.\begin{array}{l} \textbf{minimize } w(Y) = Yb \\ \text{subject to } Y \in T_d \triangleq \{Y \mid YAv \leq Cv \text{ for no } v \in \mathbb{R}^p_+\} \end{array}\right\}, \tag{17}$$

where "**minimize**" means vector minimization, $w(Y) = (w_1(Y), w_2(Y), \ldots, w_\ell(Y))$, Y is an $\ell \times m$ dual variable matrix, v is a p-dimensional column vector, and T_d means a set of Y such that there exists no $v \in \mathbb{R}^p_+$ satisfying constraints $YAv \leq Cv$. The following results on the duality of multiobjective linear programming problems are known.

Proposition 1. If x is a feasible solution to the primal problem (16) and Y is a feasible solution to the dual problem (17), then the inequality

$$w(Y) \leq z(x) \tag{18}$$

does not hold.

Proposition 2. Let x^* be a feasible solution to the primal problem (16) and let Y^* be a feasible solution to the dual problem (17) such that

$$z(x^*) = w(Y^*). \tag{19}$$

Then x^* is a Pareto optimal solution to the primal problem (16) and Y^* is a Pareto optimal solution to the dual problem (17).

Proposition 3. Consider the pair (16) and (17) of the dual problems. The following statements are equivalent:

(i) each problem has a feasible solution;

(ii) each problem has a Pareto optimal solution and there exists at least one pair $(\boldsymbol{x}^*, Y^*)$ of Pareto optimal solutions such that $z(\boldsymbol{x}^*) = w(Y^*)$.

Proposition 4. A feasible solution $\boldsymbol{x}^*$ is a Pareto optimal solution to the primal problem (16) if and only if there exists a feasible solution Y^* to the dual problem (17) such that $z(\boldsymbol{x}^*) = w(Y^*)$. Y^* is then itself a Pareto optimal solution to the dual problem (17).

Turning to the real subject, the following two problems represent a multiobjective linear programming problem with equality constraints and its dual problem, respectively:

$$\left.\begin{array}{l}\textbf{maximize } z(\boldsymbol{x}) = C\boldsymbol{x} \\ \text{subject to } \boldsymbol{x} \in T_S \triangleq \{\boldsymbol{x} \mid A\boldsymbol{x} = \boldsymbol{b}(S),\ \boldsymbol{x} \in \mathbb{R}_+^{p+m}\}\end{array}\right\}, \tag{20}$$

$$\left.\begin{array}{l}\textbf{minimize } w(Y) = Y\boldsymbol{b}(S) \\ \text{subject to } Y \in T_d \triangleq \{Y \mid YA\boldsymbol{v} \leq C\boldsymbol{v} \text{ for no } \boldsymbol{v} \in \mathbb{R}_+^{p+m}\}\end{array}\right\}, \tag{21}$$

where Y is an $\ell \times m$ variable matrix, C is an $\ell \times (p+m)$ matrix, A is an $m \times (p+m)$ matrix, and $\boldsymbol{b}(S) = (b_1(S), b_2(S), \ldots, b_m(S))$ is an m-dimensional column vector. The following theorem is shown by applying the above duality theory to the multiobjective linear production programming problem.

Theorem 3. *Let the feasible solution areas to the multiobjective linear production problem* (20) *and its dual problem* (21) *be not empty. For a Pareto optimal solution* x^* *to Problem* (20) *with* $S = N$, *there exists a Pareto optimal solution* $Y^* = \begin{pmatrix} y_{11}^* & \cdots & y_{1m}^* \\ \cdots & \cdots & \cdots \\ y_{\ell 1}^* & \cdots & y_{\ell m}^* \end{pmatrix}$ *to Problem* (21) *such that* $C\boldsymbol{x}^* = Y^*\boldsymbol{b}(N)$. *Then the following payoff vector* $\boldsymbol{u} = (\boldsymbol{u}_{1\cdot}, \boldsymbol{u}_{2\cdot}, \ldots, \boldsymbol{u}_{n\cdot}) \in \mathbb{R}^{\ell \times n}$, $\boldsymbol{u}_{i\cdot} = (u_{i1}, u_{i2}, \ldots, u_{i\ell})$ *belongs to the core* $C(N,V)$.

$$u_{ik} = b_1^i y_{k1}^* + b_2^i y_{k2}^* + \cdots + b_m^i y_{km}^*,\ i = 1,2,\ldots,n,\ k = 1,2,\ldots,\ell. \tag{22}$$

Proof In accordance with (7), for each $S \in \Lambda$, the characteristic set $V(S)$ is defined by using $\operatorname{Max} \hat{T}_S$, which is the set of all Pareto optimal values to the multiobjective linear production programming problem (20). Because both Problem (20) with $S = N$ and Problem (21) with $S = N$ have feasible solutions, from Propositions 3 and 4, Problem (20) has a Pareto optimal solution $\boldsymbol{x}^*$, and there exists at least one pair $(\boldsymbol{x}^*, Y^*)$ of Pareto optimal solutions such that $Y^*\boldsymbol{b}(N) = C\boldsymbol{x}^*$. For such a pair $(\boldsymbol{x}^*, Y^*)$, we have

$$Y^*\boldsymbol{b}(N) = C\boldsymbol{x}^* \in \operatorname{Max} \hat{T}_N,$$

where $\hat{T}_N$ is a feasible solution area in an objective space to the primal problem (20). The sum of $\boldsymbol{u}_{i\cdot}$ defined by (22) is equal to $Y^*\boldsymbol{b}(N)$ because

$$\begin{aligned}\sum_{i\in N}\boldsymbol{u}_{i\cdot} &= \sum_{i\in N}(b_1^i\boldsymbol{y}_{\cdot 1}^* + b_2^i\boldsymbol{y}_{\cdot 2}^* + \cdots + b_m^i\boldsymbol{y}_{\cdot m}^*)\\ &= b_1(N)\boldsymbol{y}_{\cdot 1}^* + b_2(N)\boldsymbol{y}_{\cdot 2}^* + \cdots + b_m(N)\boldsymbol{y}_{\cdot m}^* = Y^*\boldsymbol{b}(N),\end{aligned}$$

where $\boldsymbol{u}_{i\cdot} = (u_{i1},\ldots,u_{i\ell}), \boldsymbol{y}_{\cdot r}^* = (y_{1r}^*,\ldots,y_{\ell r}^*), r = 1,\ldots,m$. Therefore

$$\sum_{i\in N}\boldsymbol{u}_{i\cdot} = Y^*\boldsymbol{b}(N) \in \operatorname{Max}\hat{T}_N = \operatorname{Max}V(N).$$

Because Y^* is not always a Pareto optimal solution to the dual problem (21) with $S \neq N$,

$$Y^*\boldsymbol{b}(S) \in (\operatorname{Max}\hat{T}_S + \mathbb{R}_+^\ell).$$

Thus, we have

$$\begin{aligned}\sum_{i\in S}\boldsymbol{u}_{i\cdot} &= \sum_{i\in S}(b_1^i\boldsymbol{y}_{\cdot 1}^* + b_2^i\boldsymbol{y}_{\cdot 2}^* + \cdots + b_m^i\boldsymbol{y}_{\cdot m}^*)\\ &= b_1(S)\boldsymbol{y}_{\cdot 1}^* + b_2(S)\boldsymbol{y}_{\cdot 2}^* + \cdots + b_m(S)\boldsymbol{y}_{\cdot m}^*\\ &= Y^*\boldsymbol{b}(S) \notin V(S)\backslash\operatorname{Max}V(S),\end{aligned}$$

and from the definition of the stable outcome (12), we have $\boldsymbol{u} \in SO(N,V) = C(N,V)$. ■

It is shown from Theorem 3 that we can compute points in the core $C(N,V)$ by using Pareto optimal points to the dual problem of the multiobjective linear production programming problem.

The dual problems (17) and (21) are not typical multiobjective linear programming problems and therefore solving such dual problems would be far more complex than solving the primal problems (16) and (20). To resolve the difficulty, we can refer to the duality concept of multiple criteria and multiple constraint levels [19].

However, by applying an algorithm (e.g. [18]) for solving multiobjective linear programming problems to the primal problem (20), we can obtain all the Pareto optimal extreme points to the problem and can also find Pareto optimal solutions to the dual problem (21) in multiobjective simplex tableau corresponding to the Pareto optimal extreme points.

4 Conclusions

We have formulated the production problem as a multiobjective linear programming problem and have shown that a multi-commodity game, a multiobjective linear production programming game arises from the problem. It has been proven that the multiobjective linear production programming game has the nonempty core, and it has been shown that points in the core can be computed by using dual Pareto optimal solutions to the multiobjective linear production programming problem.

References

1. K. Bergstresser and P.L. Yu, "Domination structure and multicriteria problems in n-person games," *Theory and Decision* 8 (1977) 5–48.
2. G.C. Bird, "Cores of nonatomic linear production games," *Mathematics of Operations Research* 6 (1981) 420–423.
3. J.J.M. Derks and S.H. Tijs, "Stable outcome for multi-commodity flow games," *Methods of Operations Research* 55 (1986) 493–504.
4. J.J.M. Derks and S.H. Tijs, "Totally balanced multi-commodity games and flow games," *Methods of Operations Research* 54 (1986) 335–347.
5. P. Dubey and L.S. Shapley, "Totally balanced games arising from controlled programming problems," *Mathematical Programming* 29 (1984) 245–267.
6. J.G. Ecker, N.S. Hegner and I.A. Kouada, "Generating all maximal efficient faces of multiple objective linear programs," *Journal of Optimization Theory and Applications* 30 (1980) 353–381.
7. R. Engelbrecht-Wiggans and D. Granot, "On market prices in linear production games," *Mathematical Programming* 32 (1985) 366–370.
8. T. Gal, "A general method for determining the set of all efficient solutions to linear vectormaximum problem," *European Journal of Operational Research* 1 (1977) 307–322.
9. D. Granot, "A generalized linear production model: a unifying model," *Mathematical Programming* 34 (1986) 212–222.
10. H. Isermann, "The enumeration of the set of all efficient solutions for a linear multiple objective program," *Operational Research Quarterly* 28 (1977) 711–725.
11. H. Isermann, "The relevance of duality in multiple objective linear programming," *TIMS Studies in the Management Sciences* 6 North Holland Publishing Company, Amsterdam, 1977, 241–262.
12. H. Isermann, "On some relations between a dual pair of multiple objective linear programs," *Zeitschrift für Operations Research* 22 (1978) 33–41.
13. E. Kalai and E. Zemel, "Generalized network problems yielding totally balanced games," *Operations Research* 30 (1982) 998–1008.
14. G. Owen, "On the core of linear production games," *Mathematical Programming* 9 (1975) 358–370.
15. J. Rosenmüller, "L.P.-games with sufficiently many players," *International Journal of Game Theory* 11 (1982) 129–149.
16. T. Tanino, Y. Muranaka and M. Tanaka, "On multiple criteria characteristic mapping games," *Proceedings of MCDM '92*, Taipei, 1992, 63–72.
17. A. van den Nouweland, H. Aarts, and P. Borm, "Multi-commodity games," *Methods of Operations Research* 63 (1990) 329–338.
18. P.L. Yu and M. Zeleny, "The set of all non-dominated solutions in linear cases and a multicriteria simplex method," *Journal of Mathematical Analysis and Applications* 49 (1975) 430–468.
19. P.L. Yu, *Multiple-Criteria Decision Making: Concepts, Techniques and Extensions*, Plenum Press, New York, 1985.
20. M. Zeleny, *Linear Multiobjective Programming*, Springer-Verlag, Berlin, 1974.

Multiple Criteria Subset Selection Under Quantitative and Non-Quantitative Criteria

S. Rajabi, [1] K. W. Hipel[1,2] and D. M. Kilgour [1,3]

1.Department of Systems Design Engineering, University of Waterloo, Waterloo, Ontario, N2L 3G1, Canada.
2.Department of Statistics and Actuarial Science, University of Waterloo.
3.Department of Mathematics, Wilfrid Laurier University, Waterloo, Ontario, N2L 3C5, Canada.

Abstract:

Both tangible and intangible considerations must usually be taken into account in development of environmental and other public policies. Consequently, policy selection can be thought of as a multiple objective problem with mixed quantitative and non-quantitative information on criterion scores. This paper presents an efficient approach to the multiple-criterion selection of actions, or subsets of actions, from a discrete set, in the presence of mixed criteria. In particular, the approach chooses good subsets of actions when the preferences of the Decision Maker (DM) over the actions are specified ordinally, cardinally, or according to qualitative properties on the criterion. First, a procedure is presented to screen and remove actions that cannot possibly be in the best subset. In the second stage, the performance of the remaining actions is evaluated to find the best possible subset.
Keywords: Potentially optimal, ordinal criterita, cardinal criteria, qualitative criteria

1 Introduction

The difficulty of multiple criteria problems is due not only to conflicts among the criteria, but also to the diversity of measurement scales on which the criteria are measured. Consider the multiple criteria problem of purchasing a house. Criteria such as cost or distance from work can often be measured quantitatively. However, preferences over different houses on criteria such as appearance and quality of neighborhood may be expressed only ordinally. Also, there may be some qualitative criteria that can be represented as attributes that a house may or may not possess such as a pool, a sauna, natural gas heating or adequate shade trees.

Similar decision situations can be found in many multiple criteria problems, especially in environmental and public policy decision problems. The use of cardinal tools to process qualitative information such as ordinal preferences has long been criticized in the literature [5]. On the other hand, downgrading cardinal information to ordinal data risks the loss of important information.

A few multiple criteria methods are specially designed to deal with ordinal or mixed criteria. For instance, the global out-ranking method of Roubens [23] is based on the transformation of a preference table of actions and criteria into a vector of ranked actions. The voting method of Borda [2] has also been used for multiple criteria problems under ordinal information. To apply Borda's technique to a multiple criteria problem, each criterion should be considered as a voter. Bernardo's method [1] is based on solving a linear program in the form of an assignment problem to find the maximum agreement between an overall ranking of actions and ordering of actions on each criterion. Cook and Seiford [6] propose a similar assignment problem approach to minimize the total disagreement between the overall ordering of actions on each criterion.

Koksalan *et. al* [13] suggest an approach in which ordinal information about criteria is converted to cardinal values during the solution process. The method of Rietveld and Ouwersloot [22] use a random sampling approach to generate a set of qualitative values consistent with the underlying ordinal scores. This method can handle problems with mixed criteria. Cook and Kress [5] propose an extreme-point approach to obtain numerical estimates for both weights and criterion scores that are ranked ordinally. Lansdowne [16] reviews several multiple criteria approaches designed to solve problems with ordinal information. For other research related to multiple criteria problems under ordinal information, refer to [17], [15], [14], [7] and [9].

We propose an integrative approach to multiple criteria subset selection problems under mixed criteria. The procedure has three main steps. First, a model is built such that the values of actions measured using all types of criteria can be aggregated into a unified expression. This issue is discussed in Section 3. Then in the second step, a screening procedure is carried out to remove those actions that cannot possibly be in the best subset, thereby reducing the number of combinations of actions. The screening procedure is explained in Section 4. Finally, in the third step, subsets of remaining actions are evaluated to find the best subset of actions. Section 5 discusses this issue. Before defining our procedure, we introduce our notation and present some definitions.

2 Notation and Definitions

Let **A** be the set of actions, and **P** the set of criteria in a multiple crityeria decision making problem. We assume that **A** and **P** are both finite. Suppose for now that all criteria are cardinal. Then the consequence of action $a_k \in \mathbf{A}$ on criterion $p \in \mathbf{P}$ is measured by a real number $c_p(a_k) = c_p^k$. Thus, action

a_k is described by its consequences,

$$c(a_k) = (c_1^k, \ldots, c_p^k, \ldots, c_{|\mathbf{P}|}^k).$$

Also, for any set of actions $\mathbf{S} \subseteq \mathbf{A}$, define the consequence of $\mathbf{S}$ on criterion p as

$$c_p(\mathbf{S}) = \sum_{a_l \in \mathbf{S}} c_p^l.$$

(Note that this definition implies that consequences are independent.) A value is a measure of the overall worth of a subset. We assume that

$$v(\mathbf{S}) = v(c_1(\mathbf{S}), \ldots, c_p(\mathbf{S}), \ldots, c_{|\mathbf{P}|}(\mathbf{S})).$$

In other words, values are cardinally determined by the consequences on all criteria. Denote the set of all real values by $\mathbf{V}$. A subset of $\mathbf{V}$ is the set of linear values, $\mathbf{V}_L$. A value $v \in \mathbf{V}_L$ is determined by numerical criterion weights, $\lambda_1, \lambda_2, \ldots, \lambda_{|\mathbf{P}|}$, such that the v_L satisfies

$$v_L(\mathbf{S}) = \sum_p \lambda_p c_p(\mathbf{S}).$$

Throughout, we assume that $v(.)$ is a strictly monotonically increasing function in each of its $|\mathbf{P}|$ arguments. Note that the assumption of value as an increasing function of consequences is for exposition only; our definitions can easily be modified to include values that are decreasing in consequences, such as cost or damage to the environment of a large-scale engineering project.

Now suppose that the set of criteria, $\mathbf{P}$, can be divided into three disjoint subsets, $\mathbf{P}_1, \mathbf{P}_2$, and $\mathbf{P}_3$, such that $\mathbf{P}_1$ contains all criteria measured based on an ordinal scale, $\mathbf{P}_2$ all criteria measured on a cardinal scale, and $\mathbf{P}_3$ all criteria which represent attributes that each action can possess including none or some of them. We set $\mathbf{P} = \{1, 2, \ldots, p', \ldots, p'', \ldots, |\mathbf{P}|\}, \mathbf{P}_1 = \{1, 2, \ldots, p'\}$, $\mathbf{P}_2 = \{p' + 1, \ldots, p''\}$ and $\mathbf{P}_3 = \{p'' + 1, \ldots, |\mathbf{P}|\}$. We address two subset selection problems:

1. The m-best actions problem: m actions are to be selected. The number m is given a *priori.*

2. The j-constraints problem: a subset of actions satisfying j constraints is to be selected. The constraints are specified a *priori.*

Note that the j-constraints problem is a generalization of the m-best actions problem, which has $m = 1$ constraint. Our main focus in this paper is the m-best actions problems, but our procedure can be extended to j-constraints problems, as discussed in [21].

One concept we use for screening actions is the notion of Potentially Optimal (PO), defined as follows:

Definition 1 *Action* $a_i \in \mathbf{A}$ *is* potentially optimal *(PO) iff there exists at least one* $v \in \mathbf{V}$ *such that* $v(a_i) \geq v(a_l)$ *for all* $a_l \in \mathbf{A}$.

The set of potentially optimal actions in $\mathbf{A}$ is denoted $PO(\mathbf{A})$. The following mathematical program can be used to determine whether action a_i is potentially optimal:

$$(\mathbf{D1}(a_i)) \qquad \textbf{Minimize} \quad \delta$$
$$\textbf{Subject to:}$$
$$v(a_i) - v(a_l) + \delta \geq 0, \qquad a_l \in \mathbf{A} \setminus \{a_i\},$$
$$v \in \mathbf{V},$$

where $\mathbf{V}$ is the set of all possible value functions and $v(a_i)$ is value of action a_i according to value function v. The above program seeks a value function v that minimizes δ. For instance, if v is a linear value function, $v \in \mathbf{V}_L$, the program $(\mathbf{D1}(a_i))$ determines the criterion weights, λ_p, that minimize δ. If the optimal value of this problem is non-positive, then $a_i \in PO(\mathbf{A})$, because $\delta^* \leq 0$ implies that there is a value function that makes a_i at least as preferable as all other actions.

The concept of potentially optimal actions has been addressed in the context of multi-attribute decision theory (see, for example, [10], [11], [12], and [3]). This notion has been especially useful in situations where partial information on the DM's preferences is available [3]. A strictly monotonic value function is strictly monotonic in the consequence of each criterion. White [24] proves that under a strictly monotonic value function a potentially optimal action is always non-dominated. However, there may be some non-dominated actions which are not potentially optimal. Therefore,

$$PO(\mathbf{A}) \subseteq Eff(\mathbf{A}), \tag{1}$$

where $Eff(\mathbf{A})$ is the set of efficient (non-dominated) actions. In particular, when the value function is strictly monotonic, an action dominated in $\mathbf{A}$ cannot be potentially optimal in $\mathbf{A}$. In general, however, there is no relation between potentially optimal and efficient actions [12].

Our proposed approach can handle all criteria types introduced above. We first describe a method of examining each type individually, and then present a structure to aggregate them.

3 Expressing the Overall Values of Actions

3.1 Ordinal Criteria

We assume that for an ordinal criterion, $p \in \mathbf{P}_1$, actions can be ranked ordinally, in the sense that there exists an ordered partition of $\mathbf{A}$, $(\mathbf{A}_1, \mathbf{A}_2, \ldots, \mathbf{A}_{L_p})$, where L_p is defined as the number of rank positions. Note that $\mathbf{A}_i \cap \mathbf{A}_j = \emptyset$, if $i \neq j$ and $\cup_{i=1}^{L_p} \mathbf{A}_i = \mathbf{A}$. An action $a \in \mathbf{A}_i$ is strictly preferred to an action $b \in \mathbf{A}_j$ on criterion p *iff* $i < j$. Usually no $\mathbf{A}_i$ is empty, but this is not a requirement. If $a \in \mathbf{A}_i$, we say that a has rank position $l_p(a) = i$ on criterion p. Then,

$$a_k \succ_p a_j \Longleftrightarrow l_p(a_k) < l_p(a_j),$$

$$a_k \sim_p a_j \Longleftrightarrow l_p(a_k) = l_p(a_j),$$

where $\succ_p$ and $\sim_p$ represent preference and indifference on criterion p, respectively. Hence, an action with lower rank position is preferred to an action with higher rank position. Let $w_{pl}(i)$ be the worth of rank position l on criterion p for action i, and $\lambda_p(i)$ be the importance of criterion p for action i. In general, of course, $w_{pl}(i)$ and $\lambda_p(i)$ are unknown. Define $d_{pl}(i)$ as follows:

$$d_{ph}(i) = \begin{cases} 1 & \text{if } l_p(i) = h; \\ 0 & \text{otherwise.} \end{cases}$$

Assuming that the DM's value function is linear, the overall value of an action a_i according to the set of ordinal criteria, $\mathbf{P}_1$, is given by

$$v_o(i) = \sum_{p \in \mathbf{P}_1} \sum_{l=1}^{L_p} d_{pl}(i) w_{pl}(i) \lambda_p(i). \tag{2}$$

Note that, in (2), w_{pl} and λ_p are unknown and will be found during the solution process. Moreover, these parameters may have different values for different actions.

3.2 Cardinal Criteria

Recall that $c_p(i)$ is the consequence of action i on criterion p. For cardinal criteria, the consequences of actions are known, so the value of action a_i on the set of criteria P_2 can be expressed as

$$v_c(i) = \sum_{p \in \mathbf{P}_2} c_p(i) \lambda_p(i), \tag{3}$$

Note that to make (3) consistent with (2), the $c_p(i)$ is normalized to lie between 0 and 1. The only unknown parameters are the $\lambda_p(i)$.

3.3 Qualitative Criteria

This set of criteria describes qualities or characteristics that an action may or may not possess. To express the value of an action, a_i, on the set of criteria $\mathbf{P}_3$, define a binary indicator, d'_{pj} as follows:

$$d'_{pj}(i) = \begin{cases} 1 & \text{if action } a_i \text{ has attribute } x_j \text{ on criterion } p; \\ 0 & \text{otherwise.} \end{cases}$$

We assume that the DM can provide preference over individual attributes $x \in \mathbf{X}_p$ on criteria in $\mathbf{P}_3$, where $\mathbf{X}_p$ is the set of all elements on criterion p. In general, the issue of extending the rank order of individual attributes in $\mathbf{X}$ to the power set of $\mathbf{X}$ is not a trivial task and has been discussed widely in economic theory and elsewhere [8]. Selection of an extension is not an issue in our procedure, however.

Let w'_{pj} be the worth of having option x_j on criterion $p \in \mathbf{P}_3$ for action a_i and $\lambda_p(i)$ be defined as before. Then, the overall value of action i on the set of criteria $\mathbf{P}_3$ is,

$$v_q(i) = \sum_{p \in \mathbf{P}_3} \sum_{j=1}^{\mathbf{X}_p} d'_{pj} \lambda_p(i) w'_{pj}(i). \tag{4}$$

In the above expression the unknown parameters are λ_p and w'_{pj}. Therefore, the overall value of action i on all criteria is the summation of the above three expressions given in (2), (3), and (4).

4 Screening Actions

The concept of potentially optimal actions introduced in Section 2 was defined with respect to the standard problem of selecting the best action. Now we modify this concept for subset selection problems of the m-best actions type. Let $\mathbf{A}_{(n)}$ denote the collection of all subsets of $\mathbf{A}$ that contain n actions. For instance, $\mathbf{A}_{(2)}$ is the collection of all (unordered) pairs of actions in $\mathbf{A}$. Analogous to $PO(\mathbf{A})$, which denotes potentially optimal actions when only one action is to be selected, $PO(\mathbf{A}_{(n)})$ is the set of potentially optimal subsets with cardinality n within set $\mathbf{A}_{(n)}$ of all subsets containing n actions. Note that $\mathbf{A}^i_{(n)} \in PO(\mathbf{A}_{(n)})$ implies that there exists a value function such that $\mathbf{A}^i_{(n)}$ is as good as any other subset with cardinality n. In the m-best actions problem, the concept of Potentially Optimal action is defined as follows:

Definition 2 *Action $a_k \in \mathbf{A}$ belongs to the set of potentially optimal actions for m-best actions problems, $PO_m(\mathbf{A})$, iff there exists $\mathbf{A}^i_{(m)} \in \mathbf{A}_{(m)}$ such that $a_k \in \mathbf{A}^i_{(m)}$ and $\mathbf{A}^i_{(m)} \in PO(\mathbf{A}_{(m)})$.*

The above definition states that if action a_k is not in $PO_m(\mathbf{A})$, then there is no $\mathbf{A}^i_{(m)} \in \mathbf{A}_{(m)}$ that includes a_k such that $\mathbf{A}^i_{(m)} \in PO(\mathbf{A}_{(m)})$. In other words, a PO action, a_k, cannot belong to any m-best subset of actions. Clearly, $a_k \notin PO_m(\mathbf{A})$, can be removed from $\mathbf{A}$ without affecting any m-best subset. In symbols,

$$a_k \notin PO_m(\mathbf{A}) \Longrightarrow a_k \notin \mathbf{A}^*. \tag{5}$$

We utilize the following mathematical program to define whether $a_k \in PO_m(\mathbf{A})$ in a multiple criteria subset selection under mixed criteria when m

actions are to be selected:

$$
\begin{aligned}
(\mathbf{D2}(a_k)) \quad & \textbf{Minimize} \quad \delta \\
& \textbf{Subject to:} \\
& v(a_k) - v(a_l) + \delta \geq -M(1-\alpha_l), \quad a_l \in \mathbf{A} \setminus \{a_k\}, && (6) \\
& \sum_{a_l \in \mathbf{A} \setminus \{a_k\}} \alpha_l = q, && (7) \\
& \lambda_p - \lambda_{p+1} \geq \epsilon \quad p = 1, 2, \ldots, |P| - 1, && (8) \\
& w_{p,l} - w_{p,l+1} \geq \epsilon \quad \forall p \in \mathbf{P}_1, \quad l = 1, 2, \ldots, L_p - 1, && (9) \\
& w'_{p,x_i} - w'_{p,x_{i+1}} \geq \epsilon \ \forall p \in \mathbf{P}_3, \quad i = 1, 2, \ldots, |\mathbf{X}_p| - 1, && (10) \\
& L \leq w_p \leq U, && (11) \\
& \sum_{p=1}^{|\mathbf{P}|} \lambda_p = 1, && (12) \\
& \sum_{l=1}^{\mathbf{L}_p} w_{pl} = 1, \quad \forall p \in \mathbf{P}_1, && (13) \\
& \sum_{i=1}^{|\mathbf{X}_p|} w'_{px_i} = 1, \quad \forall p \in \mathbf{P}_3. && (14) \\
& \alpha_l \in \{0, 1\}, \quad a_l \in \mathbf{A} \setminus \{a_k\}, \\
& v \in \mathbf{V},
\end{aligned}
$$

where M is a sufficiently large number, $q \geq |\mathbf{A}| - m$ and v is overall value of action i. In the above program, the unknown parameters are $\delta, \alpha_l, \lambda_p, w_{pl}$, and w'_{p,x_i}. If the optimal value of the above program is non-positive, then action a_k is potentially optimal and may be included in the best subset of actions of size m. In the above program, the set of constraints in (6) is the main set of equations for comparing action a_k with all other actions. If α_l is set to one for a constraint in (6), then that equation becomes active; when α is zero, the equation is redundant because M is a very large number.

The set of constraints in (8) specifies the ordinal relationship among criteria. The term ϵ in these equations is a parameter to be set by the DM; it represents the minimum gap between two adjacent criterion weights. In constraints (9), ϵ is the minimum gap between two adjacent rank positions for ordinal criteria. Inequalities in (10) show the ordinal relationship among the qualitative criteria. The relationships in (11) constitute optional constraints which specify upper and lower bounds for the set of weights. The rest of the constraints are self-explanatory.

The issue of obtaining the extreme values of weights λ_p has been discussed elsewhere [25]. For a multiple criteria problem with three criteria and $\epsilon = 0$, if $\lambda_1 \geq \lambda_2 \geq \lambda_3$, then the extreme values are $(0, 0, 1), (0, 0.5, 0.5)$, and $(0.33, 0.33, 0.33)$.

5 Evaluating Combinations of Actions

This section discusses an approach for evaluating and selecting the best subset of actions under mixed criteria. For each subset, a separate program is solved to find its most favorable parameters. The idea stems from Data Envelopment Analysis (DEA), and follows from the work of Cook and Kress [5] and Cook and Johnston [4]. We propose that for a multiple criteria subset selection problem under mixed criteria, the following program be solved to measure the overall value of a subset, $\mathbf{S}_i \subseteq \mathbf{A}$.

$$
\begin{aligned}
Max\ v(\mathbf{S}_i) \quad = \quad & \sum_{p\in \mathbf{P}_1}\sum_{l=1}^{L_p} d_{pl}(\mathbf{S}_i) w_{pl}(\mathbf{S}_i)\lambda_p(\mathbf{S}_i) + \\
(\mathbf{D3}(a_k)) \qquad & \sum_{p\in \mathbf{P}_2} c_p(\mathbf{S}_i)\lambda_p(\mathbf{S}_i) + \\
& \sum_{p\in \mathbf{P}_3}\sum_{j=1}^{|\mathbf{X}_p|} d'_{pj}\lambda_p(\mathbf{S}_i) w'_{pj}(\mathbf{S}_i),
\end{aligned}
$$

Subject to:

$$
\begin{aligned}
& \lambda_p - \lambda_{p+1} \geq \epsilon \qquad p = 1, 2, \ldots, |P| - 1, \\
& w_{p,l} - w_{p,l+1} \geq \epsilon \qquad \forall p \in \mathbf{P}_1, \quad l = 1, 2, \ldots, L_p - 1, \\
& w'_{p,x_i} - w'_{p,x_{i+1}} \geq \epsilon \qquad \forall p \in \mathbf{P}_3, \quad i = 1, 2, \ldots, |\mathbf{X}_p| - 1, \\
& L \leq w_p \leq U, \\
& \sum_{p=1}^{|\mathbf{P}|} \lambda_p = 1, \\
& \sum_{l=1}^{\mathbf{L}_p} w_{pl} = 1, \qquad \forall p \in \mathbf{P}_1, \\
& \sum_{i=1}^{|\mathbf{X}_p|} w'_{px_i} = 1, \qquad \forall p \in \mathbf{P}_3.
\end{aligned}
$$

The above program is solved for each subset to determine the best performance of that subset. Note that the most favorable values, λ^* and w^* may have different values for different subsets. If λ^* and w^* are comparable for different subsets, then these subsets can be compared directly according to their best performances. But if not, comparison and selection of the best subset is difficult. One way to overcome this difficulty is to evaluate each subset based on the best parameters of other subsets in addition to its own best parameters. For example, the value of $\mathbf{S}_i$ in terms of the best parameters of $\mathbf{S}_j$ is

$$\begin{aligned} v_{ij} = & \sum_{p \in \mathbf{P}_1} \sum_{l=1}^{L_p} d_{pl}(\mathbf{S}_i) v_{pl}(\mathbf{S}_j) \lambda_p(\mathbf{S}_j) + \\ & \sum_{p \in \mathbf{P}_2} c_p(\mathbf{S}_i) \lambda_p(\mathbf{S}_j) + \\ & \sum_{p \in \mathbf{P}_3} \sum_{j=1}^{|\mathbf{X}_p|} d'_{pj}(\mathbf{S}_i \lambda_p(\mathbf{S}_j) v'_{pj}(\mathbf{S}_j). \end{aligned}$$

The following example demonstrates the above discussion.

Example 1 Assume that after screening actions and removing the inferior ones, five feasible subsets remain. Table 1 shows these subsets and their corresponding values on different criteria.

Table 1: The Scores of Five Feasible Subsets

Criteria	*Subsets*				
	$\mathbf{S}_1$	$\mathbf{S}_2$	$\mathbf{S}_3$	$\mathbf{S}_4$	$\mathbf{S}_5$
p_1	2	3	3	1	5
p_2	1	4	2	3	1
p_3	3	1	5	2	3
p_4	0.2	0.15	0.1	0.2	0.35
p_5	0.15	0.2	0.3	0.15	0.2
p_6	x_1, x_2	x_2, x_3	x_3, x_4	x_1, x_2, x_4	x_1, x_3, x_4

The first three criteria are measured based on the ordinal scale. For instance, $\mathbf{S}_1$ has rank position 2 on criterion 1 and $\mathbf{S}_2$ has rank position 4 on criterion 2. The number of rank positions for the first, second, and third criteria are 5, 4, and 5, respectively. The fourth and fifth criteria are measured on a cardinal scale, and the last criterion is qualitative. Hence,

$$\mathbf{P} = \{1, 2, 3, 4, 5, 6\}, \ L_1 = 5, \ L_2 = 4, \ \text{and} \ L_3 = 5;$$

$$\mathbf{P}_1 = \{1, 2, 3\}, \quad \mathbf{P}_2 = \{4, 5\} \quad \mathbf{P}_3 = \{6\}.$$

The value of subset $\mathbf{S}_1$ on the set of criteria $\mathbf{P}_1$, $\mathbf{P}_2$ and $\mathbf{P}_3$ is expressed, as

$$\begin{aligned} v_o(1) &= w_1^1 v_{12}(1) + w_2^1 v_{21}(1) + w_3^1 v_{35}(1), \\ v_c(1) &= .2 w_4^1 + .15 w_5^1, \\ v_q(1) &= w_6^1 [v'_{61}(1) + v'_{62}(1)]. \end{aligned}$$

Assume that for this problem,

$$p_1 \succeq p_2 \succeq p_3, \qquad \text{and}$$

$$x_1 \succeq x_2 \succeq x_3 \succeq x_4.$$

Then solving Problem ($\mathbf{D}_3(\mathbf{S}_k)$) for all subsets, individually, and for different values of the minimum gap, ϵ, gives Table 2. As this table shows subset $\mathbf{S}_4$ has the largest value for all possible values of ϵ. Therefore, this subset is the most favorable.

Table 2: The Overall Values of Five Feasible Subsets

Minimum	*Actions*				
Gap	$\mathbf{S}_1$	$\mathbf{S}_2$	$\mathbf{S}_3$	$\mathbf{S}_4$	$\mathbf{S}_5$
$\epsilon = 0.0$	0.82	0.53	0.67	1.0	0.66
$\epsilon = 0.01$	0.67	0.44	0.59	0.92	0.57
$\epsilon = 0.02$	0.54	0.37	0.50	0.76	0.48
$\epsilon = 0.05$	0.32	0.22	0.28	0.34	0.30

Now assume that we change the ordering of criterion importance. In other words, let

$$p_3 \succeq p_2 \succeq p_1.$$

This time, solving Problem ($\mathbf{D}_3(\mathbf{S}_k)$) shows that several subsets are very close in value. This demonstrates that it is worthwhile to evaluate the value of each subset according to the best parameters of other subsets.

6 Conclusions

An approach is developed for selecting an action or a subset of actions based on multiple mixed criteria, which may include ordinal, quantitative and qualitative scales. Our procedure is designed so that any partial information on the weights or criterion space can be included in the solution process. The proposed approach respects the integrity of ordinal and qualitative scales by avoiding conversion of ordinal or qualitative data to cardinal values. Moreover, our approach preserves cardinal values in that it does not reduce them to ordinal data. The structure of the proposed model makes extension to group decision situations feasible. Research is currently being carried out on incorporating interdependence of actions into the procedure [20].

References

[1] Bernardo, J. J., (1977) "An Assignment Approach to Choosing R & D Experiments" *Decision Sciences*, Vol. 8, pp. 489-501.

[2] Borda, J-C, (1781) Memoire Sur les Elections au Scrutin, "Histoire de l' Academie Royale das Sciences", Paris.

[3] Athanassopoulos, A.D. and Podinovski, V.V. (1997) "Dominance and Potential Optimality in Multiple Criteria Decision Analysis with Imprecise Information", *Journal of the Operational Research Society*, Vol. 48, pp. 142-150.

[4] Cook, W. D. and Johnston D. A. (1992) "Evaluating Suppliers of Complex Systems: A Multiple Criteria Approach", *Journal of Operational Research Society*, Vol. 43, pp. 1055-1061.

[5] Cook, W. D. and Kress, M. (1996) "An Extreme-Point Approach for Obtaining Weighted Ratings in Qualitative Multi-criteria Decision Making", *Naval Research Logistics*, Vol. 43, pp. 519-531.

[6] Cook, W. D. and Seiford, L. M. (1982) "On the Borda-Kendall Consensus Method for Priority Ranking Problems", *Management Sciences*, Vol. 28, pp. 621-637.

[7] Doyle, J. and Green, R. (1994) "Efficiency and Cross-efficiency in DEA: Derivations, Meaning and Uses", *Journal of the Operational Research Society*, Vol. 45, pp. 567-578.

[8] Fishburn, P. C. (1992), "Signed Orders and Power Set Extensions", *Journal of Economic Theory*, Vol. 56, pp. 1-19.

[9] Green, R. and Doyle, J. (1995) "On Maximizing Discrimination in Multiple Criteria Decision Making", *Journal of Operational Research Society*, Vol. 46, pp. 192-204.

[10] Hazen, G. B. (1986) "Partial Information, Dominance, and Potential Optimality in Multi-Attribute Utility Theory", *Operations Research*, Vol. 34, No. 2, pp. 296-310.

[11] Insua, D. R. (1990) *Sensitivity Analysis in Multiple Objective Decision Making*, Springer-Verlag, Berlin.

[12] Insua, D. R., and French, S. (1991) "A Framework for Sensitivity Analysis in Discrete Multi-Objective Decision Making, *European Journal of Operational Research*, Vol. 54, pp. 176-190.

[13] Koksalan, M., Karwan, M. H., and Zionts, S. (1988) "An Approach for Solving Discrete Alternative Multiple Criteria Problems Involving Ordinal Criteria", *Naval Research Logistics Quarterly*, Vol. 35, pp. 625-642.

[14] Korhonen, P. J. (1986) "A Hierarchical Interactive Method for Ranking Alternative Criteria", *European Journal of Operational Research*, Vol. 24, pp. 265-276.

[15] Larichev O. I., Moshkovich H. M., Mechitov, A. I., and Olson, D. L. (1993) "Experiments Comparing Qualitative Approaches to Rank Ordering of Multi-Attribute Alternatives", *Journal of Multi-Criteria Decision Analysis*, Vol. 2, pp. 5-26.

[16] Lansdowne, Z. (1996) "Ordinal Ranking Methods for Multi-Criterion Decision Making", *Naval Research Logistics*, Vol. 43, pp. 613-627.

[17] Pawlak, Z., Slowinski, R. (1994), "Rough Set Approach to Multi-Attribute Analysis", *European Journal of Operational Research*, Vol. 77, pp. 443-459.

[18] Perez, J. (1994) "Theoretical Elements of Comparison among Ordinal Discrete Multi-Criteria Methods", *Journal of Multi-Criteria Decision Analysis*, Vol. 3, pp. 157-176.

[19] Perez, J. and Barba-Romero, S. (1995) "Three Practical Criteria of Comparison among Ordinal Preference Aggregation Rules", *European Journal of Operational Research*, Vol. 85, pp. 473-487.

[20] Rajabi, S., Kilgour, D. M., and Hipel, K. W. (1998) "Modelling Action-Interdependence in Multiple Criteria Decision Making." To Appear in *European Journal of Operational Research.*

[21] Rajabi, S, Kilgour, D.M., and Hipel, K. W., (1998) Screening Actions in Multiple Criteria Subset Selection. Unpublished manuscript, Dept. of Systems Design Engineering, University of Waterloo.

[22] Rietveld, P. and Ouwersloot, H. (1992) "Ordinal Data in Multi-Criteria Decision Making, A Stochastic Dominance Approach to Siting Nuclear Power Plants", *European Journal of Operational Research*, Vol. 56, pp. 249-262.

[23] Roubens, M. (1982) "Preference Relations on Actions and Criteria in Multicriteria Decision Making", *European Journal of Operational Research*, Vol. 10, pp. 51-55.

[24] White, D. J. (1980) *Optimality and Efficiency*, Chichester, Wiley.

[25] Vgood, H. (1983), *Multicriteria Evaluation for Urban and Regional Planning*, Pion, London.

Using Block Norms in Bicriteria Optimization

Bernd Schandl Kathrin Klamroth*
Margaret M. Wiecek

Department of Mathematical Sciences
Clemson University
Clemson, SC
USA

Abstract

We propose to use block norms to generate nondominated solutions of multiple criteria programs and introduce the new concept of the oblique norm that is specially tailored to handle general problems. We show the applicability of oblique norms to deal with discrete or convex bicriteria programs and also discuss implications of using block norms in multiple criteria decision making.

Keywords: Bicriteria optimization, bicriteria programming, block norms, oblique norms, properly nondominated points.

1 Introduction

Compromise programming is based on the concept of identifying nondominated solutions of multiple criteria programs that are the closest to some utopia (ideal) point. Different norms have been used to measure the distance between the solutions and the utopia point. In particular, the family of L_p norms has been extensively studied by many researchers, including (Yu, 1973), (Zeleny, 1973), (Gearhart, 1979), (Wierzbicki, 1980), (Steuer and Choo, 1983), (Steuer, 1986), and many others. The l_∞ norm and the augmented l_∞ norm turned out to be very useful in generating nondominated solutions of general continuous or discrete multiple criteria programs and led to the well known weighted (augmented) Tchebycheff scalarization and its variations. (Kaliszewski, 1987) introduced a modified l_∞ norm and showed its applicability in generating nondominated solutions. Compromise programming was extended by (Szidarovszky et al., 1986) to composite programming

*On leave from the Department of Mathematics, University of Kaiserslautern, Kaiserslautern, Germany.

This work was partially supported by ONR Grant N00014-97-1-0784.

using more than one value of p in the l_p distance. (Ballestero and Romero, 1998) analyzed connections between compromise programming and utility theory. (Carrizosa et al., 1996) proposed a new class of norms that contains the family of L_p norms to generate the set of points that have minimal distance to the utopia point with respect to at least one norm within this class of norms. Their approach leads to solving linear programs while generating nondominated solutions.

Not only have norms been beneficial in constructing scalarization approaches to multiple criteria programs but also become suitable tools supporting decision making. The choice of the utopia point and weights usually expresses decision maker's preferences in the objective space while selecting the most preferred nondominated solution. Applications of norm-based methods can be found in structural design (Miura and Chargin, 1996), water resource management (Bárdossy et al., 1985), manpower planning (Silverman et al., 1988), transportation and location (Ogryczak et al., 1988) and many other areas.

Motivated by the success of norm-based approaches in MCDM, we propose to apply block norms to generate nondominated solutions as well as to support the decision making process. The family of block norms, also called polyhedral norms, includes all the norms whose unit ball is a polyhedral set, so that the l_1 norm and the l_∞ norm are members of this family. In this paper, we introduce the concept of the oblique norm that can be viewed as a generalization of the augmented l_∞ norm. This new norm is designed to preserve capabilities of the l_∞ norm and the augmented l_∞ norm while allowing the decision maker more freedom in the choice of a distance measure.

In the next section we define the oblique norm and derive some properties useful for finding nondominated solutions. Section 3 contains the main results of the paper. We first examine relationships between nondominated solutions of a general multiple criteria program and optimal solutions of its scalarization by means of a block norm and an oblique norm. In the second part of this section we focus on bicriteria programs. In particular, we examine relationships between (properly) nondominated solutions of (finite) discrete problems and (polyhedral) convex problems and optimal solutions of related scalarizations by means of an oblique norm. At the end of this section we discuss practical implications of using block norms in MCDM and in Section 4 we highlight future research directions.

To facilitate further discussions, the following notation is used throughout the paper. Let $u, w \in \mathbb{R}^n$ be two vectors.

- We denote components of vectors by subscripts and enumerate vectors by superscripts.

- $u < w$ denotes $u_i < w_i$ for all $i = 1, \ldots, n$. $u \leq w$ denotes $u_i \leq w_i$ for all $i = 1, \ldots, n$, but $u \neq w$. $u \leqq w$ allows equality. The symbols $>, \geq, \geqq$ are used accordingly.

- Let $\mathbf{R}^n_{\geqq} := \{x \in \mathbf{R}^n : x \geqq 0\}$. If $S \subseteq \mathbf{R}^n$, then $S_{\geq} := S \cap \mathbf{R}^n_{\geqq}$.
- $\langle u, w \rangle$ denotes the scalar product in $\mathbf{R}^n$: $\langle u, w \rangle = \sum_{i=1}^n u_i w_i$.
- $\text{conv}(S)$ denotes the convex hull of a set $S \subseteq \mathbf{R}^n$.
- $\text{int}(S)$ denotes the interior of $S \subseteq \mathbf{R}^n$.

We consider the following general multiple criteria program

$$\begin{array}{ll} \min & \{z_1 = f_1(x)\} \\ & \vdots \\ \min & \{z_n = f_n(x)\} \\ \text{s.t.} & x \in S, \end{array} \tag{1}$$

where $S \subseteq \mathbf{R}^m$ is the *feasible set* and $f_i(x), i = 1, \ldots, n$, are real-valued functions. We define the *set of all feasible criterion vectors* Z, the *set of all nondominated criterion vectors* N and the *set of all efficient points* E of (1) as follows

$$\begin{aligned} Z &= \{z \in \mathbf{R}^n : z = f(x), x \in S\} = f(S) \\ N &= \{z \in Z : \nexists \tilde{z} \in Z \text{ s.t. } \tilde{z} \leq z\} \\ E &= \{x \in S : f(x) \in N\}, \end{aligned}$$

where $f(x) = \left(f_1(x) \cdots f_n(x)\right)^T$. The set Z is assumed to be closed. The point $z^* \in \mathbf{R}^n$ with

$$z_i^* = \min\{f_i(x) : x \in S\} - \varepsilon_i \qquad i = 1, \ldots, n$$

is called the *ideal (utopia) criterion vector*, where the entries of $\varepsilon \in \mathbf{R}^n$ are small positive numbers. Without loss of generality we assume $z^* = 0$.

We define the set of properly nondominated solutions according to (Geoffrion, 1968). A point $\bar{z} \in N$ is called *properly nondominated*, if there exists $M > 0$ such that for each $i = 1, \ldots, n$ and each $z \in Z$ satisfying $z_i < \bar{z}_i$ there exists a $j \neq i$ with $z_j > \bar{z}_j$ and

$$\frac{z_i - \bar{z}_i}{\bar{z}_j - z_j} \leq M.$$

Otherwise $\bar{z} \in N$ is called *improperly nondominated.* The set of all properly nondominated points is called N_p.

2 Oblique Norms

In order to develop the new concept of oblique norms we first review some basic definitions about block norms. For a detailed introduction to norms and their properties we refer the reader to (Rockafellar, 1970), (Hiriart-Urruty and Lemaréchal, 1993a) and (Hiriart-Urruty and Lemaréchal, 1993b). An overview of basic properties of block norms is also given in (Schandl, 1998).

Definition 2.1 A norm γ with a polyhedral unit ball in $\mathbb{R}^n$ is called a *block norm.* The vectors defined by the extreme points of the unit ball are called *fundamental vectors* and are denoted by v^i. The fundamental vectors defined by the extreme points of a facet of B span a *fundamental cone.*

Definition 2.2 Let $u \in \mathbb{R}^n$. The *reflection set* of u is defined as

$$R(u) := \{w \in \mathbb{R}^n : |w_i| = |u_i| \quad \forall i = 1, \ldots, n\}.$$

Definition 2.3 (Bauer et al., 1961) A norm γ is said to be *absolute* if for any given $u \in \mathbb{R}^n$, all elements of $R(u)$ have the same distance from the origin with respect to γ, i.e.

$$\gamma(w) = \gamma(u) \quad \forall w \in R(u).$$

Note that the unit ball of an absolute norm has the same structure in every orthant, which is convenient as well as sufficient for multiple criteria programs as all nondominated solutions are located in the cone $z^* + \mathbb{R}^n_{\geqq}$ and one does not need to search the entire space $\mathbb{R}^n$.

Definition 2.4 A block norm γ with a unit ball B is called *oblique* if it has the following properties:

(i) γ is absolute.

(ii) $(z - \mathbb{R}^n_{\geqq}) \cap \mathbb{R}^n_{\geqq} \cap \partial B = \{z\} \quad \forall z \in (\partial B)_{\geqq}$.

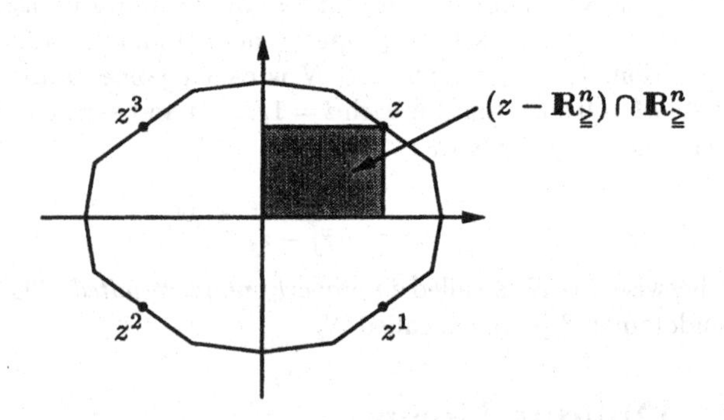

Fig. 1: Example of the unit ball of an oblique norm with $R(z) = \{z, z^1, z^2, z^3\}$

The following corollaries immediately result from Definitions 2.3 and 2.4.

Corollary 2.5 The number of fundamental vectors of an oblique norm γ in $B_{\geqq}$ is finite.

Corollary 2.6 If γ with the unit ball B is an oblique (absolute) norm, then $\tilde{\gamma}$ with the unit ball $\alpha B, \alpha > 0$ is also an oblique (absolute) norm.

The following lemmas are useful in developing our main results in the next section. Note that the condition (i) of Lemma 2.9 is identical with the condition (ii) of Definition 2.4.

Lemma 2.7 An oblique norm γ with the unit ball B has the following property:

$$(z - \mathbb{R}^n_{\geqq}) \cap \mathbb{R}^n_{\geqq} \cap \partial(\gamma(z)B) = \{z\} \quad \forall z \in \mathbb{R}^n_{\geqq}.$$

Proof. Since $z \in \partial(\gamma(z)B)$, the statement follows directly from Definition 2.4 and Corollary 2.6. □

Lemma 2.8 An absolute norm γ with the unit ball B has the following property:

$$(z - \mathbb{R}^n_{\geqq}) \cap \mathbb{R}^n_{\geqq} \subseteq \gamma(z)B_{\geqq} \quad \forall z \in \mathbb{R}^n_{\geqq}.$$

Proof. Consider first $z \in (\partial B)_{\geqq}$. It follows that $\gamma(z) = 1$. Since γ is absolute, all points in $R(z)$ are in B. Because of the convexity of B, we have $\text{conv}(R(z)) \subseteq B$. But $(z - \mathbb{R}^n_{\geqq}) \cap \mathbb{R}^n_{\geqq}$ is a subset of $\text{conv}(R(z))$ and therefore also of $B_{\geqq}$.
The general case $z \in \mathbb{R}^n_{\geqq}$ follows again from Corollary 2.6. □

Lemma 2.9 Let γ be an absolute block norm with the unit ball B. Let $\mathcal{N}$ denote the set of outer normal vectors of all the facets of B. Let e^j be the jth unit vector, $j = 1, \ldots, n$. Then the following two statements are equivalent:

(i) $(z - \mathbb{R}^n_{\geqq}) \cap \mathbb{R}^n_{\geqq} \cap \partial B = \{z\} \quad \forall z \in (\partial B)_{\geqq}$.

(ii) $\langle \mathfrak{n}, e^j \rangle \neq 0 \quad \forall j = 1, \ldots, n$ and $\forall \mathfrak{n} \in \mathcal{N}$.

Proof.

(i) $\Rightarrow$ (ii) Let F be a facet of B with the normal vector $\mathfrak{n} \in \mathcal{N}$. Assume $\langle \mathfrak{n}, e^j \rangle = 0$ for some j. Then there exists a point $z \in F$ with $z_j \neq 0$ (otherwise F would not be a facet). Since γ is absolute, we can assume without loss of generality that $z \in \mathbb{R}^n_{\geqq}$. Define a point $\tilde{z}$ as follows:

$$\begin{aligned} \tilde{z}_k &= z_k \quad \forall k \neq j \\ \tilde{z}_j &= \tfrac{1}{2} z_j. \end{aligned}$$

Then $\tilde{z}$ is in $F \subseteq \partial B$, because γ is absolute. But we also have that

$$\tilde{z} \in (z - \mathbb{R}^n_{\geqq}) \cap \mathbb{R}^n_{\geqq} \cap \partial B,$$

which is a contradiction to (i).

(ii) $\Rightarrow$ (i) Let $z \in (\partial B)_{\geq}$ and assume there exists $\tilde{z} \neq z$ with

$$\tilde{z} \in (z - \mathbb{R}^n_{\geqq}) \cap \mathbb{R}^n_{\geqq} \cap \partial B. \tag{2}$$

Because of Lemma 2.8 we have $\mathfrak{n} \geq 0$ for all normals of facets in $\mathbb{R}^n_{\geqq}$. Together with $\langle \mathfrak{n}, e^j \rangle \neq 0$ for all j we even know that $\mathfrak{n} > 0$ for these same normals. Since we assumed that both z and $\tilde{z}$ are in ∂B, they are either on the same or on two different facets.

Assume first that z and $\tilde{z}$ are on the same facet F with the normal $\mathfrak{n}$. Consequently $\langle z - \tilde{z}, \mathfrak{n} \rangle = 0$, but since $z - \tilde{z} \geq 0$ and $\mathfrak{n} > 0$ it follows that $z = \tilde{z}$, a contradiction to our assumption.

Assume now that z and $\tilde{z}$ are on different facets, say F and $\tilde{F}$ with normals $\mathfrak{n}$ and $\tilde{\mathfrak{n}}$, respectively. Since $z \in B$ and $\tilde{z} \in \tilde{F}$, the definition of the outer normal yields $\langle z - \tilde{z}, \tilde{\mathfrak{n}} \rangle \leq 0$. But since $z - \tilde{z} \geq 0$ and $\tilde{\mathfrak{n}} > 0$ it follows again that $z = \tilde{z}$, a contradiction.

Thus $\tilde{z} \notin \partial B$ and assumption (2) was wrong. □

3 Generating the Nondominated Set

3.1 General Results

We first show that for every nondominated point there exists a block norm so that this point is a unique minimizer of the related block-norm-scalarization. In the proof, to show the existence of the desired block norm we use the l_∞ norm, and thus not an oblique norm. The result gives another interpretation of the results on the weighted Tchebycheff approach in (Steuer, 1986) and illustrates the idea of introducing block norms to multiple criteria programming.

Theorem 3.1 Let $\bar{z} \in N$. Then there exists a block norm γ so that $\bar{z}$ uniquely minimizes

$$\min_{z \in Z} \gamma(z) = \min_{x \in S} \gamma(f(x)).$$

Proof. Recall that we assumed without loss of generality $z^* = 0$. Define the unit ball B of a block norm γ as $B = \text{conv}(R(\bar{z}))$. Assume there is a $\tilde{z} \in Z$, $\tilde{z} \neq \bar{z}$ with $\gamma(\tilde{z}) \leq \gamma(\bar{z})$. From the construction of γ we have that $\tilde{z} \leqq \bar{z}$. Since $\tilde{z} \neq \bar{z}$, we have $\tilde{z}_i < \bar{z}_i$ for some i, which is a contradiction to $\bar{z} \in N$. Thus $\gamma(z) > \gamma(\bar{z})$ for all $z \in Z$. □

We now focus on oblique norms and show that any optimal solution of the oblique-norm-scalarization of (1) is a nondominated solution of (1). The converse of this result is not true in general since oblique norms cannot be used to generate improperly nondominated points.

Theorem 3.2 Let γ be an oblique norm and let $\bar{z}$ be a solution of

$$\min_{z \in Z} \gamma(z) = \min_{x \in S} \gamma(f(x)).$$

Then $\bar{z} \in N$.

Proof. Assume $\bar{z} \notin N$. Then there exists $\tilde{z} \in Z$ with $\tilde{z} \leq \bar{z}$; therefore $\tilde{z} \in \left((\bar{z} - \mathbb{R}^n_{\geqq}) \cap \mathbb{R}^n_{\geqq}\right) \setminus \{\bar{z}\}$. However, according to Lemma 2.7, we have $\{\bar{z}\} = \left((\bar{z} - \mathbb{R}^n_{\geqq}) \cap \mathbb{R}^n_{\geqq}\right) \cap \partial(\gamma(\bar{z})B)$. Thus $\tilde{z} \notin \partial(\gamma(\bar{z})B)$ and from Lemma 2.8, it follows that $\tilde{z} \in \operatorname{int}(\gamma(\bar{z})B)$. Therefore $\gamma(\tilde{z}) < \gamma(\bar{z})$, which is a contradiction to the minimality of $\bar{z}$. □

3.2 The Bicriteria Case

In this section we concentrate on bicriteria problems and show that there exists an oblique norm γ for every $z \in N_p \subseteq \mathbb{R}^2$ so that z uniquely minimizes

$$\min_{z \in Z} \gamma(z) = \min_{x \in S} \gamma(f(x)).$$

We study the cases where Z is a general discrete set, a finite discrete set, a convex polyhedral set and a general convex set. In each case we prove the existence of an oblique norm with the above mentioned property by constructing its unit ball.

Theorem 3.3 (Discrete case in $\mathbb{R}^2$) Let $Z \subseteq \mathbb{R}^2$ be discrete and let $\bar{z} \in N_p$. Then there exists an oblique norm γ so that $\bar{z}$ uniquely minimizes

$$\min_{z \in Z} \gamma(z) = \min_{x \in X} \gamma\big(f(x)\big). \tag{3}$$

Proof. Since $\bar{z} \in N_p$, there exists $M > 0$ such that for every $z \in Z \subseteq \mathbb{R}^2_{>}$ with $z_1 < \bar{z}_1$ we have $z_2 > \bar{z}_2$ and $\frac{z_1 - \bar{z}_1}{\bar{z}_2 - z_2} \leq M < \infty$. Solving this for z_2 yields $z_2 \geq -\frac{1}{M} z_1 + \bar{z}_2 + \frac{1}{M}\bar{z}_1$. So all feasible points with $z_1 < \bar{z}_1$ are located above the halfline starting at $\bar{z}$ and going through $\left(0, \bar{z}_2 + \frac{1}{M}\bar{z}_1\right)$. We take $\left(0, \bar{z}_2 + \frac{1}{\alpha_1 M}\bar{z}_1\right)$ where $\alpha_1 > 1$ as the first extreme point of the unit ball. Considering the feasible points with $z_2 < \bar{z}_2$, we use $\frac{z_2 - \bar{z}_2}{\bar{z}_1 - z_1} \leq M < \infty$ to construct another extreme point at $\left(\bar{z}_1 + \frac{1}{\alpha_2 M}\bar{z}_2, 0\right)$ where $\alpha_2 > 1$. The set of extreme points of B is then defined as the union of the reflection sets of $\bar{z}$ and the two mentioned points.

Due to the chosen slope of the boundary segments of B, the unit ball is convex and satisfies both conditions of Definition 2.4, so the resulting norm is an oblique norm. Since we constructed the boundary of B so that $\bar{z}$ is the only point in $Z \cap B$, $\bar{z}$ minimizes (3) uniquely. □

Although we have given a general proof for the discrete case, it is interesting to demonstrate a construction of an oblique norm for the *finite* discrete case where $N = N_p$. The construction is described in Algorithm 3.4 while Lemma 3.5 and Theorem 3.6 show that the constructed norm is in fact an oblique norm so that $\bar{z}$ uniquely minimizes (3).

Algorithm 3.4 Let $Z \subseteq \mathbb{R}^2$ be discrete and finite, and let $\bar{z} \in N_p$.

Step 1: Finding the extreme points v of B with $v_1 \in [0, \bar{z}_1]$ and $v_2 \geq \bar{z}_2$.

If there does not exist a point $z \in N_p$ with $z_1 < \bar{z}_1$ below or on the line through $\bar{z}$ and $(0, z_2 + \alpha\bar{z}_1)$ where $0 < \alpha < 1$, then define $v^1 = \bar{z}$, $v^2 = (0, z_2 + \alpha\bar{z}_1)$ and goto Step 2.

Otherwise set $v^1 = \bar{z}$ and $i = 1$. Consider the following problem:

$$\begin{array}{ll} \min & z_2 \\ \text{s.t.} & 0 < z_1 < v_1^i \\ & z \in Z. \end{array} \tag{4}$$

Note that (4) is always feasible, because we consider it only if we have already found a point $z \in N_p$ with $z_1 < v_1^i$. Let v^{i+1} be the solution of (4).

If there does exist a point $z \in N_p$ with $z_1 < v_1^{i+1}$ below or on the line through v^i and v^{i+1}, then set $i = i + 1$ and consider again (4) to find subsequent extreme points. Otherwise redefine v^{i+1} as the intersection point of the z_2-axis and the line through v^i and v^{i+1}, i.e.

$$v^{i+1} \leftarrow \left(0, v_2^{i+1} - \frac{v_2^{i+1} - v_2^i}{v_1^{i+1} - v_1^i} v_1^{i+1}\right).$$

Step 2: Finding the extreme points v of B with $v_1 \geq \bar{z}_1$ and $v_2 \in [0, \bar{z}_2]$.

Get these extreme points in a similar way as in Step 1 by considering the following problem:

$$\begin{array}{ll} \min & z_1 \\ \text{s.t.} & 0 < z_2 < v_2^i \\ & z \in Z. \end{array} \tag{5}$$

Step 3: Finding the complete set of extreme points of B.

The entire set of extreme points of the unit ball B of γ is the union of the reflection sets of all the extreme points found in Steps 1 and 2.

Note that the procedure is finite, since Z and therefore N_p are both finite.

Lemma 3.5 The block norm constructed in Algorithm 3.4 is an oblique norm.

Proof. We first give two remarks:

(a) Each line segment between two consecutive extreme points v^{i+1} and v^i constructed in Step 1 has a negative slope, otherwise a point $z \in N_p$ with $z_1 < v_1^i$ and $z_2 \leq v_2^i$ would exist, which contradicts the construction of v^i using a nondominated point. An analogous result is valid for the points found in Step 2.

(b) The slope of the line segments between v^{i+1} and v^i constructed in Step 1 is always between 0 and -1 and increases with i. Since a slope change at v^{i+1} occurs only if there is a point $z \in N_p$ with $z_1 < v_1^{i+1}$ below the line through v^i and v^{i+1}, the slope can never decrease with i. An analogous result is valid for the points found in Step 2.

Because of remark (b), B is convex. Due to Step 3 of the algorithm, γ is an absolute norm. Due to remark (a) and Lemma 2.9, part (ii) of Definition 2.4 is satisfied, and by construction, part (i) of Definition 2.4 is satisfied as well. □

Theorem 3.6 (Finite discrete case in $\mathbb{R}^2$) Let $Z \subseteq \mathbb{R}^2$ be discrete and finite, let $\bar{z} \in N_p$. The point $\bar{z} \in N_p$ minimizes

$$\min_{z \in Z} \gamma(z) = \min_{x \in S} \gamma(f(x))$$

uniquely, where γ is the oblique norm constructed in Algorithm 3.4.

Proof. Follows directly from the construction of γ and Lemma 3.5. □

Theorem 3.7 (Convex polyhedral case in $\mathbb{R}^2$) Let $Z \subseteq \mathbb{R}^2$ be convex and polyhedral and let $\bar{z} \in N$. Then there exists an oblique norm γ so that $\bar{z}$ uniquely minimizes

$$\min_{z \in Z} \gamma(z) = \min_{x \in S} \gamma(f(x)). \tag{6}$$

Proof. Due to (Geoffrion, 1968), there exists a supporting line of Z at $\bar{z}$ with the normal vector $w > 0$. Define the two vectors $w^1 = (\alpha w_1, w_2)$ and $w^2 = (w_1, \alpha w_2)$ where $\alpha > 1$. Denote the line defined by the normal w^1 through $\bar{z}$ as l_1 and the line defined by the normal w^2 through $\bar{z}$ as l_2.

Take the intersection point of l_1 and the z_1-axis, the intersection point of l_2 and the z_2-axis, and the point $\bar{z}$ as extreme points of B in $\mathbb{R}^2_{\geqq}$ and get the entire set of extreme points of B by taking the union of the reflection sets of the three mentioned points.

Conditions (i) of Definition 2.4 is satisfied by construction. Since $w^1 > 0$ and $w^2 > 0$ and because of Lemma 2.9, B is convex and condition (ii) of Definition 2.4 is satisfied, so γ is oblique.

The point $\bar{z}$ minimizes (6) uniquely, because $\alpha > 1$ and therefore no other point of N can be in B. □

Theorem 3.8 (Convex case in $\mathbb{R}^2$) Let $Z \subseteq \mathbb{R}^2$ be convex and let $\bar{z} \in N_p$. Then there exists an oblique norm γ so that $\bar{z}$ uniquely minimizes

$$\min_{z \in Z} \gamma(z) = \min_{x \in S} \gamma(f(x)). \tag{7}$$

Proof. Since Z is convex and $\bar{z}$ is properly nondominated, there exists a supporting line of Z at $\bar{z}$ with a normal vector $w > 0$. We then proceed as in the proof of Theorem 3.7. □

3.3 Practical Implications

Having established theoretical foundations for applying block norms in bicriteria optimization we should turn our attention to the issue of enhancing the decision making process. Block norms can be viewed as a mathematical tool but also as a decision tool introducing a piecewise linear utility function in the objective space which minimized over the outcome set yields a most preferred nondominated solution. Piecewise linearity avoids computational difficulties when the utility function is nonlinear but on the other hand applies different utility to different regions of the objective space. As the number of the fundamental directions of a block norm and their length can be easily changed, the resulting utility function can be easily modified before the decision process starts or in the course of the process. This flexibility allows decision makers to change their preferences while searching for a most preferred solution.

Furthermore, block norms are dense in the set of all norms in $\mathbb{R}^n$, see (Ward and Wendell, 1985), so that any norm in $\mathbb{R}^n$ can be approximated arbitrarily close by a block norm, a feature again helpful in representing or approximating complex decision maker's preferences.

Last but not least, block norms can be helpful in exploring the objective space in several directions simultaneously, which can be beneficial in MCDM with multiple decision makers or in designing parallel algorithms for MCDM.

4 Conclusions

In this paper we introduced block norms into multiple criteria programming. We also defined oblique norms, a new class of block norms specially designed to generate properly nondominated solutions. These norms are absolute and have a unit ball whose boundary is determined by hyperplanes with normal vectors never parallel nor perpendicular to the coordinate axes of the objective space. This property makes the norms suitable to represent finite nonzero trade-offs between nondominated solutions.

We showed a general relationship between nondominated solutions and solutions of the scalarization by means of an oblique norm. Specific results are presented for bicriteria problems. We also briefly discussed the application of block norms in MCDM.

We will generalize the results of this paper for the multiple criteria case and will also study continuous nonconvex problems. In the future, we plan to develop block-norm-based approaches to MCDM which make use of these norms' flexibility and versatility.

References

Ballestero, Enrique; Romero, Carlos (1998). *Multiple Criteria Decision Making and its Applications to Economic Problems.* Kluwer Academic Publishers, Boston.

Bárdossy, Andras; Bogárdi, Istvan; Duckstein, Lucien (1985). Composite Programming as an Extension of Compromise Programming. In *Mathematics of Multiobjective Optimization* (edited by P. Serafini), pages 375–408. Springer-Verlag, New York.

Bauer, F. L.; Stoer, J.; Witzgall, C. (1961). Absolute and monotonic norms. *Numerische Mathematik*, 3:257–264.

Carrizosa, E.; Conce, E.; Pascual, A.; Romero-Morales, D. (1996). Closest Solutions in Ideal-Point Methods. In *Advances in Multiple Objective and Goal Programming* (edited by R. Caballero; F. Ruiz; R. E. Steuer), pages 274–281. Springer-Verlag, Berlin.

Gearhart, W. B. (1979). Compromise Solutions and Estimation of the Noninferior Set. *Journal of Optimization Theory and Applications*, 28:29–47.

Geoffrion, A. M. (1968). Proper Efficiency and the Theory of Vector Maximization. *Journal of Mathematical Analysis and Applications*, 22(3):618–630.

Hiriart-Urruty, Jean-Baptiste; Lemaréchal, Claude (1993a). *Convex Analysis and Minimization Algorithms I.* Springer-Verlag, Berlin.

Hiriart-Urruty, Jean-Baptiste; Lemaréchal, Claude (1993b). *Convex Analysis and Minimization Algorithms II.* Springer-Verlag, Berlin.

Kaliszewski, Ignacy (1987). A Modified Weighted Tchebycheff Metric for Multiple Objective Programming. *Computers and Operations Research*, 14:315–323.

Miura, H.; Chargin, M. K. (1996). A flexible formulation for multi-objective design problems. *AIAA Journal*, 34:1187–1192.

Ogryczak, W.; Studzinski, K.; Zorychta, K. (1988). Dynamic Interactive Network System – DINAS version 2.1. User's Manual. IIASA Working Paper WP-88-114, Laxenburg, Austria.

Rockafellar, R. Tyrrell (1970). *Convex Analysis.* Princeton University Press, Princeton, NJ.

Schandl, Bernd (1998). On Some Properties of Gauges. Tech. Rep. 662, Department of Mathematical Sciences, Clemson University, Clemson, SC.

Silverman, J.; Steuer, R. E.; Whisman, A. W. (1988). A multi-period, multiple criteria optimization system for manpower planning. *European Journal of Operational Research*, 34:160–170.

Steuer, Ralph E. (1986). *Multiple Criteria Optimization: Theory, Computation, and Application.* Wiley, New York.

Steuer, Ralph E.; Choo, E. U. (1983). An Interactive Weighted Tchebycheff Procedure for Multiple Objective Programming. *Mathematical Programming*, 26:326–344.

Szidarovszky, Ferenc; Gershon, Mark E.; Duckstein, Lucien (1986). *Techniques for Multiobjective Decision Making in Systems Management.* Elsevier Science Publishing, New York.

Ward, J. E.; Wendell, R. E. (1985). Using Block Norms for Location Modeling. *Operations Research*, 33:1074–1090.

Wierzbicki, Andrzej P. (1980). The Use of Reference Objectives in Multiobjective Optimization. *Lecture Notes in Economics and Mathematical Systems*, 177:468–486.

Yu, P. L. (1973). A Class of Solutions for Group Decision Problems. *Management Science*, 19:936–946.

Zeleny, M. (1973). Compromise Programming. In *Multiple Criteria Decision Making* (edited by J.L. Cochrane; M. Zeleny), pages 262–301. University of South Carolina, Columbia, SC.

Multiple Risk Assessment with Random Utility Models in Probabilistic Group Decision Making

Fumiko Seo

Department of Business Administration and Informatics, Setsunan University,
17-8 Ikeda-Nakamachi, Neyagawa, Osaka 572-8508, Japan.
E-mail: a50202@sakura.kudpc.kyoto-u.ac.jp

Abstract. This paper discusses a probabilistic utility approach to group decision making for risk assessment. Individual decision makers are assumed to be all anonymous and to possess their unknown and diversified preference structures. The probabilistic utility approach based on random utility models is discussed for the risk assessment. The probabilistic group utility models are defined with the probability distributions called the preference probability. The multiple risk assessment for alternative gamble prospects is discussed, which come from the incomplete information structure. The multiple risk evaluation function is constructed via the probabilistic value tradeoffs on multiple attributes under the alternative gamble prospects.

Keywords. Group decision making, Random utility model, Incomplete information, Probabilistic value tradeoffs, Multiple risk assessment.

1. Introduction

This paper discusses a probabilistic utility approach for multiple risk assessment in group decision making under incomplete information structures.

In decision analysis under uncertainty, the construction of the numerical utility functions has been major concern, because the risk attitudes of the decision maker (DM) are varied largely according to his/her particular uncertain decision environments. This approach is called the "*algebraic*" choice model where the numerical preference evaluations are assigned deterministically (c.f. Savage 1954), even though the utility functions are constructed heuristically with the probability measure in the probability space.

In realistic decision environments, however, the presentation of human preferences is usually indeterministic. In particular, in group decision making, the difficulty of the construction of group utility functions based on individual preference orders has been raised, which is known as Arrow's Possibility Theorem (1950, 1951) and leads to the difficulty of construction of the best-preferred social programs in democratic societies.

On the other hand, in the related fields such as mathematical psychology, probabilistic choice theory has been presented. The probabilistic choice model known as the Fechner-Thurstone model has been developed by Luce (1958, 1959), Marshak (1960), Block and Marshak (1960), Luce and Suppes (1965), and many others, where the random utility values are assessed at random but unequivocally and the random utility models are constructed with the probability distributions named as the preference probability. These researches, however, do not treat the group decision making intentionally, although the randomness in the preference evaluation is observed mostly in the group decision making where individual preferences are usually varied and unknown. More basically, the method to assess the utility values on which the random utility models should be constructed has not been concerned. The random utility values seem to be assessed simply as arbitrary numbers. In team theory discussed by Marschak and Radner (1972), whose major concern is with finding the optimal information structure and the optimal decision rules, the team members are assumed simply to have common, not diversified, interest and beliefs.

In this paper, we treat the group decision making for the risk assessment with multiple attributes as the multiperson decision making and discuss it based on the random utility models. The individual decision makers (IDM) are presumed to be all anonymous in the group decision making, but have their own interests and beliefs although these are all unknown. The choice behavior by the individuals is treated as latent and thus uncertain "social" events. The evaluation of group decision making is reduced to the assessment of the probability distributions of the latent individual choices. In addition, we introduce the incompleteness of the information structure which the group decision making faces. The occurrence of uncertain "natural" events, we call it *risk*, is treated in the varied risk prospects which come from the diversified information structures. Probabilistic value tradeoffs are defined on the alternative risk profiles due to the alternative information structures. Integration of the multiple risk assessments for the alternative risk prospects is performed on the probabilistic value tradeoffs. As a result, a composite risk evaluation function for the multiple risk assessments is constructed.

In Section 2, the basic concepts of the random utility models are discussed as the background, and redefined on the probabilistic choice behavior for the group decision making. In Section 3, the gamble, or risk, prospects for uncertain "natural" events are introduced and the probabilistic value tradeoffs are discussed for the alternative information structures. The construction of a composite risk evaluation function for the multiple risk assessment is presented in Section 4. Finally, a brief summary and remarks are presented in Section 5.

2. Probabilistic Choice Behavior and Random Utility Models

2.1 Probabilistic Choice Behavior

Let Ψ denote the universal set of possible information and ϖ, $\varpi \subseteq \Psi$, be its obtainable subset which defines an information structure. Denote by $\eta \in \Psi$ its element. Let A denote the universe of choice behavior, or actions, and F, $F \subseteq A$, be its subset recognized as the feasible. Let X denote the universe of objects for choice and ξ, $\xi \subseteq X$, be its known, or discriminating, subset. Denote by $a \in A$ and $x \in X$ their elements respectively. Note that A is a set of decision alternatives and X is a set of the multiple attributes evaluated as the certain variables that represent multiple objectives included in a decision problem. When the subject (IDM) is presented an information $\eta \in \varpi$ as the stimulus, the choice behavior of the subject is taken as a response to it from among the decision alternatives.

The primal choice model is defined with the attribute function as $x(a) = x(a \mid \eta)$, $x \in \xi$, on the feasible decision alternative $a \in F$, when an obtainable information $\eta \in \varpi$ is given. The information structure ϖ can be enlarged and revised, and subsequently the same for the sets of the feasible actions $F(\varpi)$ and the discriminating outcomes $\xi(\varpi)$. In this Section, however, the variations of the information structure are not treated. The diversification of the information will be introduced in Section 3.

In group decision making, the collective response of the subjects is assumed to be unknown and revealed by probabilistic mechanism. In the probabilistic choice model, the response probability by IDM is defined as the *choice probability* $p_F(a \mid \eta)$, where $p_F(a)$ denotes a probability of an action a to be chosen from the feasible set $F \subseteq A$ and which naturally obeys to the general probability rule such as

$$p_F(a \mid \eta) \geq 0, \quad \sum_{a \in F} p_F(a \mid \eta) = 1 \quad \text{for all } a \in F \subseteq A. \tag{1}$$

From our primary choice model, the choice probability is written equivalently as

$$p_\xi(x(a \mid \eta)) \geq 0, \quad \sum_{x \in \xi} p_\xi(x(a \mid \eta)) = 1 \quad \text{for all } x \in \xi \subseteq X. \tag{2}$$

Hereafter, the arguments, a and η, are omitted in the outcome function, $x(a \mid \eta)$. Then $p_\xi(x)$ is used as the choice probability, which represents a probability for a particular multiattribute value set x to be chosen in ξ.

In the case of binary preference relations,

$$p(x, y) \triangleq p_{\{x,y\}}(x) \tag{3}$$

is used in the place of $p_\xi(x)$ in *Eq.*(2), where $\xi = \{x, y\}$.

In the probabilistic choice model, the randomness occurs as the "social" uncertain events in the choice behavior among the alternative outcomes; It differs from the deterministic choice models.

2.2 Random Utility Models

Let U be a function defined on the outcome set $X(A)$ such that

$$P_r[U(x) \geq U(y);\; x, y \in \xi \subseteq X]$$
$$\triangleq \int_{-\infty}^{\infty} P_r[U(x) = t,\; U(y) \leq t;\; x, y \in \xi \subseteq X]dt. \qquad (4)$$

where the values of the function U is a vector of random variables. P_r is called the *preference probability*. A set of preference probabilities defines the *random utility model* such that

$$p_\xi(x) \;=\; P_r[U(x) \geq U(y);\; x, y \in \xi \subseteq X], \qquad (5)$$

where U is called the random utility function and represents the latent preference values of IDM.

In the binary preference relations,

$$p(x, y) = P_r[U(x) \geq U(y)]. \qquad (6)$$

The value of the preference probability P_r represents a value of the possible distribution of the preference values to be assessed by the anonymous IDM in group decision making.

The preference independence rule for the probabilistic choice behavior is presented in terms of the probability independence rule. Assume the preference evaluation as the random variables to be independent of each other. Then the random utility model is rewritten from *Eq.*(4) by the probability rule as

$$P_r[U(x) \geq U(y);\; x, y \in \xi \subseteq X] \triangleq$$
$$\int_{-\infty}^{\infty} P_r[U(x) = t] \prod_{y \in \xi - \{x\}} P_r[U(y) \leq t]dt. \qquad (7)$$

The irrelevance rule for the probabilistic choice behavior shows the independence of the preference evaluation from irrelevant objects and is represented in terms of the choice probability as

$$p_Y(x) \;=\; p_\xi(x \mid Y), \quad \text{for all } x \in Y \subseteq \xi \subseteq X, \qquad (8)$$

These rules correspond to the equivalent rules in the "algebraic" decision theory (e.g. Luce and Raiffa 1957).

2.3 Stochastic Concepts of "Algebraic" Utility Functions

The alternative utility models embodying the stochastic utility functions are discussed with the binary choice probability in *Eq.(3)*. The strong and strict random utility models are defined (Luce 1959, Block and Marshak 1960, Luce and Suppes 1965).

Definition 1. The *strong binary utility model* is defined with a set of the binary preference probabilities $p(x, y)$ on $\xi \subseteq X$ such that

(i) $p(x, y) = \phi[u(x) - u(y)]$ for all $x, y \in F$ for which $p(x, y) \neq 0$ or 1 (9)

(ii) $\phi(0) = \frac{1}{2}$. (10)

The function u is a real-valued function which defines the *Strong Utility Function* on an interval scale, that is, u is unique up to positive linear transformations. Condition (ii) reflects the indifference preference.

The existence of the strong utility model in *Eqs.*(9) (10) is assured only as the necessary condition of that of the strict utility model (Block and Marshak, Marshak).

Definition 2. The *strict binary utility model* is defined with a set of the binary preference probabilities $p(x, y)$ on $\xi \subseteq X$ such that

$$p(x,y) = \frac{v(x)}{v(x)+v(y)} \quad \text{for all } x, y \in \xi \text{ for which } p(x, y) \neq 0 \text{ or } 1. \qquad (11)$$

The function v is a positive real-valued function which defines the *Strict utility function* on a ratio scale, that is, v is unique up to multiplications by a positive constant.

The strict utility model is defined also in a general form on a finite set $\xi \subseteq X$ such that

$$p_\xi(x) = \frac{v(x)}{\sum_{y \in \xi} v(y)} \quad \text{for } p_\xi(x) \neq 0,\ 1. \qquad (12)$$

The following are the practical implications of the random utility models.

Proposition 1. When the function ϕ is the logistic distribution function $\phi(t) = 1/(1+e^{-t})$, using the definition $u = \log v + b$,

$$\begin{aligned} p(x,y) &= \frac{1}{1+v(y)/v(x)} = \frac{1}{1+\exp\{-[u(x)-u(y)]\}} \\ &= \phi[u(x)-u(y)], \end{aligned} \qquad (13)$$

which is represented with the logistic curves familiar to practitioners.

Note that any binary strict utility model *Eq.*(11) implies a strong utility model Eqs.(9) (10), but not reversely.

In a practical point of view, it will not be unreasonable that the difference of the utility values is taken, for example, as a positive linear function of the most preferred outcome x^* as a variable. Then we can use $\phi(t) = \phi[u(x) - u(y)] = \phi(\alpha x^* + \beta)$ in *Eq.*(13). As an alternative example, we can also use an exponential function $e^{\alpha x^* + \beta}$ or its variations for assessing the difference.

Proposition 2 (Luce and Suppes 1965). Define the random utility function U whose values are independent random variables with probability density functions such that

$$P_r[U(x)=t] = \begin{cases} v(x)e^{v(x)t} & \text{if } t \le 0 \\ 0 & \text{if } t > 0 \end{cases} . \tag{14}$$

Then, for the choice probabilities different from 0 and 1, any strict utility model (*Eq.*(12)) implies an independent random utility model (*Eq.*(5)(7)).

This proposition assures that the random utility model can be constructed with an algebraic form. For the proof, consider that

$$\begin{aligned} P_r[U(x) \ge U(y),\ y \in \xi \subseteq X] &= \int_{-\infty}^{\infty} P_r[U(x)=t] \prod_{y \in Y-\{x\}} P_r[U(y) \le t]dt \\ &= \int_{-\infty}^{0} v(x)e^{v(x)t} \prod_{y \in Y-\{x\}} [\int_{-\infty}^{t} v(y)e^{v(y)\tau}d\tau]dt \\ &= \int_{-\infty}^{0} v(x)e^{v(x)t} \prod_{y \in Y-\{x\}} e^{v(y)t}dt \\ &= \int_{-\infty}^{0} v(x)\exp(\sum_{y \in Y} v(y)t)dt \\ &= \frac{v(x)}{\sum_{y \in Y} v(y)}. \end{aligned} \tag{15}$$

From the practical point of view, when the strict utility function $v(x)$ is taken as $v(x) = e^{u(x)-b}$, the random utility model is evaluated such that

$$P_r[U(x) \ge U(y), x, y \in \xi \subseteq X] = \frac{e^{u(x)-b}}{\sum_{y \in Y} e^{u(y)-b}} . \tag{16}$$

The next proposition for the strong utility model is also useful in practice.

Define the random utility function $U(x) = u(x) + \varepsilon(x)$ with the strong utility function u. Let ϕ be the distribution function of $\varepsilon(x) - \varepsilon(y)$ where $x \neq y$ and $\varepsilon(x)$ and $\varepsilon(y)$ are random variables which are independent of each other and identically distributed. Then by the definition of the random utility model Eqs.(4) (5),

$$\begin{aligned} P_r[U(x) \geq U(y)] &= P_r[u(x) - u(y) \geq \varepsilon(y) - \varepsilon(x)] \\ &= \phi[u(x) - u(y)] \\ &= p(x, y). \end{aligned} \tag{17}$$

Proposition 3. Any strong utility model satisfying *Eq*.(9) (10) implies a random utility model in *Eq*.(7) as a binary choice model.

When the group utility function is assumed to be additively separable with the probabilistic term $\varepsilon(x)$ whose values are independently and identically distributed random variables with the Weibull distributions, then the random utility model Eq.(16) is satisfied (*c.f.* McFadden 1974).

For establishing the rationality of evaluations in the random utility models, the stochastic transitivity conditions for the probabilistic choice model have also been established.

Definition 3. *Quadruple condition* If $p(w,x) \geq p(y,z)$, then $p(w,y) \geq p(x,z)$.

Definition 4. *Stochastic transitivity.* If $p(x,y) \geq \frac{1}{2}$ and $p(y,z) \geq \frac{1}{2}$, then $p(x,z) \geq \frac{1}{2}$.

Definition 5. *Strong stochastic transitivity.* If $p(x,y) \geq \frac{1}{2}$ and $p(y,z) \geq \frac{1}{2}$, $x, y, z \in \xi \subseteq X$, then $p(x,z) \geq \max\,[p(x,y), p(y,z)]$.

Proposition 4. (Block and Marshak, Luce and Suppes). The quadruple condition for the binary random choice is assured in the strict random utility model and, as its consequence, also in the strong random utility model. When the quadruple condition is satisfied, then the strong stochastic transitivity and, as its consequence, the stochastic transitivity are satisfied.

This proposition assures that the strict random utility model and subsequently the strong utility model satisfy, via the quadruple condition, the strong stochastic transitivity and then the stochastic transitivity condition. Then the behavioral propriety of the strict random utility model is assured.

The Propositions 1-4 present the practical prospects for applicability of the random utility models. It should be noted, however, that the heuristic construction of the numerical utility values as the arguments in the probability distributions has not been cared at all in these discussions. We are concerned with the construction of these assessments on proper base.

In the next section, we discuss the method to evaluate the random utility function as the risk function without depending on the particular individual preference evaluations. In this process, we also introduce the incompleteness of information in the risk assessment.

3. Probabilistic Value Tradeoffs for Multiple Risk Assessment

First, we introduce the uncertainty in the "natural" state of the world in the stochastic choice model. Let Ω denote the universe of the state of the world and be its observable subset. Denote by $\theta \in \Omega$ its element. Note that Ω is a set of possible events which forms the statistical populations. The observable subset Θ (ϖ) is defined on an information structure ϖ. An outcome $x_j \in \xi \subseteq X$ is obtained when an event $\theta_j \in \Theta \subseteq \Omega$, $j = 1, ..., n$, occurs under the information structure ϖ. Let $\pi_j \in \Pi$, $j = 1, ..., n$, be a "natural" probability with which a "natural" event θ_j occurs and then a certain outcome x_j is obtained. Denote a gamble, or a lottery, by $\pi \in \Pi$. An n-chance fork gamble i is written by $\pi^i = (\pi_1^i, \pi_2^i, \cdots, \pi_n^i)$, $\pi^i \in \Pi$.

The incompleteness of available information forms alternative information structures. Thus the multiple risk assessments for the alternative gamble prospects should be taken into account in the incomplete information structures.

π^ALet A, B, C, ... denote the alternative gambles which assign the different "natural" probabilities for the uncertain prospects depending on the diversified information structures. Define a gamble $\pi^A = (\pi_1^A, \cdots, \pi_n^A)$, $\pi^A(\varpi^A) \in \Pi(\Psi)$, on the multiattribute value set $x = (x_1, ..., x_m)$ whose elements x_i, $i = 1, ..., n$, describe the multiple objects in group decision making.

The assessment of the gamble prospect $\pi(\varpi) \in \Pi(\Psi)$ is assumed to be a technical problem independent of the subjective evaluations.

Define the expectation for a multiattribute gamble π^A as

$$E\pi^A \triangleq \frac{1}{m} \sum_{i=1}^{m} \sum_{j=1}^{n} \pi_j^A x_{ij} \,, \tag{18}$$

or alternatively, if it is preferable,

$$E\pi^A \triangleq \sqrt[m]{\prod_{i=1}^{m} \left(\sum_{j=1}^{n} \pi_j^A x_{ij}\right)} \quad \text{for } A, B, C, \ldots . \tag{19}$$

The *risk evaluation function* $\Re$ *for a multiattribute gamble* is defined on the expectation $E\pi^A$ of a gamble π^A such that

$$\Re(E\pi^A(\varpi^A)) \triangleq \Re[(\pi^A{}_1, x_1 \mid \theta^A{}_1(\varpi^A)), (\pi^A{}_2, x_2 \mid \theta^A{}_2(\varpi^A)), \ldots, (\pi^A{}_n, x_n \mid \theta^A{}_n(\varpi^A))], \quad (20)$$

where $x_j \triangleq (x_{1j}, \ldots, x_{mj}), j = 1, \ldots, n$, is a multiattribute value set to be obtained as the outcome of the occurrence of an uncertain event $\theta^A{}_j(\varpi^A), j = 1, \ldots, n$.

Note that, although the risk evaluation function *Eq.*(20) for a gamble prospect π^A can be derived as a utility evaluation, say on the negative scale, for the gamble expectation, the function forms are constructed from the results of scientific research as a technical evaluation problem, not as the preference evaluation problem.

The risk evaluation function *Eq.*(20) can be used in the random utility models in group decision making such as in the place of the strong utility functions in *Eq.*(13) or *Eq.*(16) and in the strict utility function in *Eq.*(14).

In the incomplete information structure, however, we should face alternative risk profiles, or alternative gamble prospects. We intend to construct more possible risk profile on an integration of the alternative risk prospects. For this purpose, we introduce the probabilistic value tradeoffs among the alternative risk prospects.

The probabilistic value tradeoff T_R is defined on the tradeoff between expectations of the alternative gambles, $A, B, \ldots$, as

$$T_R = \Delta E\pi^A / \Delta E\pi^B \qquad \text{for } A, B, C, \cdots. \quad (21)$$

Note that the expectation $E\pi^A$ is a certain value defined on the multiple attributes and is the object for the numerical ranking. Thus, although the assessment of the value tradeoffs is still preferential, the numerical evaluation can be performed as a technical problem based on the scientific knowledge for the risk assessment. Thus the evaluation forms a *pseudo* preference problem.

4. Construction of the Multiple Risk Evaluation Functions

A composite risk evaluation function defined with the alternative gambles can be derived, via the probabilistic value tradeoffs, on the component risk evaluation functions (*Eq.*(20)). The probabilistic value tradeoffs are executed for the assignment of the scaling constants in the derivation of the composite risk evaluation function.

The composite risk evaluation function $\mathscr{R}$ for the multiple risk evaluation, which we call the ***multiple risk evaluation function for alternative gambles*** (MRFG), can be constructed in the alternative representation forms.

Multiple risk evaluation functions for alternative multiattribute gambles (MRFG):
(Multiplicative)

$$\mathscr{R}(\pi) = \frac{1}{K}\Big[\prod_{A,B,C,\cdots}(Kk^A\Re(E\pi^A)+1)-1\Big], \text{ when } \sum_{A,B,C,\cdots} k^A \neq 1, \tag{22}$$

(Additive)

$$\mathscr{R}(\pi) = \sum_{A,B,C,\cdots} k^A\Re^A(E\pi^A), \quad \text{when} \quad \sum_{A,B,C,\cdots} k^A = 1, \tag{23}$$

where k is the scaling constant, $0 \le \Re(E\pi^A) \le 1$ and $0 \le \mathscr{R}(\pi) \le 1$.

$K > -1, K \neq 0$, is a unique solution to $1 + K = \prod_{A,B,C\cdots} (1+Kk^A)$.

The assessment of the scaling constant k in these forms can be performed with the same procedures as that for the nonprobabilistic composite utility function, named as the multiattribute utility function (MUF), which is defined directly on the multiple attributes x (*c.f.* Keeney and Raiffa 1976).

In the derivation of MRFG, however, the evaluation of the "natural" probability for the alternative gambles is performed as the technical problem. This work does not introduce the direct evaluation problems of the preferences for the multiple attributes. Only mathematical expectations are used for the evaluation of the uncertain prospects. Although a preference evaluations are still necessary in the derivation of the component risk function, the evaluations has been treated as the scientific problem. In other words, DM's preferential risk attitudes can be assumed to be linear, for example, to a data base for the attributes.

The MRFG can be evaluated for alternative value sets of the multiple attributes x^Q, $Q, R, S, \ldots$, which represents alternative policy programs, and can be compared with each other for selecting the most preferred attribute set, say S, for construction of the revised policy programs. Sensitivity analysis can be performed for selecting the revised policy programs more effectively from the points of view of the group decision making. According to alternative policy aims, alternative value sets of selected attributes can be constructed.

5. Concluding Remarks

In group decision making where individuals as IDM are supposed to be all anonymous, the individual preference structures are usually latent and unknown. We show that the probabilistic choice models based on the random utility models can treat this situation properly, where the identification of the group utility function as well as the individual utility functions is not necessary to be assessed directly. On the other hand, the risk evaluation in uncertain environments needs to assess the "natural" probabilities for the chance of the occurrence of uncertain events. This task should not be preferential and should be executed by a cooperation with the technological specialists on the scientific knowledge basis. In addition, the preference probabilities to be evaluated in the random utility models also should not the subjective probability, but be assessed on the objective basis. In this process, we should also take the incomplete information structure into account, which introduces alternative gamble prospects, or risk profiles, in the risk evaluation.

These problems which we intend to discuss in the present paper have not been treated in the preceding works.

In this paper, the method for the evaluation of MRFG has been presented in the context of the group decision making, which is defined on the alternative gamble prospects and provides the domain of the preference probability to be assessed in the random utility models. With this device, the probabilistic choice model for group decision making will be constructed on more persuasive bases.

It has been shown that for example, in the binary case, the strict utility model (*Eq.*(13)) can be constructed with the logistic distributions. In the general case, another strict utility model (*Eq.*(16) can be constructed. In both cases, the strong utility functions can be used in the place of the strict utility functions. These properties provide practical applicability in the probability assessment for group decision making.

It may be discussed that, when the construction of the composite risk evaluation functions is not intended, the probabilistic choice model can be directly assessed by the general form of the strict expected utility models as found in the original developers (Luce and Suppes 1965, & Others). We are concerned, however, even in this case with the incompleteness of the information structure in assessing the uncertain prospects. Subsequently, the total risk evaluation based on the alternative gamble prospects should be intended. MRFG will meet this need.

Application fields of this approach are various. One example is the plant-site selection problem, such as for the waste disposal treating the toxic substance, say, dioxin which may bring serious adverse effects on the local residents. The numerical comparison of alternative site selection programs can be performed with the construction of MRFG.

The use of computer-aided decision support systems for the construction of MRFG is recommended. The new computer programs MAP and IDASS have already been developed for the multiattribute and expected utility analyses, respectively (Seo, Nishizaki and Park 1999, Seo and Nishizaki 1997). A revised and extended version for treating the probability evaluations and constructing the multiple expected utility functions is now under development, which can be applied to the works discussed in this paper.

The probabilistic choice model for group decision making as discussed in the present paper, however, still includes a critical shortcoming. The group decision making usually includes the conflict of interest among group members, or IDM. The present probabilistic approach is not able to treat this problem. The further development of the logic for the negotiation processes in the conflict resolution should be expected.

Acknowledgment: The author is indebted to the Japan Ministry of Science, Culture and Education for the financial support to this research in 1998-99.

References:

Arrow, K. J. (1950). A difficulty in the concepts of social welfare, *Journal of Political Economy*, **58**. 328-346.

Arrow, K. J. (1951). *Social Choice and Individual Values*, Wiley. 2nd ed. 1963.

Becker, G. M., DeGroot, M. H., and J. Marschak (1963). Stochastic models of choice behavior, *Behavioral Science*, **8**, 41-55.

Block, H. D.. and J. Marschak (1960), Random orderings and stochastic theories of responses, in I. Olkin, S. G. Ghurye, W. Hoeffding, W. G. Madow and H. B. Mann (eds.) *Contributions to Probability and Statistics*, Stanford University Press, Stanford. 97-132.

Keeney, R. L. and H. Raiffa (1976). *Decisions with Multiple Objectives; Preferences and Value Tradeoffs*, Wiley, New York.

Luce, R. D. (1958). A probability theory of utility, *Econometrica*, **26**, 193-224.

Luce, R. D. (1959). *Individual Choice behavior*, Wiley, New York.

Luce, R. D. And H. Raiffa. *Games and Decisions, Wiley 1957,*

Luce, R. D. and P. Suppes (1965). Preferences, Utility, and Subjective Probability, in R. D. Luce, R. R. Bush, and E. Galanter (eds.) *Handbook of Mathematical Psychology*, **III**, Chap. 19. Wiley, New York. 249-410.

Luce, R. D. and J. W. Tukey (1964). Simultaneous Conjoint measurement: a new type of fundamental measurement, *Journal of Mathematical Psychology*, **I**, 1-27.

Marschak, J. (1960). Binary-Choice Constraints and Random Utility Indicators, in K. J. Arrow, S. Karlin and P. Suppes (eds.) *Mathematical methods in Social Sciences*, Stanford University Press, Stanford. Chap. 21. 312-329.

Marschak, J. And R. Radner (1972). *Economic Theory of Teams*, Yale University Press, New Haven.

McFadden, D. (1974). Conditional logit analysis of qualitative choice behavior, in P. Zarembka (ed.) *Frontiers in Econometrics*, Academic Press, New York.. Chap. 4. 105-142.

Savage, L. J. (1954). *The Foundations of Statistics*, Wiley, New York.

Seo, F. and I. Nishizaki (1997). On development of interactive decision analysis support systems (IDASS) in the IDSS environments. Presented paper at the IIASA Workshop, to appear in *the proceeding*, IIASA, Laxenburg Austria.

Seo, F., Nishizaki, I. and S. Park. (1999). Object-oriented programming for multiattribute utility analysis MAP and its application, *Annals of Department of Business Management and Informatics*, Setsunan University. Osaka.

On a Behavioral Model of Analytic Hierarchy Process for Modeling the Legitimacy of Rank Reversal

Hiroyuki Tamura, Satoru Takahashi*, Itsuo Hatono**
and Motohide Umano#

Department of Systems and Human Science
Graduate School of Engineering Science
Osaka University
1-3 Machikaneyama, Toyonaka, Osaka 560-8531, JAPAN
FAX:+81-6-850-6341; tamura@sys.es.osaka-u.ac.jp

Abstract: This paper deals with a behavioral extension of a conventional Analytic Hierarchy Process (AHP), called behavioral AHP, such that rank reversal phenomena are legitimately observed and explanatory. In AHP, the main causes of rank reversal are either inconsistency in pairwise comparison or change in hierarchical structure, or both. Without these causes AHP should not lead to rank reversal. But if we use inappropriate normalization procedure such that entries sum to 1, the method will lead to rank reversal even when the rank should be preserved. Behavioral AHP proposed in this paper will preserve rank when it should be preserved, and will lead to rank reversal when the rank reversal is legitimate. Some numerical examples obtained by the behavioral AHP are included which could explain the legitimacy of the rank reversal.

Keywords: Multiple criteria methods; weighting methods; Analytic Hierarchy Process (AHP); behavioral AHP; legitimacy of rank reversal

---

* Presently with Mitsubishi Electric Corporation, Amagasaki, Hyogo, Japan

** Presently with Kobe University, Nada-ku, Kobe, Japan

Presently with Osaka Prefecture University, Sakai, Osaka, Japan

1. Introduction

Analytic Hierarchy Process (AHP) (Saaty, 1980) has been widely used as a powerful tool for deriving priorities or weights which reflect the relative importance of alternatives for multiple criteria decision making problems, because the method of ranking items by means of pairwise comparison is easy to understand and easy to use compared with other methods of multiple criteria decision making. For example, although multiattribute utility theory (Keeney and Raiffa, 1993) is deeper and stronger in theoretical basis compared with AHP, it is laborious and hard to identify a proper multiattribute utility function of a decision maker even when utility independence property is postulated to hold. That is, AHP is appropriate as a normative approach which prescribes optimal behavior how decision should be made. However, there exist difficult phenomena to model and to explain by using conventional AHP. Rank reversal is one of these phenomena. That is, conventional AHP is an inappropriate model as a behavioral model which is concerned with understanding how people actually behave when making decisions.

In AHP rank reversal has been regarded as inconsistency in the methodology. When a new alternative is added to an existing set of alternatives, several attempts have been made to preserve the rank (Belton and Gear, 1983; Barzilai, et al., 1987; Dyer, 1990). However, the rank reversal could occur in real world as seen in the well-known example of a person ordering his/her meal in a restaurant shown by Luce and Raiffa (1957) and in other works (Corbin and Marley, 1974; Cox and Grether, 1996).

In this paper we propose a behavioral extension of a conventional AHP, called *behavioral AHP*, such that the rank reversal phenomena are legitimately observed and explanatory. In general, it is pointed out that the main causes of rank reversal are violation of transitivity and/or change in decision making structure (Tversky, et al., 1990). In AHP these causes correspond to inconsistency in parwise comparison and change in hierarchical structure, respectively. Without these causes, AHP should not lead to rank reversal. But if we use inappropriate normalization procedure such that the entries sum to 1, the method will lead to rank reversal even when the rank should be preserved (Dyer, 1990; Salo and Hamalainen, 1992). Some numerical examples which show the inconsistency in the conventional AHP and which show the legitimacy of the rank reversal in our behavioral AHP, are included.

2. Two Characteristics to Construct Behavioral AHP

We propose to use two characteristics; *preference characteristics* and *status char-*

acteristics, in behavioral AHP. The preference characteristics represent the degree of satisfaction of each alternative with respect to each criterion. The status characteristics represent the evaluated value of a set of alternatives. The evaluation of each alternative for multiple criteria decision making is performed by integrating these two characteristics.

2.1 Preference Characteristics

In a conventional AHP it has been regarded that the cause of rank reversal lies in inappropriate normalization procedure such that entries sum to 1 (Belton and Gear, 1983). In this paper we propose to add a hypothetical alternative such that it gives aspiration level of the decision maker (DM) for each criterion, and the (ratio) scale is determined by normalizing the eigenvectors so that the entry for this hypothetical alternative is equal to 1. Then, the weighting coefficient for satisfied alternative will become more than or equal to 1, and the weighting coefficient for dissatisfied alternative will become less than 1. That is, weighting coefficient of each alternative under a concerning criterion represents the DM's degree of satisfaction. Unless the aspiration level of DM changes, the weighting coefficient for each alternative does not change even if a new alternative is added or an existing alternative is removed from a set of alternatives.

2.2 Status Characteristics

The status characteristics represent the evaluated value of a set of alternatives under a criterion. If the average importance of all alternatives in the set is far from aspiration level 1 under a criterion, the weighting coefficient for this criterion is increased. Furthermore, the criterion which gives larger consistency index can be regarded that the DM's preference is fuzzy under this criterion. Thus, the importance of such criterion is decreased.

Let A be an $n \times n$ pairwise comparison matrix with respect to a criterion. Let $A = \left(a_{ij}\right)$, then

$$1/\rho \leq a_{ij} \leq \rho \tag{1}$$

Usually, $\rho = 9$. Since $a_{ij} = w_i / w_j$ for priorities w_i and w_j

$$1/\rho \leq w_i / w_j \leq \rho \tag{2}$$

Equation (2) satisfies when item j is at the aspiration level. In this case $w_j = 1$, then

$$1/\rho \leq w_i \leq \rho \tag{3}$$

Then, we obtain

$$-1 \leq \log_\rho w_i \leq 1. \tag{4}$$

Taking geometric mean of w_i' s, we still obtain

$$1/\rho \leq \left(\prod_{i=1}^{n} w_i\right)^{1/n} \leq \rho, \tag{5}$$

$$-1 \leq \log_\rho \left(\prod_{i=1}^{n} w_i\right)^{1/n} \leq 1. \tag{6}$$

Let

$$C := \left|\log_\rho \left(\prod_{i=1}^{n} w_i\right)^{1/n}\right| \tag{7}$$

then we obtain

$$0 \leq C \leq 1. \tag{8}$$

We call C the status characteristics which denotes the average importance of n alternatives. If $C = 0$, the average importance of n alternatives is at the aspiration level. For larger C the importance of the concerning criterion is increased.

3. Integration of Two Characteristics

Let w_i^B be basic weight obtained from preference characteristics, $C.I.$ be consistency index, and $f(C.I.)$ be a function of $C.I.$, which is called reliability function. We evaluate the revised weight w_i by integrating preference characteristics w_i^B and status characteristics C as

$$w_i = w_i^B \times C^{f(C.I.)} \tag{9}$$

$$0 \leq C \leq 1$$

$$0 \leq f(C.I.) \leq 1$$

where

$$f(C.I.) = 0, \quad \text{for} \quad C.I. = 0$$

If $\sum_{i=1}^{n} w_i \neq 1$, then w_i is normalized to sum to 1. The same procedure is repeated when there exist many levels in the hierarchical structure.

If the priority of an alternative is equal to 1 under every criterion, the alternative is at the aspiration level. In this case the overall priority of this alternative is obtained as 1. Therefore, the overall priority of each alternative denotes the satisfaction level of each alternative. If this value is more than or equal to 1, the corresponding alternative is satisfactory, and conversely, if it is less than 1, the corresponding alternative is unsatisfactory. Behavioral AHP gives not only the ranking of each alternative, but it gives the level of satisfaction.

4. Algorithm of Behavioral AHP

Step 1. Multiple criteria and multiple alternatives are arranged in a hierarchical structure.

Step 2. Compare the criteria pairwise which is arranged in the one level higher level of alternatives. Eigenvector corresponding to the maximum eigenvalue of the pairwise comparison matrix is normalized to sum to 1. The priority obtained is set to be preference characteristics which represent basic priority.

Step 3. For each criterion, aspiration level is asked to DM. A hypothetical alternative which gives aspiration level for all the criteria is added to a set of alternatives. Including this hypothetical alternative pairwise comparison matrix for each criterion is evaluated. Eigenvector corresponding to the maximum eigenvalue is normalized so that the entry for this hypothetical alternative is equal to 1.

Step 4. If $C.I. = 0$ for each comparison matrix, preference characteristics, that is, basic priority is used as the weighting coefficient for each criterion. If $C.I. \neq 0$ for some criteria the priority for these criteria is revised by using eqn.(9) taking into account the status characteristics.

Step 5. If some priorities are revised taking into account the status characteristics, the priority for each criterion is normalized to sum to 1.

Step 6. Overall weight is evaluated. If there exists upper level in the hierarchy, go to Step 7. Otherwise, stop.

Step 7. Evaluate pairwise comparison matrix of criteria with respect to each criterion in the higher level. If some pairwise comparison matrices are not consistent, evaluate status characteristics and revise the priority. Go to Step 6.

5. Numerical Examples

Case 1. Rank reversal should not be observed

When pairwise comparisons are consistent and there exists no change in hierarchical structure of decision making, rank reversal should not be observed. This example deals with such a case, where by using conventional AHP, rank reversal will be observed.

Suppose there exist 4 criteria (C_1, C_2, C_3, C_4) and 4 alternatives (A_1, A_2, A_3, A_4) where each criterion is equally weighted. **Table 1** shows the result of direct rating of each alternative under each criterion.

Table 1. Direct rating of alternatives

	C_1	C_2	C_3	C_4
A_1	1	9	1	3
A_2	9	1	9	1
A_3	8	1	4	5
A_4	4	1	6	6

From this table we can describe a consistent pairwise comparison matrix, and thus overall weighting and ranking of 4 alternatives can be obtained by using AHP. **Table 2** shows overall weighting and ranking obtained by Saaty's AHP and our behavioral AHP where in the behavioral AHP it is assumed that alternative A_4 is at the aspiration level for all the 4 criteria.

Suppose we eliminate alternative A_4 from a set of alternatives. **Table 3** shows overall weighting and ranking obtained by Saaty's AHP and our behavioral AHP.

Table 2. Overall weighting and ranking before eliminating an alternative

	Saaty' s AHP		Behavioral AHP	
	Weight	Rank	Weight	Rank
A_1	0.261	1	2.479	1
A_2	0.252	2	1.229	2
A_3	0.245	3	1.125	3
A_4	0.241	4	1.000	4

Table 3. Overall weighting and ranking after eliminating an alternative

	Saaty' s AHP		Behavioral AHP	
	Weight	Rank	Weight	Rank
A_1	0.320	3	2.479	1
A_2	0.336	2	1.229	2
A_3	0.344	1	1.125	3

In Saaty's conventional AHP rank reversal is observed even though the rank reversal should not be observed.

Case 2. Rank reversal could be observed (1)

When a pairwise comparison matrix is inconsistent under a criterion, consistency index will become large. Under this criterion DM's preference is fuzzy. In this case it is considered that DM's decision making process will be proceeded taking into account the preference characteristics and the status characteristics. This example deals with such case.

Suppose 3 alternatives (a,b,c) are evaluated under 2 criteria (X,Y). **Table 4** shows pairwise comparison of these two criteria.

Table 4. Pairwise comparison of multiple criteria

	X	Y	Weight
X	1	1/2	0.333
Y	2	1	0.667

As this result basic weight for criteria (X,Y) is obtained as

$$w_X^B = 0.333, \qquad w_X^B = 0.667$$

Suppose the aspiration levels for the criteria (X,Y) are (s_X, s_Y). Suppose we obtained pairwise comparison matrices as shown in **Table 5** in which the aspiration level is included for each criterion.

Table 5. Pairwise comparison matrix of each alternative

X	a	b	c	s_X	Weight
a	1	1/6	1/2	1/2	0.50
b	6	1	3	3	3.00
c	2	1/3	1	1	1.00
s_X	2	1/3	1	1	1.00
$C.I. = 0$					

Y	a	b	c	s_Y	Weight
a	1	7	2	3	2.97
b	1/7	1	1/3	1/3	0.57
c	1/2	3	1	2	1.56
s_Y	1/3	3	1/2	1	1.00
$C.I. = 0.014$					

Pairwise comparison matrix under criterion X is completely consistent, therefore the status characteristics does not affect the preference characteristics at all. But pairwise comparison matrix under criterion Y is somewhat inconsistent, therefore we need to take into account the status characteristics. If we compute status characteristics C based on eqn.(7), we obtain 0.147. Suppose a reliability function is written simply as $f(C.I.) = 10 \times C.I.$, we obtain revised weight for criteria X and Y

as

$$w_X = 0.333$$

$$w_Y = 0.667 \times 0.147^{10\times 0.014} = 0.510,$$

respectively. Since pairwise comparison matrix under criterion Y is somewhat inconsistent, weight for criterion Y is decreased. Then, weights for criteria are normalized to sum to 1 as

$$w_X = 0.395, \qquad w_Y = 0.605.$$

As the result the weight for consistent criterion is increased and the weight for inconsistent criterion is decreased.

Table 6 shows overall weighting and ranking for 4 alternatives including a hypothetical alternative s such that it gives aspiration level for each criterion.

Table 6. Overall weighting and ranking

	X	Y	Weight	Rank
	0.395	0.605		
a	0.500	2.976	2.00	1
b	3.000	0.423	1.44	2
c	1.000	1.560	1.34	3
s	1.000	1.000	1.00	4

Suppose DM was told that pairwise comparison matrix under criterion Y is somewhat inconsistent. If he revised pairwise comparison matrix to decrease the consistency index as shown in **Table 7**, weight for each criterion would be revised as

$$w_X = 0.333$$

$$w_Y = 0.667 \times 0.163^{10\times 0.005} = 0.609$$

and the normalized weights to sum to 1 are obtained as

$$w_X = 0.354, \qquad w_Y = 0.646.$$

Table 7. Revised result of pairwise comparison matrix under criterion Y

Y	a	b	c	s_Y	Weight
a	1	8	2	3	3.11
b	1/8	1	1/4	1/3	0.38
c	1/2	4	1	2	1.68
s_Y	1/3	3	1/2	1	1.00

$C.I. = 0.005$

The overall weighting and ranking are obtained as shown in **Table 8**.

Table 8. Overall weighting and ranking after revising the pairwise comparison matrix

	X 0.354	Y 0.646	Weight	Rank
a	0.500	3.111	2.19	1
b	3.000	0.377	1.27	3
c	1.000	1.679	1.45	2
s	1.000	1.000	1.00	4

In this example we can find that rank reversal could be observed when there exist inconsistency and ambiguity of pairwise comparison.

Case 3. Rank reversal could be observed (2)

In general criteria included in hierarchical structure of AHP are such that alternatives to be evaluated are discriminated and ranked under each criterion. A common criterion such that it takes the same value for all the alternatives cannot be included in the hierarchical structure. Existence of such criterion can be found in the example of a man who orders his meal in a restaurant as shown in Luce and Raiffa (1957).

Suppose a man ordered beef steak when he found beef steak and salmon steak in the menu. But when he found that escargot is added in the menu, he changed his mind and ordered salmon steak. This is a typical rank reversal phemenon when an alternative is added in the existing set of alternatives. How could we explain his preference?

The reason why rank reversal is observed in his preference is as follows: By recognizing that the restaurant could serve escargot, he found that the quality of the restaurant is very high. As the result what he wants in this restaurant has changed. By adding an alternative "escargot" in the menu a new attribute "quality of restaurant" is added in the criteria. If we would add this new attribute in the criteria of conventional AHP, could we model a proper decision making process of rank reversal? Answer to this question is no, since the attribute "quality of restaurant" is common to all the alternatives "beef", "salmon" and "escargot". That is, it is not possible to do pairwise comparison of these three alternatives under the criterion "quality of the restaurant".

By recognizing the change in quality of restaurant, his preference has changed that he found better taste for all the alternatives, he could increase his budget, it must be tasty even for an inexpensive alternative, and so forth. Change of aspiration level could model these phenomena.

Suppose he evaluates two alternatives "beef steak" and "salmon steak" under two criteria "taste" and "price" at the beginning. He describes his aspiration level for each criterion. Then, he will start to do pairwise comparison. Suppose **Table 9** is obtained for pairwise comparison of two criteria and **Table 10** is obtained for pairwise comparison of two alternatives under each criterion where it is assumed that the aspiration level for taste is same as salmon steak and prices for beef steak, salmon steak and aspiration level are 2000yen, 1000yen and 800yen, respectively.

Table 9. Pairwise comparison of criteria

	Taste	Price	Weight
Taste	1	1/2	0.333
Price	2	1	0.667

Table 10. Pairwise comparison of alternatives under each criterion

(a) For criterion "Taste"

Taste	Beef	Salmon	A.L.	Weight
Beef	1	2	2	2.0
Salmon	1/2	1	1	1.0
A.L.	1/2	1	1	1.0

(b) For criterion "Price"

Price	Beef	Salmon	A.L.	Weight
Beef	1	0.5	0.4	0.4
Salmon	2	1	0.8	0.8
A.L.	2.5	1.25	1	1.0

A.L.: Aspiration Level

Taking into account that each pairwise comparison matrix is completely consistent, we will obtain overall weighting and ranking for beef steak and salmon steak as shown in **Table11**.

Suppose he found a new alternative escargot with 4800yen and found that this restaurant is a high quality restaurant. Under this circumstance suppose his aspiration level is changed that the aspiration level for taste is same as beef steak and the aspiration level for price is raised up to 1200yen. **Table 12** shows the resulting weighting and ranking for beef steak, salmon steak and escargot. Rank reversal is observed comparing Tables 11 and 12. This rank reversal phenomenon can be interpreted that, by the change of aspiration level, the degree of predominance of beef steak in taste is decreased and the degree of predominance of salmon steak in price is increased. As the result overall weight of salmon steak is increased and that of beef steak is decreased.

Table 11. Overall weighting and ranking before adding an alternative

	Taste	Price	Weight	Rank
	0.333	0.667		
Beef	2.000	0.400	0.933	1
Salmon	1.000	0.800	0.867	2

Table 12. Overall weighting and ranking after adding an alternative

	Taste	Price	Weight	Rank
	0.333	0.667		
Beef	1.000	0.600	0.733	3
Salmon	0.500	1.200	0.967	1
Escargot	2.000	0.250	0.833	2

6. Concluding Remarks

It is shown that a conventional AHP is not an adequate behavioral model of multiple criteria decision making, although it has been used as a powerful tool of a normative model. In this paper a behavioral analytic hierarchy process, called behavioral AHP is developed. Behavioral AHP could properly model the legitimacy of rank reversal when an alternative is added or removed from an existing set of alternatives. Key ideas are to include a hypothetical alternative which gives aspiration level to each criterion when pairwise comparison matrix is composed and to take into account the status characteristics to modify the preference characteristics.

Without increasing the load for a decision maker so much we could model various decision making processes, and the flexibility of modeling is increased enormously.

For further research we will try to model a behavioral ANP (Analytic Network Process) to avoid the irrational rank reversal phenomena that may arise in usual ANP (Saaty, 1996).

References

Barzilai, J., W.D. Cook and B. Golany (1987): Consistent weights for judgements matrices of relative importance of alternatives, *Operations Research Letters*, Vol. 6, No. 3, pp. 131-134.

Belton, V. and T. Gear (1983): On a shortcoming of Saaty's method of analytic hierarchies, *OMEGA The International Journal of Management Sciences*, Vol. 11, No. 3, pp. 228-230.

Cobin, R. and A.A.J. Marley (1974): Random utility models with equity: An Apparent but not actual generalization of random utility models, *Journal of Mathematical Psychology*, Vol.11, pp. 274-293.

Cox, J. and D.M. Grether (1996): The preference reversal phenomenon: Response mode, markets and incentives, *Economic Theory*, Vol. 7, pp. 381-405.

Dyer, J.S. (1990): Remarks on the analytic hierarchy process, *Management Sciences*, Vol. 36, No. 3, pp. 249-258.

Keeney, R.L. and H. Raiffa (1993): *Decisions with Multiple Objectives*, Cambridge University Press, Cambridge. (First published in 1976 by Wiley)

Luce, R.D. and H. Raiffa (1957): *Games and Decisions*, Wiley, New York.

Saaty, T.L. (1980): *The Analytic Hierarchy Process*, McGraw-Hill, New York.

Saaty, T.L. (1996): *Decision Making with Dependence and Feedback: The Analytic Network Process*, RWS Publications, Pittsburgh.

Salo, A.A. and R.P. Hamalainen (1992): Preference assessment by imprecise ratio statements, *Operations Research*, Vol. 40, No. 6, pp. 1053-1061.

Tversky, A., P. Slovic and D. Kahneman (1990): The cause of preference reversal, *The American Economic Review*, Vol. 80, No. 1, pp. 204-217.

Resolution of Variational Inequality Problems by Multiobjective Programming

Hsiao-Fan Wang

Department of Industrial Engineering, National Tsing-Hua University
Hsinchu, Taiwan, Republic of China, 30043
Hfwang@ie.nthu.edu.tw

Abstract. In this study, we consider the equivalence between a Generalized Complementarity Problem (GCP) and a Variational Inequality (VI) Problem over a convex cone. A new approach based on multiple objective programming for solving VI problems in a convex cone is proposed of which an existence theorem is presented. We have shown that the proposed method solves a VI in a linear case by $O(n^4)$ arithmetic operations and solves an Linear Complementarity Problem (LCP), a special case of GCP, by the order of $O(n^3)$. An example is provided for illustration.

Keywords: Variational Inequality Problems, Generalized Complementarity Problems, Linear Complementarity Problems, Nonlinear Complementarity Problems, Multiple Objective Programming.

1. Introduction

Over the past decade, the issues regarding the finite-dimensional Variational Inequality and Complementarity Problems have drawn much attention as concerns both theory and applications [Kinderdehrer and Stampocchia, 1980]. In the simplest format a Variational Inequality, denoted by VI(X, f) or simply VI, consists in finding a vector $x^* \in X$, such that

$$< f(x^*), x - x^* > \geq 0, \forall x \in X \tag{1}$$

where <• , •> denotes the inner product, X is a nonempty subset of $\mathbf{R}^n$ and f is a mapping from into itself $\mathbf{R}^n$. Normally, the set X is assumed to be closed and convex, and in practice, X is often a polyhedron. Therefore, let us divide the convex set X into three subsets of interior points I, extreme points E, and the boundary B, with $X = I \cup E \cup B$. Now, we can observe that a solution x* of VI(X, f) should satisfy (a) f(x*)=0, when $x^* \in I$; (b)that f(x*) is inward normal to X at x*, when $x^* \in B$; (c) that whenever C(x*) is generated by the vectors which are inward normal to X at x*, then it is a convex cone with respect to the point x* but not the origin, and $f(x^*) \in C(x^*)$, when $x^* \in E$. Denote $X^*_I = \{x | x \in I, I \subset X\}$, $X^*_B = \{x | x \in B, B \subset X\}$, $X^*_E = \{x | x \in E, \quad E \subset X\}$, then, the solution set is $X^* = X^*_I \cup \ X^*_B \cup X^*_E$. Since the three cases may happen for the same problem, we have to test all of them. However it is not easy to find those solutions from the above observation.

A convex cone is a convex set with an additional property that $\lambda x \in X$ for each $x \in X$ and for each $\lambda > 0$. Hence if X is a convex cone, it is a convex set that consists of all rays emanating from the origin. It has been shown that, when X is a convex cone, the VI is equivalent to the Generalized Complementarity Problem defined below. Therefore, in this study we shall propose an alternative way to solve Generalized Complementarity Problems under convexity assumption. Then, the problem of VI will consequently be solved.

A Generalized Complementarity Problem [Karamardian, 1971], denoted by GCP(X,F) or simply GCP, consists in finding x*, such that

$$x^* \in X, \ f(x^*) \in X^d, \ \langle f(x^*), x^* \rangle = 0, \tag{2}$$

where $X^d = \{ y \in \mathbf{R}^n : \langle y, x \rangle \geq 0, \ \forall \, x \in X\}$.

In Kostreva and Wiecek [1993] is presented a Multiple Objective Program (MOP) to solve a Linear Complementarity Problem, denoted by LCP(q,M) or simply LCP, which consists in finding $x^* \in \mathbf{R}^n$, such that

$$f(x^*) \geq 0,\ x^* \geq 0,\ \langle f(x^*), x^* \rangle = 0 \tag{3}$$

where $f(x^*)=Mx^*+q$, M is an n×n matrix and q is an n×1 column vector. Because using MOP can characterize an LCP, and LCP is a special case of GCP, therefore we shall extend the concept to solve GCP.

In Section 2, we shall formulate a GCP in terms of an MOP. In Section 3 an existence theorem for the solutions to a linear GCP and a linear VI is presented. Besides, an LCP is discussed. In Section 4, we propose a new algorithm to solve a linear VI with an example. Evaluation of computation complexity is provided. Finally, summary and conclusions are drawn in Section 5.

2. Problem Formulations

Consider a Multiple Objective Problem as follows:

$$\begin{aligned} &\text{minimize } [y_1x_1,\ y_2x_2,\ \ldots,\ y_nx_n] \\ &\text{Subject to } x \in X,\ y = f(x),\ y \in X^d \end{aligned} \tag{4}$$

where X is a convex cone and $X^d := \{ y \in \mathbf{R}^n : \langle y, x \rangle \geq 0\ \forall\ x \in X\}$.

Model (4) is a kind of Vector Optimization problem in which if X_{eff} denotes the set of all efficient solutions of (4), X_{eff} consistes of all pareto optimal solutions with the following property:

Theorem 2.1. Let X be a convex cone. The point $x^* \in X_{eff}$ is a solution of the MOP such that $y^*_1x^*_1 + y^*_2x^*_2 + \ldots + y^*_nx^*_n = 0$ if and only if x^* is a solution of GCP(X,F).

This theorem shows the equivalence in solutions between MOP in (4) and GCP in (2). Thus, once the MOP is solved, GCP can be solved too.

Furthermore, let a convex cone $X=\{ x : Dx\geq 0, x\in \mathbf{R}^n, D=[d_i], D$ is a m×n matrix and d_i's are row vectors of D, $i=1,2,...,m\}$, then $y^*\in X^d=\{ y\in \mathbf{R}^n : \langle y, x \rangle \geq 0$ for all $x\in X \}$ if and only if there exists a nonnegative vectors $v\in \mathbf{R}^m$ such that $y^*=v_1d_1+v_2d_2+...+v_md_m$, i.e., $y^*=D^T v$.

Since X^d is a convex cone generated by the row vectors of the matrix D, therefore, with $y = f(x)$, we can apply the concept of weighted-sum method for the MOP [Steuer, 1986] defined in (4) into

$$\begin{aligned} &\text{Minimize } v^T Dx \\ &\text{Subject to } Dx\geq 0 \\ &\qquad v\geq 0 \\ &\qquad f(x)= D^T v \end{aligned} \tag{5}$$

Corollary 2.1. $x^*\in X_{eff}$ is a solution of the MOP such that $y^*_1x^*_1+ y^*_2x^*_2+...+y^*_nx^*_n=0$ if and only if x^* is a solution of model (5) such that $v^TDx^* = 0$.

Corollary 2.2. x^* is a solution of model (5) such that $v^T Dx^*=0$ if and only if x^* solves GCP in (2) with $X=\{ x : Dx\geq 0\}$.

3. Solutions for Linear GCP, Linear VI and LCP

From the previous section, we know the relation between GCP and MOP. In this section, we apply the reformulated model (5) to solve linear GCP, linear VI, and also LCP. First, we shall solve MOP in (4) under the linear case.

Theorem 3.1. x* is a solution of model (5) with $f(x)=Mx+q$, $M\in\mathbf{R}^{n\times n}$, $q\in\mathbf{R}^{n}$, $X=\{x : Dx\geq 0,\ D\in\mathbf{R}^{m\times n}\}$, such that $v^T Dx^* = 0$. Then,

(1) x*=0, if and only if $q = D^T v$ and $v\geq 0$; or,

(2) x*≠0, f(x*)= 0 if and only if $Dx^*\geq 0$, $Dx^*\neq 0$; or,

(3) x*≠0, f(x*)≠0 if and only if

(a). there exists a $k\in\{1,2,...,m\}$ such that $d_k x^*=0$, $d_i x^*>0$ for all $i=1,2,...,m$, $i\neq k$, where d_i's are row vectors of D, and there exists a $l>0$, $l\in\mathbf{R}$, such that $Mx^*+q = l\cdot d_k$; or

(b). if $m>2, n>2$, there exist k_1 and k_2, $k_1\neq k_2$ and k_1, $k_2\in\{1,2,...,m\}$ such that $d_{k_1}x^*=0$, $d_{k_2}x^*=0, d_i x^*>0$ for all $i=1,2,...,m$, $i\neq k_1,k_2$, and there exist some $l_1>0$ and $l_2>0$, $l_1,l_2\in\mathbf{R}$, such that $f(x^*)=l_1 d_{k_1}+l_2 d_{k_2}$.

Remarks.

(1). The above theorem states that the solutions may occur individually or simultaneously on thre situations of x*=0, f(x*)=0, and f(x*) normal to x*:

(a). If q belongs to X^d, then x*=0 is a solution and $x^*\in X^*_E$.

(b). If the intersection of f(x*)=0, x*≠0 and the convex cone X is not empty, then this intersection is a solution set with $x^*\in X^*_I\cup X^*_B$.

(c). On the boundary of X, if we can find an f(x*)≠0 that is inward normal to X at x*, then x* is a solution and $x^*\in X^*_B$.

(2). (a) Since X is a convex cone represented by $\{x : Dx\geq 0, D\in\mathbf{R}^{m\times n}\}$, therefore, in Theorem 3.1(1) if m=n, then D is of full row rank and its inverse must exist.

(b). Since $M\in\mathbf{R}^{n\times n}$, if M^{-1} does not exist, then from $f(x)=Mx+q=0$ at Theorem 3.1(2), we know that either $f(x)=0$ has no solution or we have an infinite number of solutions to the equations $f(x)=0$.

In other words, if M^{-1} exists, then $f((x)=0$ has an unique solution.

Corollary 3.1. Let $X=\{x : Dx\geq 0\}$ be a convex cone and $f(x)=Mx+q$. If $x^*\in X$ solves the problem of GCP(X,F), then

(1) $x^*=0$ and $x^*\in X^*_E$., if and only if $q=D^T v$, $v\geq 0$; or,

(2) $x^*\neq 0$, $f(x^*)=0$ and $x^*\in X^*_I\cup X^*_B$ if and only if $Dx^*\geq 0$, $Dx^*\neq 0$; or,

(3) $x^*\neq 0, f(x^*)\neq 0$ and $x^*\in X^*_B$ if and only if

(a) there exists a $k\in\{1,2,...,m\}$ such that $d_k x^*=0$, $d_i x^*>0$ for all $i=1,2,...,m$, $i\neq k$, where d_i's are row vectors of D, and there exists a $l>0$, $l\in\mathbf{R}$, such that $Mx^*+q=l\cdot d_k$; or

(b) if $m>2, n>2$, there exist k_1 and k_2, $k_1\neq k_2$ and k_1, $k_2\in\{1,2,...,m\}$ such that $d_{k_1}x^*=0$, $d_{k_2}x^*=0, d_i x^*>0$ for all $i=1,2,...,m$, $i\neq k_1,k_2$, and there exist some $l_1>0$ and $l_2>0$, $l_1,l_2\in\mathbf{R}$, such that $f(x^*)=l_1 d_{k_1}+l_2 d_{k_2}$.

Corollary 3.2. Let $X=\{x : Dx\geq 0\}$ be a convex cone and $f(x)=Mx+q$. If $x^*\in X$ solves the problem VI(X,F) (1), then

(1) $x^*=0$ and $x^*\in X^*_E$., if and only if $q=D^T v$, $v\geq 0$; or,

(2) $x^*\neq 0$, $f(x^*)=0$ and $x^*\in X^*_I\cup X^*_B$ if and only if $Dx^*\geq 0$, $Dx^*\neq 0$; or,

(3) $x^*\neq 0, f(x^*)\neq 0$ and $x^*\in X^*_B$ if and only if

(a) there exists a $k\in\{1,2,...,m\}$ such that $d_k x^*=0$, $d_i x^*>0$ for all $i=1,2,...,m$, $i\neq k$, where d_i's are row vectors of D, and there exists a $l>0$, $l\in\mathbf{R}$, such that $Mx^*+q=l\cdot d_k$; or

(b). if $m>2, n>2$, there exist k_1 and k_2, $k_1\neq k_2$ and $k_1, k_2\in\{1,2,...,m\}$ such that $d_{k_1}x^*=0$, $d_{k_2}x^*=0, d_i x^*>0$ for all $i=1,2,...,m$, $i\neq k_1,k_2$, and there exist some $l_1>0$ and $l_2>0$, $l_1,l_2\in\mathbf{R}$, such that $f(x^*)=l_1 d_{k_1}+l_2 d_{k_2}$.

Corollary 3.3. If x* solves **LCP-MOP** of the problem (5), then
(1) x*=0, if and only if $q^T \geq 0$; or,
(2) x*≠0, *f*(x*)= 0 if and only if x*>0; or,
(3) x*≠0, *f*(x*)≠0 and $x^* \in X^*_B$ if and only if
(a) there exists a $k \in \{1,2,...,m\}$ such that $x^*_k=0$, $x^*_i>0$, $m_k x^* + q_k > 0, m_i x^* + q_k = 0$ for all $i = 1,2,...,m$, $i \neq k$, where $m_i^{'}$s are row vectors of *D*, and there exists a $l>0$, $l \in \mathbf{R}$, or
(b) if $m > 2, n > 2$, there exist k_1 and k_2, $k_1 \neq k_2$ and $k_1, k_2 \in \{1,2,...,m\}$ such that $x^*_{k1}=0, x^*_{k2}=0, x^*_i>0$, for all $i=1,2,...,m$, $i \neq k_1, k_2$, and there exist some, $l_1, l_2 \in \mathbf{R}$.

4. Algorithm and Examples

In this section, based on the theorems developed in the previous section we propose an algorithm for solving linear VI(*X*,*f*) (1). A numerical example is provided for illustration.

Algorithm for Linear VI:

STEP 0. Let X*=∅ be the solution set of the VI.
STEP 1. If *D* is a rectangular matrix, then solve the following system

$$q = D^T v. \tag{6}$$

If $v \geq 0$, $\mathbf{0} \in X^*_E$; otherwise, compute $q^T D^{-1}$. If $q^T D^{-1} \geq 0$, then $\mathbf{0} \in X^*_E$.

STEP 2. If M^{-1} does not exist, then solve the following system:

$$\begin{cases} Mx + q = 0 \\ Dx \geq 0, Dx \neq 0 \end{cases} \tag{7}$$

If the solution x* satisfies that some element of *D*x* is zero, then

$x^* \in X^*_B$; otherwise, $x^* \in X^*_I$. Hence, the solutions $x^* \in X^*_I \cup X^*_B$.

Otherwise, compute $-DM^{-1}q$. If $-DM^{-1}q \geq 0$, then $-M^{-1}q$ is a single solution and belongs to $X^*_I \cup X^*_B$.

STEP 3.

STEP 3.1. Let $k=1$.

STEP 3.2. Let $d_k x=0$, $d_i x>0$ for all $i=1,2,...,m$, $i \neq k$, where d_i's are row vectors of D, and $l>0$.

STEP 3.3. Solve the following system to obtain the solution x*.

$$\begin{aligned} & d_k x=0 \\ & d_i x>0 \quad i \neq k \\ & Mx+q = l \cdot d_k, \end{aligned} \tag{8}$$

Then, $x^* \in X^*_B$.

STEP 3.4. If $k<m$, let $k = k+1$, go back to STEP 3.2; otherwise, if $n>2$ and $m>2$, go to STEP 4; else go to STEP 5.

STEP 4.

STEP 4.1 Let $k_1=1$, $k_2=2$.

STEP 4.2. Solve the following system (9) and find the solution x*.

$$\begin{aligned} & d_{k_1} x=0, d_{k_2} x=0 \\ & d_i x>0 \quad i \neq k_1, k_2 \\ & Mx+q = l_1 d_{k_1} + l_2 d_{k_2} \\ & l_1, l_2 > 0 \end{aligned} \tag{9}$$

Then $x^* \in X^*_B$.

STEP 4.3. If $k_2 \neq m$, then $k_2 = k_2+1$ and go to STEP 4.2. If $k_2 = m$ and $k_1 < m-1$, let $k_1 = k_1+1, k_2 = k_1+1$ and $k_2 = m$ go to STEP 4.2. If $k_2 = m$ and $k_1 = m-1$, $k_2 = m$ then go to STEP 5.

STEP 5. Let $X^* = X^*_I \cup X^*_B \cup X^*_E$ and output the solution set X*, then STOP.

With this algorithm, the computation complexity for linear VI has been shown to be $O(n^4)$, and for LCP, it is $O(n^3)$.

Now, we demonstrate the proposed algorithm by the following example:

Example 4.1. Solve a VI with
X={ $x: x_1 + x_2 \geq 0$, $x_2 \geq 0$, $x=(x_1, x_2) \in \mathbf{R}^2$ } and

$$f(x)=\begin{bmatrix} 1 & -3 \\ -2 & 5 \end{bmatrix}\begin{bmatrix} x_1 \\ x_2 \end{bmatrix}+\begin{bmatrix} 1 \\ 3 \end{bmatrix}=\begin{bmatrix} x_1 - 3x_2 + 1 \\ -2x_1 + 5x_2 + 3 \end{bmatrix}.$$

Reformulate the VI(X, f) into the form of the model (5), then

$$D=\begin{bmatrix} 1 & 1 \\ 0 & 1 \end{bmatrix}, \quad M=\begin{bmatrix} 1 & -3 \\ -2 & 5 \end{bmatrix}, \ q=\begin{bmatrix} 1 \\ 3 \end{bmatrix} \text{ and}$$

$$D^{-1}=\begin{bmatrix} 1 & -1 \\ 0 & 1 \end{bmatrix}, \ M^{-1}=\begin{bmatrix} -5 & -3 \\ -2 & -1 \end{bmatrix}, \ d_1=(1 \ \ 1), \ d_2=(0 \ \ 1).$$

STEP 0. Let X*=∅ be the solution set of the VI.

STEP 1. $q^T D^{-1} = [1 \ \ 3]\begin{bmatrix} 1 & -1 \\ 0 & 1 \end{bmatrix} = [1 \ \ 2] > 0, \ 0 \in X^*_E$.

STEP 2. $-DM^{-1}q = -\begin{bmatrix} 1 & 1 \\ 0 & 1 \end{bmatrix}\begin{bmatrix} -5 & -3 \\ -2 & -1 \end{bmatrix}\begin{bmatrix} 1 \\ 3 \end{bmatrix} = \begin{bmatrix} 19 \\ 5 \end{bmatrix} > 0$, then

$$x^* = -M^{-1}q = \begin{bmatrix} 14 \\ 5 \end{bmatrix} \in X^*_I \cup X^*_B.$$

STEP 3.
3.1. k=1.
3.2.Suppose $d_1x=0$, $d_2x>0$, $l>0$.

3.3. $\begin{cases} d_1 x = 0 \\ Mx + q = l \cdot d_1 \end{cases} \Rightarrow \begin{cases} x_1 + x_2 = 0 \\ x_1 - 3x_2 + 1 = l \quad ,l > 0 \\ -2x_1 + 5x_2 + 3 = l \end{cases}$

The solution set of above system is empty.
3.4. k=2.
3.2. Suppose $d_1x>0$, $d_2x=0$, $l>0$.

3.3. $\begin{cases} d_2 x = 0 \\ Mx + q = l \cdot d_2 \end{cases} \Rightarrow \begin{cases} x_2 = 0 \\ x_1 - 3x_2 + 1 = 0 \\ -2x_1 + 5x_2 + 3 = l \end{cases}$.

The solution set of above system is empty.

STEP 4. Output the solution set X*={(0, 0),(14, 5)}and STOP.

In this problem, two types of solutions exist simultaneously.

5. Summary and Conclusion

This study proposes an alternative approach to solving variational inequality problems in a convex cone by a multiple objective programming model. A polynomial-time algorithm is designed to find the exact solutions of the linear VI and the LCP. Theoretical evidences are illustrated by an numerical example. It is concluded that by investigating the properties of GCP, the transformed structure of an MOP provides an efficient tool to find the solutions.

Acknowledgement: This work was supported by National Science Council, Taiwan Republic of China with the project number NSC83-0415-E007-006.

References

1. Karamardian, S., Generalized Complementarity Problem, *Journal of Optimization Theory and Applications*, 8, 161-167, (1971).

2. Kinderdehrer, D. and Stampocchia, G., *An Introduction to Variational Inequalities and their Applications*, Academic Press, (1980).

3. Kostreva, M. M., and Wiecek, M. M., Linear Complementarity Problems and Multiple Objective Programming, *Mathematical Programming*, 60, 349-359, (1993).

4. Steuer, R. E., *Multiple Criteria Optimization: Theory, Computation, and Application*, John Wiley & Sons, Inc., (1986).

An Outer Approximation Method for Optimization over the Efficient Set

Syuuji Yamada, Tetsuzo Tanino, and Masahiro Inuiguchi

Graduate School of Engineering, Osaka University, Japan.

Abstract. In this paper, we consider an optimization problem which aims to minimize a convex function over the efficient set of a multi-objective programming problem constrained by a compact convex set X. In case when X is not a polytope, From a computational viewpoint, we may compromise our aim by getting an approximate solution of such a problem. To find an approximate solution, we propose an outer approximation method for a dual problem.

Keywords: Efficient Set, Global Optimization, Dual Problem, Outer Approximation Method.

1 Introduction

We consider the following multi-objective programming problem:

$$(P)\begin{cases} \text{maximize } \langle c^i, x\rangle, & i = 1, \ldots, K, \\ \text{subject to } x \in X \subset R^n, \end{cases}$$

where X is a compact convex set and $\langle \cdot , \cdot \rangle$ denotes the Euclidean inner product in R^n. The objective functions $\langle c^i, x\rangle$, $i = 1, \ldots, K$, express the criteria which the decision-maker wants to maximize. A feasible vector $x \in X$ is said to be efficient if there is no feasible vector y such that $\langle c^i, x\rangle \leq \langle c^i, y\rangle$ for every $i \in \{1, \ldots, K\}$ and $\langle c^j, x\rangle \leq \langle c^j, y\rangle$ for at least one $j \in \{1, \ldots, K\}$. The set X_e of all feasible efficient vectors is called the efficient set. Let $C = \{x \in R^n : \langle c^i, x\rangle \leq 0 \text{ for all } i \in \{1, \ldots, K\}, \text{ and } \langle c^j, x\rangle < 0 \text{ for some } j \in \{1, \ldots, K\}\}$. Then X_e can be formulated as $X_e = X \backslash (X + C)$. For problem (P), we shall assume the following throughout this paper:

(A1) $X = \{x \in R^n : p(x) \leq 0\}$ where $p : R^n \to R$ is a convex function satisfying $p(0) < 0$ (whence $0 \in \text{int } X$),
(A2) $\text{int } C \neq \emptyset$.

In this paper, we consider a cost function minimization problem over the efficient set where the cost function is convex. An example of such a problem is furnished by the portfolio optimization problem in capital markets. A fund manager may look for a portfolio which minimizes the transaction cost on the efficient set. In case X is a polytope, Konno, Thach and Tuy [5] have proposed a cutting plane method for solving the problem. In order to solve the problem in case X is a compact convex set and is not necessarily a polytope, we propose an outer approximation method for solving the problem.

The organization of this paper is as follows: In Section 2, we explain a convex function minimization over the efficient set in R^n. Moreover, we describe a subproblem having an approximate solution of the problem, and the dual problem for the subproblem. In Section 3, we formulate an outer approximation algorithm for the dual problem, and make sure of convergence of the algorithm. In Section 4, to be efficient in executing the algorithm, we propose a procedure in identifying redundant constraints for the subproblem.

Throughout the paper, we use the following notation: int X, bd X and co X denote the interior set of $X \subset R^n$, the boundary set of X and the convex hull of X, respectively. $\bar{R} = R \cup \{-\infty\} \cup \{+\infty\}$. Given a convex polyhedral set (or polytope) $X \subset R^n$, $V(X)$ denotes the set of all vertices of X. For a subset $X \subset R^n$, $X^\circ = \{u \in R^n : \langle u, x\rangle \leq 1, \ \forall x \in X\}$ is called the polar set of X. For a subset $X \subset R^n$, the indicator of $X : \delta(\ \cdot\ |X)$ is an extended-real-valued function defined as follows:

$$\delta(x|X) = \begin{cases} 0 & \text{if } x \in X \\ +\infty & \text{if } x \notin X. \end{cases}$$

Given a function $f : R^n \to R$, $f^H : R^n \to \bar{R}$ is called the quasi-conjugate of f if f^H is defined as follows:

$$f^H(u) = \begin{cases} -\sup\{f(x) : x \in R^n\} & \text{if } u = 0 \\ -\inf\{f(x) : \langle u, x\rangle \geq 1\} & \text{if } u \neq 0. \end{cases}$$

2 Minimizing a Convex Function over the Efficient Set

Let us consider the following problem which minimizes a function f over the weakly efficient set of (P):

$$(OES) \begin{cases} \text{minimize } f(x) \\ \text{subject to } x \in X_e = X \backslash (X + C), \end{cases}$$

where $f : R^n \to R$ satisfies the following assumptions:

(B1) f is a convex function,
(B2) f is regular at the origin, that is, $f(0) = \inf\{f(x) : x \in R^n \backslash \{0\}\}$,
(B3) $\arg\min\{f(x) : x \in R^n\} = \{0\}$.

By using the indicator of X, problem (OES) can be reformulated as

$$\begin{cases} \text{minimize } g(x) \\ \text{subject to } x \in R^n \backslash (X + C) \end{cases} \tag{1}$$

where $g(x) := f(x) + \delta(x|X)$.

For any positive number s, we define

$$c^i(s) = c^i + s\sum_{j=1}^{K} c^j,\ i = 1,\ldots,K,$$
$$C_s = \{x \in R^n : \langle c^i(s), x\rangle \leq 0,\ i = 1,\ldots,K\},$$
$$X_s = X\backslash \text{int } (X + C_s).$$

Since $C \subset \text{int } C_s$, C_s has a nonempty interior from (A3). Moreover, since $0 \in \text{int } X$ (assumption (A1)), $0 \in \text{int } (X + C_s)$.

Proposition 1. *(see Konno, Thach and Tuy [5]) The following assertions are valid.*

(i) X_s *is a compact subset of* X_e,
(ii) $X_s \subset X_t$ *if* $s \geq t$ *and* $X_e \subset \text{cl } (\cup_{s>0} X_s)$,
(iii) $\inf\{f(x) : x \in X_s\} \downarrow \inf\ (OES)$ *as* $s \downarrow 0$

where $\inf\ (OES)$ *is the infimum of the feasible values of the objective function in* (OES).

From Proposition 1, a relaxed problem of problem (OES) is

$$(MP)\begin{cases}\text{minimize } g(x)\\ \text{subject to } x \in R^n\backslash\text{int } (X + C_s).\end{cases}$$

The dual problem of problem (MP) is formulated as

$$(DP)\begin{cases}\text{maximize } g^H(x)\\ \text{subject to } x \in (X + C_s)^\circ.\end{cases}$$

Since g^H is a quasi-convex function and $(X + C_s)^\circ$ is a compact convex set, we note that problem (DP) is a quasi-convex maximization problem over a compact convex set in R^n. Denote by $\inf\ (MP)$ and $\sup\ (DP)$ the optimal values of the objective functions in (MP) and (DP), respectively. It follows from the duality relation between problem (MP) and problem (DP) that $\inf\ (MP) = -\sup\ (DP)$ (cf., Konno, Thach and Tuy [5]).

3 An Outer Approximation Method for (DP)

3.1 An Outer Approximation Method for (DP)

We propose an outer approximation method for problem (DP) as follows:

Algorithm OAM-(DP)

Initialization. Generate a polytope S_1 such that $S_1 \subset X$ and that $0 \in \text{int } S_1$. Set $k \leftarrow 1$ and go to Step 1.

Step 1. Consider the following problem (P_k):

$$(P_k)\begin{cases}\text{minimize } g(x)\\ \text{subject to } x \in R^n \setminus \text{ int } (S_k + C_s).\end{cases}$$

Choose $v^k \in V((S_k+C_s)^\circ)$ such that v^k solves the following dual problem of problem (P_k):

$$(D_k)\begin{cases}\text{maximize } g^H(x)\\ \text{subject to } x \in (S_k + C_s)^\circ.\end{cases}$$

Let $x(k)$ be an optimal solution of the following convex minimization problem:

$$\begin{cases}\text{minimize } f(x)\\ \text{subject to } x \in X \cap \{x \in R^n : \langle v^k, x\rangle \geq 1\}.\end{cases} \tag{2}$$

Step 2. Solve the following problem:

$$\begin{cases}\text{minimize } \phi(x; v^k) = \max\{p(x), h(x, v^k)\}\\ \text{subject to } x \in R^n\end{cases} \tag{3}$$

where $h(x, v^k) = -\langle v^k, x\rangle + 1$. Let z^k and α_k denote an optimal solution of problem (3) and the optimal value, respectively. It will be proved later in Theorem 2 and Lemma 4 that there exists the optimal value in problem (3) and that $z^k \in X$, respectively.

a. If $\alpha_k = 0$, then stop; v^k solves problem (DP). Moreover, $x(k)$ solves problem (MP) and the optimal value of problem (2) is the optimal value of problem (MP).

b. Otherwise, set $S_{k+1} = \text{co}\ (\{z^k\} \cup S_k)$. Set $k \leftarrow k+1$ and go to Step 1.

At every iteration k of the algorithm, problems (2) and (3) are convex minimization problems.

Let $\{S_k\}$ be generated by the algorithm. Then $S_k + C_s$, $i = 1, 2, \ldots$, are convex polyhedral sets and satisfy that $0 \in \text{int } (S_k + C_s)$. Hence, from the principle of duality, $(S_k + C_s)^\circ$ is a polytope. Moreover, the following assertions are valid.

- for any k,

$$\begin{aligned}(S_k + C_s)^\circ &= (S_k)^\circ \cap (C_s)^\circ\\ &= \{x \in R^n : \langle z, x\rangle \leq 1\ \forall z \in V(S_k),\\ &\qquad\quad \langle u, x\rangle \leq 0\ \forall u \in E(C_s)\}\end{aligned}$$

  where $E(C_s)$ is a finite set of extreme directions of C_s satisfying $C_s = \{x \in R^n : x = \sum_{u \in E(C_s)} \lambda_u u,\ \lambda_u \geq 0\}$,
- for any k, $S_k + C_s = \{x \in R^n : \langle v, x\rangle \leq 1,\ \forall v \in V((S_k + C_s)^\circ)\}$.

Since problem (D_k) is a quasi-convex maximization problem over $(S_k + C_s)^\circ$, we can choose an optimal solution of problem (D_k) from the set of all vertices of $(S_k + C_s)^\circ$. Denote by inf (P_k) and sup (D_k) the optimal values in problem (P_k) and problem (D_k), respectively. Since (D_k) is the dual problem of problem (P_k), inf $(P_k) = -$sup (D_k) (Konno, Thach and Tuy [5]).

3.2 A Relationship between Problem (P_k) and Problem (2)

Assume that $S_k \subset X$ at iteration k of Algorithm OAM-(DP). The validity of this assumption will be proved later in Lemma 4. Under this assumption it follows that an optimal solution $x(k)$ of problem (2) solves (P_k) (see Theorem 1 described later):

Remark 1. If $S_k \subset X$, then $S_k + C_s \subset X + C_s$. Moreover, by the principle of duality, $(S_k)^\circ \supset X^\circ$ and $(S_k + C_s)^\circ \supset (X + C_s)^\circ$.

Lemma 1. *At iteration k of Algorithm OAM-(DP), $v^k \neq 0$.*

Proof. For any $y \in (\text{bd } X^\circ) \cap (C_s)^\circ$, there is $x' \in X$ such that $\langle y, x' \rangle = 1$. Therefore, $g^H(y) > -\infty$ for any $y \in (\text{bd } X^\circ) \cap (C_s)^\circ$. Since $(S_k + C_s)^\circ \supset (X + C_s)^\circ \supset (\text{bd } X^\circ) \cap (C_s)^\circ$ and v^k is an optimal solution of (D_k), we obtain $g^H(v^k) > -\infty$. Moreover, since $g^H(0) = -\infty$, we have $v^k \neq 0$.

Lemma 2. *At iteration k of Algorithm OAM-(DP), for any $v \in V((S_k + C_s)^\circ) \setminus \{0\}$, $v \notin \text{int } X^\circ$.*

Proof. Suppose to the contrary that there exists $v \in V((S_k + C_s)^\circ) \setminus \{0\}$ satisfying $v \in \text{int } X^\circ$. Then, since v is a vertex of $(S_k + C_s)^\circ$ and $(S_k + C_s)^\circ = \{x \in R^n : \langle z, x \rangle \leq 1 \ \forall z \in V(S_k), \ \langle u, x \rangle \leq 0 \ \forall u \in E(C_s)\}$,

$$\exists a^1, \ldots, a^r \in V(S_k) \cup E(C_s) \text{ such that } \dim\{a^1, \ldots, a^r\} = n \text{ and } \langle a^i, v \rangle = b_i \ i = 1, \ldots, r \tag{4}$$

where for all $i \in \{1, \ldots, r\}$,

$$b_i = \begin{cases} 1 \text{ if } a^i \in V(S_k) \\ 0 \text{ if } a^i \in C_s. \end{cases}$$

Note that $y \notin \text{int } X$ if a point $y \in R^n$ satisfies that $\langle z, y \rangle \leq 1$ for some $z \in V(S_k)$, because $X^\circ \subset (S_k)^\circ = \{x \in R^n : \langle z, x \rangle \leq 1 \text{ for all } z \in V(S_k)\}$. Hence, by the assumption of v, $\{a^1, \ldots, a^r\} \subset E(C_s)$. Then v is the origin of R^n because $\cap_{i=1}^r \{x \in R^n : \langle a^i, x \rangle = 0\} = \{0\}$. This is a contradiction and hence, we have $v \notin \text{int } X^\circ$ for any $v \in V((S_k + C_s)^\circ) \setminus \{0\}$.

Lemma 3. *At iteration k of Algorithm OAM-(DP), $v^k \notin \text{int } X^\circ$.*

Proof. We remember that v^k is a vertex of $(S_k + C_s)^\circ$. By Lemma 1 and Lemma 2, we get that $v^k \notin \text{int } X^\circ$.

Theorem 1. *At iteration k of Algorithm OAM-(DP), let v^k be an optimal solution for problem (D_k) and $x(k)$ an optimal solution for problem (2). Then*

(i) $X \cap \{x \in R^n : \langle v^k, x \rangle \geq 1\} \neq \emptyset$,
(ii) $x(k)$ is contained in the feasible set of problem (P_k),
(iii) $x(k)$ solves problem (P_k).

Proof. (i) By Lemma 3, $v^k \notin \text{int } X^\circ$. Therefore, there is $x' \in X$ such that $\langle v^k, x' \rangle \geq 1$, that is, $X \cap \{x \in R^n : \langle v^k, x \rangle \geq 1\} \neq \emptyset$.

(ii) Since $v^k \in (S_k + C_s)^\circ$, $x(k) \in \{x \in R^n : \langle v^k, x \rangle \geq 1\}$ and $((S_k + C_s)^\circ)^\circ = S_k + C_s$, we get that $x(k) \notin \text{int } (S_k + C_s)$. Therefore $x(k)$ is contained in the feasible set of problem (P_k).

(iii) Since v^k is an optimal solution and (D_k) is the dual problem of problem (P_k),

$$\begin{aligned}\inf\ (P_k) &= -\sup\ (D_k) \\ &= -g^H(v^k) \\ &= \inf\{g(x) : \langle v^k, x \rangle \geq 1\} \quad \text{(by Lemma 1)} \\ &= \inf\{f(x) : \langle v^k, x \rangle \geq 1,\ x \in X\} \\ &= \inf\ (2)\end{aligned}$$

where inf (2) is the optimal value of problem (2). Therefore, we have $f(x(k)) = g(x(k)) = \inf\ (P_k)$.

3.3 Stopping Criterion of Algorithm OAM-(DP)

In this section, we examine the suitability of the stopping criterion of Algorithm OAM-(DP).

Theorem 2. *For any $v \in R^n$, there exists the minimal value of the objective function $\phi(x; v)$ of problem (3) over R^n.*

Proof. Since p is a continuous function and $X = \{x \in R^n : p(x) \leq 0\}$ is compact, the minimal value of p over R^n exists (Hestenes [2]). Let $\alpha := \min\{p(x) : x \in R^n\}$. Then $\inf_{x \in R^n} \phi(x; v) = \inf_{x \in R^n} \max\{p(x), h(x, v)\} \geq \inf_{x \in R^n} p(x) = \alpha$. This implies that $\phi(x; v)$ has the infimum over R^n. Let $\beta := \inf_{x \in R^n} \phi(x; v)$. Since p is a proper convex function and X is compact, we obtain that for any $\gamma \geq \alpha$, the level set $L_p(\gamma) = \{x \in R^n : p(x) \leq \gamma\}$ is compact (Rockafellar [6], Corollary 8.7.1). We note that $\beta \geq \alpha$ and that $L_p(\gamma) \supset L_\phi(\gamma) \neq \emptyset$ for any $\gamma \geq \beta$. Hence, $L_\phi(\gamma)$ is compact for any $\gamma \geq \beta$. Consequently, the minimal value of $\phi(x; v)$ over R^n exists (Hestenes [2], Theorem 2.1).

Lemma 4. *At iteration k of Algorithm OAM-(DP), assume that $S_k \subset X$. Then*

(i) $\alpha_k \leq 0$,
(ii) $z^k \in X$.

Proof. By Lemma 3, $v^k \notin \text{int } X^\circ$. Therefore, there is $\hat{x} \in X$ such that $\langle v^k, \hat{x} \rangle \geq 1$. Furthermore,

$$\alpha_k = \min_{x \in R^n} \phi(x; v^k) \leq \phi(\hat{x}; v^k) = \max\{p(\hat{x}), -\langle v^k, \hat{x} \rangle + 1\} \leq 0.$$

Since $\alpha_k \leq 0$, we obtain $p(z^k) \leq \alpha_k \leq 0$. Consequently, $z^k \in X$.

From Lemma 4, we have

- $S_1 + C_s \subset S_2 + C_s \subset \ldots \subset S_k + C_s \subset \ldots \subset X + C_s$,
- $(S_1 + C_s)^\circ \supset (S_2 + C_s)^\circ \supset \ldots \supset (S_k + C_s)^\circ \supset \ldots \supset (X + C_s)^\circ$.

Moreover, we note that sup $(D_{k-1}) \geq$ sup (D_k) for any $k \geq 2$, that is,

$$g^H(v^1) \geq g^H(v^2) \geq \cdots \geq g^H(v^k) \geq \cdots \geq \sup\ (DP), \tag{5}$$

and that inf $(P_{k-1}) \leq$ inf (P_k) for any $k \geq 2$, that is,

$$f(x(1)) \leq f(x(2)) \leq \cdots \leq f(x(k)) \leq \cdots \leq \inf\ (MP). \tag{6}$$

If the algorithm terminates after finite iterations, then it satisfies that we have an optimal solution of problem (DP), as follows:

Theorem 3. *At iteration k of Algorithm OAM-(DP), $\alpha_k = 0$ if and only if $v^k \in X^\circ$.*

Proof. We shall show that $v^k \in X^\circ$ if $\alpha_k = 0$. Suppose that $v^k \notin X^\circ$. Then, there is $\hat{x} \in X$ such that $\langle v^k, \hat{x}\rangle > 1$. Moreover, since $\{x \in R^n : \langle v^k, x\rangle > 1\}$ is an open set, there exists $\varepsilon > 0$ such that $B(\hat{x}, \varepsilon) \subset \{x \in R^n : \langle v^k, x\rangle > 1\}$ where $B(\hat{x}, \varepsilon) = \{y \in R^n : \|y - \hat{x}\| < \varepsilon\}$. This implies that $(\text{int}\, X) \cap B(\hat{x}, \varepsilon) \neq \emptyset$. Let $x' \in (\text{int}\, X) \cap B(\hat{x}, \varepsilon)$, then

$$\begin{aligned} \alpha_k &= \min_{x \in R^n} \phi(x; v^k) \\ &= \min_{x \in R^n} \max\{p(x), -\langle v^k, x\rangle + 1\} \\ &\leq \max\{p(x'), -\langle v^k, x'\rangle + 1\} \\ &< 0. \end{aligned}$$

Therefore, we get that $\alpha_k < 0$ if $v^k \notin X^\circ$. Consequently, $v^k \in X^\circ$ if $\alpha_k = 0$.

Next, we shall show that $\alpha_k = 0$ if $v^k \in X^\circ$. Suppose that $v^k \in X^\circ$. Then, since $(X^\circ)^\circ = X$, we obtain $X \subset \{x \in R^n : \langle v^k, x\rangle \leq 1\}$. Therefore, $X \cap \{x \in R^n : \langle v^k, x\rangle > 1\} = \emptyset$, that is,

$$\nexists x \in R^n \text{ such that } p(x) < 0 \text{ and } -\langle v^k, x\rangle + 1 < 0.$$

Hence, for any $x \in R^n$, $\phi(x; v^k) \geq 0$, that is, $\alpha_k \geq 0$. Consequently, by Lemma 4, $\min_{x \in R^n} \phi(x; v^k) = \alpha_k = 0$.

Theorem 4. *At iteration k of Algorithm OAM-(DP), if $\alpha_k = 0$, then*

(i) v^k is an optimal solution of problem (DP),
(ii) $x(k)$ is an optimal solution of problem (MP).

Proof. Suppose that $\alpha_k = 0$. Then, by Theorem 3, $v^k \in X^\circ$. Furthermore, we obtain $v^k \in X^\circ \cap (C_s)^\circ = (X + C_s)^\circ$ because $v^k \in (S_k + C_s)^\circ \subset (C_s)^\circ$. Therefore, $g^H(v^k) \leq \sup(DP)$. Since v^k is an optimal solution of (D_k) and

$(S_k + C_s)^\circ \supset (X + C_s)^\circ$, we get $g^H(v^k) \geq \sup(DP)$. Hence, $g^H(v^k) = \sup(DP)$. Consequently, v^k is an optimal solution of problem (DP).

Since $\langle v^k, x(k)\rangle \geq 1$ and $v^k \in (X + C_s)^\circ$, we have $x(k) \notin \text{int}\,(X + C_s)$. Therefore, $x(k)$ is contained in the feasible set of problem (MP). By Theorem 1,

$$f(x(k)) = -g^H(v^k) = -\sup(DP) = \inf(DP).$$

Consequently, $x(k)$ is an optimal solution of problem (MP).

At iteration k of Algorithm OAM-(DP), $\langle v^k, z^k\rangle > 1$ if $\alpha_k < 0$. Hence, $S_{k+1} + C_s = \text{co}\,(S_k \cup \{z^k\}) + C_s \neq S_k + C_s$ because $S_k + C_s \subset \{x \in R^n : \langle v^k, x\rangle \leq 1\}$. Moreover, since $V(S_{k+1}) \subset V(S_k) \cup \{z^k\}$, we have

$$(S_{k+1} + C_s)^\circ = (S_k + C_s)^\circ \cap \{x \in R^n : \langle z^k, x\rangle \leq 1\} \neq (S_k + C_s)^\circ \quad (7)$$

(see Fig. 1).

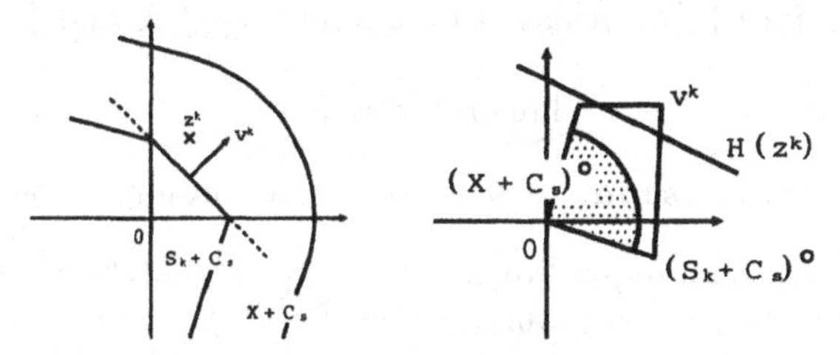

Fig. 1. Construct $S_{k+1} + C_s$ and $(S_{k+1} + C_s)^\circ$; $H(z^k) = \{x : \langle z^k, x\rangle = 1\}$.

Remark 2. At iteration k of the algorithm, for any $v \in V((S_{k+1} + C_s)^\circ)$ such that $v \notin V((S_k + C_s)^\circ)$, $\langle z^k, v\rangle = 1$.

3.4 Convergence of Algorithm OAM-(*DP*)

Algorithm OAM-(DP) using the stopping criterion discussed in section 3.3 doesn't necessarily terminate after finite iterations. In this section, we consider the case that an infinite sequence $\{v^k\}$ is generated by the algorithm.

Lemma 5. *Assume that $\{v^k\}$ is an infinite sequence such that for all k, v^k is an optimal solution of (D_k) at iteration k of Algorithm OAM-(DP). Then, there exists an accumulation point of $\{v^k\}$.*

Proof. Since $\{v^k\} \subset (S_1 + C_s)^\circ$ and $(S_1 + C_s)^\circ$ is compact, there exists an accumulation point of $\{v^k\}$.

It follows from the following theorem that every accumulation point of $\{v^k\}$ belongs to the feasible set of problem (DP).

Theorem 5. *Assume that $\{v^k\}$ is an infinite sequence such that for all k, v^k is an optimal solution of (D_k) at iteration k of Algorithm OAM-(DP) and that $\bar{v}$ is an accumulation point of $\{v^k\}$. Then $\bar{v}$ belongs to $(X+C_s)^\circ$.*

Proof. Let a subsequence $\{v^{k_q}\} \subset \{v^k\}$ converge to $\bar{v}$. Then, there is the sequence $\{z^{k_q}\}$ such that z^{k_q} is an optimal solution of problem (3) at iteration k_q of the algorithm. Since $\{z^{k_q}\}$ belongs to the compact set X, there are an accumulation point $\bar{z}$ and a subsequence $\{z^p\} \subset \{z^{k_q}\}$ such that $z^p \to \bar{z}$ as $p \to \infty$. Moreover, for $\{z^p\}$, there is the subsequence $\{v^p\} \subset \{v^{k_q}\}$. Obviously, $\{v^p\}$ converges to $\bar{v}$. Since $\{v^p\} \not\subset X^\circ$, by Theorem 3,

$$0 > \alpha_p = \max\{p(z^p), h(z^p, v^p)\} \geq -\langle v^p, z^p\rangle + 1, \qquad \text{for all } p.$$

Therefore, $\lim_{p\to\infty}\langle v^p, z^p\rangle = \langle \bar{v}, \bar{z}\rangle \leq 1$. On the other hand, since $v^{p'} \in (S_{p+1}+C_s)^\circ$ for all $p' > p$, and $(S_{p+1}+C_s)^\circ = (S_p+C_s)^\circ \cap \{x \in R^n : \langle z^p, x\rangle \leq 1\}$, we obtain $\lim_{p\to\infty}\langle v^{p+1}, z^p\rangle = \langle \bar{v}, \bar{z}\rangle \leq 1$. Hence, $\lim_{p\to\infty}\langle v^p, z^p\rangle = \langle \bar{v}, \bar{z}\rangle = 1$. Furthermore, we get that $\lim_{q\to\infty}\langle \bar{v}, z^{k_q}\rangle = 1$ (Aubin [1], Proposition 1, Section 6, Chapter 1).

Moreover, since $\{z^{k_q}\}$ belongs to the compact set X and $\{v^{k_q}\}$ converges to $\bar{v}$,

$$\lim_{q\to\infty}\langle v^{k_q}, z^{k_q}\rangle = 1. \tag{8}$$

By Lemma 4, $\lim_{q\to\infty}\sup \alpha_{k_q} \leq 0$. Moreover, according to condition (8),

$$\begin{aligned}\lim_{q\to\infty}\inf \alpha_{k_q} &= \lim_{q\to\infty}\inf \max\{p(z^{k_q}, h(z^{k_q}, v^{k_q})\}\\ &\geq \lim_{q\to\infty} \in h(z^{k_q}, v^{k_q})\\ &= 0.\end{aligned}$$

Consequently, $\lim_{q\to\infty}\alpha_{k_q} = 0$.

In order to obtain contradiction, suppose that $\bar{v} \notin X^\circ$. Then, we have

$$\exists x' \in X \text{ such that } h(x', \bar{v}) = -\langle \bar{v}, x'\rangle + 1 < 0.$$

Since $h(\ \cdot\ , \bar{v})$ is a continuous function over R^n,

$$\exists \varepsilon > 0 \text{ such that } B(x', \varepsilon) \subset \{x \in R^n : h(x, \bar{v}) < 0\}$$

where $B(x', \varepsilon) = \{x \in R^n : \|x - x'\| < \varepsilon\}$. This implies that for any $\bar{x} \in (\text{int } X) \cap B(x', \varepsilon)$, $p(\bar{x}) < 0$ and $h(\bar{x}, \bar{v}) < 0$ because int $X \neq \emptyset$. Then, we obtain

$$\exists \delta > 0 \text{ such that} h(\bar{x}, v) < \frac{1}{2}h(\bar{x}, \bar{v}) < 0, \qquad \forall v \in B(\bar{v}, \delta)$$

and, for any $v \in B(\bar{v}, \delta)$,

$$\begin{aligned}\min_{x\in R^n} \phi(x; v) \ \min_{x\in R^n} &\max\{p(x), h(x, v)\}\\ &\leq \max\{p(\bar{x}), h(\bar{x}, v)\}\\ &\leq \max\left\{p(\bar{x}), \frac{1}{2}h(\bar{x}, \bar{v})\right\} < 0.\end{aligned}$$

Consequently, $\lim_{q\to\infty} \alpha_{k_q} \leq \max\{p(\bar{x}), \frac{1}{2}h(\bar{x}, \bar{v})\} < 0$. This is a contradiction. Hence $\bar{v} \in X^\circ$. Moreover, since $\{v^{k_q}\} \subset (S_1 + C_s)^\circ \subset (C_s)^\circ$ and $(C_s)^\circ$ is a closed set, we have $\lim_{q\to\infty} v^{k_q} = \bar{v} \in (C_s)^\circ$. Therefore, we get that $\bar{v} \in (X + C_s)^\circ = (X^\circ) \cap ((C_s)^\circ)$.

Corollary 1. *Assume that $\{v^k\}$ is an infinite sequence such that for all k, v^k is an optimal solution of problem (D_k) at iteration k of Algorithm OAM-(DP) and that $\bar{v}$ is an accumulation point of $\{v^k\}$. Then $\bar{v} \notin \text{int } X^\circ$.*

Proof. Let a subsequence $\{v^{k_q}\} \subset \{v^k\}$ converge to $\bar{v}$. By Theorem 4, $v^{k_q} \notin \text{int } X^\circ$ for all q. Since $R^n \backslash \text{int } X^\circ$ is a closed set, we have $\lim_{q\to\infty} v^{k_q} = \bar{v} \in R^n \backslash \text{int } X^\circ$.

Moreover, we get from the following theorem that every accumulation point of $\{v^k\}$ solves problem (DP).

Theorem 6. *Assume that $\{v^k\}$ is an infinite sequence such that for all k, v^k is an optimal solution of (D_k) at iteration k of Algorithm OAM-(DP) and that $\bar{v}$ is an accumulation point of $\{v^k\}$. Then $\bar{v}$ solves problem (DP).*

Proof. Let a subsequence $\{v^{k_q}\} \subset \{v^k\}$ converge to $\bar{v}$. Since that f is continuous over R^n, h is continuous over $R^n \times R^n$, X is a compact set and $\{x \in R^n : \langle v, x\rangle \geq 1,\ x \in X\} = \{x \in R^n : -h(x, v) \geq 0,\ x \in X\} \neq \emptyset$ for any $v \in (C_s)^\circ \backslash (\text{int } X^\circ)$, we obtain that g^H is upper semi-continuous over $(C_s)^\circ \backslash (\text{int } X^\circ)$ (Hogan [3]). Therefore, by condition (5),

$$g^H(\bar{v}) \geq \lim_{q\to\infty} \sup g^H(v^{k_q}) \geq \sup(DP).$$

By Theorem 5, $\bar{v} \in (X + C_s)^\circ$. Hence, $g^H(\bar{v}) \leq \sup(DP)$. Consequently,

$$g^H(\bar{v}) = \lim_{q\to\infty} g^H(v^{k_q}) = \sup(DP).$$

By Theorem 5 and 6, we get that every accumulation point of $\{v^k\}$ belongs to the feasible set of problem (DP) and solves problem (DP).

Remark 3. At iteration k of the algorithm, since $f(0) = \inf\{f(x) : x \in R^n \backslash \{0\}\}$, $\arg\min\{f(x) : x \in R^n\} = \{0\}$ and $0 \in \{x \in R^n : \langle v^k, x\rangle < 1\}$, every optimal solution of problem (2) belongs to $\{x \in R^n : \langle v^k, x\rangle = 1\}$.

Remark 4. For the feasible set $R^n \backslash \text{int}(X + C_s)$ of (MP), we have

$$R^n \backslash \text{int}(X + C_s) \supset X \backslash \text{int}(X + C_s) \neq \emptyset.$$

Theorem 7. *Assume that $\{x(k)\}$ is an infinite sequence such that for all k, $x(k)$ is an optimal solution of problem (P_k) at iteration k of Algorithm OAM-(DP) and that $\bar{x}$ is an accumulation point of $\{x(k)\}$. Then $\bar{x}$ belongs to $R^n \backslash \text{int } (X + C_s)$ and solves problem (MP).*

Proof. Let a subsequence $\{x(k_q)\} \subset \{x(k)\}$ converge to $\bar{x}$. Then, there is a sequence $\{v^{k_q}\}$ such that v^{k_q} is an optimal solution of (D_{k_q}) at iteration k_q of the algorithm. By Remark 3, $\langle v^{k_q}, x(k_q)\rangle = 1$ for all q. Therefore, $\lim_{q\to\infty}\langle v^{k_q}, x(k_q)\rangle = \lim_{q\to\infty}\langle v^{k_q}, \bar{x}\rangle = 1$. Moreover, for every accumulation point $\bar{v}$ of $\{v^{k_q}\}$, $\langle \bar{v}, \bar{x}\rangle = 1$. By Theorem 5, since $\bar{v} \in (X + C_s)^o$, we obtain $\bar{x} \in \text{bd}\,(X + C_s)$. Consequently, $\bar{x} \notin \text{int}\,(X + C_s)$.

Since $\{x(k_q)\} \subset X$ and for any q, $x(k_q)$ is an optimal solution of problem (2) at iteration k_q of the algorithm, we get $g^H(v^{k_q}) = -g(x(k_q)) = -f(x(k_q))$ for all q. Therefore, by Theorem 6 and continuity of f,

$$\inf(MP) = -\sup(DP) = -\lim_{q\to\infty} g^H(v^{k_q}) = \lim_{q\to\infty} f(x(k_q)) = f(\bar{x}).$$

The proof is complete.

4 Conclusion

In this paper, instead of solving problem (OES) directly, we have considered a relaxed problem and presented an outer approximation method for its dual problem. By the algorithm, we can obtain an approximate solution contained in the feasible set of problem (OES).

To execute the algorithm, a convex minimization problem (3) is solved at each iteration. However, we note that it is not necessary to obtain an optimal solution for problem (3) at each step. At iteration k of the algorithm, it suffices to get a point which is contained in X and is not contained in $S_k + C_s$. That is, at each step, we can compromise solving problem (3) by getting a point z^k satisfying $\phi(z^k; v^k) < 0$, because z^k belongs to $X\backslash(S_k + C_s)$ if $\phi(z^k; v^k) < 0$.

Moreover, problem (D_k) is solved at each step. To solve problem (D_k), for every vertex v of the feasible set of problem (D_k), it is necessary to get the objective function value $g^H(v^k)$. This means that a convex minimization problem (2) is solved for every vertex of $(S_k + C_s)^o$. Hence, by solving problem (D_k), $x(k)$ is obtained.

As is mentioned above, by solving two kinds of convex minimization problems successively, it is possible to obtain an approximate solution of problem (OES). These convex minimization problems are fairly easy to solve and therefore the proposed algorithm is practically useful.

References

1. Aubin, J.P., *Applied Abstract Analysis*, John Wiley, New York (1977).
2. Hestenes, M.R., *Optimization Theory: The Finite Dimensional Case*, John Wiley, New York (1975).
3. Hogan, W.W., "Point-To-Set Maps in Mathematical Programming", *SIAM Review*, Vol. 15, No. 3 (1973).
4. Horst, R. and H. Tuy, *Global Optimization*, Springer-Verlag, Berlin (1990).

5. Konno, H., P.T. Thach and H. Tuy, *Optimization on Low Rank Nonconvex Structures* Kluwer Academic Publishers, Dordrecht (1997).
6. Rockafellar, R.T., *Convex Analysis* Princeton University Press, Princeton, N.J. (1970).
7. Sawaragi, Y., H. Nakayama and T. Tanino, *Theory of Multiobjective Optimization*, Academic Press, Orland (1985).

On Efficiency of Data Envelopment Analysis

Y.B. YUN[1], H. NAKAYAMA[2] and T. TANINO[1]
[1] Department of Electronics and Information Systems
Graduate School of Engineering, Osaka University
2-1 Yamada-oka, Suita, Osaka 565-0871, Japan
[2] Department of Applied Mathematics, Konan University
8-9-1 Okamoto, Higashinada, Kobe 658-8501, Japan

Abstract

In this paper, we suggest a new concept of "Value Free Efficiency" which does not introduce any value judgment for outputs and inputs. That is, similarly to the usual multiple criteria decision analysis, a Decision Making Unit (DMU) is defined to be efficient if there is no unit that consumes less inputs and produces more outputs than the DMU. In addition, we propose a generalized DEA model for estimating value free efficiency, ratio value efficiency proposed by Charnes, Cooper and Rhodes [4], and sum value efficiency proposed by Belton [2] and Belton and Vickers [3] as special cases. An illustrative example compares these concepts of efficiency.

Keywords : Data Envelopment Analysis, Value Free Efficiency, Multiple Criteria Decision Analysis.

1 Introduction

Data Envelopment Analysis (DEA) which is originally proposed by Charnes, Cooper and Rhodes [4] is a method to measure the relative efficiency of comparable entities called Decision Making Units (DMUs) essentially performing the same task using similar multiple inputs to produce similar multiple outputs. Since then, theoretical studies as well as practical applications of CCR model have been developed and extended : the BCC model by Banker, Charnes and Cooper [1] ; the model by Belton [2], or Belton and Vickers [3] ; the REF model by Joro, Korhonen and Wallenius [8], or Hamel et *al.* [7] ; the CCWH model by Charnes et *al.* [6], etc.

In the following section, a brief outline on several DEA models is demonstrated. In section 3, we suggest the concept of "Value Free Efficiency", which appears usually in multiple criteria decision analysis, and formulate a generalized DEA model for estimating the value free efficiency. It will be seen that the ratio value efficiency and the sum value efficiency can be evaluated as special cases. In section 4, an illustrative example compares these concepts of efficiency.

2 Data Envelopment Analysis

We assume that

$n,\ p,\ m$: numbers of DMUs, outputs, inputs
y_{kj} : amount of output k generated by DMUj,
$\boldsymbol{y}_j := [y_{kj}] \in \mathbb{R}^p_+$ x_{ij} : amount of input i generated by DMUj,
$\boldsymbol{x}_j := [x_{ij}] \in \mathbb{R}^m_+$ μ_k : weight associated with output k
ν_i : weight associated with input i
$Y := [y_{kj}] \in \mathbb{R}^{p\times n}_+, \quad X := [x_{ij}] \in \mathbb{R}^{m\times n}_+$

First, the measure of efficiency for unit o as ratio of weighted sum of outputs to weighted sum of inputs can be given by solving the following:

$$\begin{aligned} \text{Maximize} \quad & \frac{\sum_{k=1}^{p} \mu_k y_{ko}}{\sum_{i=1}^{m} \nu_i x_{io}} \\ \text{subject to} \quad & \frac{\sum_{k=1}^{p} \mu_k y_{kj}}{\sum_{i=1}^{m} \nu_i x_{ij}} \leqq 1, \ j = 1, \cdots, n, \\ & \mu_k \geqq \varepsilon, \ \nu_i \geqq \varepsilon, \ k = 1, \cdots, p, \ ; \ i = 1, \cdots, m, \\ & \varepsilon \text{ is a non-Archimean infinitesimal.} \end{aligned}$$

The above fractional linear program can be reformulated as the following linear program:

$$\begin{aligned} \text{Maximize} \quad & \sum_{k=1}^{p} \mu_k y_{ko} \\ \text{subject to} \quad & \sum_{i=1}^{m} \nu_i x_{io} = 1, \\ & \sum_{k=1}^{p} \mu_k y_{kj} - \sum_{i=1}^{m} \nu_i x_{ij} \leqq 0, \ j = 1, \cdots, n, \\ & \mu_k \geqq \varepsilon, \ \nu_i \geqq \varepsilon, \ k = 1, \cdots, p, \ ; \ i = 1, \cdots, m, \\ & \varepsilon \text{ is a non-Archimean infinitesimal.} \end{aligned}$$

For computational convenience, it is usually solved via the following dual problem:

$$
\begin{aligned}
\text{Minimize} \quad & \theta - \varepsilon(\mathbf{1}^T \boldsymbol{s}_x + \mathbf{1}^T \boldsymbol{s}_y) \\
\text{subject to} \quad & -\theta \boldsymbol{x}_o + X\lambda + \boldsymbol{s}_x = 0, \\
& Y\lambda - \boldsymbol{y}_o - \boldsymbol{s}_y = 0, \\
& \lambda \geqq 0, \ \lambda \in \mathbb{R}^n, \\
& \varepsilon \text{ is a non-Archimean infinitesimal.}
\end{aligned}
$$

where $\boldsymbol{s}_x \in \mathbb{R}^m_+$, $\boldsymbol{s}_y \in \mathbb{R}^p_+$ are slack variables.

Definition 1 (Charnes et *al.*, [4]). DMUo is *efficient* if the optimal value θ^* to the above problem is equal to one and the slack variables $\boldsymbol{s}_x^*$ and $\boldsymbol{s}_y^*$ are all zero. In this paper, we call this efficiency *ratio value efficiency.*

On the other hand, Belton [2] suggested the measure of efficiency as weighted sum to aggregation measures of input and output:For any non-negative weight set μ_k, ν_i, let

$\sum_{k=1}^{p} \mu_k y_{kj} = O_j$ (aggregate output for unit j)

$\sum_{i=1}^{m} \nu_i x_{ij} = I_j$ (aggregate input for unit j)

β is constant, $0 \leqq \beta \leqq 1$

$V_j = (1-\beta)O_j - \beta I_j$

Definition 2 (Belton, [2]). DMU$_o$ is *efficient* if for given weight μ_k and ν_i, there exists β such that $V_o \geqq V_j$ for all units $j : j \neq o$. In this paper, we call this efficiency *sum value efficiency.*

Theorem 1 (Belton, [2]). *If DMUo is sum value efficient, then for some value β, the optimal solution of the following linear program has the value of the objective function $\sum D_j^-$ is equal to zero.*

$$
\begin{aligned}
\text{Minimize} \quad & \sum D_j^- \\
\text{subject to} \quad & V_o - V_j + D_j^- - D_j^+ = 0, \quad (j \neq o) \\
& \mu_k, \ \nu_i, \ D_j^-, \ D_j^+ \geqq 0,
\end{aligned}
$$

where D_j^-, D_j^+ are deviation variables, as used in goal programming.

Because the above problem is nonlinear, Belton and Vickers [3] introduced the following variables;

$$p_k = (1-\beta)\,\mu_k \ \text{ and } \ q_i = \beta\nu_i .$$

As a result, we have the following linear problem:

$$
\begin{aligned}
\text{Minimize} \quad & \sum D_j^- \\
\text{subject to} \quad & \sum p_k d_{koj} - \sum q_i e_{ioj} + D_j^- - D_j^+ = 0, \quad (j \neq o) \\
& \sum p_k = 1 - \beta, \\
& \sum q_i = \beta, \\
& \epsilon \leqq \beta \leqq 1 - \epsilon, \\
& p_k,\ q_i,\ D_j^-,\ D_j^+ \geqq 0,
\end{aligned}
$$

where $d_{koj} = y_{ko} - y_{kj}$ and $e_{ioj} = x_{io} - x_{ij}$.

3 Generalized DEA Model

The following conventions for vectors in $\mathbb{R}^n$ will be used:

$$
\begin{aligned}
& \boldsymbol{x} < \boldsymbol{y} \Leftrightarrow x_i < y_i, \quad i = 1, \cdots, n; \\
& \boldsymbol{x} \leqq \boldsymbol{y} \Leftrightarrow x_i \leqq y_i, \quad i = 1, \cdots, n; \\
& \boldsymbol{x} \leq \boldsymbol{y} \Leftrightarrow x_i \leqq y_i, \quad i = 1, \cdots, n \text{ but } \boldsymbol{x} \neq \boldsymbol{y}; \\
& \boldsymbol{x} \not\leq \boldsymbol{y} \text{ is the negation of } x \leq y.
\end{aligned}
$$

Next we define the concept of value free efficiency.

Definition 3. DMUo is *Value Free Efficient* if there does not exist another DMUj, $j = 1, \cdots n$ such that $(\boldsymbol{y}_j, -\boldsymbol{x}_j) \geq (\boldsymbol{y}_o, -\boldsymbol{x}_o)$.

The value free efficiency is evaluated by solving the following problem:

(GDEA)

$$
\begin{aligned}
\underset{(\Delta,\, \mu_k,\, \nu_i)}{\text{maximize}} \quad & \Delta \\
\text{subject to} \quad & \Delta \leqq \tilde{d}_j, \quad j = 1, \cdots, n,
\end{aligned}
$$

where $\tilde{d}_j$ is the optimal value of the following.

$$
\begin{aligned}
\underset{(d_j, d_{kj}^+, d_{kj}^-, d_{ij}^+, d_{ij}^-)}{\text{minimize}} \quad & d_j + \alpha \left(\sum_{k=1}^{p} (d_{kj}^+ - d_{kj}^-) + \sum_{i=1}^{m} (d_{ij}^+ - d_{ij}^-) \right) \\
\text{subject to} \quad & \mu_k (y_{ko} - y_{kj}) = d_{kj}^+ - d_{kj}^-, \quad k = 1, \cdots, p, \\
& \nu_i (-x_{io} + x_{ij}) = d_{ij}^+ - d_{ij}^-, \quad i = 1, \cdots, m, \\
& d_{kj}^+ - d_{kj}^- \leqq d_j, \quad k = 1, \cdots, p, \\
& d_{ij}^+ - d_{ij}^- \leqq d_j, \quad i = 1, \cdots, m, \\
& d_{kj}^+,\ d_{kj}^-,\ d_{ij}^+,\ d_{ij}^- \geqq 0, \quad k = 1, \cdots, p;\ i = 1, \cdots, m, \\
& \mu_k,\ \nu_i \geqq \varepsilon, \quad k = 1, \cdots, p;\ i = 1, \cdots, m, \\
& \varepsilon \text{ is a non-Archimedean infinitesimal,}
\end{aligned}
$$

where d^+_{kj}, d^-_{kj}, d^+_{ij}, d^-_{ij} are deviation variables.

We study the relations between the DEA efficiencies and the optimal value of GDEA under a assumption.

Theorem 2. *DMUo is value free efficient if and only if for sufficiently small* $\alpha > 0$, *the optimal value of GDEA is equal to zero.*

Proof. (Necessity) Suppose that the optimal value solution of GDEA is not equal to zero, i.e. there exists another DMUj such that for sufficiently small $\alpha > 0$,

$$\tilde{d}_j = \min\left\{d_j + \alpha\left(\sum_{k=1}^{p}(d^+_{kj} - d^-_{kj}) + \sum_{i=1}^{m}(d^+_{ij} - d^-_{ij})\right)\right\} < 0.$$

Therefore, from the constraints, the following inequalities hold. For $k = 1, \cdots, p,\ i = 1, \cdots, m$,

$$\mu_k(y_{ko} - y_{kj}) \leqq \min d_j < -\min\left\{\alpha\left(\sum_{k=1}^{p}(d^+_{kj} - d^-_{kj}) + \sum_{i=1}^{m}(d^+_{ij} - d^-_{ij})\right)\right\},$$

$$\nu_i(-x_{io} + x_{ij}) \leqq \min d_j < -\min\left\{\alpha\left(\sum_{k=1}^{p}(d^+_{kj} - d^-_{kj}) + \sum_{i=1}^{m}(d^+_{ij} - d^-_{ij})\right)\right\}$$

and since $\mu_k,\ \nu_i > 0$, the above inequalities imply that

$$(\boldsymbol{y}_j, -\boldsymbol{x}_j) \geq (\boldsymbol{y}_o, -\boldsymbol{x}_o).$$

This contradicts the value free efficiency of DMUo.
(Sufficiency) Suppose that DMUo is not value free efficient, i.e. there exists another DMUj such that $(\boldsymbol{y}_j, -\boldsymbol{x}_j) \geq (\boldsymbol{y}_o, -\boldsymbol{x}_o)$. Then $\min d_j \leqq 0$ and for all $\mu_k,\ \nu_i > 0$,

$$\begin{aligned}
&\sum_{k=1}^{p}\mu_k(y_{ko} - y_{kj}) + \sum_{i=1}^{m}\nu_i(-x_{io} + x_{ij}) \\
&= \sum_{k=1}^{p}(d^+_{kj} - d^-_{kj}) + \sum_{i=1}^{m}(d^+_{ij} - d^-_{ij}) \\
&< 0.
\end{aligned}$$

Therefore, since for sufficiently small $\alpha > 0$, $\tilde{d}_j = \min\left\{d_j + \alpha\left(\sum_{k=1}^{p}(d^+_{kj} - d^-_{kj}) + \sum_{i=1}^{m}(d^+_{ij} - d^-_{ij})\right)\right\} < 0$, which contradicts the assumption. □

Theorem 3. *DMUo is sum value efficient if and only if for sufficiently large $\alpha > 0$, the optimal value of GDEA is equal to zero .*

Proof. (Necessity) Suppose that the optimal value of GDEA is not equal to zero, i.e. there exists another DMUj such that for sufficiently large $\alpha > 0$, $\max \Delta = \max_{(\mu_k,\nu_i)} \tilde{d}_j < 0$. It means that

$$\max_{(\mu_k,\nu_i)} \left\{ \alpha \left(\sum_{k=1}^{p} (d_{kj}^{+} - d_{kj}^{-}) + \sum_{i=1}^{m} (d_{ij}^{+} - d_{ij}^{-}) \right) \right\} < 0$$

and we have that for $k = 1, \cdots, p$ and $i = 1, \cdots, m$,

$$\mu_k (y_{ko} - y_{kj}) \leqq 0 \text{ and } \nu_i(-x_{io} + x_{ij}) \leqq 0.$$

Thus for all $\beta : 0 < \beta < 1$,

$$(1-\beta) \sum_{k=1}^{p} \mu_k (y_{ko} - y_{kj}) + \beta \sum_{i=1}^{m} \nu_i (-x_{io} + x_{ij}) < 0.$$

This contradicts the assumption of sum value efficiency.
(Sufficiency) Suppose that DMUo is not sum value efficient. Then there exists another DMUj such that for all $\beta : 0 < \beta < 1$ and for some $\mu_k,\ \nu_i > 0$,

$$(1-\beta) \sum_{k=1}^{p} \mu_k (y_{ko} - y_{kj}) + \beta \sum_{i=1}^{m} \nu_i (-x_{io} + x_{ij}) < 0.$$

Taking $\beta = \frac{1}{2}$ and multiplying by $\alpha > 0$, it yields that

$$\alpha \left(\sum_{k=1}^{p} \mu_k (y_{ko} - y_{kj}) + \sum_{i=1}^{m} \nu_i (-x_{io} + x_{ij}) \right) < 0$$

and since α is sufficiently large,

$$\tilde{d}_j = \min \left\{ d_j + \alpha \left(\sum_{k=1}^{p} (d_{kj}^{+} - d_{kj}^{-}) + \sum_{i=1}^{m} (d_{ij}^{+} - d_{ij}^{-}) \right) \right\} < 0.$$

This contradicts the assumption. □

Theorem 4. *Let μ_k , ν_i be satisfied $\sum_{k=1}^{p} \mu_k y_{ko} / \sum_{i=1}^{m} \nu_i x_{io} = 1$ in GDEA. DMUo is ratio value efficient if and only if for sufficiently large $\alpha > 0$, the optimal value of GDEA is equal to zero.*

Proof. (Necessity) Assume that DMUo is ratio value efficient. Then there exist μ_k, $k = 1, \cdots, p$ and ν_i, $i = 1, \cdots, m$ such that

$$1 = \frac{\sum_{k=1}^{p} \mu_k y_{ko}}{\sum_{i=1}^{m} \nu_i x_{io}} \geqq \frac{\sum_{k=1}^{p} \mu_k y_{kj}}{\sum_{i=1}^{m} \nu_i x_{ij}}, \quad \text{for all } j = 1, \cdots, n.$$

It follows that for all $j = 1, \cdots, n$

$$\sum_{k=1}^{p} \mu_k y_{ko} = \sum_{i=1}^{m} \nu_i x_{io} \text{ and } \sum_{k=1}^{p} \mu_k y_{kj} \leqq \sum_{i=1}^{m} \nu_i x_{ij}$$
$$\Rightarrow \sum_{k=1}^{p} \mu_k (y_{ko} - y_{kj}) + \sum_{i=1}^{m} \nu_i (-x_{io} + x_{ij}) \geqq 0$$

and then for sufficiently large $\alpha > 0$,

$$\tilde{d}_j = \min \left\{ d_j + \alpha \left(\sum_{k=1}^{p} (d_{kj}^{+} - d_{kj}^{-}) + \sum_{i=1}^{m} (d_{ij}^{+} - d_{ij}^{-}) \right) \right\} \geqq 0.$$

Thus the results holds.
(Sufficiency) Suppose that DMUo is not ratio value efficient. Then there exists another DMUj such that

$$\frac{\sum_{k=1}^{p} \mu_k y_{ko}}{\sum_{i=1}^{m} \nu_i x_{io}} < \frac{\sum_{k=1}^{p} \mu_k y_{kj}}{\sum_{i=1}^{m} \nu_i x_{ij}},$$

which yields that

$$\sum_{k=1}^{p} \mu_k y_{ko} = \sum_{i=1}^{m} \nu_i x_{io} \text{ and } \sum_{k=1}^{p} \mu_k y_{kj} > \sum_{i=1}^{m} \nu_i x_{ij}$$
$$\Rightarrow \sum_{k=1}^{p} \mu_k (y_{ko} - y_{kj}) + \sum_{i=1}^{m} \nu_i (-x_{io} + x_{ij}) < 0.$$

Thus for sufficiently large $\alpha > 0$,

$$\tilde{d}_j = \min \left\{ d_j + \alpha \left(\sum_{k=1}^{p} (d_{kj}^{+} - d_{kj}^{-}) + \sum_{i=1}^{m} (d_{ij}^{+} - d_{ij}^{-}) \right) \right\} < 0.$$

It contradicts the assumption. □

4 Numerical Example

We illustrate our value free efficiency along an example. Assume that there are six DMUs, consuming one input and producing one output in Table 1.

DMU	A	B	C	D	E	F
input	2	3	4.5	4	5.5	6
output	1	3	3.5	2	5	4

Table 1: 1-input and 1-output

Figure 1 shows that DMU C is dominated by DMU B in terms of ratio value efficiency. On the other hand, Figure 2 shows that DMU C is dominated by a convex combination of points B and E in terms of sum value efficiency. However, DMU C is not dominated by any other Decision Making Units in terms of value free efficiency as can be seen in Figure 3. The reason is that linear combination of DMUs in some form can be also feasible in CCR model and Belton's model, while the feasible region of GDEA model consists of DMUs themselves.

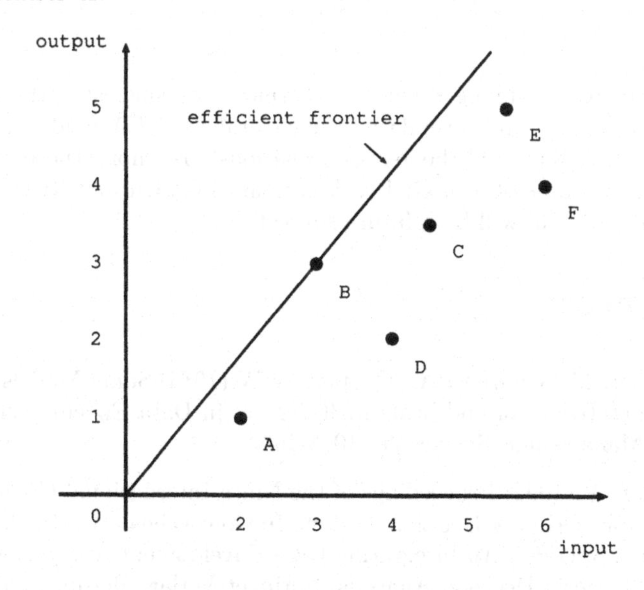

Figure 1: Ratio Value Efficiency

5 Conclusion

In this paper, we examined several concepts of efficiency, for example, ratio value efficiency in CCR model and sum value efficiency in Belton's model

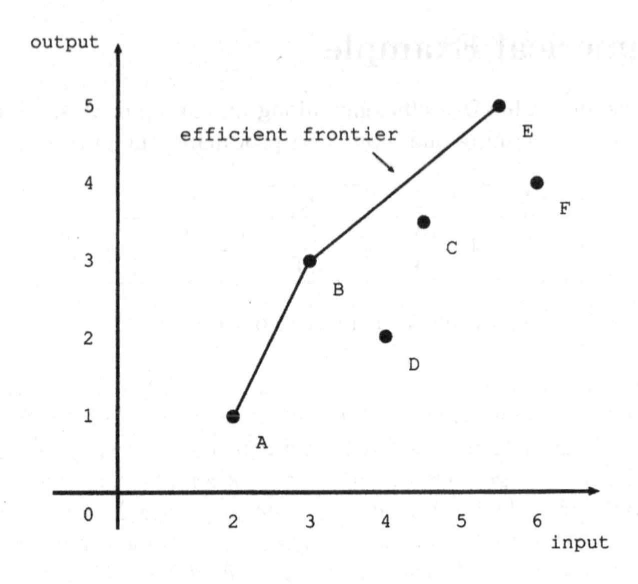

Figure 2: Sum Value Efficiency

and some relations among them. Furthermore, we suggested the notion of value free efficiency and formulated the generalized DEA model to evaluate it. We established several theorems on relationships among those efficiencies in terms of the value of α in GDEA. Relations of GDEA and REF model [7], or CCWH model [6] will be a future subject.

References

[1] Banker, M., Charnes, A., Cooper, W.W.(1984):Some Models for Estimating Technical and Scale Inefficiencies in Data Envelopment Analysis. Management Science 30, 1078-1092.

[2] Belton, V.(1992):Proceedings of the Ninth International Conference on Multiple Criteria Decision Making.In:Goicoechea, A., Duckstein, L., Zoints, S. (Eds.):An Integrating Data Envelopment Analysis with Multiple Criteria Decision Analysis. Springer-Verlag, Berlin, 71-79.

[3] Belton. V., Vickers, S.P.(1993):Demystifying DEA–A Visual Interactive Approach Based on Multiple Criteria Analysis. Journal of Operational Research Society 44, 883-896.

[4] Charnes, A., Cooper, W.W., Rhodes, E.(1978):Measuring the Efficiency of Decision Making Units. European Journal of Operational Research 2, 429-444.

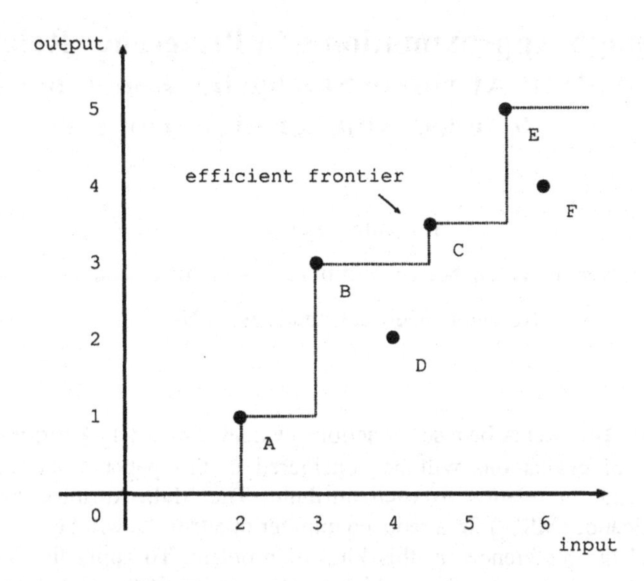

Figure 3: Value Free Efficiency

[5] Charnes, A., Cooper, W.W., Rhodes, E.(1979):Measuring Efficiency of Decision Making Units. European Journal of Operational Research 3, 339.

[6] Charnes, A., Cooper, W.W., Wei, Q.L., Huang, Z.M.(1989): Cone Ratio data Envelopment analysis and Multi-objective Programming. International of Systems Science 22, 1099-1118.

[7] Halme, M., Joro, T., Korhonen, P., Salo, A., Wallenius, J.(1996):A Value Efficiency Approach to Incorporating Preference Information in Data Envelopment Analysis. Helsinki School of Economics and Business Administration, Working Papers W–171.

[8] Joro, T., Korhonen, P., Wallenius, J.(1995):Structural Comparison of Data Envelopment Analysis and Multiple Objective Linear Programming. Helsinki School of Economics and Business Administration, Working Papers W–144. (Forthcoming in Management Science)

[9] Stewart, T.J.(1993):Relationships between Data Envelopment Analysis and Multiplecriteria Decision Analysis. Journal of Operational Research Society 47, 654-665.

Rough Approximation of a Preference Relation by a Multi-Attribute Stochastic Dominance for a Reduced Number of Attributes

Kazimierz Zaras

Université du Québec en AbitibiTémiscamingue, Rouyn-Noranda, Québec, Canada J9X 6N9

Abstract: Let A be a set of actions evaluated by a set of attributes. Two kinds of evaluations will be considered in this paper : determinist or stochastic in relation to each attribute. The Multi-Attribute Stochastic Dominance (MSD_r) for a reduced number of attributes will be suggested to model the preferences in this kind of problem. To apply the MSD_r the subset R of attributes from which approximation of the global preference is valid should be known. The theory of Rough Sets gives us an answer on this issue allowing us to determine a minimal subset of attributes that enables the same classification of objects as the whole set of attributes. In our approach these objects are pairs of actions. In order to represent preferential information we shall use a pairwise comparison table (PCT). This table is built for subset $B \subset A$ described by Stochastic Dominance relations for particular attributes and a total order for the decision attribute given by the decision maker (DM). Using a Rough Sets approach for the analysis of the subset of preference relations, a set of decision rules is obtained, and these are applied to a set A\B of potential actions. The Rough Sets approach of looking for the reduction of the set of attributes gives us the possibility of operating on Multi-Attribute Stochastic Dominance for a reduced number of attributes.

Keywords: Preferences; Multiple attributes; Rough sets theory; Multi-attribute Stochastic Dominance; Decision making

1. Introduction

Let there be a multi-attribute decision problem where the evaluations of actions can be determinist or stochastic, and the global preference is approximated from the subset of attributes. To solve this kind of problem the Multi-Attribute Stochastic Dominance (MSD_r) for a reduced number of attributes can be applied, but this application raises the question about which collection of attributes should be used to build the MSD_r. The Rough Sets approach deals with this problem by looking for the reduced set of attributes which provides the same quality of approximation as the original set of attributes. This paper attempts to show that this way of modelling can be matched by the Multi-Attribute Stochastic Dominance approach.

In the Rough Sets approach the point of departure is the global evaluation of the actions chosen as examples by a Decision Maker (DM). The consequence of the ambiguity of these evaluations is that some rules are non-deterministic, i.e. they can not be univocally described by means of "granules" of the representation of preferential information. A formal framework for dealing with the granularity of information has been given by Pawlak (1991), and called Rough Sets theory. The Rough Sets theory assumes a representation of the information in a table form where rows correspond to objects and columns to attributes. If the description of objects by a set of attributes is produced by the DM, these attributes are called decisions, and the remaining ones, conditions, and all together these form the decision table. This decision table is a particularly appropriate form for the description of decision sorting problems . As was shown by Greco, Matarazzo and Slowinski (1995), a direct application of the Rough Sets approach to multi-criteria decision analysis is not possible when the ordinal properties of the criteria have to be taken into account. In this paper, and following this way of reasoning, a decision table is supposed whose objects are pairs of actions. For each pair of actions the partial evaluations of the preferences in respect to each attribute from the Stochastic Dominance (SD) relations are given. To determine the type of the SD we need additional information : about the distribution of evaluations of each of the actions in relation to these attributes which are usually made by experts and about the attitude of the DM to facing risk. This paper is structured as follows. The problem is formulated in section 2. In this section we also present the results emanating from Stochastic Dominance conditions for each attribute. In section 3, these results are incorporated to build the rough sets table. In section 4, suggested approach is applied to solve a famous example of the selection of Nuclear Power Plant Sites described in Keeney and Nair (1977).

2. Formulation of the problem

The decision situation may be conceived as a problem (A, X, E) where the sets Action, Attribute, and Evaluation are defined as follows :

A - a finite set of actions a_i, $i= 1,2, \dots, n$;

X - a finite set of attributes X_k, $k= 1,2, \dots m$;

E - a set of distribution of evaluations $f(x_{ik})$,

where $f(x_{ik})= f_{ik}$, is the distribution of the evaluations associated with the performance of each action a_i with respect to the attribute X_k. The attributes are supposed to be probabilistically independent, and are also supposed to satisfy the preference independence condition. Thus, the overall comparison is between two actions a_i, $a_j \in A$, the first leading to a comparison of the two vectors of probability distributions on each attribute X_k, and the second to globally concluding that an action a_i, is at least as good as another action a_j where each $(a_i, a_j) \in A \times A$. In relation to each attribute the preferences can be modelled by using the Stochastic Dominance (SD) rules. We can consider, one of two groups of the SD for two classes of utility functions. The first one is for increasing concave utility functions (FSD, SSD, TSD), which means First, Second, Third Degree Stochastic Dominances, and the second one presented in Zaras (1989) is for increasing convex utility functions (FSD, SISD, TISD1 and TISD2) which means Second and Third Degree Inverse Stochastic Dominance of the first and second kind. The Stochastic Dominances are also defined for the discrete cases (see Hadar and Russel, 1969). In the case of determinist evaluations the cumulated distribution function $F(x_{ik})$ is a unit step function, and the preferences can be modelled based only on FSD which is proved for a class of increasing utility functions.

Generally speaking we assume one of two classes of utility functions : concave, which expresses that the DM is risk averse in relation to the attribute X_k, or convex, which expresses that the DM is risk prone in relation to the attribute X_k. In reality, the rules of the SD are more or less restrictive. For example the FSD rule which is proved for the increasing utility functions ($u' > 0$), according to Levy and Sarnat (1984) is very often observed in practice. The SSD rule is limited to the DM who is risk averse (concave increasing utility function, $u'>0$ and $u''\leq 0$). The TSD rule is still more restrictive, imposing non increasing absolute risk aversion. Then, the preference between two actions is not always clear from the SD rules because the attitude of DM which doesn't correspond to the risk situation which he is facing. We have the same difficulties modelling preferences from the rules of the SD for the class of convex utility functions Consequently, when using the SD rules in relation to each attribute X_k we may have the following situations :

- a_i SD_k a_j and not a_j SD_k a_i ;
- not a_i SD_k a_j and not a_j SD_k a_i ;

The last situation is possible for example if we have two distributions of evaluations between two actions exactly the same or the preference can not be explained by the SD. This last situation will generate non deterministic rules in the Rough Sets approach.

The problem consists of modelling global preferences. In 1978, Huang et all (1978) suggested the Multi-Attribute Stochastic Dominance (MSD). The rule of the MSD can be expressed in the following manner : " the action a_i is at least as great as the action a_j in the sense of the MSD, if and only if the action a_i dominates a_j by the SD in relation to each attribute". In practice this rule is very rarely verified. Based on psychological observation that people tend to simplify the multi-attribute complex problem by choosing a better action from the minimal subset of the most important attributes, we suggested the Multi-Attribute Stochastic Dominance for a reduced number of attributes (see Zaras and Martel (1994)). Given a_i, $a_j \in A$ and $R \subseteq Q \subseteq X$ this kind of the SD noted MSD_r can be defined as follows :

Definition 1

a_i MSD_r,a_j if and only if f_{ik} SD_k f_{jk} for all $X_k \in R$ where $r = |R|$ and $r \geq 2$, (1)

We can identify the same idea in the theory of Rough Sets. Looking for the reduced set of attributes which provides the same quality of an object's classification as the original set of attributes is one of the basic concepts of this theory. The core, which is defined as an intersection of minimal subsets of attributes, is a collection of the most significant attributes in the multi-attribute decision table. Application of Rough Sets to the concept of the MSD_r will allow us to find the subsets of attributes X_k from which base it will be possible to model global preferences.

3. Decision table and basic concepts of the Rough Sets theory

In order to represent the preferential information provided by the DM in the decision table, we will use the pairwise comparison of some actions as objects. It will be similar to the pairwise comparison table introduced in Greco, Matarazzo and Slowinski (1995).
Let B be a finite set of actions , considered by the DM as the basis for exemplary pairwise comparisons. Also let $C \subseteq X$ be the set of attributes (condition attributes) describing the actions, and D the decision attribute. The decision table is defined as 4-tuple $T = \langle H, C \cup D, V_C \cup V_D, g \rangle$, where $H \subseteq BxB$ is a finite set of pairs of actions, $C \cup D$ are two sets of attributes, called condition and decision attributes, $V_C \cup V_D$ are the values of the function g which is defined as follows :
$g : H \times (C \cup D) \rightarrow V_C \cup V_D$ is a total function where $V_C = \cup V_k$. This function is such that :

1) $g[(a_i, a_j),k] = 1$, if f_{ik} SD_k f_{jk} is verified for $\forall X_k \in C$, and $\forall (a_i, a_j) \in H$;

(2)

2) $g[(a_i, a_j),k] = 0$, if f_{ik} not SD_k f_{jk} if is verified $\forall X_k \in C$, and $\forall (a_i, a_j) \in H$.

In our decision table $g[(a_i, a_j),D]$ can also have two values on $H \subseteq BxB$:

1) $g[(a_i, a_j),D] = P$, if a_i is preferred to a_j;

(3)

2) $g[(a_i, a_j),D] = N$, if a_i is not preferred to a_j,.

These two situations are motivated by the hypothesis that the comprehensive preference model can be inferred by studying comprehensive evaluations made by the DM when presented with a set of representative pairs of actions H. Then, we suppose that the DM is able to express his preferences on the small number of

actions. The appeal of this approach is that the DM is typically more confident in exercising his evaluations than explaining them.
The set H is composed of two subsets: H_p which represents the preferences and H_N which represents non preferences on the set of actions B. The indifference is usually noted if the actions are not preferred in two directions : $g[(a_i, a_j),D] = N$ and $g[(a_j, a_i),D] = N$. The relation of incomparability can be identified during the SD verification. In the case where $g[(a_i, a_j),k] = 0$ and $g[(a_j, a_i),k] = 0$, it can be possible that one of the two indifference or incomparability relations is true. This situation introduces the elements of imprecision in the Rough Sets approach. In general the decision table can be presented as follows :

Table 1. Decision Table

		X_1	**X_2**	.....	**X_m**	**D**
	(a_i, a_j)	$g[(a_i, a_j),1]$	$g[(a_i, a_j),2]$	.....	$g[(a_i, a_j),m]$	$g[(a_i,a_j),D]=P$
	.	.	.	.....	.	.
H_p	.	.	.	.....	.	.
	.	.	.	.....	.	.
	.	.	.	.....	.	.
H_N	.	.	.	.....	.	.
	.	.	.	.....	.	.
	(a_s,a_t)	$g[(a_s, a_t),1]$	$g[(a_s, a_t),2]$	.....	$g[(a_s, a_t),m]$	$g[(a_s,a_t)D]=N$

Let $Q \subseteq C$ be a subset of condition attributes. We say that a pair of actions (a_i, a_j) and a pair of actions (a_s, a_t) are indiscernible by the subset of attributes Q in T iff $g[(a_i, a_j),k] = g[(a_s, a_t),k]$ for every $X_k \in Q$. Then every $Q \subseteq C$ generates a binary relation on H which is called an indiscernible relation, denoted by IND(Q). IND(Q) is an equivalent relation for any Q. Equivalence classes of IND(Q) correspond to "granules" of knowledge representation which will be denoted $Q(a_i, a_j)$, and called the Q-elementary set. Let $Q \subseteq C$ and $H_P \subseteq H$. The next step in the Rough Sets methodology is to determine $Q_*(H_P)$-lower approximation of H_P, and $Q^*(H_P)$-upper approximation of H_P, which are defined as follows :

$$Q_*(H_P)=\{ (a_i, a_j) \in H : Q(a_i, a_j) \subseteq H_P\}, \text{ and}$$
$$Q^*(H_P)= \{ (a_i, a_j) \in H : Q(a_i, a_j) \cap H_P \neq \emptyset\}. \quad (4)$$

Each decision table $T=< H, C \cup D, V_C \cup V_D , g>$ can be uniquely decomposed into two decision tables
$T_1= < H_P , C \cup D, V_C \cup V_D , g>$ such that $C \rightarrow_1 D$, and
$T_2 =< H_N , C \cup D, V_C \cup V_D , g>$ such that $C \rightarrow_2 D$,
where $H_P = POS_C(D)$ is a part of H where a_i is preferred to a_j. The Q-boundary (doubtful region) of set H_P is defined as

$$BN_Q(H_P) = Q^*(H_P) - Q_*(H_P) \qquad (5)$$

$Q_*(H_P)$ is the set of all pairs of actions from H which can be certainly classified as elements of H_P, employing the set of attributes Q. $Q^*(H_P)$ is the set of objects from H which can be possibly classified as elements of H_P using the set of attributes Q. The set $BN_Q(H_P)$ is the set of objects which cannot be certainly classified to H_P using the set of attributes Q. If $BN_Q(H_P) \neq \varnothing$, then H_P is the rough set. With every set $H_P \subseteq H$, we can associate an accuracy of approximation as follows :

$$\alpha_Q(H_P) = |Q_*(H_P)|/|Q^*(H_P)|. \qquad (6)$$

We also need an approximation of a partition of H, which is called the *quality of approximation of partition H*, and which is defined as follows :

$$\gamma_Q(H_P) = |Q_*(H_P) \cup Q_*(H_N)|/|H|. \qquad (7)$$

An important issue is that of attribute reduction, in such a way that the reduced set of attributes provides the same quality of classification as the original set of attributes. The minimal subset $R \subseteq Q \subseteq C$ such that $\gamma_C(H_P) = \gamma_Q(H_P) = \gamma_R(H_P)$will be denoted by RED_l (C).Let us notice that the decision table may have more than one reduct. The intersection of all reducts is called the core of C, i.e.

$$CORE(C) = \cap RED_l(C). \qquad (8)$$

The core is a collection of the most significant attributes in the decision table. From a reduced set of attributes we generate the decision rules. Any decision reduced rule, corresponding to a pair of actions (a_i, a_j) can be viewed as an implication $\varphi(a_i, a_j)$: $R \rightarrow V_D$ where $R \subseteq Q \subseteq C$. $\varphi(a_i, a_j)$ is a description of the pair of actions in terms of condition attributes from a reduced subset of attributes, and V_D is the value of the decision attribute.

4. Application

To illustrate the application of the Rough Sets approach modelling the preferences relation using Multi-Attribute Stochastic Dominance for a reduced number of attributes, let us consider the famous example of the selection of Nuclear Power Plant Sites given by Keeney and Nair (1977). The selection is described by two certain attributes :
X_1 – Site Population Factor,X_5 – Environmental Impact,and by four uncertain attributes, X_2 – Loss of Salmonids, X_3 – Biological Impact, X_4 – Socio-economic Impact,X_6 – Differential System Cost.

In relation to these attributes the probability assessments were given implicitly assuming that probabilistic independence exists among the attributes. The first step in our approach is the identification of the type of Stochastic Dominance by comparing the probability distribution of evaluations of sites two by two in respect to uncertain attributes. The results of this comparison in relation to the attribute X_4-Socioeconomic Impact are presented in Table 2.

Table 2. Stochastic Dominance relations identified in respect to the attribute X_4-Socioeconomic Impact

Sites	S1	S2	S3	S4	S5	S6	S7	S8	S9
S1	-	SSD	SSD	FSD	FSD	FSD	FSD	FSD	FSD
S2	0	-	SSD	FSD	FSD	FSD	FSD	FSD	FSD
S3	0	TD2	-	FSD	FSD	FSD	FSD	FSD	FSD
S4	0	0	0	-	0	FSD	FSD	FSD	0
S5	0	0	0	FSD	-	FSD	FSD	FSD	FSD
S6	0	0	0	0	FSD	-	0	FSD	0
S7	0	0	0	0	0	SID	-	FSD	0
S8	0	0	0	0	0	0	0	-	0
S9	0	0	0	FSD	0	FSD	FSD	FSD	-

The TD2 in table 2 means TISD2 and SID means SISD which are for the DM who is risk seeking. In this example the attitude of DM is suppose to be risk aversion, then in the first case S2 SSD S3 explain the preferences, in the second case SISD can not explain the preferences between two sites S7 and S6. Four first sorting sites given by the DM create a training set to build the decision table as a pairwise comparison table presented in Table 3.

Table 3. Decision Table

	X_1	X_2	X_3	X_4	X_5	X_6	D
(S1,S2)	0	0	0	1	1	1	P
(S1,S3)	0	0	1	1	1	1	P
(S1,S4)	0	1	1	1	1	1	P
(S2,S3)	0	0	1	1	1	1	P
(S2,S4)	1	1	1	1	1	1	P
(S3,S4)	1	1	1	1	0	0	P
(S2,S1)	1	0	0	0	1	0	N
(S3,S1)	1	0	0	0	0	0	N
(S4,S1)	1	0	0	0	0	0	N
(S3,S2)	1	0	0	0	0	0	N
(S4,S2)	0	0	0	0	0	0	N
(S4,S3)	0	0	0	0	1	1	N

In the table, the evaluations in relation to the decision attribute D make a dichotomic partition of the set of pair sites : H_P , where P means preference, H_N , where N means non preference. The partition of the set of pair sites in relation to the whole set of conditional attributes $X_k \in C$ is as follows :

$$I(C) = \{(1,2), (1,4), \{(1,3),(2,3)\}, \{(3,1),(3,2,),(2,1),(4,1)\}, (2,4), (4,2),(3,4),(4,3)\}$$

The accuracy of sets H_P and H_N corresponding to the classes of preference and non preference relations on the pair sites respectively, is equal to one, and the quality of approximation of the decision by the whole set C of attributes is also equal to one. It means that using all the condition attributes one can perfectly approximate the decision. The next step of the Rough Sets analysis is the construction of minimal subsets of independent attributes ensuring the same quality of classification as the whole set C, i.e. the reducts of C. There is one such reduct:

$$Red_1(C) = \{X_4\}$$

It can be said that the DM made his decision by taking into account only one attribute. The decision rules generated from the reduced decision table have the following form:

Rule No.1: IF S_i MSD_1 S_j in relation to the attribute $(X_4,)$ THEN $S_i \succ S_j$.
Rule No. 2:IF S_i not SD S_j in relation to the attribute $\{X_4\}$ THEN not $S_i \succ S_j$.

$S_i \succ S_j$ means that S_i is globally preferred to S_j, and not $S_i \succ S_j$ means that S_i is not globally preferred to S_j. The last step of the suggested methodology is to apply the Mullti-Attribute Stochastic Dominances for a reduced number of attributes to order the set of nine sites. This is very often a partial preference order on the set of potential actions. The overall binary preference relation noted ($\succ$) is identified if, the rule 1 is fulfilled between sites. If the rule 2 is fulfilled, the overall non-preference is identified which is noted (N$\succ$). The others which are not explained, will be noted ($\sim$). The results of this identification by comparison of all sites to each other are presented in Table 4.

Table 4. The global preference relation in the multi-attribute problem

MSD	S1	S2	S3	S4	S5	S6	S7	S8	S9
S1	N≻	≻	≻	≻	≻	≻	≻	≻	≻
S2	N≻	N≻	≻	≻	≻	≻	≻	≻	≻
S3	N≻	N≻	N≻	≻	≻	≻	≻	≻	≻
S4	N≻	N≻	N≻	N≻	N≻	≻	≻	≻	N≻
S5	N≻	N≻	N≻	≻	N≻	≻	≻	≻	≻
S6	N≻	N≻	N≻	N≻	N≻	N≻	N≻	≻	N≻
S7	N≻	N≻	N≻	N≻	N≻	~	N≻	≻	N≻
S8	N≻	N≻	N≻	N≻	N≻	N≻	N≻	N≻	N≻
S9	N≻	N≻	N≻	≻	N≻	≻	≻	≻	N≻

The preference relation from Table 4 allows us to determine the following order of sites :

S1→S2→S3→S5→S9→S4→{S6,S7}→S9→S8.

The stochastic dominance between two sites S6 and S7 can not explain the preferences, then we consider the relationship between these two sites as situation of incomparability.

5. Conclusions

I have been using the Rough Sets approach for the analysis of preferential information concerning multi-attribute uncertain choice and ranking problems. This information is given by the DM as a set of pairwise comparisons among some reference actions taking into account the preferences deduced from the SD rules in relation to each conditional attribute. The SD rules deal with partial uncertainty on the level of each attribute. The Rough Sets approach deals with global uncertainty on the level of aggregation. Finally, the global preference on the set of actions is approximated by means of Multi-Attribute Stochastic Dominance rules for a reduced number of attributes. They represent the preference model of the DM, which can be applied to a new set of potential actions.

References

- Greco, S.Matarazzo,B., and Slowinski, R.(1997) "Rough Approximation of a Preference Relation by Fuzzy Dominance Relation" *Proceedings of International Conference on Methods and Applications of Multicriteria Decision Making* , FUCAM, Mons, May 14-16.

- Greco, S.Matarazzo,B., and Slowinski (1997) "Rough Set Approach to Multi-Attribute Choice and Ranking problems", *ICS Research Report 38/95*, Warsaw University of Technology, Warsaw, 1995, and in : G.Fandel and T. Gal (Eds), *Multiple Criteria Decision Making*, Springer-Verlag, Berlin.

- Hadar, J. and Russel, W.(1969) "Rules for Ordering Uncertain Prospect", *American Economic Review 59.*

- Huang, C.C., Kira, D., Vertinsky, I. (1978) "Stochastic Dominance Rules for Multiattribute Utility Functions", *Review of Economics Studies,* vol. 41.

- Keeney, R.L. and Nair, K. (1977) "Selecting Nuclear Power Plant Sites in the Pacific Northwest Using Decision Analysis" in D.E.Bell, R.L. Keeney and H. Raiffa (Eds), *Conflicting Objectives in Decisions,* John Wiley&Sons, Chichester-New York-Brisbane-Toronto.

- Levy, H. and Sarnat, M.(1984) *Portfolio and Investment Selection: Theory and Practice,* Prentice Hall.

- Pawlak Z.(1997) "Rough set approach to kowledge-based decision support" *European Journal of Operational Research*, 99.

- Pawlak, Z, and Slowinski, R.(1994) "Rough Set Approach to Multi-Attribute Decision Analysis" *European Journal of Operational Research*, 72.

- Pawlak, Z.(1991) *Rough Sets. Theoretical Aspects of Reasoning about Data* Kluwer Academic Publishers, Dordrecht.

- Slowinski, R.(ed.)(1992) *Intelligent Decision Support. Handbook of Applications and Advances of the Rough Sets Theory* Kluwer Academic Publishers, Dordrecht.

- Zaras, K, and Martel J.M.(1994) "Multi-Attribute Analysis Based on Stochastic Dominance", in Munier B., and Machina M.J.(eds.) ,*Models and Experiments in Risk and Rationality* Kluwer Academic Publishers.

- Zaras, K.(1989) "Dominances stochastiques pour deux classes de fonctions d'utilité: concaves et convexes", *RO/OR, Recherche Opérationnelle*, vol.23.

Acknowledgment

Research was done in the framework which was directed by prof. Jean-Marc Martel for Defence Research Establishment Valcartier (DREV) in Canada.

An Evaluation of Multicriteria Decision-Making Methods in Integrated Assessment of Climate Policy

Michelle L. Bell, Benjamin F. Hobbs, Emily M. Elliott, Hugh Ellis, Zachary Robinson
Department of Geography and Environmental Engineering
Johns Hopkins University
313 Ames Hall, 3400 North Charles Street, Baltimore, Maryland, USA

Abstract: Those who conduct integrated assessments (IAs) are increasingly aware of the need to explicitly consider uncertainty and a range of criteria when evaluating alternative policies for preventing global warming. Multi-criteria decision-making (MCDM) methods provide a useful set of tools for understanding tradeoffs and gaining insight into policy alternatives. A difficulty facing potential MCDM users is the multitude of different techniques, each with distinct advantages and disadvantages. Methods differ widely in terms of their ease of use and appropriateness to the issue under consideration. Most importantly, different methods can yield strikingly different rankings of alternatives. A workshop was held to expose climate change experts, IA researchers, and policy makers to a range of MCDM methods and to evaluate and compare their potential usefulness to IA. Participants applied several methods in the context of a hypothetical greenhouse gas policy decision and evaluated each method. Analysis of method results and participant feedback through questionnaires and discussion provide the basis for conclusions regarding the use of MCDM methods for climate change policy and IA analysis.

Keywords: multi-criteria decision-making, climate change, experiment, weight selection, method evaluation, integrated assessment

1. Introduction

"Greenhouse gases" such as carbon dioxide, nitrous oxide, and water vapor raise the earth's temperature by altering the earth's radiation balance. Anthropogenic emissions of greenhouse gases have increased significantly since the industrial revolution, and may enhance the greenhouse effect. If such climate change occurs, potential impacts include sea-level rise, stressed ecosystems, and jeopardized human health. Climate change experts do not agree on the magnitude, distribution, or timeframe of global warming impacts and are faced with numerous uncertainties (Morgan & Keith 1995). Integrated assessment (IA) aids the understanding of climate change consequences through the development of complex models with interrelationships and feedbacks among system components (Dowlatabadi & Morgan 1993). The basic functions of IA are system modeling (simulation of physical, biological, and/or social systems) and decision evaluation (comparison of alternatives in terms of their risks and performance on important objectives). IAs are most useful to policy makers if they are explicitly linked to decision-making (NAPAP 1991). Given the importance and complexity of climate change, analysis of IA could be an important application of multicriteria decision-making (MCDM) methods. These methods can be used to improve the quality of decisions by providing information on tradeoffs, increase confidence in the decision, provide insight into the decision and alternatives, facilitate negotiation, and document the process.

This paper presents results from a workshop in which climate change experts used MCDM methods to evaluate climate change policies in the context of IA. The experiment is unique in that methods were applied and assessed by *experts*, rather than inexperienced subjects. The purposes of the workshop were to: (1) test the appropriateness, ease of use, and validity of each method for IA; (2) assess visualization techniques for displaying tradeoffs and risks; and (3) expose workshop participants to MCDM methods. Method performance was assessed through participant feedback (discussions and evaluation questionnaires) and analysis of method results. Limitations such as small sample size prevent definitive conclusions regarding the relative merits of the methods. Nonetheless, such case studies or quasi-experiments can provide useful information (Adelman 1991). For instance, such studies often possess an ecological validity (realism of problem setting and sophistication of participants) lacked by better controlled experiments such as those involving large numbers of undergraduates (e.g., Stillwell *et al.* 1987). Results of quasi-experiments with real practitioners combined with results from controlled experiments can yield more definitive conclusions than each type of study alone (Elmes *et al.* 1995). Many field studies of this nature have been conducted with beneficial results (Hobbs 1986, von Winterfeldt & Edwards 1986).

2. Experimental Design and Process

The workshop explored policy decisions for limiting greenhouse gas emissions: base case (no policy); global tax of $75, $150, or $300 per ton of carbon emitted; relaxed sulfur dioxide (SO_2) emission standards (which could have a cooling effect); promotion of nuclear power through nuclear fuel subsidies; and promotion of biomass energy. Policy attributes considered were temperature increase (1990 to 2050), sea-level rise (1990 to 2050), annualized SO_2 emissions (1990 to 2050), annualized nuclear waste generation (1990 to 2050), ecosystem stress (in 2050), and annualized cost (in 2050). Attribute values were global aggregate estimates obtained through use of the Holmes/Ellis IA Model (Holmes & Ellis 1998). Uncertain scenarios were generated with Monte Carlo simulation using probabilistic inputs for climate sensitivity, SO_2 cooling effect, energy efficiency, labor productivity, natural gas reserves, and population growth. Participants were informed that model runs were used to provide plausible attribute values for the purpose of evaluating MCDM methods, not definitive values for policy making.

Numerous MCDM methods have been proposed (e.g., Stewart 1992). The methods used for this research can be divided into three categories: weighting methods, deterministic ranking methods, and uncertainty ranking methods. A listing and brief description of the methods is provided in Table 1. For more detailed descriptions of the methods, see Hobbs & Meier (1998).

Table 1. MCDM Methods Used

METHOD	NOTES
Weighting Methods	w_i = weight for criterion *i*
Point allocation	Performed twice: a) 6 attributes, b) 4 attributes
Hierarchical point allocation	6 attributes
Swing weighting/Analytical Hierarchy Process (AHP)	Used swing method to compare two attributes at a time. Inconsistencies resolved via AHP eigenvector method (Saaty 1980)
Tradeoff weighting	Tradeoff annualized cost
Revision of weights	Participants were given weighting results for each of the above methods, and asked to provide a final set of weights.
Deterministic Ranking Methods	x_{ij} = value for criterion *i* for alternative *j*; x_i^{**} = best value for criterion *i*, among all alternatives; x_i^{*} = worst value for criterion i, among all alternatives; *n* = number of criteria. Revised weights used except where specified.
Initial Holistic Assessment	Alternatives ranked from most desirable (1) to least desirable (7), and rated from most desirable (100) to least desirable (0). Performed twice: a) 6 attributes, b) 4 attributes
Additive Linear Value Function V(x)	Repeated for all sets of weights. $v_i(x_{ij})$ = single criterion value function, $v_i(x_i^{**}) = 1$, $v_i(x_i^{*}) = 0$ $\underset{j}{MAX} \quad V(x_j) = \sum_{i=1}^{n} w_i v_i(x_{ij})$
Non-Linear Value Function V(x)	a) Mid-value splitting: $x_{i0.5}$ = user-specified value for each attribute *i* that is halfway in desirability between the best and worst values $v_i(x_{i0.5}) = 0.5$ b) For each attribute, participants drew $v_i(x_i)$ on a graph.
Goal programming	a) p = 2, b) p = ∞ g_i = user-specified max acceptable value $\underset{j}{MAX} \sum_{i=1}^{n} w_i \left(MAX(0, v_i(g_i) - v_i(x_{i,j}))\right)^p$
ELECTRE I	Set of alternatives that are not outranked defines a "kernel."
Revision of ranks & ratings (final holistic assessment)	Participants were given results for the deterministic ranking methods, and asked to provide a final set of ranks and ratings.

Table 1. MCDM Methods Used (Continued)

Uncertainty Ranking Methods	x_{ijk} = value for criterion *i* for alternative *j* for simulation *k*, *K* = number of simulations, x_i^{**} and x_i^{*} = best and worst values, respectively, for criterion *i* among all alternatives and simulations. Revised weights used.
Initial Holistic Assessment	
Linear Utility Function U(x)	$u_i(x_{ijk})$ = single criterion utility function, $u_i(x_i^{**}) = 1$, $u_i(x_i^{*}) = 0$ $U(x_j) = \sum_{i=1}^{n} w_i \sum_{k=1}^{K} u_i(x_{ijk})/K$
Non-Linear Utility Function *U(x)*	Gamble method: User-specified value $x_{i0.5}$ so that he/she is indifferent between a deterministic alternative, $x_{i0.5}$, and a gable, 50/50 change of best and worst values, $u_i(x_i^{**}) = 1$, $u_i(x_i^{*}) = 0$. Repeated for all attributes.
Regret	Regret (R_{jk}): Loss in utility under scenario *k* if policy *j* is chosen rather than the best alternative under that scenario. Used linear *U(x)*. $R_{jk} = \underset{h=1,\dots,n}{\mathrm{MAX}} U(x_{hk}) - U(x_{jk})$ Minimize maximum regret: $\mathrm{MIN} \underset{k=1,\dots,K}{\mathrm{MAX}} R_{jk}$
Stochastic dominance	Results are whether one alternative dominates another a) 1st order dominance, b) 2nd order dominance
Revision of ranks & ratings (final holistic assessment)	Participants were given results for the above uncertainty ranking methods, and asked to provide a final set of ranks and ratings.

The workshop was held June 1 and 2, 1998, at Johns Hopkins University, Baltimore, Maryland. Participants included 20 climate change experts, policy-makers, and IA practitioners, from academic, governmental, national laboratory, and corporate organizations. Each method was explained to provide participants with the conceptual understanding necessary to answer questionnaires, which were used to elicit the information needed to apply the methods. Results were calculated and provided for participants at the workshop. Nominal group discussions were held to elicit participant views on the methods and their application to IA and to gain insight into the thought processes used to answer questionnaires. Additional evaluation questionnaires asked participants to rate the appropriateness and ease of use of each method.

All experiments must be viewed in light of their limitations. Our results' internal validity may suffer from limitations (e.g., small sample size) that preclude control for alternate hypotheses (Adelman 1991). Various methods' results may differ due to fundamental differences in the types of responses they elicit, or due to an order effect, as the process of completing MCDM exercises and related discussions can provide insight into the decision. While this constrains our comparisons of methods, such insights are a benefit of MCDM and help people reflect upon their choices and focus on objectives and tradeoffs. Although a rigorous experimental design was not possible due to sample size and time limitations, our results are useful because they can aid the application of MCDM to IA. The experiment's external validity was increased by the use of participants who may actually use MCDM methods in climate change decision-making. Qualitative insights from the experts, which does not suffer from the aforementioned limitations, was an important outcome of this research.

3. Hypotheses, Results, and Discussion

Hypothesis 1. MCDM methods have different predictive validity. MCDM results should be valid (i.e., reflect actual preferences). Because preferences are subjective and often imprecise, there exists no universally accepted measure of validity (Larichev 1992). "Predictive validity" measures a method's ability to predict the holistic assessment (unaided judgment). We examined Spearman's correlations between each method's alternative rankings and the final holistic rankings (deterministic final holistic assessment for deterministic methods, uncertainty final holistic assessment for uncertainty methods) (Figure 1). Such an "intermethod correlation" is defined as the correlation between two methods' results for a specific user, averaged across all users: $\frac{1}{W}\sum_{a=1}^{W}\frac{\text{cov}(r_{sa}, r_{ta})}{\sigma_{r_{sa}} \cdot \sigma_{r_{ta}}}$ where r_{sa} and r_{ta} = participant a's ranks for methods s and t, respectively; W = number of users; and cov( ) = covariance.

No method's ranks are highly correlated with the holistic assessment's ranks, which is consistent with previous research (e.g., Hobbs *et al.* 1992, Hobbs 1986). Predictive validity of the linear V(x) with cost tradeoff weights is statistically lower than that of the linear V(x) using any other weighting method (Wilcoxon signed-rank test p-values 0.001). The mid-value splitting, non-linear V(x)'s predictive validity is significantly lower than that of the linear V(x) with revised weights (p-value 0.006). All uncertainty methods have approximately equal predictive validity. Our hypothesis that methods have different validity holds for the deterministic methods, but not for uncertainty methods. Revised weights were used for all uncertainty methods, and it appears that differences in rankings resulted more from different weights than method differences. Method results may differ because methods frame the problem differently, or because users learn and change their opinions throughout the process. The implications for MCDM use are the same in either event: inconsistencies between methods, whether from ordering effects or inherent method differences, can help people focus on their objectives and provide an opportunity to reflect further on the decision. In this sense, the use of several MCDM approaches is preferable to a single approach.

Figure 1. Average Intermethod Correlations Final Holistic Assessments

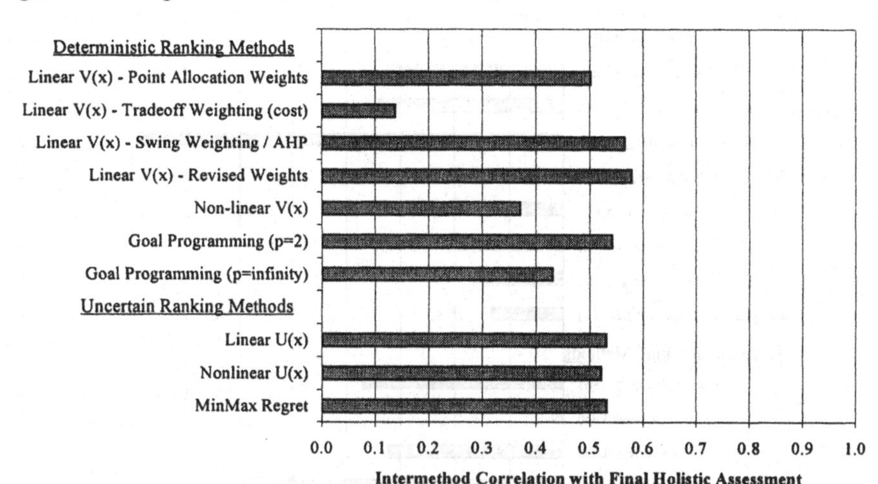

Hypothesis 2. Methods differ in the convergence of different persons' results. Differences in interperson correlations have implications for consensus building through MCDM, as users can explore different methods' results and framing of the decision. Mean interperson correlations for different methods were compared to determine if some methods produce more similar results across all persons than others. We define "interperson correlation" as the correlation between a pair of users' results for a given method, averaged across all pairs of users: $\frac{1}{\binom{W}{2}} \cdot \sum_{a=1}^{W} \sum_{\substack{b=1 \\ (b>a)}}^{W} \frac{\mathrm{cov}(r_{sa}, r_{sb})}{\sigma_{r_{sa}} \cdot \sigma_{r_{sb}}}$ where r_{sa} and r_{sb} = ranks from method s for participants a and b, respectively. Formal MCDM methods generally have greater convergence of different persons' results than holistic evaluation (e.g., Hobbs & Meier 1994). Surprisingly, this was not the case with this experiment. Interperson correlations ranged from 0.02 [linear *U(x)*] to 0.79 [linear *V(x)* with cost tradeoff weighting] (Figure 2). Most participants had similar weights for linear *V(x)* with cost tradeoff weighting, with high weight on cost (average cost weight = 53%), causing the high interperson correlation (statistically different from other deterministic methods', Wilcoxon signed-rank test p-value < 0.000). Interperson correlations for goal programming ($p = 2$ and $p = \infty$) are statistically lower than correlations for the holistic assessment and the linear *V(x)* with point allocation weights, cost tradeoff weights, or revised weights (p-values < 0.000 to 0.03). The linear *U(x)* and min max regret methods' interperson correlations are statistically different from other uncertainty methods' interperson correlations (p-values < 0.000).

Figure 2. Interperson Correlations for MCDM Methods

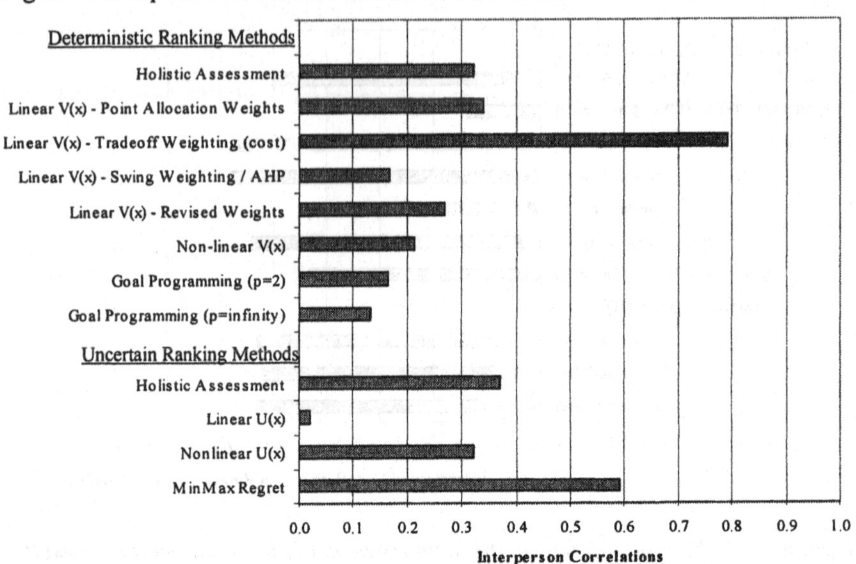

Hypothesis 3. Climate change experts are subject to classic weighting biases. Direct weighting methods have been repeatedly shown to suffer from various biases. For example, experiments using inexperienced subjects found that less weight is given to an attribute when is split into several attributes (a hierarchical approach, the user first assigns weights to broad categories, then allocates each category's weight among subcategories) than when the original criterion is categorized as a single attribute (direct approach, attributes presented simultaneously) (Eppel 1992, Stillwell *et al.* 1987). We hypothesize that this "splitting bias" also applies to experts. Participants weighted six attributes by direct point allocation and by a hierarchical approach [allocate weights: (1) between cost (x_1) and an environmental category; (2) then among the environmental categories of SO_2 emissions (x_2), nuclear waste (x_3), and climate; and (3) finally among the climate attributes of temperature increase (x_4), sea-level rise (x_5), and ecosystem stress (x_6)]. As splitting bias would predict, the average sum of weights for x_4, x_5, and x_6 is higher for direct point allocation (42%) than the hierarchical approach (22%) (t-test p-value 0.008).

Hypothesis 4. Weights chosen by different methods differ significantly. We hypothesize that *how* weights are chosen (the method used) can be as important as *who* chooses the weights. Comparison of weighting methods' intermethod and interperson correlations tested this hypothesis (Table 2). The cost tradeoff weighting method almost uniformly produced higher weight for cost, causing this method to have low intermethod correlations and the highest interperson correlation (0.46, statistically higher than other interperson correlations: Wilcoxon signed-rank test p-values < 0.000). Intermethod correlations of weights can be divided into two groups: high (0.78, 0.88, 0.89) and low (0.39, 0.49, 0.58) (p-values 0.001 to 0.04). This indicates that tradeoff weighting elicits very different weights. Comparison of intermethod (range: 0.39 to 0.89) and interperson correlations (range: 0.03 to 0.46) indicates that *who* performs the exercise and *which method* is used both have significant impact on the resulting weights. This analysis is subject to the limitations discussed earlier. We cannot determine the specific cause of variations in weights. Use of multiple weighting methods may be preferred to a single approach, either because methods *frame* the problem differently, thereby providing a learning opportunity, or because decision-makers learn from the *process* of applying methods.

Table 2. Intermethod Correlations of Weights from Weighting Methods

Weighting Method	Point allocation	Swing weighting/ AHP	Tradeoff weighting (cost)	Revision of weights
Average Intermethod Correlations of Ranks for the Same Person				
Point allocation		0.78	0.58	0.89
Swing weighting / AHP			0.39	0.88
Tradeoff weighting (cost)				0.49
Average Interperson Correlation of Ranks for the Same Method				
	0.05	0.03	0.46	0.06

4. Insights into the Application of MCDM to Climate Change and Integrated Assessment

The workshop provided the opportunity to learn from climate change experts how MCDM methods may best be applied to climate change policy evaluation. Five key points were derived from participants' feedback, and are applicable to decisions other than climate change.

1. Decision-makers must have confidence in attribute values. Whether estimates of attribute values are *believed* was crucial for some participants. The consequences of particular policies are uncertain, as can be seen from the current heated debates on climate change science. Each decision-making entity must have a basis for determining what will or might happen if a certain action is taken, even if they disagree on that outcome, in order to compare and input values for MCDM methods. Incorporation of uncertainty can help address this issue.

2. Attributes and objectives must be clearly defined and must encompass the decision-makers' values. To allow flexibility, we did not specify general objectives; participants operated according to their own preferences. Some participants found the exercises difficult without explicit statements of *who* is the decision-maker (e.g., government agency) and what they care about (e.g., individual versus social goals). Information some participants considered valuable, such as distribution of cost, was omitted. Although actual decision-making would involve identifying appropriate criteria, the participant feedback stressed the importance of explicit and complete objectives and attributes. Determining what the decision-makers value may be especially complex for climate change, as goals are often divergent or unspecific.

3. *Anchoring, although a potential source of bias, may be helpful or necessary.* To prevent "anchor and adjust" bias (anchoring on the numbers presented), we did not provide "anchors." However, several workshop participants commented that exercises were difficult without a reference point. In such cases, anchors, although a source of bias, may be necessary. Offering a range of anchors limit the bias.

4. *MCDM methods that involve decomposition of the problem require special attention.* Several methods require users to decompose the problem by valuing changes in some attributes, holding other attributes constant (e.g., tradeoff weighting). For experts who are highly familiar with feedbacks between attributes, this was difficult to impossible, indicating that such methods require additional time to allow users to feel comfortable or that other methods may be more appropriate for IA.

5. *Decision-makers gain insights from the process of applying MCDM methods.* Insights can be gained from the process, which is a primary benefit of MCDM. This sentiment was echoed in our workshop. Many participants felt they learned about *how* they think about the problem by completing the exercises and examining results. For example, one person said he wanted to do the "right" thing when he completed the holistic assessment, so he choose a costly alternative, whereas with other methods he placed more emphasis on cost. This caused him to reconsider his thought process.

5. Summary

A workshop with 20 climate change experts and IA practitioners explored the application of MCDM methods to a climate change policy and IA. Participants applied MCDM methods (weighting, deterministic ranking, and uncertainty ranking) to a hypothetical climate change decision. Methods were compared and evaluated through analysis of method results and participant feedback (questionnaires and nominal group discussions). Differences between methods were identified and found to present opportunities to learn about the decision process. Additional hypotheses explored by this research (participant evaluation of methods and visualization methods) will be addressed in later papers. We would like to thank workshop participants. This research was supported by the National Science Foundation (SBR9634336).

References

Adelman, L. (1991): Experiments, Quasi-Experiments, and Case Studies: A Review of Empirical Methods for Evaluating Decision Support Systems. *IEEE Trans. Systems Man & Cybernetics* 21(2), 293-301

Dowlatabadi, H., Morgan, M. G. (1993): Integrated Assessment of Climate Change. *Science* 259, 1813-1814

Elmes, D. G., Kantowitz, B. H., Roediger, III, H. C. (1995): *Research Methods in Psychology*, 5th Edition, West Publishing Co., St. Paul, Minnesota

Eppel, T. (1992): Description and Procedure Invariance in Multiattribute Utility Measurement. School of Management, Purdue Univ., West Lafayette, Indiana

Hobbs, B. (1986): What Can We Learn From Experiments in Multiobjective Decision Analysis? *IEEE Trans Syst. Man Cyber.* SMC-16, 384-394

Hobbs, B. F., Chankong, V., Hamadeh, W., Stakhiv, E. Z. (1992): Does Choice of Multicriteria Method Matter? An Experiment in Water Resources Planning. *Water Resources Research* 28(7), 1767-1780

Hobbs, B. F., Meier, P. M. (1998): *Energy Decisions and the Environment: A Guide to the Use of Multicriteria Methods*, in preparation.

Hobbs, B. F., Meier, P. M. (1994): Multicriteria Methods for Resource Planning: An Experimental Comparison, *IEEE Transactions on Power Systems* 9(4), 1811-1817

Holmes, K. J., Ellis, J. H. (1998): An Integrated Assessment Modeling Framework for Assessing Primary and Secondary Impacts from Carbon Dioxide Stabilization Scenarios, *Environmental Modeling and Assessment*, accepted pending revisions.

Larichev, O. I. (1992): Cognitive Validity in Design of Decision-Aiding Techniques. *Journal of Multi-Criteria Decision Analysis* 1, 127-138

Morgan, M. G., Keith, D. W. (1995): Subjective Judgments by Climate Experts. *Environmental Science & Technology* 29(10), 468A-476A

National Acid Precipitation Assessment Program (NAPAP), Oversight Review Board (1991): The Experience and Legacy of NAPAP, Report to the Joint Chairs Council of the Interagency Task Force on Acidic Deposition, Washington DC

Saaty, T. L. (1980): *The Analytical Hierarchy Process*. McGraw-Hill, New York

Stewart, T. J. (1992): A Critical Survey on the Status of Multiple Criteria Decision Making Theory and Practice. *OMEGA, The International Journal of Management Science* 20(5/6), 569-586

Stillwell, W. G., von Winterfeldt, D., John, R. S. (1987): Comparing Hierarchical and Nonhierarchical Weighting Methods for Eliciting Multiattribute Value Models. *Management Science* 33(4), 442-450

von Winterfeldt, D., Edwards, W. (1986): *Decision Analysis and Behavioral Research*. Cambridge University Press, New York

A Revised Minimum Spanning Table Method for Optimal Expansion of Competence Sets

C. I. Chiang[1], J. M. Li[2], G. H. Tzeng[3], and P. L. Yu[4]

1, 2, 3: Visiting scholars at School of Business, University of Kansas. 1 and 3 from Institute of Traffic and Transportation, National Chiao Tung University, Taipei, Taiwan, and 2 from Institute of Applied Mathematics, Guizhoou University of Technology, Guiyang, China.
4: C. A. Scupin Distinguished Professor, School of Business, The University of Kansas, Lawrence, Kansas.

Abstract. Competence set analysis is a new approach to solve an import aspect of challenging MCDM problems. Thus, many researchers have developed efficient methods for solving the optimal expansion of competence sets (please see the references). The minimum spanning table method (MST) proposed by Feng and Yu (1997) is an efficient algorithm using spreadsheet tableaus to solve optimal expansion problems. In order to develop a computer program, in this paper, we propose the revised algorithm of MST, which can efficiently solve the optimal expansion problems. We develop a user-friendly program called MINST, which can solve fairly large-scale expanding problems even with PC of pentium 133.

Keywords. Competence set expansion, habitual domains, minimum spanning table.

1 Introduction

There are many ways to attack challenging MCDM problem. Traditionally, we first try to find the sets of criteria, the alternatives and then identify the possible good or nondominated solutions. While this approach can produce good results, it can't solve many challenging MCDM problems. Thus Yu and Zhang (1989) propose a new concept of competence set analysis based on habitual domains theory. The main idea is first to identify what competence sets (including knowledge, skills, resource, information) are needed to solve a given challenging problem, then try to identify the way to effectively and/or efficiently acquire the needed competence. For the details, please see Yu (1990) and Yu and Zhang (1990). Many articles based on this concept have provided insight of efficient method to acquire the needed competence set.

Given the cost functions of acquiring j from i, denoted by $c(i, j)$, among the skills of the needed competence set (CS), the problem of how to expand from a subset of CS to the entire CS has been studied analytically and mathematically by Yu and Zhang (1990) when $c(i, j)$ is symmetric, and by Shi and Yu (1996) when

$c(i, j)$ is asymmetric. Using the deduction graph without cycles, Li and Yu (1994) reported a method of how to solve the problem of competence set expansion when there are compound skills, intermediate skills and skills of the multilevel proficiency. The above methods eventually convert the problem into mathematical programming problems.

The minimum spanning table method (MST) of Feng and Yu (1997) is an efficient algorithm using relevant spreadsheet tableaus to solve optimal expansion problems. In order to develop a computer code to solve large-scale problems, we propose a revised minimum spanning table (RMST) method by introducing a bookkeeping method and modifying some technical operations of the MST method. Based on the RMST method, we developed a user-friendly computer program, called MINST, which can solve easily expansion problems with table up to 100×100 that covers 100 skills and 9900 connections, which can be further expanded without difficulty.

This paper is organized as follows: Section 2 briefly describes the terminology and procedure of MST method. Section 3 elaborates the procedures of the RMST method and gives a numeric example. Section 4 a conclusion and some further research are offered.

2 Minimum Spanning Table (MST) Method

Let us first briefly describe some terminology in the MST method. An *expansion table*, as shown in table 1, is a matrix representation of a digraph, in which the component of row i and column j stands for the cost function denoted by $c(i, j)$ of acquiring from skill x_i to skill x_j.

Table 1 Expansion table

	x_1	...	x_j	...	x_m
x_1					
:					
x_i			$c(i, j)$		
:					
x_m					

Except for complete graphs, an expansion table may have empty cells indicating that the corresponding arc, say $< x_i, x_j >$, does not exist in the digraph. When $c(i, j)$ is well defined, the corresponding element of the table and, for simplicity, the cost $c(i, j)$ is called a *connecting element*, or simply *conn-element*. Note, a conn-element $c(i, j)$ is connecting x_i from x_j . Thus, the node x_i in the rows is the *out-node*, and node x_j in the column is the *into-node*.

The MST method has the following procedures:

Step 0. Initializing condition.

Step 1. Selecting and marking procedure.
Step 2. Cycle detecting procedure.
Step 3. Crossing out procedure.
Step 4. Stopping rule.
Step 5. Compressing procedure.
Step 6. Unfolding procedure.
For a detailed description of the MST method, refer to Feng and Yu (1997).

3 Revised Minimum Spanning Table (RMST) Method

3.1 An Overview

The RMST method consists of eight procedures: *Initializing*, *Choosing*, *Constructing candidate list*, *Sorting*, *Marking and Cycle Detecting*, *Compressing*, *Unfolding* and *Output* procedures. This method, as compared with MST, has two new features: bottom line shortcut and bookkeeping. The bottom line shortcut is a heuristic but efficient way to find a candidate for a minimum spanning tree, which may be further refined. The bookkeeping allow us to detect a cycles in the forward procedure and track down a minimum spanning tree in the unfolding (backward) procedure.

The RMST method starts with appending two rows, *marked index(MI)* and *cycle index(CI)*, to the expansion table. Then, by bottom line shortcut, the minimal conn-element of each column is chosen from the augmented expansion table. The chosen conn-element, its corresponding out-node, into-node, *MI*, and *CI* form a *record*, and the collection of such records forms a so-called *candidate list*. After the formation of the candidate list, the list is sorted by the *key values* (that is, the chosen conn-elements), so that the list is in nondecreasing order of the chosen conn-elements.

In the marking and cycle detecting procedure, RMST applies *MI* to indicate if a conn-element is marked, and *CI* to indicate if a conn-element forms a cycle with conn-elements that are already marked. In the compressing procedure, we store some data for tracing compression to *six stacks*, so that the unfolding procedure can be more easily done. The above procedures are based on bookkeeping method. In addition, the output procedure can rearrange the output of the unfolding procedure as to easily draw the minimum spanning tree.

The technical details of RMST method will be elaborate in section 3.2, and a numerical example will be given in section 3.3.

3.2 Procedures of RMST

For convenience, assume that there are *m* skills in an expansion table.

Initializing procedure: Augment the expansion table by appending two rows to the original expansion table, and label them *marked index* (*MI*) and *cycle index*

(CI), respectively. Let all elements of row MI and CI be 0. An augmented expansion table is shown in Table 2.

Table 2 Augmented Expansion Table

	x_1	x_2	...	x_j	...	x_m
x_1						
x_2						
:						
x_i				$c(i,j)$		
:						
x_m						
MI	0	0	...	0	...	0
CI	0	0	...	0	...	0

Choosing procedure: Select the minimal conn-element from each column in the augmented expansion table. Randomly choose, if there are more than one minimal conn-elements in each column. The chosen minimal conn-element of column j is denoted by $C(j)$, and the corresponding *out-node* and *into-node* are denoted by $O(j)$ and $I(j)$, respectively. In other words,

$$C(j) = \underset{i}{Min}\ c(i,j),\ i,j=1,2,\ldots,m, \tag{1}$$

$$O(j) \in \{k \mid \underset{i}{Min}\ c(i,j) = c(k,j)\}, \tag{2}$$

$$I(j) = j,\ j=1,2,\ldots,m. \tag{3}$$

Constructing the candidate list procedure: Construct the candidate list, which is a collection of records ($R_1, R_2, \ldots, R_m$) that consists of 5 fields: *conn-element*, *out-node*, *into-node*, *MI*, and *CI*. The conn-element field stores the chosen conn-elements; the out-node and into-node fields store out-nodes and into-nodes, respectively. The *MI* field stores marked index, and the *CI* field stores cycle index. The structure of the candidate list is shown as Table 3.

Table 3 Candidate list

Record	*conn-element*	*out-node*	*into-node*	*MI*	*CI*
R_1					
R_2					
:					
R_j					
:					
R_m					

Sorting procedure: Rearrange the records ($R_1, R_2, \ldots, R_m$) of the candidate list. Let the conn-element field be the *key value*. Sort and reorder the records according to the key value in nondecreasing order. That is, find a permutation, σ, such that $C(\sigma(j)) \le C(\sigma(j+1)),\ 1 \le j \le m-1$. This procedure produces a *sorted candidate list.*

Marking and cycle detecting procedure: Sequentially check each record in the sorted candidate list. For instance, check the ith record, denoted by R_i. First, set *MI* of R_i as 1. That is, set *MI(i)*=1, and let temporary variable, *temp*, be out-node of R_i. Next, select the record R_j whose into-node is that corresponding to *temp*. As R_j is found, check whether *MI(j)* equals 0 or 1. If *MI(j)* equals 0, the inclusion of R_i's conn-element does not form a cycle. In this case, clear all cycle indexes in the list (i.e. reset all *CI* of marked records corresponding to *MI*=1 to be 0). Then, add *count*, the number of marked record, by 1, i.e. *count*=*count*+1. After that, continue to check the next record R_{i+1}. If *MI(j)* equals 1, set *CI(j)*=1 and let *temp* be out-node of R_j. Then continually select the record whose into-node is that corresponding to *temp*, and check its *MI* with 1.

Repeat the above selecting and checking *MI* process. If the newly selected record is identical to the starting record R_i, then a *cycle* is detected. Once a cycle is detected, append the records with *MI*=1 to the unfolding list and go to the *compressing procedure*.

Stopping Rule: If no cycle exists after checking m-1 records, i.e. *count*=m-1, then we have found the minimum spanning tree. In this case, append the records with *MI* =1 to the unfolding list and go to the *output procedure*.

Compressing procedure: Assume the cycle detected has n nodes. Compress the nodes in the cycle, denoted by **C**, into a compressed node, denoted by x_{r+m}, where r stands for the rth compression. Then transform the expansion table to the one in the next stage, which has m-n+1 nodes, and the corresponding cost function as follows:

$$c(x_i, x_j) = c_{ij} \quad if \quad x_i \notin \mathbf{C} \text{ and } x_j \notin \mathbf{C} \tag{4}$$

and for any node x_i not in **C**, define

$$c(x_{r+m}, x_i) = \min\{c(y, x_i),\ y \in \mathbf{C}\}, \tag{5}$$

$$c(x_i, x_{r+m}) = \min\{c(x_i, y) + c(x_s, x_t) - c(x_y, y),\ y \in \mathbf{C}\}, \tag{6}$$

where $c(x_s, x_t)$ is the largest conn-element in cycle and x_y is such that (x_y, y) is the conn-element in the cycle **C**. The equations (5) and (6) are called the *transformation equations* (Feng and Yu, 1997).

In order to facilitate the tracking of compression for the Unfolding procedure, six stacks, named S^1, S^2, S^3, S^4, S^5, and S^6, are employed for storing $c(x_i, x_{r+m})$, x_i, y, $c(x_{r+m}, x_i)$, y, and x_i in this procedure. The way to implement these stacks is by using a two-dimensional array, say $S_{p,q}$, in which p stands for the total number of compressing, up to this point and q for the number of nodes not in the cycle. Let y be the component which solves equation (6) for $c(x_i, x_{r+m})$. Store $c(x_i, x_{r+m})$, x_i, and y respectively in S^1, S^2, and S^3. Thus, $S^1_{r,k} = c(x_i, x_{r+m})$, $S^2_{r,k} = x_i$, $S^3_{r,k} = y$, where the subscript (r, k) indicates stack's row and column number respectively, with r indicating the rth compression and k indicating kth node that is not in the cycle. That is, $(S^2_{\bullet k}, S^3_{\bullet k})$ is the arc which solve (6) with $c(x_i, x_{r+m}) = S^1_{\bullet k}$.

Similarly, let y be the component which solves equation (5) for $c(x_{r+m}, x_i)$. Store $c(x_{r+m}, x_i)$, x_i, and y respectively in S^4, S^5, and S^6. Thus, $S^4_{r,k} = c(x_{r+m}, x_i)$, $S^5_{r,k} = y$, $S^6_{r,k} = x_i$. That is, $(S^5_{\bullet k}, S^6_{\bullet k})$ is the arc which solve (5) with $c(x_{r+m}, x_i) = S^4_{\bullet k}$. After creating the expansion table of a new stage and storing the data to stacks, go to the choosing procedure and continue the choosing procedure to the marking and cycle detecting procedure.

Unfolding procedure: If there are p times of compression for us to find the final minimum spanning tree (i.e. the stopping rule of the forward procedure is reached after p times of compression), we must operate the Unfolding procedure p times. So, we sequentially and backwardly unfold the compressed nodes x_{r+m}, where $r=p, p-1,...,1$. There are two steps to complete the Unfolding procedure.

Step 1. Determine whether or not the compressed node, x_{r+m}, is the root. For this purpose, check if there is a record whose into-node of the unfolding list is x_{r+m}. Denote such a record by $\overline{R}$. If $\overline{R}$ does not exist, it means that x_{r+m} is the root. In this case, we first discard the maximum conn-element in the cycle that is detected in stage r-1. To do this, select the record whose stage number equals r-1, and whose conn-element is the largest in such a stage. Then, let all fields of that record be "*."

Thereafter, select the record whose out-node is x_{r+m}. Denote such a record by $\underline{R}$. Note that the $S^6_{r,k}$'s element is identical to $\underline{R}$'s into-node. Then replace $\underline{R}$'s conn-element, out-node, and into-node with $S^4_{r,k}$, $S^5_{r,k}$, and $S^6_{r,k}$, respectively.

If the x_{r+m} is not the root, then $\overline{R}$ exists. Note, the element of $S^2_{r,k}$ is $\overline{R}$'s out-node. Replace $\overline{R}$'s conn-element, out-node, and into-node by $S^1_{r,k}$, $S^2_{r,k}$, and $S^3_{r,k}$, respectively.

Continue the above processes of Step 1 until r=1, and then go to the step 2.

Step 2. Check the records of the unfolding list backwardly, starting from the last record to the first one. When there are two or more records that have the same into-node, delete all records except the one with the highest index of r (i.e. the compression index).

Output procedure: Since we cannot easily draw a minimum spanning tree in accordance with the unfolding list after the unfolding procedure, we use the following output procedure to rearrange the unfolding list so that the minimum spanning tree can be easily drawn.

Step 1. Let i=1, j=1, and l=1. Locate the root of the minimum spanning tree by searching the node that does not appear in the into-node field of the unfolding list. Denote such a node by x_{root}. Then let the root(i, j)= x_{root}.

Step 2. Search all of the records of the unfolding list to locate the records whose out-node is root(i, j). Once such a record is located, we have three sub-steps as follows: (i) append this record to the output table; (ii) let the root(i+1, l) be this record's into-node; (iii) let l=l+1.

Step 3. Let j=j-1 and check whether or not j>0. If so, continue Step 2; otherwise, let i=i+1, j=l . Check whether u<v or not, where u represents the total number of the records in the output table, v represents the total number of the records in the unfolding list. If the condition is satisfied, continue Step 2 and Step 3; otherwise, we have completed the output procedure.

The flowchart of RMST is shown in figure 1.

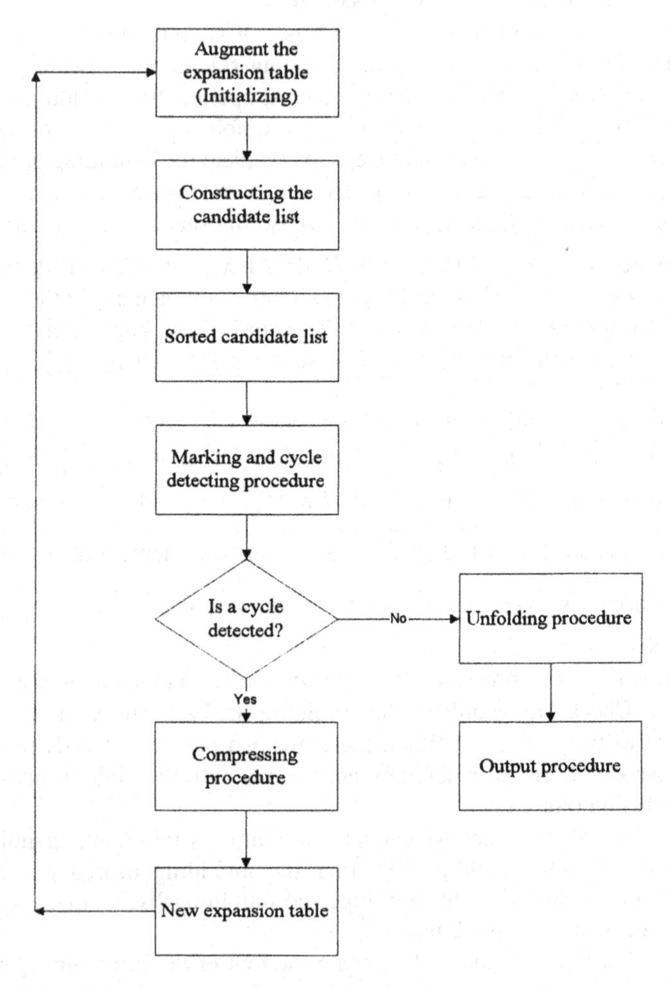

Figure 1. Flowchart of RMST

According to the above procedures, we developed the MINST program for the purpose of computing the optimal expansion competence set by the Visual Basic for Application (VBA) on the EXCEL 7.0. It is a user-friendly program, which can compute the minimum spanning tree with the RMST method. The maximum size of the expansion table is 100×100, which can be applied to real-word applications. The running time of the program depends on the elements of the table and your computer. The running time of various table sizes (with each $c(i, j)$, $i \neq j$ randomly chosen from 1 to 9) tested on a Pentium 133 PC are shown in the Table 4.

Table 4 Running Time of MINST

Size of the expansion table	Running time(seconds)
10×10	1
20×20	12
50×50	240
75×75	2358
100×100	4858

3.3. Numerical Example

We use the example 4.2 in Feng and Yu (1997), which is illustrated in Table 5. By the MINST program, the result, i.e. the output table, is shown in Table 6. For details, see the unabridged version of this paper.

Table 5 Expansion table

	x_1	x_2	x_3	x_4	x_5	x_6	x_7
x_1	*	9	8	4	5	9	4
x_2	9	*	4	6	4	9	9
x_3	3	4	*	2	3	5	4
x_4	4	8	5	*	1	7	8
x_5	5	9	8	8	*	1	4
x_6	3	4	2	7	3	*	6
x_7	3	4	8	5	9	4	*

Once the output table is established, the minimum spanning can be easily drawn, as shown in Figure 2.

Table 6 Output table

conn-element	*out-node*	*into-node*
1	x_4	x_5
4	x_5	x_7
1	x_5	x_6
3	x_7	x_1
4	x_6	x_2
2	x_6	x_3

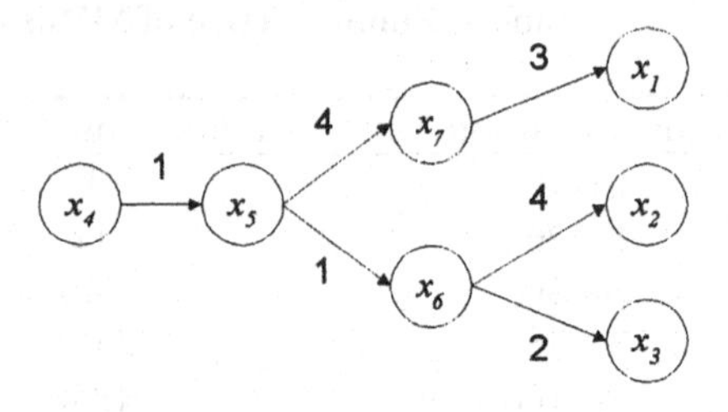

Figure 2. Minimum spanning tree of numerical example

4 Conclusions

Because traditional MCDM method sometimes can not effectively solve the challenging MCDM problem, Yu proposed a new approach: competence set analysis. In this paper, we focus on the MST method, and provide a revised algorithm, RMST method. Because of the two features: bottom line shortcut and bookkeeping, we can write user-friendly computer program, MINST, to solve optimal expansion of competence sets. By using MINST, we can handle large-scale optimal expansion problems that mathematical programming methods cannot manage.

Many research problems remain to be solved. For instances, how to solve expansion problems using RMST when there are compound nodes or skills (see Li and Yu, 1994) in the expansion process. When the cost function have uncertainty or unknowns, how to use RMST to solve the problems effectively.

References

Chiang, C. I., Li, J. M., Tzeng, G. H., and Yu, P. L., A Revised Minimum Spanning Table Method for Optimal Expansion of Competence Sets (1998).

Feng, J. W., and Yu, P. L., Minimum Spanning Table and Optimal Expansion of Competence Set, accepted for publication by *Journal of Optimization Theory and Applications* (1997).

Horowitz, E., and Sahni, S., *Fundamentals of Data Structures in Pascal*, 4th Edition. Computer Science, New York, New York (1994).

Li, H. L., and Yu, P. L., Optimal Competence Set Expansion Using Deduction Graphs, *Journal of Optimization Theory and Applications* **80**, 75-91 (1994).

Li, H. L., Incorporating Competence Sets of Decision Makers by Deduction Graphs, accepted for publication by *Operations Research* (1997).

Shi, D. S., and Yu, P. L., Optimal Expansion and Design of Competence Set with Asymmetric Acquiring Costs, *Journal of Optimal Theory and Applications* **88**, 643-658 (1996).

Yu, P. L., and Zhang, D., Competence Set Analysis for Effective Decision Making, *Control Theory and Advanced Technology* **5**, 523-547 (1989).

Yu, P. L., *Forming Winning Strategies: An Integrated Theory of Habitual Domains*, Springer, Berlin, Germany (1990).

Yu, P. L., and Zhang, D., A Foundation for Competence Set Analysis, *Mathematical Social Sciences* **20**, 251-299 (1990).

Yu, P. L., and Zhang, D., Marginal Analysis for Competence Set Expansion, *Journal of Optimization Theory and Applications* **76**, 87-109 (1991).

Yu, P. L., and Zhang, D., Optimal Expansion of Competence Set and Decision Support, *Information System and operational Research* **30**, 68-84 (1992).

Constructing and Implementing a DSS to Help Evaluate Perceived Risk of Accounts Receivable

Leonardo **Ensslin**, Gilberto **Montibeller** Neto and Marcus Vinicius A. de **Lima**

LabMCDA, Department of Production Engineering, Federal University of Santa Catarina
CEP 88040-900, C.P. 476, Florianópolis, SC, Brazil

Abstract. A factoring company is a means to provide working capital for small and medium sized companies that sell goods and/or services on a delayed-payment basis, by selling their accounts receivable to that factoring company. This paper describes the real life process of multicriteria model construction for Plaza Invest Factoring Co. (Brazil), using a decision-aid approach. A multi-attribute value function was developed, in order to measure the credit manager's perceived risk generated by the transaction, thus helping him to decide if Plaza Invest should buy or not buy a given account receivable.

Keywords: perceived risk, factoring business, decision-aid, MAVT.

1 Introduction

Factoring companies are means to provide working capital for those companies that sell goods and/or services on a delayed-payment basis. This paper describes a real life process of a multicriteria model construction for Plaza Invest Factoring Co. (Brazil), using a decision-aid approach. This company asked our team of consultants to design a decision support system to help its credit manager in deciding if he should buy or not buy a given account receivable from its client. In this paper we present the different phases of the decision-aid process, from problem construction through to evaluation of accounts receivable performed by the multicriteria model.

2 A Framework for Consulting

In our consulting we adopt a *cognitive approach*, and this is the basic assumption of our work as facilitators. Following this approach means giving high importance to understanding how decision-makers think (gathering and processing information about the decisional situation) when they take their decisions.

We feel it is important to make explicit the scientific paradigm underpinning a decision support process (Roy, 1993) for real decision-makers (DMs). A paradigm serves to define the legitimate problems and methods of a research field for those practitioners engaged in it. We adopt a *constructivist paradigm*, following Roy (1993).

This paradigm considers that each DM has his/her own value system that *should* be taken into account when we are aiding his/her decision. And each one has a subjective view of reality in accord with his/her objectives and interests. Theories and models are simple constructs, *tools* that the DMs agree using for aiding their decision. The solutions are dependent on the model because we are not modeling the "truth", but only those aspects of reality perceived and valued by the DMs. The aim of the process of modeling *is not* to provide an optimum solution, but to *generate knowledge* for the DMs, guiding them in the decision process. We *believe* that this paradigm is more appropriate for providing *decision aiding* for our clients, and seems fit the cognitive approach we follow.

3 The Decisional Context

In September 1996, the credit manager of Plaza Invest Co. requested a meeting with our consulting group. The company is positioned between the top five factoring companies in Florianópolis, a town located in South of Brazil. In our initial meeting with him, he explained how the factoring system works and proposed his idea: developing a system "to evaluate the risk" of accounts receivable (AsR), so that he could decide whether to buy or not buy them.

Our client company's organization is very small (4 employees). Its portfolio is about US$ 500,000 (70% of Equity Capital). The spread level is 5% per month (Brazilian interest rate is one of the biggest in the world) and its total revenue was about US$ 350,000 (1997 data). Our decision-maker (DM) was the credit manager. His main function is the management of risk in operations involving buying AsR, clearly essential for the success of its business.

Figure 1 describes the Brazilian factoring system in more detail (with cash flows of each company). First (day 0), the seller company sells goods/services (priced at M_{AR} monetary units) and issues an account receivable (AR), that is endorsed by the buyer company (BC). This AR has to be honored at day t_x. Second, the factoring company buys the AR from the seller company at de day t_1. For this it charges a spread S. Third, the buyer company pays the AR to the factoring company at day t_x.

Suppose now that at day t_x, the buyer company, which would have to pay the AR, does not honor it. This generates an uncertainty, in the factoring company's cash flow, about *when* (t_y) that AR will be honored.

After buying an account receivable from the seller company, the factoring company will have the entire onus of collecting this AR from the buyer company. But in Brazil, the judiciary system is slow to judge cases of not honoring the AsR. The result is a high level of AsR not honored (about 6% of the whole annual transactions). Another problem is the absence of reliable data about the financial performance of those companies. This occurs because (1) Brazil had a long period of high inflation rates (that distorted the balance sheets); (2) buyer and seller companies are usually small ones (monthly total revenue between U$20,000 and US$500,000), and therefore less supervised by governmental control.

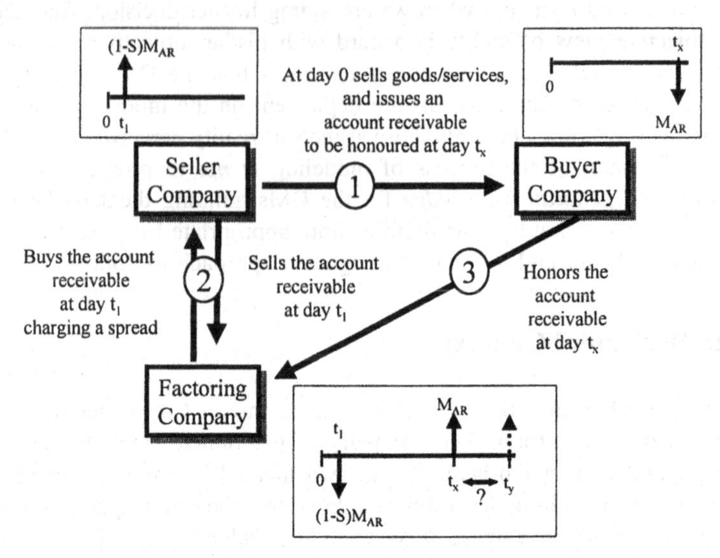

Figure 1. The factoring business system in Brazil.

4 Perceived Risk

When we had the initial meeting with the DM, we did not have a clear understanding of what was *his* concept of risk. Following the cognitive approach, we have to understand first the ways in which our DM thinks about and responds to risk. Studies have shown that the concept "risk" means different things to different people and perceptions of risk are determined by a variety of quantitative and qualitative characteristics (Slovic, 1987).

In our second meeting, together with the DM, we decided to construct a model for performing the evaluation of *his perceived* risk related to AsR. We define *perceived risk* here as *the degree or unattractiveness that a pair of buyer-seller companies that transacted an AR generates for the credit manager, due to their characteristics, following his value system.* Also, we understood that there were a set of aspects to be taken into consideration and, hence, we chose to construct a multicriteria model which would make this possible. Then we decided, following Fischhoff (1990), to use as our *tool* of evaluation a multi-attribute value function (Keeney, 1992) to measure the DM's perceived risk concerning each aspect. A value function will be called here a *perceived risk function.*

To propose a decision support system (DSS) was a natural conclusion of our analysis about the decisional context. The reduced number of employees, associated with a large number of AsR (which have a small period of duration, usually between 30 to 90 days, and a small unitary monetary value, typically no more than US$1,000) have generated an overload of work for the credit manager. Only an automatic system for decision *aiding* could reduce this overload.

As far as we know, a similar DSS for this kind of decisional problem has not been reported in literature, especially in the absence of reliable financial data (the usual situation in developing countries) and in a framework of perceived risk. The credit granting decision, a relatively common problem for multicriteria analysis (*eg*, Mareschal and Brans (1991)) can be considered a related concern. However, it *is not* the same problem. First, in factoring business there are two companies (buyer and seller) in the transaction (instead of one bank's borrower). Second, the AR is honored by the buyer company, and usually it is impossible to have a direct contact with, or require more information from, this company (instead of getting the information from a well-known bank's client).

5 Which Are Manager's the Points of View?

A *problem belongs to a person*: it is always a *personal construction* that a DM places on events (Eden, 1988). A cognitive map is a tool that provides a graphical representation, in linguistic terms, of the DM's problem. During the process of construction, the articulation of his ideas about the decisional context is helpful in structuring the situation (see also Eden (1988) and Belton *et al.* (1997)).

We made use, in our third meeting with the decision-maker, of a cognitive map. The aim was to define which aspects he considered relevant in evaluating the perceived risk of an AR. The mapping scheme proposed by Montibeller (1996) was applied, where a cognitive map is a network of nodes (text blocks) linked by arrows that represent influence relationships between means-ends concepts. Each text block represents a concept, with a presented pole and a contrast pole. The two labels are linked with '...' (read as "rather than").

Following Bana e Costa *et al.* (1999) we will call *fundamental points of view* (FPsV) those aspects that are considered fundamental, following the DM's value system, in evaluating the perceived risk of an AR. The process described below (during our fourth meeting with the DM) had as final aim to find these FPsV.

Using the *traditional* analysis of cognitive maps (Eden, 1988) we detected the hierarchical structure and performed a cluster identification. Three clusters were identified about the buyer company's characteristics: *behavior in the market, financial performance, honoring pattern of the previous financial obligations*.

The *dimension* analysis proposed by Montibeller (see Bana e Costa *et al.* (1997) for an overview), tries to identify, from the cognitive map, which are the fundamental points of view to be taken into account in the evaluation model. First, the reader can observe that an acyclic cognitive map (*i.e.*, without loops) is composed by a series of lines of reasoning. A *line of reasoning* (see also Axelrod

(1976)) is defined here as a chain or arguments, starting in a tail concept (only out-arrows) and ending in a head concept (only in-arrows). Second, performing a content analysis, we group the lines of reasoning which have similar ideas. Each group becomes a *branch.* Figure 2 presents the branches B_1, B_2, B_3 and B_8 of the cluster *behavior in the market.*

Each branch represents a specific concern of the DM. The method was to search for and elicit a FPV in each branch, which we did by trying to answer the following question: which is the main concern in this branch that is, at the same time, *essential* for the DM's system of values and *controllable.* (A FPV is controllable if it addresses those consequences which are influenced only by the AR – see Keeney (1992).) The same kind of analysis was done in each branch of the map. Table 1 presents, for each branch, the candidate FPV elicited.

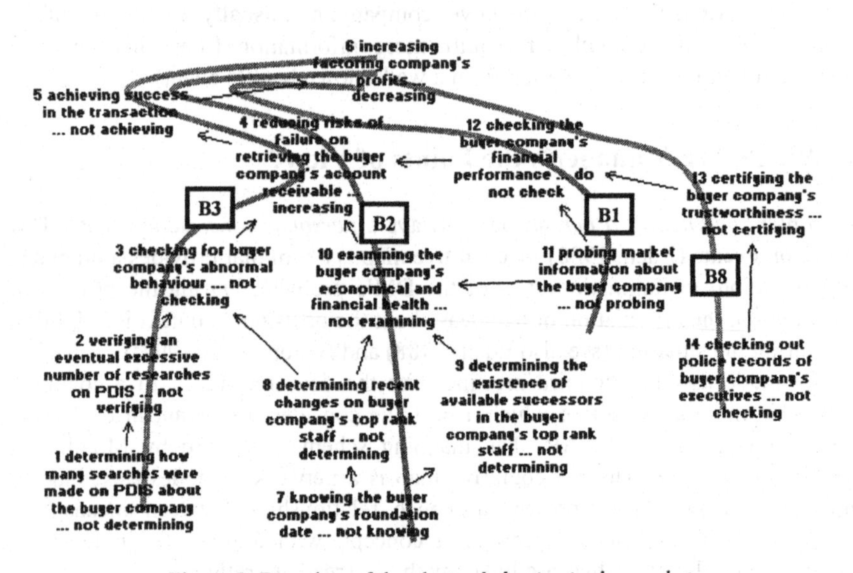

Figure 2. Branches of the cluster *behavior in the market.*

When we tested the properties of the FPsV (Keeney, 1992) we discovered that the set was not considered complete by the DM. An untrustworthy seller company and/or a seller company with a bad economic and financial performance could reduce the quality of its products/services and, thus, provoke the buyer company to not honor the AR. Then we add two FPsV which concern the seller company: "Trustworthiness" and "Economical and Financial Performance". On the other hand, "Police Records" (branch B_8) was eliminated because it was not operational (Keeney, 1992): it was impossible to obtain this information speedly.

Finally, the decision-maker decided to eliminate the FPV "Loans by Financial Companies" (branch B_9). It became a *rejection criterion* of the model, that is: if the monetary value of the AR was bigger than the biggest monetary value lent in the past to the buyer company by another financial institution, the AR will be rejected *before* being evaluated by the model.

Table 1. Candidate FPsV.

Branch #	Candidate FPV	Branch #	Candidate FPV
1	Market Information	6	Payment Profile
2	Company Administration	7	PDIS Data Reliability
3	Pattern of Behavior	8	Police Records
4	Business' Field Insertion	9	Loans by Financial Institutions
5	Checks, Insolvency & Bankruptcy	PDIS = Past Data Information System	

6 Constructing the Multicriteria Model

6.1 Developing the Criteria

For each FPV we developed a criterion (Bouyssou, 1990). It is a real-valued function on the set of potential actions (the AsR), such that it appears meaningful to compare two AsR *a* and *b* according to this fundamental point of view on the sole basis of the two numbers v(a) and v(b). These numbers will reflect the local unattractiveness (following the DM's system of values) of *a* and *b*, respectively.

Two devices are needed to develop a criterion. First, it is necessary to construct a descriptor (Bana e Costa *et al.*, 1999). Second, it is required to define a perceived risk function, measured in terms of unattractiveness. Following the constructivist paradigm, there is no the "correct" descriptor or perceived risk function (Roy, 1993), Descriptors and perceived risk functions must only be considered by the DM as suitable tools to evaluate the AsR.

We spent one month (2 meetings per week) developing the criteria, together with the DM. Two of them will be presented here. Wherever it was possible we constructed quantitative descriptors ("Payment Profile" and "PDIS Data Reliability"): the others are qualitative due the nature of information required.

To perform the evaluation of AsR we developed a software called *RiskInvest*. It was designed to be extremely user-friendly, using visual data input. It consists basically of four modules: seller company database; buyer company database; account receivable database, and perceived risk analyzer.

6.1.1 Market Information Criterion

The DM was concerned with the informal information about economic and financial performance of the buyer company (during the short run) in two ways: inside information (only available to the factoring company), and market information (available from other companies, media, *etc*.). Because this concern deals with "soft" data (usually neglected in formal decision models) we decided to construct a qualitative descriptor. Even with a lot of ambiguity, we considered it better to take this into account than neglect an aspect considered important by the DM just because "there are no hard data to measure". For each of these two variables we asked him to define three states (see Figure 3). The states were then combined and ranked in terms of DM's preference (Bana e Costa *et al*., 1999). Each combination was considered as an impact level (Table 2).

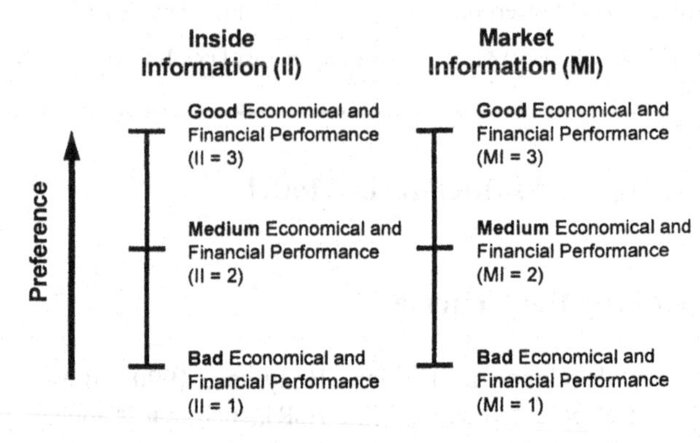

Figure 3. States for the descriptor "Market Information".

The reader may note, in the last column of Table 2, that the DM stated two impact levels as the *good* (N_8) and *neutral* (N_4). They represent the impact of two dummy AsR considered, respectively, as good and neutral in this descriptor. These two actions delimit the frontiers of perceived risk.

Table 2. Descriptor and Perceived Risk Function for "Market Information".

Impact Level	Inside Information	Market Information	Perceived Risk Function v_1	Risk Ranges
N_9	II = 3	IM = 3	120	*low risk*
N_8	II = 3	IM = 2	100	good level
N_7	II = 2	IM = 3	80	
N_6	II = 2	IM = 2	73	*medium risk*
N_5	II = 3	IM = 1	27	
N_4	II = 2	IM = 1	0	neutral level
N_3	II = 1	IM = 3	-60	
N_2	II = 1	IM = 2	-154	*high risk*
N_1	II = 1	IM = 1	-207	

The multicriteria model that we were constructing needed numerical parameters. But humans think in a qualitative way. This is a serious cognitive constraint of the DM for building a quantitative model. That is why we adopt the Macbeth software (Bana e Costa *et al.*, 1999) to construct perceived risk functions (and also to determine the substitution rates - see section 6.2). Using the software we developed the perceived risk function v_1 (Table 2) from the DM's qualitative judgments (the greater the value, the smaller the unattractiveness.)

The *RiskInvest* screen allows the user to input this information about the buyer company (*Casas da Água*) in a visual data way. The DM can choose one of three states in each variable. In this example: II=2, IM = 2, v_1(AR) = 73 (see Table 2).

6.1.2 Payment Profile Criterion

The DM was concerned here with the honoring of previous financial obligations by the buyer company. Frequent large delays ($t_y - t_x$ in Figure 1) mean for him higher risk. When dealing with Plaza Invest, the buyer company's conduct may differ from its conduct when dealing with other financial companies in the market. So we developed an index for the payment profile in the market (r_m) and in factoring (r_f). The index is the same in both cases, but the source of information are different (from the PDIS and from the Plaza Invest, respectively). The construction of r_m is described below.

Our choice was create an index, using a weighted sum, which could take into account (via the substitution rates) the degree of unattractiveness that large delays cause for the DM (*eg*, 20 days of delay is more repulsive than 3 days). Then:

$$r_m = \alpha \left[\frac{c_0 p + c_1 a_1 + c_2 a_2 + c_3 a_3}{p + a_1 + a_2 + a_3} \right] - \beta \qquad [1]$$

where:

p, a_1, a_2, a_3= number of payments in the past: without delay (p), delay between 5 and 15 days (a_1), delay between 16 and 30 days (a_2) and delay greater than 30 days (a_3);

c_0, c_1, c_2, c_3 = the substitution rates of each delay range;

α e β = parameters of a linear transformation.

The substitution rates cannot be arbitrarily defined, as discussed by Keeney (1992). For this reason we used the swing procedure (von Winterfeld and Edwards, 1986) to help the decision-maker in defining them (see also Section 6.2 below). At first we created four reference actions (dummy BCs) that have the best performance in one sub-criterion and the worst performance in all others sub-criteria, and one (a_0) that is worst in all sub-criteria.

Now it was possible to find a decision-maker's perceived risk function for those actions with the help of Macbeth, where p (the best BC) has 100 points and a_0 (the worst BC) has 0 points. After inputting his value judgments in terms of qualitative difference of unattractiveness for a pair of actions, the software outputs a_1 = 72 points, a_2 = 56 points, and a_3 = 22 points. To find the weights we had to just normalize them by the total sum of points (250): c_0 = 0.40; c_1 = 0.29; c_2 = 0.22; c_3 = 0.09.

If $\alpha = 1$ and $\beta = 0$, then r_m range, as defined by [1], is from 40% (the best case when p = 100% and $a_1 = a_2 = a_3$ = 0%) to 9% (the worst case when a_3 = 100% and p = $a_1 = a_2$ = 0%). This is a cognitively unsuitable scale. We then set: α =3.2258 and β = 29.0323 with the purpose of generate r_m ranging from 0 to 100. This index has not yet a perceived risk function that shows the desirability of a given AR with an established r_m performance. Again we used the Macbeth software to construct a perceived risk function as shown in Figure 4, assuming a linear interpolation between each pair of points.

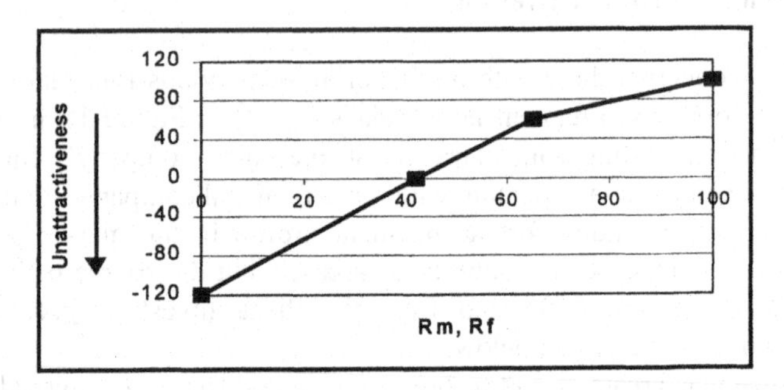

Figure 4. Payment Profile - Perceived Risk Function.

Using the same procedure described above, we found the substitution rates for the market (45%) and the factoring (55%) indexes. Then we can calculate the local perceived risk in this criterion by: v_6 (AR) = $0.45v(r_m) + 0.55v(r_f)$. Figure 5 shows RiskInvest evaluating the payment profile of *Casas da Água*. It has a good payment profile, better for the factoring company (r_f(AR) = 100%, $v_6(r_f)$ = 100 – see Figure 4) than for the market (r_m(AR) = 98%, $v_6(r_f)$ = 95 – see Figure 4). Thus the local performance was v_6(AR) = 0.45(95) + 0.55(100) = 98 points.

6.2 Finding the Substitution Rates

The global perceived risk V of an AR_i will be calculated by a weighted sum of the local perceived risk v_j in the j-th criterion:

$$V(AR_i)=\sum_{j=1}^{9} w_j v_j(AR_i) \qquad [2]$$

Determining the substitution rates w_j is a critical concern of this kind of aggregation (von Winterfeld and Edwards, 1986). For the constructivist paradigm the numerical values obtained are not considered as the "true ones" but just as tools accepted by the DM as adequate to guide his decision (Roy, 1993).

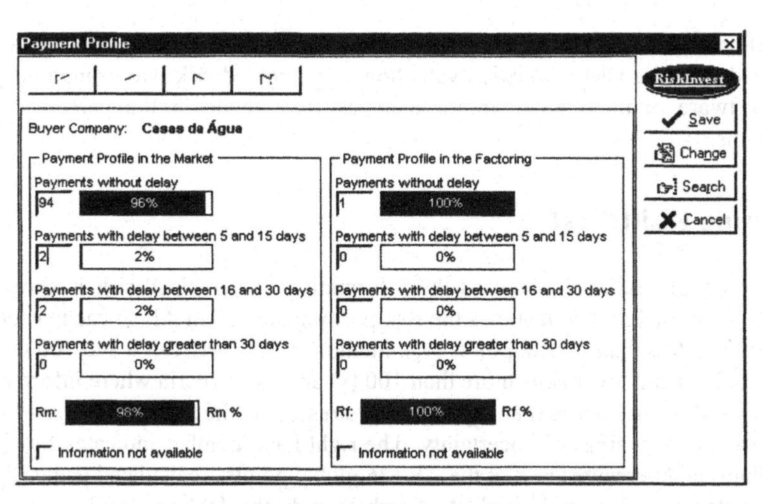

Figure 5. Payment Profile - RiskInvest Screen.

We adopted the same procedure described in the Section 6.1.2 to find w_j. For each j-th criterion, we created a reference AR that is at the *good* level in this j-th criterion and is at the *neutral* level in all other criteria. And also a reference AR a_0 that is at the neutral level in all criteria. (Using *good* and neutral *level*, instead of the usual *best* and *worst*, we tried to eliminate extremes of unattractiveness in a given criterion that could affect largely the substitution rates.) With these references AsR it was possible, to use Macbeth to obtain the substitution rates (Figure 6). The reader may note that the criteria related to the buyer company have a major influence in the global perceived risk (darker bars).

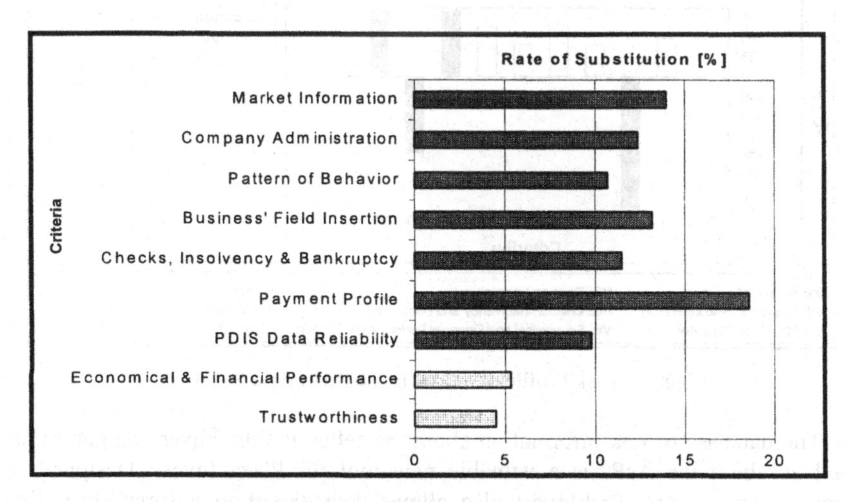

Figure 6. Rates of Substitution of the Criteria.

After developing the software, there was a period of tests. A set of 24 AsR was selected, and the DM's holistic evaluation of perceived risk was compared with the software outputs. Some minor changes were made in the perceived risk functions.

7 Using RiskInvest

The way that *RiskInvest* displays the information of perceived risk is shown in Figure 7. In the left side it shows the risk performance of an AR in each criterion. High risk is less than 0 local points (v_8), medium risk is between 0 and 100 (v_1, v_3, v_4, v_5 and v_6) and low risk is more than 100 (v_7 and v_9). Criteria where information is unavailable (also lesser than 0 points) are stressed by the software (v_2), and is an indication of key areas of uncertainty. The right hand number indicates risk level (medium) and perceived risk of the AR (26 global points – calculated using [2]).

Besides perceived risk evaluation, which aids the DM in deciding whether should buy or not buy the AR, *RiskInvest* has other potential tools for supporting the decision-making procedure. One of the most interesting is the management of the perceived risk level in a portfolio. Another decision support tool is to identify *where* Plaza Invest must directs its information gathering efforts.

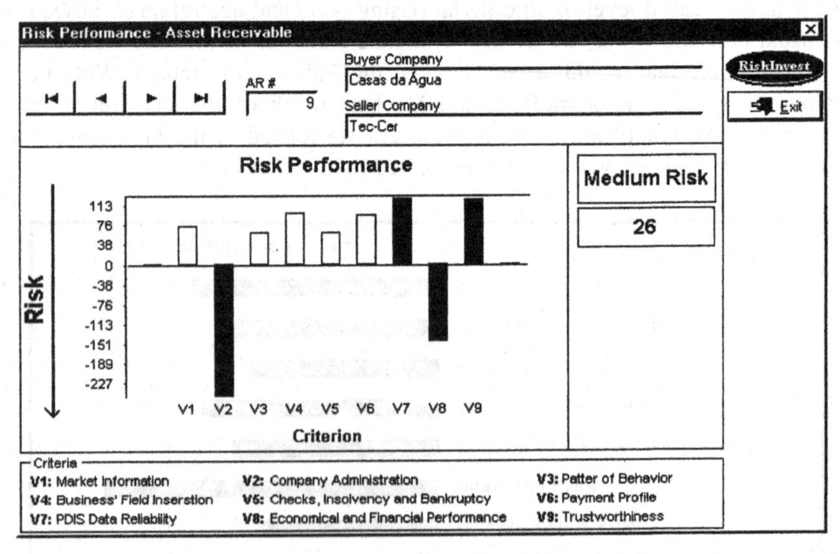

Figure 7. Risk Profile of an AR in the RiskInvest Software.

The database of risk information about the seller and the buyer companies, as well as about the AsR, is a valuable new tool for Plaza Invest. Designed to generate knowledge, RiskInvest also allows data export to a spreadsheet like

Excel. This data can be used to generate correlation analysis between variables.

8 Conclusions

This paper presented a real life multicriteria decision aid (MCDA) application in the financial area, more specifically in the factoring business. We constructed and developed a software, *RiskInvest*, that allows a DM to perform his perceived risk evaluation of accounts receivable. It is a decision support system that helps in deciding whether to buy or not buy an account receivable; control the portfolio risk; and last but not least, *learn* about the variables that effectively influence the risk of delay of payment in accounts receivable.

References

Axelrod, R. (Ed.) (1976): Structure of Decision. Princeton University Press, Princeton

Bana e Costa, C.A., Ensslin, L., Corrêa, E.C., Vansnick, J.C. (1999): DSS in Action - Integrated Application in a Multicriteria Decision Aid Process. European Journal of Operational Research 113, 315-335

Bana e Costa, C.A., Ensslin, L., Montibeller, G.N. (1997): From Cognitive Maps to Multicriteria Models. Proceedings of the International Conference on Methods and Applications of MCDM. Mons, Belgium, 247-250

Bouyssou, D. (1990): Building Criteria: A Prerequisite for MCDA. In Bana e Costa, C.A. (Ed.): Readings in Multiple Criteria Decision Aid. Springer, Berlin, 58-80

Belton, V., Ackermann, F., Shepherd, I. (1997): Integrated Support from Problem Structuring Through to Alternative Evaluation Using COPE and VISA. Journal of Multicriteria Decision Analysis 6, 115-130

Eden, C. (1988): Cognitive Mapping. European Journal of Operational Research 36, 1-13

Fischhoff, B., Watson, S.R., Hope, C. (1990): Defining Risk. In Glickman, T.S., Gough, M. (Eds.): Readings in Risk. Resources for the Future, Washington, 30-41

Keeney, R. L. (1992): Value-Focused Thinking: A Path to Creative Decision-Making. Harvard University Press, Cambridge

Mareschal, B., Brans, J.P. (1991): BANK ADVISER - An Industrial Evaluation System. European Journal of Operational Research 54, 318-324

Montibeller, G.N. (1996): Cognitive Maps - A Tool in Aiding Problem Structuring. M. Sc. Thesis (in Portuguese). Federal University of Santa Catarina, Florianópolis, Brazil

Roy, B. (1993): Decision Science or Decision Aid Science?. European Journal of Operational Research 66, 184-203

Slovic, P. (1987): Perception of Risk. Science 236, 280-285

von Winterfeld, D., Edwards, W. (1986): Decision Analysis and Behavioral Research, Cambridge University Press, Cambridge

Acknowledgements. We would like to thank the software programmer Clodaldo Federle for his help with the English version of *Riskinvest*, and our colleagues Sandro Noronha, Alec Morton, and Tatiana Souza for their revision on our draft.

IMPROVING WEIGHTING INFORMATION IN INTERACTIVE DECISION PROCEDURES. A VISUAL GUIDE

Férnandez F.R[1]., A. Mármol[2] and J. Puerto[1]
[1]Departamento de Estadística e I.O. Universidad de Sevilla. Spain.
[2]Departamento de Economía Aplicada III. Universidad de Sevilla. Spain.

Abstract

In this paper we address multicriteria decision problems when only partial information is given in the decision making process. We develop a methodology to sequentially incorporate preference information in these processes. A geometrical representation is presented, providing assistance to visualize the quality of the given partial information and to recognize the need of further improvement of this information.

Keywords: Multicriteria Decision Making, Partial Information, Weights.

1. INTRODUCTION

Decision processes with multiple criteria, usually in conflict, have been widely studied in the literature due to their practical interest (see Huang and Masud(1979) [6], Huang and Yoon(1981) [7], Chankong and Haimes(1983) [2], Vinke(1989) [14]). Multiple Criteria Decision Making processes can be classified according to the cardinality of the set of feasible decisions. When this set is finite, we have multiple attribute decision problems. Otherwise we have multiple objective decision problems studied by vector programming. In this paper we deal with the first case.

Let D be a finite set of possible decisions. For each $d \in D$, let $f_j(d)$ be the value of the criterion $j (j = 1, ..., m)$ for this decision. The choice of the best decision, $\hat{d} \in D$, depends on the criteria and on the additional information the decision maker (D-M) has about the decision process. We assume that the D-M provides "a priori" information about the importance of the criteria given by special structured relations among the weighting coefficients.

Information about weights can be specified in different ways. A first approach consists of establishing intervals for the individual variation of weights (Steuer, (1976) [13]). In other cases, the D-M may be able to give certain linear relations which express partial information about marginal substitution rates between the criteria (Carrizosa et al.(1995) [1]). The ordinal approach (see Hannan(1981) [5], Kirkwood and Sarin(1985)[8], Paelinck(1975) [12]) is a particular case of the former one where the D-M is requested to estimate only the rank order of the criteria.

In general, if the information about weights is given by a set of constraints that define a polyhedron P, containing the information

$$P \subset \Lambda^+ = \{\lambda \in I\!R^m : \lambda_i \geq 0, \sum_{i=1}^{m} \lambda_i = 1\}$$

we define the preference between alternatives by the binary relation R_P as:

$$d_i R_P d_j \Leftrightarrow \lambda^t (f(d_i) - f(d_j)) \geq 0, \forall \lambda \in P$$

where $f(.) \equiv (f_1(.) \ldots f_m(.))$

Since P has an infinite number of elements, this binary relation is difficult to handle, but if the extreme points of P, $\lambda^1, \ldots, \lambda^h$, are known, the relation R_P can be evaluated equivalently as

$$d_i R_P d_j \Leftrightarrow (\lambda^s)^t (f(d_i) - f(d_j)) \geq 0, \forall s = 1, \ldots, h$$

which makes possible to manage this relation in certain contexts.

Example 1. A consulting firm has evaluated four members of staff according to three skills strongly related to the job to be done. Their evaluation is given in the following table

	c_1	c_2	c_3
d_1	3	4	3
d_2	3	1	3.5
d_3	1	3	4
d_4	3	2	5

The company, interested in employing the best of the candidates, considers that the attribute c_1 is more important than c_2. Therefore the consulting firm must transform the evaluation assigned to the candidates. Using the extreme points of the set of information the new values are

	c_1	c_2	c_3
d_1	3	3.5	3
d_2	3	2	3.5
d_3	1	2	4
d_4	3	2.5	5

As the best candidate cannot yet be chosen, the company supplies a new piece of information stating that attribute c_3 is more important than attribute c_2.

Then the consulting firm transforms the evaluation again getting the following

	c_1	c_2	c_3
d_1	3	10/3	3
d_2	3	7.5/3	3.5
d_3	1	8/3	4
d_4	3	10/3	5

Therefore under this information, candidate number four is the best suited for the job.

The aim of the paper is to provide a sequential procedure that permits to compute easily the extreme points, or at least a set of generators, of the polyhedron of weights which is the result of adding a new relation at each step. The procedure begins from a situation of partial uncertainty where the set of weights is defined by a set of generators (preferably extreme points) represented by the columns of a matrix. When the D-M provides additional information in the form of linear relations on weights, the information is processed in order to get the characterization of the new polyhedron in the form of a matrix whose columns are the generators (in many cases the extreme points). If the initial situation is of no-information the matrix of extreme points is the identity.

We also provide a procedure to visualize the process in two dimensions. This procedure will help the D-M to understand how the uncertainty is being reduced. The larger the set of weights the lower the level of information it provides. The shape of the projected set of weights in relation to the projection of the alternatives also gives information about the degree of the incomparabilities that those weighting coefficients produce on the set of alternatives.

The paper proceeds as follows: in Section 2 we expose the iterative procedure to characterize the set of weights when sequential information is added and it is illustrated with examples. In Section 3 we explain the visual representation of the process. Section 4 is devoted to the conclusions.

2. IMPROVING THE INFORMATION

Assume that the D-M's preference structure is represented by a polyhedron of weights, $P_L \subset \Lambda^+$, where $L \in I\!R^{m \times p}$ stands for a matrix whose columns are a set of generators of the set P_L. Let $L = L^{(k)}$ and $L^1 = I$.

Let $P_{L^{(k+1)}}$ be the new preference polyhedron obtained when a new constraint of the form $b^{(k)} \leq a^{(k)}w \leq c^{(k)}$ is added to $P_{L^{(k)}}$ by the D-M. Hence we have

$$P_{L^{(k+1)}} = P_{L^{(k)}} \cap \{\lambda \in I\!R^m : \lambda \geq 0, b^{(k)} \leq a^{(k)}\lambda \leq c^{(k)}\}$$

$b^{(k)}, c^{(k)} \in I\!R, a^{(k)} \in I\!R^m$.

The applicability of the new structure of preferences given by $P_{L^{(k+1)}}$ depends on the difficulty to obtain a set of generators of the new polyhedron.

A way to simplify this problem consists of using the previously obtained set of generators and sequentially recompute the sets generated by the new constraints.

Let us define the set

$$\Omega^{(k)} = \{w \in I\!R^p : e^t w = 1, b^{(k)} \leq v^{(k)} w \leq c(k)\},$$

where

$$v^{(k)} = a^{(k)} L^{(k)}$$

and consider the linear mapping

$$T_{L^{(k)}} : \Omega^{(k)} \longrightarrow P_{L^{(k+1)}}$$

$$w \longrightarrow \lambda = L^{(k)} w$$

The introduction of the set $\Omega^{(k)}$ is due to the fact that we will transform the problem of finding the set of generators of $P_{L^{(k+1)}}$ in that of finding the generators of $\Omega^{(k)}$. It is worth noting that $\Omega^{(k)}$ has a simpler structure than $P_{L^{(k+1)}}$, and what is more important, there exists tractable enumerative procedures (see Mármol et al. (1998) [11]) to compute its set of extreme points.

Let $W^{(k)}$ be the matrix whose columns are a set of generators of $\Omega^{(k)}$, then the following result holds

Theorem 2.1. *[11] The columns of the matrix $L^{(k)} W^{(k)}$ are a set of generators of the set $P_{L^{(k+1)}}$.*

Since Theorem 1 only assures that generators (not necessarily extreme points) are obtained, in order to determine which of those points are extreme, one can use Theorem 8.11 in Edelsbrunner's book (1987) [4]. This procedure has complexity O(mlogm+m) which is not too large because, in general, the number of criteria, m, is small.

Nevertheless, if the mapping $T_{L^{(k)}}$ is injective, there exists a one-to-one correspondence between the extreme points of $\Omega^{(k)}$ and those of $P_{L^{(k+1)}}$. Therefore, any extreme of $P_{L^{(k+1)}}$ is obtained by $T_{L^{(k)}}$ applied to the extreme points of $\Omega^{(k)}$.

Theorem 2.2. *[11] If the rank of matrix $L^{(k)}$ is p, and w_0 is an extreme point of $\Omega^{(k)}$, then $\lambda_0 = L^{(k)} w_0$ is an extreme point of $P_{L^{(k+1)}}$.*

When this condition holds, since the extreme points of $\Omega^{(k)}$ are easily obtained, it is straightforward to obtain those of $P_{L^{(k+1)}}$. When it is possible, it is also important to work with the set of extreme points instead of a set of generators because the size of the matrix affects strongly the applicability of the approach.

The above results lead us to a direct manner to describe a sequential procedure to incorporate new information about weights. The iterative procedure is given by $L^{(k+1)} = L^{(k)}W^{(k)}$ in the k-th step, thus

$$L^{(k+1)} = L^{(1)}W^{(1)}W^{(2)}\dots W^{(k)}$$

It is worth noting that for restrictions given as $a^{(k)}\lambda \geq b^{(k)}$ we can obtain explicitly the expression for the set of extreme points of $\Omega^{(k)}$. We denote

$$\bar{P}_{L^{(k+1)}} = \{\lambda \in I\!R^m : \lambda = L^{(k)}w, w \geq 0, e^t w = 1, b^{(k)} \leq v^{(k)}w\}$$

$$\bar{\Omega}^{(k)} = \{w \in I\!R^p, w \geq 0, e^t w = 1, b^{(k)} \leq v^{(k)}w\}$$

where $v^{(k)} = a^{(k)}L^{(k)}$.

Theorem 2.3. *[11] If $v_i^{(k)} > b^{(k)}$ and $v_j^{(k)} < b^{(k)}$ then*

$$(0,\dots,\overset{i)}{\frac{b^{(k)}-v_j^{(k)}}{v_i^{(k)}-v_j^{(k)}}},\dots,\overset{j)}{\frac{v_i^{(k)}-b^{(k)}}{v_i^{(k)}-v_j^{(k)}}},0,\dots,0)$$

is an extreme point of $\Omega^{(k)}$.

If $v_i^{(k)} \geq b^{(k)}$ then e^i is an extreme point of $\Omega^{(k)}$.

We use these results to obtain an exact description of the whole set of extreme points of $\Omega^{(k)}$, $W^{(k)}$, when the information given by the D-M is of the form $a^{(k)}\lambda \geq b^{(k)}$. Without loss of generality, assume that

$$v^{(k)} = a^{(k)}L^{(k)} = (v_1,\dots,v_s,v_{s+1},\dots,v_{s+r},v_{s+r+1},\dots,v_{s+r+t})$$

where $v_i = b^{(k)}, \forall i = 1,\dots,s,\quad v_i > b^{(k)}, \forall i = s+1,\dots,s+r,\quad$ and $v_i < b^{(k)}, \forall i = s+r+1,\dots,s+r+t$.

Then $\Omega^{(k)}$ has exactly s+r+(rxt) extreme points and they are

$$e^i = (0,\dots,\overset{i)}{1},\dots,0), i = 1,\dots,s+r$$

$$(0,\dots,\overset{i)}{\frac{b^{(k)}-v_j^{(k)}}{v_i^{(k)}-v_j^{(k)}}},\dots,\overset{j)}{\frac{v_i^{(k)}-b^{(k)}}{v_i^{(k)}-v_j^{(k)}}},0,\dots,0),\quad \forall i,j, v_i > b^{(k)}, v_j < b^{(k)}$$

Example 2. In a MCDM problem with four criteria, consider an initial polyhedron of weights which is the result of providing ordinal information in the form

$$\lambda_4 \geq \lambda_3 \geq \lambda_2 \geq \lambda_1$$

Its extreme points are the columns of the matrix

$$L^{(1)} = \begin{pmatrix} 1/4 & 0 & 0 & 0 \\ 1/4 & 1/3 & 0 & 0 \\ 1/4 & 1/3 & 1/2 & 0 \\ 1/4 & 1/3 & 1/2 & 1 \end{pmatrix}.$$

Assume that the D-M provides also the following relation

$$\lambda_1 \geq 0.5\lambda_2 + 0.5\lambda_3$$

The incorporation of this relation to the model is as follows $a^{(1)} = (1, -0.5, -0.5, 0)$, $v^{(1)} = a^{(1)}L^{(1)} = (0, -1/3, -1/4, 0)$.

The extreme points of $\Omega^{(1)}$ are e^1, e^4, hence

$$W^{(1)} = \begin{pmatrix} 1 & 0 \\ 0 & 0 \\ 0 & 0 \\ 0 & 1 \end{pmatrix}$$

and as the condition of Theorem 2.2 holds the new matrix of extreme points is

$$L^{(2)} = L^{(1)}W^{(1)} = \begin{pmatrix} 1/4 & 0 \\ 1/4 & 0 \\ 1/4 & 0 \\ 1/4 & 1 \end{pmatrix}.$$

Now consider the new relation

$$\lambda_1 + \lambda_2 \geq \lambda_4 + 0.1,$$

$a^{(2)} = (1, 1, 0, -1)$, $v^{(2)} = a^{(2)}L^{(2)} = (1/4, -1)$ The extreme points of $\Omega^{(2)}$ are $(1, 0)^t$ and $(1.1/1.25, 0.15/1.25)^t$, thus

$$W^{(2)} = \begin{pmatrix} 1 & 1.1/1.25 \\ 0 & 0.15/1.25 \end{pmatrix}, \text{ and } L^{(3)} = L^{(2)}W^{(2)} = \begin{pmatrix} 1/4 & 1.1/5 \\ 1/4 & 1.1/5 \\ 1/4 & 1.1/5 \\ 1/4 & 1.7/5 \end{pmatrix}.$$

Example 3. Consider a polyhedron of weights whose extreme points are the columns of the matrix

$$L^{(1)} = \begin{pmatrix} 1 & 1/2 & 1/2 & 0 \\ 0 & 1/2 & 0 & 1/2 \\ 0 & 0 & 1/2 & 1/2 \end{pmatrix}$$

and assume that the new additional information is $\lambda_1 \geq \lambda_2$.

Now $v^{(1)} = (1, 0, 1/2, -1/2)$ and

$$W^{(1)} = \begin{pmatrix} 1 & 0 & 0 & 1/3 & 0 \\ 0 & 1 & 0 & 0 & 0 \\ 0 & 0 & 1 & 0 & 1/2 \\ 0 & 0 & 0 & 2/3 & 1/2 \end{pmatrix}, L^{(2)} = L^{(1)}W^{(1)} = \begin{pmatrix} 1 & 1/2 & 1/2 & 1/3 & 1/4 \\ 0 & 1/2 & 0 & 1/3 & 1/4 \\ 0 & 0 & 1/2 & 1/3 & 1/2 \end{pmatrix}.$$

.Notice that $(1/3, 0, 0, 2/3)^t$ is an extreme point of $\Omega^{(1)}$, but $L^{(1)}(1/3, 0, 0, 2/3)^t = (1/3, 1/3, 1/3)^t$ is not an extreme point of $P_{L^{(2)}}$ because it is a convex combination of $(1/2, 1/2, 0)^t$ and $(1/4, 1/4, 1/2)^t$. Notice also that all the remainder extreme points of $P_{L^{(2)}}$ are the images of the extreme points of $\Omega^{(1)}$ by the linear mapping represented by $P_{L^{(1)}}$.

3. VISUAL GUIDE

The sequential procedure given in Section 2 may end providing a complete ranking of the alternatives. Nevertheless, it may also end not showing which is the best decision. In these cases we can offer the D-M a way to avoid this inconvenience. The goal of this section is to develop a tool to visualize the incomparabilities that his/her current partial information still keeps. In general, the larger the set of weights is, the lower the level of information it provides. However, only two-dimensional representations are useful for the D-M, and the m-dimensional attribute space should be reduced to a two-dimensional plane. This procedure is usual in statistical analysis and it is done projecting the matrix of values of the alternatives onto the plane defined by its two principal components (see Mardia et al.(1979) [9]for further details). Any software which implements this projection can be used. In particular, we will use Promethee-Gaia [10] in our applications because it allows to do this projection.

Since each alternative is evaluated with respect to the current set of weights we also have to project this set onto the considered plane. It is worth noting that the lengths and directions of these projections will be used to discriminate between the actions in our decision problem. If a particular vector λ is almost orthogonal to the plane, it has a short projection, and the evaluation of the alternatives by means of λ, i.e. $\sum_{j=1}^{m} \lambda_j f_j(d)$ for all $d \in D$, are all close together. Therefore, even if a total ranking between the alternatives exists the result may depend on the choice of λ.

The directions of the projected weights also give us information on the degree of opposition of the alternatives with the current partial information.

The problem of determining whether such a set helps to discriminate between the alternatives or more information is needed is similar to checking whether the origin is in the convex hull of the projected set of weights. Weights pointing in opposite directions give rise to conflicts in the evaluation of the alternatives.

Let us consider the following example proposed by Mareschal and Brans (1988) [10] and also analyzed in Conde et al.(1996) [3]. In the context of a power plant location, they compare the advantages and disadvantages of six locations which have been measured on six different criteria as it is shown in table 1.

Criteria	A1	A2	A3	A4	A5	A6	Type	Parameter
f1	min	80	65	83	40	94	II	$q = 10$
f2	max	90	58	60	80	96	III	$p = 30, q = 0.5$
f3	min	6	2	4	10	7	V	$p = 5$
f4	min	5.4	9.7	7.2	7.5	3.6	IV	$q = 1, p = 6$
f5	min	8	1	4	7	5	I	-
f6	max	5	1	7	10	8	VI	$\sigma = 5$

(Table 1)

The second column refers to the positive or negative influence of the criterion in the preference of the D-M and the data associated to the columns denoted by type and parameter refer to the preference measure used in the evaluation of each criterion on every alternative. In order not to waste effort explaining the details of a well-known MCDM we only present the matrix of the standardized values associated to the criteria (in columns) and alternatives (in rows). These values are free of scales and can be mixed via the vector of weights to obtain a global value for each alternative (further details can be found in [10]. Assume we have no-information on the weights, hence only nonnegativity and additive normalization can be assumed on this vector. We use its representation on a two-dimensional plane built up by the first two principal components u, w of

	f1	f2	f3	f4	f5	f6
A1	-0.40	0.55	-0.05	0.10	-1.00	-0.08
A2	0.20	-0.65	0.78	-0.70	1.00	-0.52
A3	-0.40	-0.60	0.38	-0.20	0.20	0.09
A4	1.00	0.16	-0.82	-0.20	-0.60	0.34
A5	0.60	-0.16	-0.05	0.60	0.60	0.16
A6	-1.00	0.70	-0.24	0.40	-0.20	0.01

(Table 2)

the alternative matrix. This kind of representation is in essence what is used by Mareschal and Brans(1988) [10] for defining their Gaia-plane. Figure 3.1 is the projection on the Gaia plane of the alternatives and the criteria, where the vectors u, v are

$$u = (0.184, -0.514, 0.417, -0.248, 0.665, -0.163)$$

$$v = (-0.897, 0.104, 0.374, 0.034, 0.059, -0.197)$$

The fact that the convex hull of the projected criteria (shown in figure 3.2) contains the origin of coordinates leads us to think that nothing is clear about the preference assigned to the different alternatives.

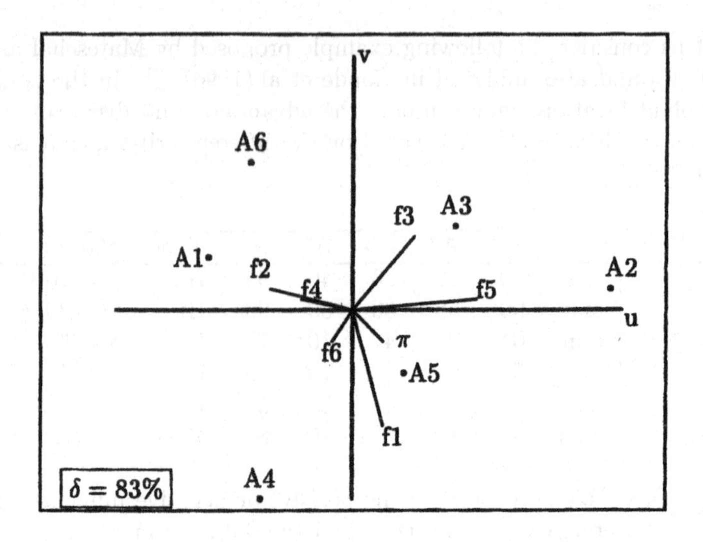

Figure 3.1: Alternatives and criteria on a plane

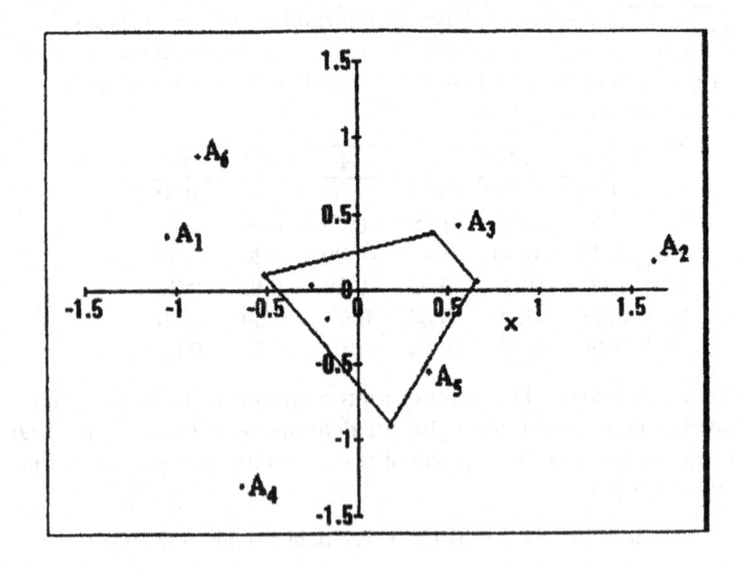

Figure 3.2: Convex hull of the projected weights

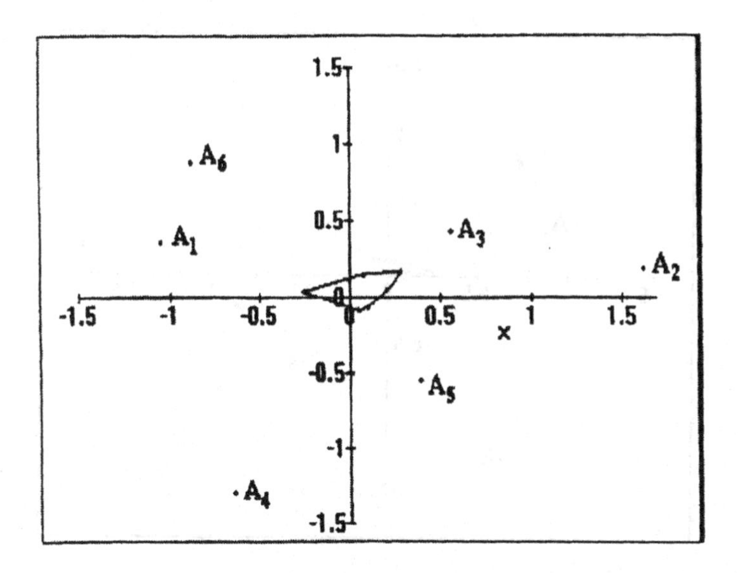

Figure 3.3: Reduction of the projected weights

It is not hard to assume some extra relationships between weights given the economic interpretation of the criteria. Let us impose the complete ranking

$$\lambda_4 \geq \lambda_5 \geq \lambda_3 \geq \lambda_2 \geq \lambda_1 \geq \lambda_6$$

Although under this assumption a total ranking among the alternatives is not obtained one can see (figure 3.3) the drastic reduction on the projection of the normalized extreme directions.

However, the origin of coordinates still belongs to the convex hull of the projected set of weights. This fact can be interpreted as a need of providing additional information in order to rank the alternatives.

To reduce the projected set of weights the D-M considers a new constraint

$$5\lambda_4 \geq \lambda_1 + \lambda_2 + \lambda_3 + \lambda_5 + \lambda_6$$

and from Theorem 2.2 and Theorem 2.3 we obtain the set of extreme points of the new polyhedron of weights. Projecting its normalized columns on the our plane the projected set of weights does not contain the origin of coordinates (figure 3.4).

This procedure can be seen as an interactive method for enrichments of the preference structures given by the successive set of weights in use.

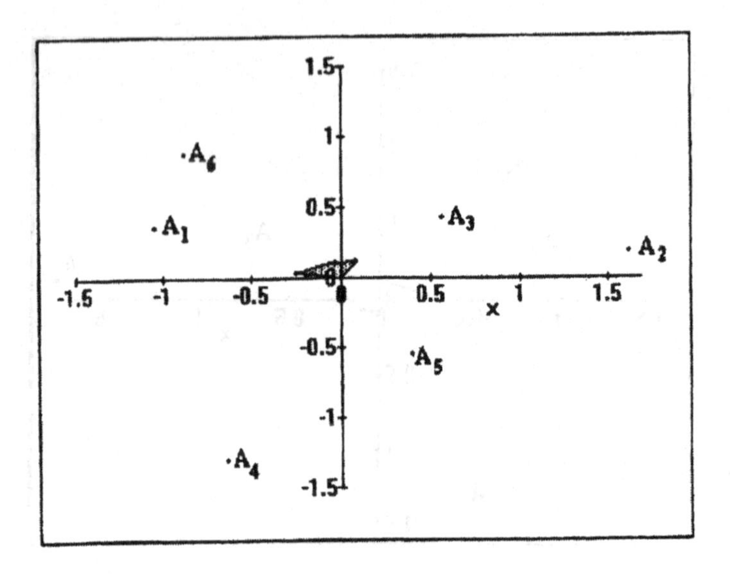

Figure 3.4: A case not containing the origin

4. CONCLUSIONS

The preference information available in multiple criteria decision processes must be used in order to reduce the set of nondominated alternatives to be considered. Since the natural way to provide information is in sequential order we analyze the iterative incorporation of information that comes from interaction of the decision-maker with the decision problem. One of the main advantages of this approach is that the incorporation of new information does not augment the difficulty of the problem to be solved and even in many cases it leads to simpler models.

The results that we present also allow to test graphically the quality of the information provided. To this end, we use a projection of the set of weights onto a two-dimensional space given by the two most significant principal components of the matrix of the evaluation of the criteria.

Acknowledgments

The research of the first and third authors has been partially supported by Spanish Ministerio de Educación y Cultura grants no. PB97-0707 and HA1997/0123.

References

[1] Carrizosa E., Conde E., Fernández F.R., Puerto J. (1995). Multicriteria Analysis with Partial Information About the Weighting Coefficients. *European Journal of Operational Research,* 81(2), 291-301.

[2] Chankong V., Haimes J.(1983). *Multiobjective Decision Making.* North-Holland.

[3] Conde E., Fernández F.R., Puerto J. (1996): Handling linear constraints in interactive weighting based decision procedures. Prepublicación n.2/96 Marzo (1996). Facultad de Matemáticas. Universidad de Sevilla.

[4] Edelsbrunner H. (1987). *Algorithms in Geometrical Geometry.* Springer-Verlag.

[5] Hannan E.L.(1981). Obtaining Nondominated Priority Vectors for Multiple Objective Decision making Problems with Different Combinations of Cardinal and Ordinal Information. *IEEE Transactions on Systems, Man. and Cybernetics,* SMC-11(85), 38-543.

[6] Huang Ch., Masud A.(1979). *Multiple Objective Decision Making.* Springer-Verlag.

[7] Huang Ch., Yoo K.(1981). *Multiple Attribute Decision Making.* Springer-Verlag.

[8] Kirkwood C.W. and Sarin R.K.(1985). Ranking with Partial Information. *Operations Research,* 34, 296-310.

[9] Mardia K.V., Kent J.T.,Bibby J.M. (1979). *Multicriteria Analysis.* Academic Press.

[10] Mareschal B. and Brans J.P. (1988). Geometrical Representation for MCDA. *European Journal of Operational Research,* 34,.69-77.

[11] Mármol A.M., Puerto J., Fernández F.R. (1998). Sequential incorporation of imprecise information in multiple criteria decision processes. Prepublicación.1/98 . Facultad de Matemáticas. Universidad de Sevilla.

[12] Paelinck J.(1975). Qualitative Multiple Criteria Analysis, Environmental Protection and Multiregional Development. European Meeting of the Regional Science Association.

[13] Steuer R.E. (1976). Multiple Objective Linear Programming with Interval Criterion Weights. *Management Science,* 23, 305-316.

[14] Vincke P. (1989). *L'aide Multicritere a la Decision.* E. de l'Université de Bruselles.

A MULTICRITERIA FRAMEWORK FOR RISK ANALYSIS

Winfried Hallerbach & Jaap Spronk[1]

1. Introduction

Many of the theoretical developments in Multicriteria Decision Analysis took off in the seventies and eighties. In the meantime, the field has attracted many academic researchers from literally all over the world and from a great variety of disciplines. Also, many of the approaches proposed to tackle multicriteria decision problems have been and are being used in practice. Some approaches have even gained a rather wide acceptance in practice. One may wonder what is the reason for the enormous attention for and success of multicriteria decision analysis. At the same time one may also wonder why the use of multicriteria decision analysis in practice has not yet been much larger, where so many researchers are strongly convinced of its great potential and practical relevance.

Reasons for the attention for multicriteria decision analysis may be found in the dissatisfaction with more traditional decision support methodologies in which decision makers were pressed into iron harnasses of assumptions being hardly acceptable in many if not most real-life situations. Other reasons may be found in the cultural developments that originated around and during the seventies, by which many value systems were shaken and reshaped into more multifarious value systems.

The question why multicriteria decision analysis has not yet been used even more widely cannot be simply answered by stating that it takes time for new decision technologies to become accepted. Of course it takes time, but some methodologies are accepted very quickly while others are accepted only very slowly or even not accepted at all. We believe that many methodologies have difficulties in finding acceptance because they are still based on many too rigid assumptions with respect to the availability of data, the information processing capabilities of decision makers and with respect to the preferences and choice behaviour of decision makers. One notable example is the way multicriteria decision analysis deals with uncertainty and risk. Either uncertainty and risk are

[1] Both Erasmus University Rotterdam, Rotterdam Institute of Business Economic Studies, and Tinbergen Institute Graduate School of Economics.

neglected all together or they are dealt with by making rigid distribution assumptions. In the first case, risk factors are sometimes introduced by decision makers through the definition of criteria that are somehow intended to control the risk inherent in the decision problem. In the second case, one often finds that it is simply assumed that a probability distribution is available which is then used as input for the multicriteria decision analysis. We believe that multicriteria decision analysis will gain much wider acceptance if it pays considerably more attention to uncertainty, risk analysis and risk management.

In this paper we describe a multicriteria framework for risk analysis that makes use of both concepts borrowed from multicriteria analysis and of concepts borrowed from modern financial economic theory. Over the years, this framework has proven its use in applications such as portfolio management, capital investment selection, financial planning and performance evaluation.[2] The use of the presented framework is certainly not limited to financial decisions. Especially applications in which uncertainty and risk play an important role may benefit from the framework. The framework is based on three pillars:

1. It makes use of all available information. At the same time, it does not make irrealistic assumptions with respect to the availability of information. So in our approach, decision makers are not squeezed out to provide information they do not possess. Actually, we do not torture them at all.
2. Borrowing from modern financial economic theory, the framework incorporates a rich description of uncertainty and risk. It includes the multi-factor approach, contingent claims, game elements and combinations – depending on the situation at hand.
3. The framework integrates the above elements in a process- oriented approach towards financial decisions.

The paper is organized as follows. Section 2 presents five snapshots of uncertainty and risks involved in problem solving. Section 3 concentrates on the modelling of uncertainty and risk given a particular decision context. Section 4 concludes the paper and outlines some directions for future research.

2. Five snapshots of uncertainty and risk in problem solving

In conventional micro-economic theory and in the 'traditional' decision methodologies that apparently have borrowed many of their assumptions from micro-economices, simplifying assumptions are made with respect to the

---

[2] For applications and further references, see Vermeulen, Spronk & Van Der Wijst [1996] and Spronk & Hallerbach [1997].

preferences of the decision maker and with respect to the representation of choice alternatives. Assuming that all necessary information is available, decision rules can be derived that are *normative* and *conditional* on the above two sets of assumptions, cf. Keynes [1891]. One of the charms of multicriteria decision analysis is that most of the time not too many rigid preference assumptions are being made but, as argued above, the uncertainty and risk aspects are often treated inadequately. For most decision makers, dealing with uncertainty and risk is an integral part of daily life. To illustrate this point, we will give five snapshots, including commentary, of problem solving and the associated uncertainties and risks. These snapshots relate to *(1)* problem awareness, *(2)* problem identification, *(3)* different ways of describing problems, *(4)* risks in modelling problems and *(5)* modelling uncertainty.

(1) Problem awareness means that one recognizes there *is* a problem to be solved. Problems can be of different nature. For instance, a potential labour strike is a threat whereas the possibility of a take-over of another firm may be a great opportunity. Systematical scanning and monitoring of one's activities may reduce the risk of not recognizing problems or recognizing problems too late (e.g. the intriguing Y2K problem that has surprised already many and will probably surprise many more at the dawn of the next century!).

(2) Even if one is aware that there are problems to be solved one has to decide which problem(s) is (or are) to be addressed and in which order. The apparent risk in this *problem identification stage* is addressing the wrong problem.

(3) There are many *different ways of describing problems.* As many philosophers and scientists from P.L. Ato to P.L. Yu [1990] have noted, the problem formulation depends on the person(s) who formulate the problem and on the language (defined in a broad sense) they speak. As such you do not describe 'the' problem but 'your projection of the problem against the wall'.

(4) Even if there is no doubt with respect to the way in which a particular problem can be described, there are still *many risks in modelling the problem.* Important choices have to be made regarding – among others:

-the identification of the decision maker(s)
-the problem horizon
-the static,comparatively static or dynamic nature of the problem
-the description of the preferences
-the description of the opportunities
-the description of the relationships between preferences and opportunities.

The choices one can make and the 'correctness' of these choices will depend on the quality of both the available set of information (e.g. are all data relevant, are there any biases?) and the body of knowledge of the decision maker and his/her helpers (e.g. what do we know for sure, is there anything ignored, do we suffer from illusions?). Fortunately, depending on one's learning potential, analytical skills and imagination people do learn. People learn to get a better view on an existing problem on the basis of new information, new insights and falsification of old insights. People do also learn in situations in which the problem setting is changing over time. In terms of the work of Yu [1990], the possibility to learn given a fixed problem setting can be summarized with the help of Figure 1.

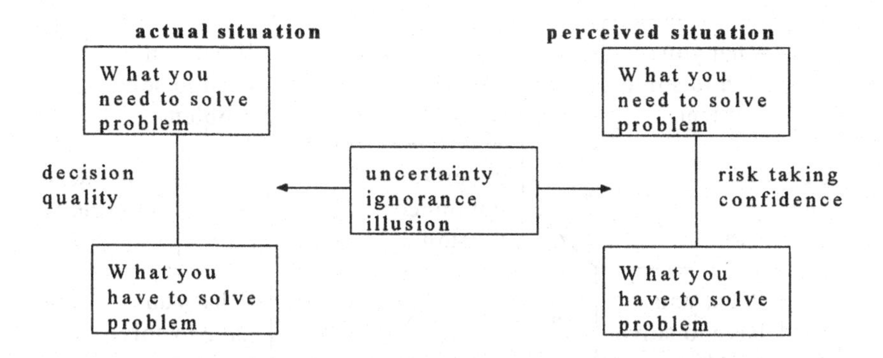

Figure 1: Uncertainty and learning in a fixed problem setting

On the left there is the (unobservable) set of 'what you need' and of 'what you have' to solve a problem. Obviously, the better the 'need' set is covered by the 'have' set, the better the decision quality. On the right are the same two sets, but now as perceived by the decision maker. In this case, the decision maker can make a 'calculated risk' and feel more or less confident about the decision made. Also, the decision maker can try to expand the perceived 'have' set by acquiring new information and by learning – thus getting a better coverage of the 'need' set. The bridge between the actual and the perceived situation is formed by intrinsic uncertainty, ignorance and illusion.

Figure 2 shows the same relationships for the case the problem setting is changing. A dynamic version of the comments to Figure 1 is easily given and thus left to the reader.

(5) The fifth snapshot concerns the *modelling of uncertainty and risk*. Conventional approaches normally assume that uncertainty can be caught in well-defined probability distributions that can be specified because of the implicit assumption of complete information. The above discussion already suggests that often in real-life problems, incompleteness of information is the rule rather than exception. Ignorance and illusions do exist, together with biases in estimation and flaws in data. The following section shows how to cope (or to live) with some of these problems.

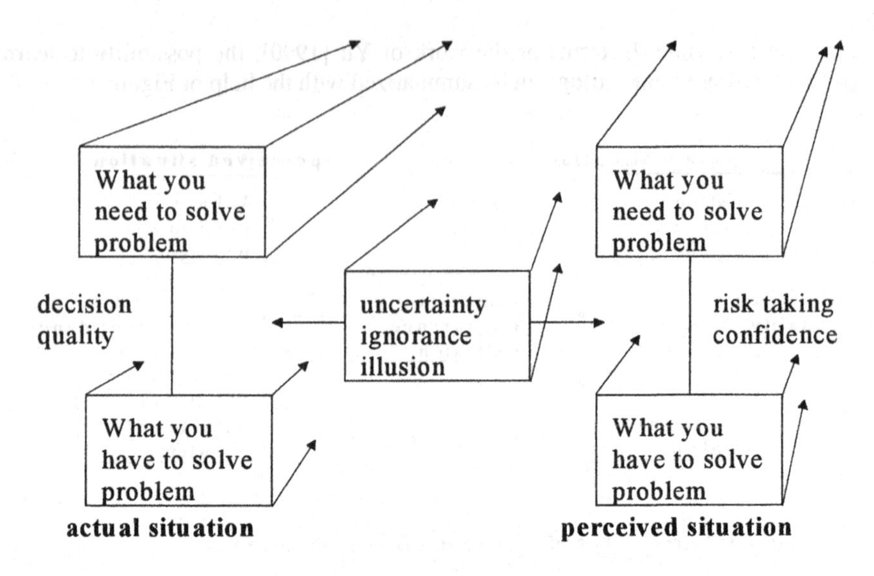

Figure 2: Uncertainty and learning in a changing problem setting

3. Describing uncertainty and risk in decision problems

After the uncertainty and risks involved in recognizing, identifying and describing decision problems, this section addresses the modelling of uncertainty and risk given a particular decision problem. Basically, there are two perspectives to this intrinsic risk: *ex-ante*, related to prospective choice problems, and *ex-post*, related to retrospective performance evaluation issues. For the time being, we'll ignore the latter and confine ourselves to the former.

In section 3.1 we first briefly summarize two conventional representations of risk and uncertainty in decision problems. In section 3.2, we then outline the key concepts of a competing way of representation, which leads to a prototypical classification of decision problems. Section 3.3 discusses these prototypical problems in more detail, making a first step from risk analysis towards risk management and incorporating some important insights from finance theory.

3.1 Conventional representations

In the conventional representation of decision making under uncertainty, all decision alternatives are given and fixed. The alternatives are first described in terms of their possible outcomes. Next, for each decision alternative, the decision maker must assess a probability distribution defined over the outcomes of the alternative. In this way, the decision problem is translated into one of choosing between alternative probability distributions defined over all possible outcomes. Uncertainty is incorporated by expressing the preference structure in terms of a preference functional defined over the stochastic outcomes. The optimal decision is identified by maximizing the expected value of the preference functional.

In an alternative representation of risk in choice problems, uncertainty is incorporated in a somewhat more explicit way. First, the choice alternatives – the probability distribution(s) of the outcomes – are described in terms of distribution characteristics. For example, the distribution is described by means of its expected value,variance and perhaps higher order statistical moments (provided that they exist). Then a preference functional is specified over the distribution characteristics. This is a Lancaster [1966]-type of preference functional. Maximizing the value of this function identifies the optimal decision.

The problem with both of these approaches is that the decision maker is assumed to have sufficient information as to be able to specify the complete probability distributions, either directly, or indirectly through specifying the distributions' moments. Furthermore, the decision maker is assumed to be able to specify the preference functional concerned. The latter assumption, or rather its high degree of irrealism, has given rise to the development of multicriteria decision analysis. Here, we address the first problem, the often irrealistic assumption of sufficient information to be able to specify probability distributions. Below, we present a competing approach, intended to support the modelling of uncertainty and risk in the case of partial information.

3.2 An alternative way of describing uncertainty and risk in decision problems

In decision problems under uncertainty and risk, *two questions* play a crucial role:

- where does the uncertainty stem from or, in other words, what are the sources of risk?
- when and how can this uncertainty be changed?

The answer to the first question leads to the decomposition of uncertainty. This involves attributing the inherent risk (the potential variability in the outcomes) to the variability in several underlying state variables or factors. We can thus view the outcomes as being *generated* by the factors. Conversely, the stochastic outcomes are *conditioned* on these factors. The degree in which fluctuations in the factors propagate into fluctuations in the outcomes can be measured by response coefficients. These sensitivity coefficients can then be interpreted as exposures to the underlying risk factors and together they constitute the multi-dimensional risk profile of a decision alternative.

The answer to the the second question leads to three prototypes of decision problems:

A. The decision maker makes and implements a final decision and waits for its outcome. This outcome will depend on the evolution of external factors, beyond the decision maker's control.
B. The decision maker makes and implements a decision and observes the evolution of external factors (which are still beyond the decision maker's control). However, depending on the value of these factors, the decision maker may make and implement additional decisions. For example, a decision maker may decide to produce some amount of 'a new and spectacular software package' and then, depending on the development of the market, he may decide to stop, decrease or increase production.
C. As in B, but the decision maker is not the sole player and thus has to take account of the potential impact of decisions made by others sometime in the future (where the other(s) are of course confronted with a similar type of decision problem). The interaction between the various players in the field gives rise to dynamic *game situations*.

Below, we will discuss each of these three prototype decision problems in somewhat more detail, paying attention to the situation that information is less than perfect. In prototype A problems the importance of multi-dimensional risk profiles is sketched. Prototype B problems involve the use of contingent claims whereas in prototype C problems game aspects are added.

3.3 Three prototypical decisions in the case of imperfect information

Multi-dimensional risk profiles in "prototype A" problems

The complexity of the decision process stems from assessing probability information regarding the outcomes (the "information problem") and from confronting this probability information with the preference structure of the decision maker (or incorporating it in some preference functional, the "criterion problem"). In general, the process of conditioning and decomposition greatly reduces the complexity of the decision process.[3] With respect to the information problem, decomposition reduces the complexity of risk judgments because it allows shifting attention from the uni-dimensional probability distribution of the outcomes to the exposures to the multi-farious underlying factors, summarized in

[3] The general idea of decomposition for the resolution or reduction of uncertainty can be traced to Simon [1962]. The process of conditioning described here is in the spirit of the stochastic hierarchies of Raiffa [1968].

the risk profile.[4] In this setting, the decision maker is relieved of the burden to explicitize the probability distribution of the outcomes. Moreover, it isn't even necessary that he or she has detailed information regarding the distributions of the factors. When a decision alternative shows a larger sensitivity (in absolute sense) to some factor j, it has a larger exposure to factor j risk, meaning that potential fluctuations in that factor propagate to a larger extent into potential fluctuations of the outcomes. So when two decision alternatives have the same risk profile except for the exposure to factor j, the alternative with the smaller exposure is less risky. This brings us to the criterion problem. Instead of the need to incorporate either the whole probability distribution of the outcomes or its relevant moments in the decision process, it suffices to express preferences with respect to the risk profile as summarized by the factor exposures. These sensitivity coefficients thus become the relevant risk attributes of the decision alternatives. Now, for example, again a Lancaster [1966]-type of preference functional can be specified, but now over the various risk exposures and other relevant attributes. An even less demanding alternative is to employ an interactive choice procedure. Asking the decision maker to evaluate the trade-offs between the risk attributes does not entail heroic preference assumptions.

A multidimensional view on risk allows multidimensional attitudes towards risk. A decision maker can show different degrees of aversion towards variability in the outcomes depending on the specific type of the underlying risk generating factor. Aside from the type of risk source, also the size of the exposure matters. A small exposure combined with a large factor variability can generate the same degree of risk as a large exposure combined with a small factor variability. However, the decision maker can face restrictions on risk exposures or possess smaller or larger buffers to absorb different types of risks. An example of the former is a restriction placed by a firm on the currency risk attached to its sales abroad; an example of the latter is the stock of input resources a firm holds in order to maintain production continuity.

Risk and variability are by no means synonymous. In general, there is a clear dichotomy between downside variability and upside variability. When a decision maker strives to maximize an alternative's outcome, downside risk can be separated from upside potential. When a decision maker instead strives to minimize the outcome, the asymmetry is reversed. This observation not only holds in general for the potential variability in the outcomes, but also more specifically for the potential fluctuations of the factors. Not only restrictions on risk exposures play a role, or the presence of buffers to absorb the unwanted part of variability, but also the availability of specific instruments that can be used to mitigate or eliminate the exposure to some part of factor variability. This leads us to the issue

---

[4] In effect, combining the conditional distributions of the outcomes with respect to the factors with the marginal (joint) distributions of the factors yields the unconditional distribution of the outcomes. Ravinder, Kleinmuntz & Dyer [1988] analyze the measurement error that can be attributed to the use of decomposition when compared to direct assessment.

of risk management and the application of derivative instruments, as discussed below.

A multidimensional view on risk quite naturally allows for performing sensitivity analyses. Given the factor exposures, it can be analyzed how the outcomes are influenced by various changes of the factor values. Likewise, the decision maker can determine how the outcomes are influenced by hypothesized changes in the exposures to the factors. This implies a dynamic instead of a static approach to problem solving. The decision maker can *(i)* evaluate the different risk profiles that are generated by the set of available alternatives; by examining the trade-offs offered by the alternatives he can *(ii)* obtain new insights and shape his preferences in more detail, or *(iii)* realize that he needs more information on a specific part of the problem. Because of this interdependence between the characteristics of the choice alternatives and the decision maker's preferences, the aspect of *learning* in a complex decision situation gains importance. This leads us to sutuations where the decision maker may take additional decision, depending on the course of the environment.

The use of contingent claims in "prototype B" problems

It is important to distinguish between potential variability and risk, and also between *sources* of risk and *exposures* to risk. The former distinction is important because of the potential asymmetric nature of risk, whereas the latter follows directly from the principle of conditioning underlying a multi-factor representation. In a capital budgeting context, an investment project derives its present value from the future cash flows it generates. The project's cash flows –and hence its value– can be conditioned on factors describing its economic context, like commodity prices, exchange rates, interest rates and inflation. The tuplet of its exposures for these factors constitute the project's characterizing risk profile. This multi-dimensional representation enriches the project's description and enhances its evaluation. This applies even when information about the type and nature of risk sources or about the exposures is *incomplete*. By pasting as many pieces of the puzzle as possible together, one can obtain an impression of the weak and strong sides of the project, especially when comparing it to other competing projects.

When evaluating a portfolio of investment projects, also the *diversification effect* is relevant. Not the sum of the risks attached to the separate projects is relevant, but the risks attached to the sum of the projects. The lower the degree of dependence between the underlying risk factors, the greater the benefits reaped from the diversification effect. Aside from diversification, a firm may control risks by *matching* factor exposures. Combining a project with an exposure to factor j of +1 with a project with an exposure of –1 will in effect eliminate the exposure to factor j risk.

When evaluating risk profiles of decision alternatives, not only the trade-off between the various risk exposures is relevant, but also the trade-off between risk on the one hand and value or return on the other. This *risk-return trade-off* implies that there is a reward for bearing risks. For example, lowering a project's risk exposures can at the same time decrease its value or profitability. A firm must then compare the risk-return trade-off offered by the portfolio of investment opportunities to the risk-return trade-off implied by its degree of risk aversion. So, generally speaking, there is a reward for risk offered by the decision alternatives, and there is premium for risk required by the decision maker – the subjective price of risk. To complicate things in a necessary way, there is in addition a reward for risk offered and required by the "outside world". When this so-called market price of risk exceeds the subjective price of risk, the firm clearly has a comparative advantage – i.e., compared to the outside world – in bearing the risk. The firm then has specific and valuable expertise in handling the risk in the decision context at hand . Selling the risk to third parties in the outside world implies paying a market compensation that is too high from the firm's point of view.

Above, we already mentioned the possibilities of diversification and matching in order to change risk profiles. But what opportunities does a firm have to sell risks to third parties? This question opens a whole toolbox of financial instruments, designed for hedging purposes. These instruments can be used in many other (multicriteria) decision problems. At this point, we like to stress the importance of derivative instruments or *"contingent claims"* to transfer risk exposures from parties that do not want or are not able to bear these risks, to parties who are better equipped to handle them. Option contracts, forward contracts, futures, and swaps are some examples of derivatives that allow a firm to hedge, i.e. to adjust individual risk exposures by shifting them to others.[5] Option contracts take a special position, since they work in an asymmetric way, like insurance contracts: they allow selling the downside risk of an exposure (by paying a market price) while maintaining its full upside potential. By chosing the right kind of contingent claims contract, risk exposures can be eliminated with surgical precision – provided the decision maker has sufficient knowledge of the nature of the contracts involved. With the increasing complexity of the available contracts, a detailed and sophisticated contingent claims analysis is a prerequisite to successful risk management.

Contingent claims analysis not only focuses on explicit options (i.e. the options that are traded on financial markets) but also on implicit options. The latter options are embedded in decision alternatives and manifest themselves as "flexibility". For example, during the life of an investment project the firm may expand the project when profitable or abandon it when it proves disappointing. A firm may even have the option to postpone the decision and start a project when and if market conditions prove favorable.[6]

The above discussion urges us to expand the description of decision alternatives. Of course, the exposures to the underlying sources of risks are still an

[5] Smithson & Smith [1995] provide a good introduction to financial risk management and the use of derivatives for hedging.

[6] For an overview, see for example Brealey & Myers [1996].

essential part of a risk profile. To this, we have to add the aspects of flexibility, relating to the possibility to make decisions somewhen in the future depending on the evolution of the environment.

Contingent claims and games in "prototype C" problems

The decision maker is of course not the sole player in this environment. This implies that the potential impact of decisions taken by others sometime in the future must also be taken into account. The interaction between the various players in the field gives rise to *game situations*. For example, suppose we have two competitors, firm A and firm B. Firm A expects B to enter the market and is trying to understand B's likely pricing strategy. Firm A can then construct a payoff matrix, summarizing the payoffs (sales or net profit, e.g.) under various pricing strategies. Firms A and B have perverse incentives to lower prices: maintaining prices at the current level while the competitor cuts prices will lower the payoff. Like in the prisoner's dilemma, both firms are inclined to cut prices, thus both lowering their payoff. As a result, we can imagine that each firm then will separately try to compete on other factors, like product features, service levels or advertising. An analysis of this pricing game can only be made when the firm has information about costs to enter and exit the market, about demand functions, about revenue, cost structures etc. So the decision process has various dynamic aspects, stemming on the one hand from flexibility in the own decision context and on the other hand from the impact of other parties' decisions on this decision context. Because of these dynamic aspects, the role of learning – from your own decisions and from others – cannot be underestimated.

4. Summary and conclusions

We described a multicriteria framework for risk management that makes use of concepts borrowed from both multicriteria analysis and modern financial economic theory. Over the years, this framework has proven its use in applications such as portfolio management, capital investment selection, financial planning and performance evaluation. However, its scope is much broader: especially applications in which uncertainty and risk play an important role may benefit from the framework.

The framework is based on three pillars. First, it makes use of all available information. At the same time, it does not make irrealistic assumptions with respect to the availability of information. So in our approach, decision makers are not expected to provide information they do not possess. Second, borrowing from modern financial economic theory, the framework incorporates a rich description of uncertainty and risk. It includes the multi-factor approach, contingent claims, game elements and combinations – depending on the situation at hand. Finally, the framework integrates the above elements in a process-oriented approach towards financial decisions.

In the present paper we concentrated on the modelling of uncertainty and risk. Starting from five snapshots of problem solving and the associated uncertainties and risks (problem awareness and identification, problem description and modelling, and modelling uncertainty), we focused on how uncertainty in a decision context can be changed. Depending on the flexibility to take additional decisions when new information arrives and the possibility to interact with third parties, we distinguish between three prototypical decisions. We discussed each of these decisions and provided some examples.

A number of the ideas and concepts set out above (e.g. the concept of multifactor models and that of contingent claims), were borrowed from finance. During the last decade, finance has embraced risk analysis and risk management very strongly, both in theory and practice. We believe that many of these ideas and concepts can also be applied fruitfully in all other areas of (multicriteria) decision making under uncertainty. We encourage the reader to get acquainted with some of the standard references in risk management that have emerged in the financial discipline (Smithson & Smith [1995], e.g.) so that these insights can fertilize efforts in other areas.

References

Brealey, R.A, S.C. Myers (1996): *Principles of Corporate Finance*, 5th Edition, McGraw-Hill, New York.

Keynes, J.N. (1891): *The Scope and Method of Political Economy*, Macmillan, London.

Lancaster, K.J. (1966): A New Approach to Consumer Theory. *Journal of Political Economy* 74, 132-157.

Raiffa, H. (1968): *Decision Analysis*, Addison-Wesley, Reading.

Ravinder, H.V., D.N. Kleinmuntz, J.S. Dyer (1988): The Reliability of Subjective Probabilities Obtained Through Decomposition. *Management Science* 34, 186-199.

Simon, H.A. (1962): The Architecture of Complexity. *Proceedings of the American Philosophical Society* 106, 467-482.

Smithson, C.W., C. Smith (1995): *Managing Financial Risk*. Irwin, Burr Ridge.

Spronk, J., W.G. Hallerbach (1997): Financial Modelling: Where to Go? With an Illustration for Portfolio Management. *European Journal of Operational Research* 99, 113-125.

Vermeulen, E.M., J. Spronk, N. Van Der Wijst (1996): Analyzing Risk and Performance Using the Multi-Factor Concept. *European Journal of Operational Research* 93, 173-184.

Yu, P.L. (1990): *Forming Winning Strategies: An Integrated Theory of Habitual Domains*. Springer Verlag, Heidelberg.

Intelligent User Support in Multicriteria Decision Support

Julie Hodgkin, Valerie Belton, and Iain Buchanan, University of Strathclyde, Glasgow, UK

Abstract:

In an attempt to enhance the relationship between user and system, Multicriteria Decision Support Systems (MCDSS) which incorporate some form of intelligent support have been developed. Intelligent MCDSS (IMCDSS) are potentially useful to the multicriteria practitioner in (at least) two respects. Firstly, they can act as an "intelligent assistant" to the multicriteria analyst both when working interactively with decision makers in a decision making environment and when carrying out "backroom" analysis. Secondly, by supporting the D.I.Y user they may enable models to be handed over to clients with less need for ongoing support. There is also the potential to make modelling tools more available for D.I.Y use by managers, whilst guarding against their misuse. Intelligence takes on many guises, both in how it is generated and how it is relayed to the user. This paper discusses our research in the area of intelligent user support over the last two years. We begin by taking a brief look at initial studies, which examined a number of decision support systems (*namely,* V•I•S•A , *SIMUL8, Frontier Analyst, and Decision Explorer).* This highlighted possible support areas as well as analysing different methods of defining, capturing and relaying intelligence. The paper then goes on to present and discuss a prototype intelligent MCDSS designed and implemented by taking an existing multicriteria package and building in different forms of "intelligence". The next stage of the research is to evaluate the effectiveness of the IMCDSS and we will discuss ways in which this might be done.

Keywords: Multicriteria analysis, decision support, intelligence, user support

1 Introduction

In recent decades, many modelling packages have been developed for a wide variety of application areas. Through the use of such packages, it is hoped that the user is able to attain a better understanding of the problems they face and is assisted in the decisions they must make. In the continuous process of re-evaluation and development of the packages, one aspect that is of increasing importance is the concept of user support. How do packages fare in terms of their ability not only to act as a tool for model building, but also in providing a means of empowering the user to gain a better understand of the problem they are solving? One potential way of achieving this is by the incorporation of "intelligent help". A general approach to this topic is described by Angehrn and Luthi (1990). They suggest that intelligence can go some way in replicating a human consultant, supporting the decision maker in tasks such as expressing, defining, and ultimately, better understanding their problems. Belton and Vickers (1993) suggest that it is such support that can ultimately determine whether the user regards a system as a black-box, and thus probably an incomprehensible and unusable package, or if they regard a system as user-friendly, one with which they are comfortable, in control of, and perhaps most importantly, one in which they feel they can place their confidence as an aid to decision making.

This paper will begin by describing initial work performed examining existing DSS systems. Research examining the concept of intelligence and discussing ways in which it has been associated with MCDSS will then be given. We will then go on to give a brief introduction to V•I•S•A, followed by a description of Intelligent V•I•S•A, the prototype that has been developed to explore the usefulness of intelligent user support. This will be supported by an example showing potential ways in which the intelligent user support might be used. The paper will then conclude with a discussion of the evaluation which is currently taking place followed by a brief look at future work.

2 Background to Research

Research in the field of providing intelligent user support began by analysing four DSS: V•I•S•A - a Visual Interactive Sensitivity Analysis tool designed to support the multicriteria decision making process: SIMUL8 - a visual interactive simulation package: Decision Explorer - a cognitive mapping tool designed to help users structure their ideas and thoughts: and Frontier Analyst - a tool which uses the method of Data Envelopment Analysis to determine the efficiency of given units.

The investigation of these four systems addressed the adequacy of the current support offered and identifying areas for potential incorporation of intelligent user support. This research indicated that although the packages provided some form of on-line support, this was very limited. There was a lack of more in-depth support designed to help the non-expert user in less trivial aspects of decision making. Based on these initial ideas and impressions of user support it was apparent that there was substantial opportunity for further research in this area. As well as seeking to generate and implement generic ideas, it was considered necessary to test such ideas in the context of an actual DSS, preferably one that already existed. This would require an understanding of the process and techniques associated with the DSS, a clear need within the chosen system for advances of the sort being developed, and open access to its implementation. For these reasons, it was decided that future practical work would centre on V•I•S•A and the potential for incorporating intelligent support.

3 V•I•S•A and Intelligent V•I•S•A

3.1 V•I•S•A

V•I•S•A is a decision support system designed to aid in MultiCriteria Decision Making (Belton and Vickers,1989; Belton and Vickers,1990). It is based on the use of a multi-attribute value function. One of V•I•S•A 's main strengths is its "visual interactivity" (Bell, 1985; Belton and Elder,1994). It's acronym stands for Visual Interactive Sensitivity Analysis, and as this suggests, the software provides a range of visual interactive facilities to support the user in their decision making. Further details regarding V•I•S•A and its use in the decision making process can be found in Belton, 1997.

3.2 Intelligent V•I•S•A

Although the concept of intelligence is not new, the implementation and evaluation of intelligent systems is relatively little studied in the field of MCDA. At this point it is worthwhile briefly examining more closely what is meant by "intelligence". It is a particularly nebulous concept which has been addressed by scholars from many disciplines. In the field of computer support it is perhaps most often associated with the notion of "artificial intelligence", stemming from the work of Turing in the 1950's (Turing, 1992). In this context "intelligence" is taken to be the ability of a machine to replicate human behaviour. However, in the context of decision support systems, intelligence is also interpreted as effective user support - essentially replacing the role of a human analyst or facilitator. Such a system is not intended to replace the decision maker, but to offer guidance, to prompt deeper thinking, or highlight specific issues. The intelligence may be context specific, or process oriented. Triplce C (Angehrn and Dutta (1992) is an example of a process oriented MCDSS, incorporating intelligence to support the user in model building and analysis. Systems described by Bayaud and Pomerol(1992), O'Keefe et al (1986), Srivinsan and Ruparel(1990), all incorporate some form of context specific knowledge-base to support the decision maker. The incorporation of intelligence to support the user in the use and understanding of MCDA is described by Pasche(1991) in reference to his system EXTRA, and by Klein(1994), in describing the system VIRTUS.

The study of intelligent user support can be viewed as a three-part process; generation of the idea, implementation of the idea, and testing and evaluation of the idea. Extensive research has been performed in the first of these areas, but few of the ideas and theories developed have been followed through to the implementation and testing stages. It is these two areas which this research focuses on. However, as the process is cyclic, it should be the case that in building and evaluating intelligent support, the development of new ideas and theories will be developed.

The prototype we have developed differs from the systems described above in two ways. Firstly, the intelligent support offered is process based; focusing on the processes involved in MCDA decision making, with the intent of supporting the user in understanding and applying the MCDA approach. Secondly, the intelligent support is generic; the information offered is generated from the data entered by the user and therefore is not restricted to a particular problem domain.

There are many different ways in which intelligent support can be used and provided (Belton and Hodgkin, 1997; Wensley, 1989; Silverman, 1995). Different users can be supported, different environments can be accommodated, and different stages of the decision process can be addressed. In order to begin investigating the effect of intelligence, it was decided that initial experiments should focus on one particular type of user, environment and decision making stage. The current version of the prototype, referred to as "Intelligent V•I•S•A", incorporates facilities to support the user in performing sensitivity analysis. It was felt that this focus combined ease of implementation with potentially high impact. The intelligent support prompts the users to highlight and debate issues, whilst promoting deeper thinking about the problem under consideration, and outlining the significance of the results obtained.

This intelligent support is designed primarily to support the naïve user; someone who may be inexperienced in the use of MCDA and/or the use of DSS in general. The motivation for this was derived from experiences working with MBA students. These students are introduced briefly to the method and software before having to work on a problem. They bring differing levels of expertise in the use of DSS. Often the students are too quick to accept the initial results of the model and the aim of the intelligent support is to encourage more in depth exploration of the problem and solution by highlighting interesting issues and providing different forms of feedback. Such feedback is particularly important in the decision making process, both in providing reassurance to the user that their analysis is well founded and in ensuring that caution is practised against overconfidence of results (Jelassi,1986).

The intelligent facilities currently available are highlighted in figure 1, as shown in the pull-down menu. In one sense the intelligent facilities act as an information processor, highlighting such things as "good-allrounder" alternatives; alternatives that rely on particular strengths counter-balancing weaknesses elsewhere; and strong and weak alternatives in terms of the scores they receive. Such facilities are particularly useful if the decision maker is examining a large problem; requiring the evaluation of many alternatives or the use of a particularly detailed and complex set of criteria. Other features the intelligent information offers include the ability to identify potentially optimal solutions, similar to the work conducted by Rios Insua (1990). Here the aim is that by using optimisation to identify changes which could effect the overall ranking of alternatives, the decision maker will be stimulated further to reflect in more depth on results.

3.3 Example using Intelligent V•I•S•A

In order to demonstrate the intelligent user support offered by Intelligent V•I•S•A, an example problem will be used. The problem chosen is that of choosing a holiday destination in the UK. Figure 1 shows the relevant criteria, the alternatives being considered, the alternatives' initial overall ranking, as well as the intelligent facilities menu.

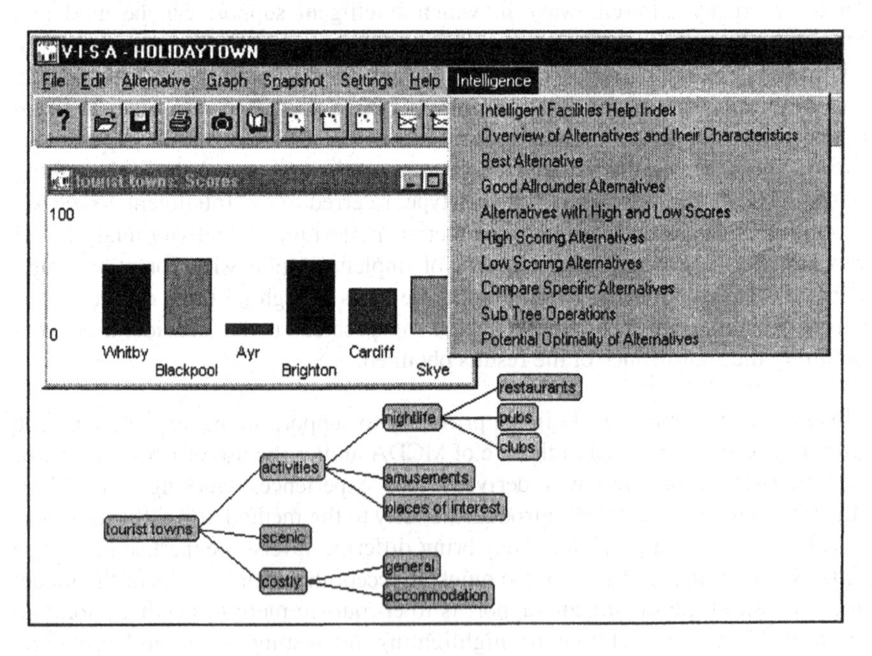

Figure 1

We join the decision making process at the stage of sensitivity analysis. The naïve user may think that the bar chart showing the initial ranking of alternatives is sufficient information on which to base their decision, however, this does not reveal the whole picture. For example, how robust is this ranking of alternatives? What kind of solution are we looking for? Do we want a holiday destination that scores well on all criteria or is this not important, so long as it scores highly on criteria which are the most important to us in our decision? This sort of information can only be determined by examining the model further, and performing investigative analysis of the alternatives and their scores and weights.

Perhaps a good place to start would be to examine how our top scoring alternative fares on each of the lower level criteria. By choosing the "best alternative" option from the intelligence menu, as shown in figure 2, we see how the best alternative performs on the lower level criteria. As is clear from the profile graph, Blackpool has a mixture of high and low scores for the different criteria. Indeed if we were to click on the intelligent option "high/low scoring alternatives" Blackpool would be highlighted, in that it is clear that it has high scores counterbalancing low scores. Along with the weights applied, these high scores ensure that Blackpool, despite scoring low on some criteria, still results in the "best" alternative.

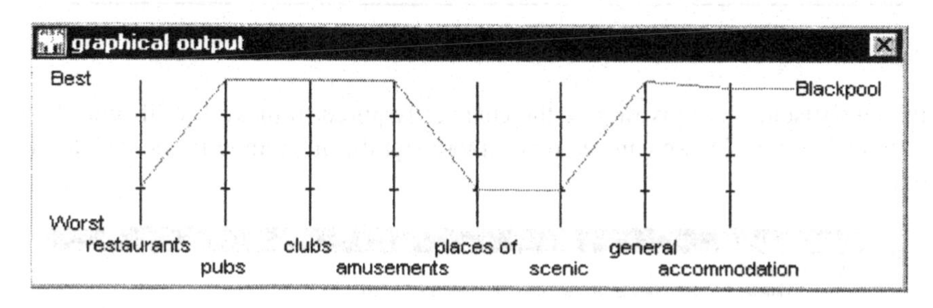

Figure 2

Perhaps the decision maker is wanting an alternative that achieves a reasonable score on all alternatives. Such alternatives can be identified by choosing the "good allrounder" option as shown in figure 3 . As can be seen, both the second and third ranked alternatives; Brighton and Whitby, are classed as "good allrounders". Given that these alternatives are only just ranked below Blackpool in the overall ranking, such alternatives may be better options to consider.

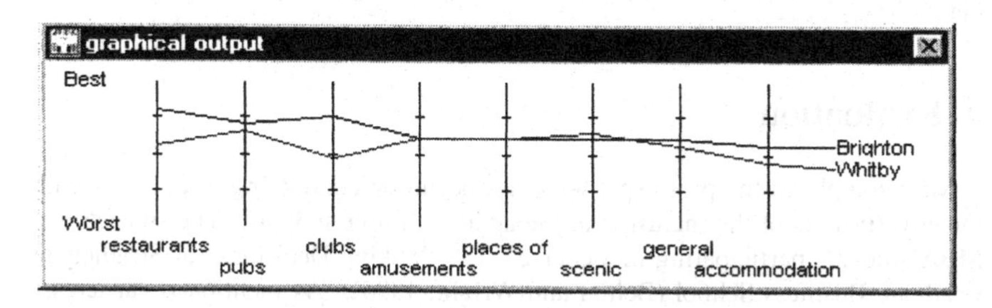

Figure 3

In addition, we may wish to investigate the robustness of the overall rankings. How sensitive to weight changes is the overall ranking of alternatives? By selecting the "potentially optimal alternative option" we can find out how much of a deviation from the original weights would be required for an alternative to be ranked top. Figure 4 shows the outcome of this for Brighton. It is clear that only a small change in the weights would result in Brighton being preferred, indicating a lack of robustness in our initial solution.

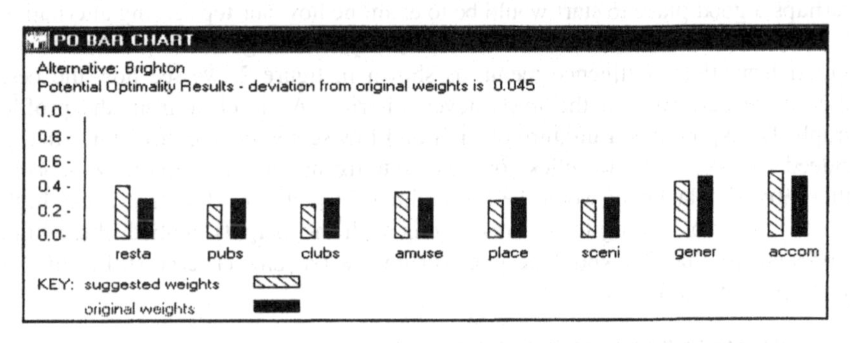

Figure 4

In comparison, if we investigate the changes required to make Cardiff our "best solution" (figure 5), we can see that a more significant change in the weights is required.

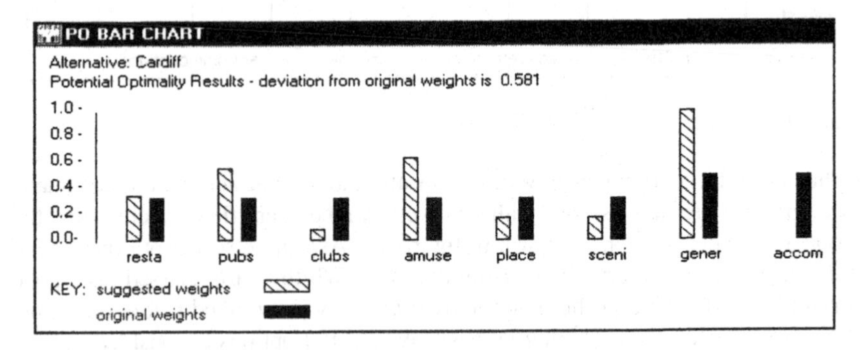

Figure 5

4 Evaluation

Having completed the prototype, work has begun on conducting tests to evaluate the effectiveness of the inclusion of the additional user support. The subjects are MBA students participating in a course titled "Making Decisions" at Strathclyde Graduate Business School (Belton and Wright, 1995). As mentioned earlier, as part of their course the students are required to work on a substantial multicriteria problem, which forms the basis of the assessment for the course. It was therefore felt that allowing them to use Intelligent V•I•S•A (in place of the standard version) would be a good method of attaining user views on the software, whilst at the same time, providing a tool for the students to complete their work. In addition, working with MBA students has the added advantage of obtaining feedback from a broad range of user types, as the level of experience with DSS, and of multicriteria analysis differs between individuals.

Before embarking on their project, all students are introduced to the SMART approach for multicriteria analysis (Goodwin and Wright, 1998), and to the V•I•S•A software. This introduction includes a discussion of the importance of sensitivity analysis and hints at what to look out for when exploring a solution. The students work in groups of 4-6 and are encouraged to adopt the roles of decision makers; facilitators; technical support and observer.

4.1 Methods of Tracking

To assess the students' views and experiences, the following methods were adopted:

1. The Full-Time MBA students - using V•I•S•A

(control group consisting of 15 sub-groups)

- All groups were given the standard version of V•I•S•A
- Three groups were selected at random and monitored in their use of the methodology and software. This was performed by sitting in on their meetings; taping their discussion sessions and making general observations about how they used the software and any problems that arose.
- All groups were asked to complete a questionnaire asking them about their thoughts on V•I•S•A and their ideas on potential ways in which V•I•S•A could better support the user.

2. *The Part-Time MBA students - using Intelligent V•I•S•A*

(15 sub-groups)

- All groups were given Intelligent V•I•S•A which automatically has a tracking device to see how the user navigates round the facilities.
- Three groups were selected at random and monitored as before.
- All groups were asked to complete questionnaires asking for their thoughts on Intelligent V•I•S•A , whether they made use of the "intelligent" facilities, suggestions, etc.

Currently, only the initial stages of the testing process have been completed. The final stages of attaining questionnaires and tracking information will not be available until students have finished their assignments.

4.2 Interesting questions

As a result of the evaluations, it is hoped that many interesting questions will be raised and perhaps answered. Anticipated questions are:

1. Does the intelligent user support provided match the intelligent user support required?
This question asks whether the intelligent user support that the full-time students suggested they might benefit from is in-line with the intelligent user support the part-time students had access to. As software development is by nature a cyclic process, the answer to this question will influence how the facilities offered will evolve in the next stage of software development.

2. What is the role of the intelligence?
This question asks not only who is the current intelligence aimed at and in what environment is it best put to use, but also how is it made use of? Does it simply reassure the decision makers suppositions? Does the intelligence highlight something new? Does its use promote more in-depth sensitivity analysis? Do the facilities encourage the decision maker to think about the problem more carefully, or perhaps do the opposite by replacing any deeper thinking the decision maker might have performed?

3. As a result of the intelligent user support, did the user perform more in-depth analysis?
Did the presence of the intelligent user support lead the part-time students to perform more detailed analysis or otherwise. The answer to this question will go some way in indicating how the intelligence is used, whether it is used as an information provider or as a confirmation facility, as highlighted by question 2.

4. Is the intelligent user support relayed to the user in the most effective way?
This question not only addresses the effectiveness of different ways of using text and graphics to display data, but also examines the method by which the information is offered to the user. Currently the intelligent facilities have to be manually chosen by the user via a pull-down menu. Would it be more useful if the intelligent information were to be made available to the user when deemed "appropriate" by the software?

5. How can the intelligent support be expanded?
Hopefully the answers to the above questions will provide some indication of possible ways forward. Should intelligent facilities be made available at other stages of the decision making process? Would intelligent user support be useful in a group networked decision making process? Which type of users should be the focus for future research?

5 Future Work Extending the Nature of Intelligent Support

5.1 Visualising Information

Although there is much inconclusive evidence about the effect of graphics on the user's ability to understand information (Belton and Vickers, 1989), many practitioners believe that viewing data graphically does in fact greatly enhance the ability of a human to acquire insight into a problem (Dawkes et al,1996; Robertson et al,1993). In addition, the ability of a package to represent data in different ways and from different perspectives has been argued to be beneficial to the user's understanding of the problem (Tweedie et al,1996; Schmid and Hinterberger,1994). Linking such visualisation and interface ideas to MCDA in particular, it is clear that as suggested by O'Keefe and Davis(1987) and O'Keefe and Bell(1995), this could only seek to advance the user's ability to not only generate alternatives and score and weight them accordingly, but also to evaluate them more fully through in-depth investigation of the problem.

In line with this thinking, we are currently undertaking a project dealing with the representation of n-dimensional weight space, similar to the work conducted by Brans and Mareschal(1994). It is the intention that the user will be able to investigate visually the effect of weight changes on the ranking of alternatives. In addition, we are currently implementing code which by using a graphical image of weight space, will relay to the user information indicating the proportion of weight space a potentially optimal solution occupies, thus giving indication of how robust such a solution is. Similar work has been reported by Antunes et al(1992), Vetschera(1997), Swain Tan and Fraser(1998).

5.2 Group Decision Support Systems

Currently, the intelligent user support implemented is designed to provide support for the individual decision maker. It is our intention that work will begin examining the feasibility of incorporating intelligent support for a networked decision making environment. Recently a networked version of V•I•S•A has been developed within the Department of Management Science (Belton and Elder,1998). By using networked computing, this facility can allow keyboard entry from any or all members of a group, allowing each decision maker to work on the same model but enter their individual evaluations, which can then be collated and compared. In this context the facilitator is faced with the need to quickly assimilate and interpret a large amount of data. It is anticipated that intelligent support can contribute to this task.

6 References

Angehrn, A.A., Dutta, S. (1992): Integrating Case-Based Reasoning in Multi-Criteria Decision Support Systems. In : Jelassi, T., Klein, M.R., Mayon-White, W.M. (Eds.): Decision Support Systems: Experiences and Expectations, Elsevier Science, 133-150

Angehrn, A.A., Luthi, H.J. (1990): Intelligent Decision Support Systems : A Visual Interactive Approach. Interfaces, 20, 17-28

Antunes, C.H., Paulo Melo, M., Climaco, J.N.(1992): On the integration of an interactive MOLP procedure. European Journal of Operational Research, 61, 135-144

Bayaud, R., Pomerol, J-Ch. (1992): An "Intelligent" DSS for the reinforcement of urban electrical power networks. In: Jelassi, T., Klein, M.R., Mayon-White, W.M. (Eds:): Decision Support Systems: Experiences and Expectations, Elsevier Science, 122-150

Bell, P.C. (1985): Visual Interactive Modelling as an Operations Research Techniques. Interfaces 15, 26-33

Belton, V. (1997): Implementing Multicriteria Decision Analysis. University of Strathclyde, Working Paper Series in the Theory, Method and Practice of Management Science

Belton, V., Elder, M.D. (1994): Decision Support Systems : Learning from Visual Interactive Modelling. Decision Support Systems, 12, 355-364

Belton, V., Elder, M.D. (1998): Focusing discussion using Group V•I•S•A for multicriteria decision support. University of Strathclyde, Working Paper Series in the Theory, Method and Practice of Management Science.

Belton, V., Hodgkin, J. (1999): Facilitators, Decision Makers, D.I.Y Users: Is Intelligent Multicriteria Decision Support for All Feasible or Desirable?. European Journal of Operational Research, 113,247-260

Belton, V., Vickers, S.P. (1989): V•I•S•A - VIM for MCDA. In : Lockett, G. and Islei, G. (Eds.): Improving Decision Making in Organisations, Springer Verlag, Berlin Heidelberg, 287-304

Belton, V., Vickers, S.P. (1990): Use of a Simple Multi-Attribute Value Function Incorporating Visual Sensitivity Analysis for Multiple Criteria Decision Making. In : Bana e Costa, C.A. (Ed): Readings in Multiple Criteria Decision Aid, Springer-Verlag, Berlin Heidelberg

Belton V., Vickers, S.P. (1993): Demystifying DEA - A Visual Interactive Approach Based on Multiple Criteria Analysis. Journal of the Operational Research Society, 44, 9, 883-896

Belton V., Wright G (1995): Making Decisions. Open Learning Unit, Strathclyde Graduate Business School, University of Strathclyde.

Brans, J.P., Mareschal, B., Vincke, P.H. (1994): The PROMCALC & GAIA Decision Support systems. Decision Support Systems, 12/4,14-

Dawkes, H., Tweedie, L.A., Spence, R. (1996): VICKI - The VIsualisation Construction Kit. Proceedings of the Workshop on Advanced Visual Interfaces, ACM Press, 257-259

Goodwin, P., Wright, G. (1998): Decision analysis for management judgement. Second Edition, Wiley.
Jelassi, M.T., (1986): MCDM: From 'Stand-Alone' Methods to Integrated and Intelligent DSS. Proceedings of VIIth International Conference on MCDM, Kyoto, Vol1, 250-262.
Klein, D.A. (1994): Decision-analytic Intelligent Systems: Automated Explanation and Knowledge Acquisition. Lawrence Erlabum associates, Inc.
O'Keefe, R.M., Bell, P.C. (1995): An Experimental Investigation into the Efficay of Visual Interactive Simulation. Management Science, 41, 6, 1018-1038
O'Keefe, R.M., Belton, V., Ball, T. (1986): Experiences with using expert systems and OR. Journal of the Operational Research, 37, 657-668
O'Keefe, R.M., Davis (1987): A Microcomputer system for Simulation Modelling. European Journal of Operational Research, 24, 23-29
Pasche, C. (1991): EXTRA : An expert system for multiple criteria decision making. European Journal of Operational Research, 45, 293-308
Rios Insua, D. (1990): Sensitivity Analysis in Multi-Objective Decision Making. Lecture Notes in Mathematical systems, Springer-Verlag, Berlin Heidelberg
Robertson, G.C., Card, S.K., Mackinlay, J.D., (1993): Information Visualization Using 3D Interactive Animation. Communications of the ACM, 36, 4, 57-71
Schmid, C., Hinterberger, H. (1994): Comparative Multivariate Visualisation Across Conceptually Different Graphical Displays. Proceedings of the SSDBM VII, IEEE Computer Society Press, September 1994
Silverman, B.G. (1995): Knowledge-Based Systems and the Decision Sciences. Interfaces, 25, 6, 67-82
Srivinsan, V., Ruparel, B. (1990): CGX: An expert support system for credit granting. European Journal of Operational Research, 49, 293-308
Swian Tan, Y., Fraser, N.M. (1998): The Modified Star Graph and the Petal Diagram: Two New Visual Aids for Discrete Alternative Multicriteria Decision Making. Journal of Multi-Criteria Decision Analysis , 7, 20-33
Turing, A.M., (1992): Intelligent Machinery. In Ince, D.C. (Ed.): Collected Works of A. M. Turing : Mechanical Intelligence, North Holland, 107-128
Tweedie, L.A., Spence, R., Dawkes, H., Su H., (1996): Externalising Abstract Mathematical Models. Proceedings of CHI'96, Vancouver, ACM Press
Vetschera, R. (1997): Volume-Based Sensitivity Analysis for Single- and Multiobjective Linear Programming Problems. Paper presented at DGOR Working Group on Decision Theory and Applications, Bad Liebenzell, March 12-14, 1997
Wensley, A. (1989): Research Directions in Expert Systems. In: Doukidis, G.I., Land, F., Miller, G. (Eds.): Knowledge Based Management Support Systems, 248-

Softwares

V•I•S•A , Visual Thinking International, Glasgow, UK
SIMUL8, Visual Thinking International, Glasgow, UK
Decision Explorer, Banxia Software Ltd, Glasgow, UK
Frontier Analyst, Banxia Software Ltd, Glasgow, UK

Data Envelopment Analysis with Fuzzy Input-Output Data

Masahiro Inuiguchi and Tetsuzo Tanino

Osaka University, Suita, Osaka 565-0871, Japan

Abstract. In this paper, we develop a DEA (Data Envelopment Analysis) with fuzzy input-output data. There are several approaches to extend the DEA to the case of fuzzy input-output data. We chose the most natural approaches among them. In one of these approaches, a linear programming problem solved in the conventional DEA is regarded as a mapping from an input-output data set to the efficiency score set. Applying the extension principle to the mapping, we obtain a fuzzy set of efficiency scores from given fuzzy input-output data. We also propose an efficiency analysis based on possibility theory. The relations between the fuzzy set of efficiency scores and the possibilistic efficiency analysis are investigated.

Keywords: Data envelopment analysis, fuzzy input-output data, efficiency, possibility, necessity, efficiency score

1 Introduction

Data Envelopment Analysis (DEA) has been developed in order to evaluate the efficiency of systems with multiple inputs and outputs by given input-output data [2][3]. Because of its tractability and usefulness, it has been applied to various fields of organizations, hospitals, banks, factories and offices. In the conventional DEA, the input-output data are assumed to be obtained as crisp numbers. However, in the real world, we are sometimes force to analyze the efficiency under the uncertain data, e.g., estimated data, predicted data, data with noise and so on. The efficiency in DEA is sensitive to data change and then, it is important to take care of data fluctuation. From this point of view, sensitivity and stability analysis [4][15][16] is developed in order to obtain the input/output stability region within which the efficiency of a specific efficient decision making unit (DMU) remains unchanged. The techniques except [16] can treat changes only in the specific efficient DMU's data. Moreover, only a region satisfies a sufficient condition for the preservation of efficiency is obtained by those methods except [15].

Such uncertain data can be treated as random variables or fuzzy numbers. Stochastic approaches to DEA with random input-output data have been proposed by several researchers [9][12][14]. Chance constrained programming approaches [9][14] are applied to DEA to examine whether the probability of the event that a specific DMU is efficient is not less than a given probability degree. This approach can treat simultaneous changes in all DMUs' data but, generally, the reduced problem is a non-convex programming problem and difficult to solve it. Morita and Seiford [12] have proposed a quadratic programming approach to obtain the amount of stochastic variations that remain a specific efficient DMU being efficient to a given probability degree restricting changes only in the data of the specific efficient DMU. They have discussed some statistics such as an expected efficiency score, a probability being efficient and α-percentile of efficiency score distributon. Those are obtained by numerical simulations.

In this paper, we propose a possibilistic approach to DEA with fuzzy input-output data. As is known in fuzzy mathematical programming literature [10], a possibilistic approach will be more tractable than stochastic approaches. Fuzzy set approaches already introduced into DEA for treating not only uncertain data [6][13][17] but also vaguely classified categorical input variable [11] and introduction of human experts' efficiency evaluation [8]. In the previous approaches [6][13][17] to deal with fuzzy input/output data, the linear and the fractional programming problems for the calculation of a efficiency score are replaced with the corresponding fuzzy mathematical programming problems in order to treat the fuzziness of input/output data. Fuzzy mathematical programming approaches are applied to the problems. However, a constraint, which requires the weighted sum of input values to be 1 or ratios of the total input value to the total output value not to be larger than 1, cannot be fully treated by a constraint with fuzzy parameters. As the results, a part of the obtained fuzzy set of efficiency scores may exceed 1 without any connection to superefficiency score [1]. We feel embarrassed by this result and redefine the fuzzy set of efficiency scores by cutting the part exceeds 1 (see [13]).

To avoid such embarrassment, we discuss a natural extension of DEA to the case of fuzzy input-output data. We regard the linear programming problem in the conventional DEA as a multi-valued function which maps crisp input-output data to efficiency score and slack values, and apply the extension principle [5] in this paper. Thus, we obtain a fuzzy set of efficiency scores. Moreover, we extend the efficiency concept to possible and necessary efficiencies based on the possibility theory [5]. An improvement guide for a system to be possibly/necessarily efficient to a given degree are discussed in the case of fuzzy input-output data. Relations between fuzzy set of efficiency scores and possible and necessary efficiencies are investigated. From results of this paper, we realize the advantages in tractability of fuzzy DEA over stochastic DEA.

2 Data Envelopment Analysis

The efficiency of a system such as corporation, enterprise, hospital, library, etc. are often evaluated by a ratio of the amount of output such as benefit, services, goods, etc. to the amount of input such as capital, materials, manpower, etc. When we consider single output and single input, the efficiency can easily be defined by the ratio, (the amount of output)/(the amount of input). However, when we consider multiple outputs and/or multiple inputs, it is not straightforward to evaluate the efficiency.

The data envelopment analysis (DEA) is proposed to evaluate the efficiency of systems with multiple inputs and outputs in a ratio form. In DEA, systems to be analyzed are called decision making units (DMUs). Given the input-output data of n DMUs with the same inputs and outputs, it is known that the efficiency of the Kth DMU, say DMU_K, is analyzed by solving the following programming problem with decision variables θ_K, $\boldsymbol{\lambda} = (\lambda_1, \lambda_2, \ldots, \lambda_n)^{\mathrm{T}}$, $\boldsymbol{u}^K = (u_1^K, u_2^K, \ldots, u_m^K)^{\mathrm{T}}$ and $\boldsymbol{v}^K = (v_1^K, v_2^K, \ldots, v_s^K)^{\mathrm{T}}$:

$$
\begin{aligned}
&\text{lex-min} \left(\theta_K, \ -\sum_{i=1}^{m} u_i^K - \sum_{j=1}^{s} v_j^K \right), \\
&\text{subject to } \theta_K x_{iK} = \sum_{k=1}^{n} \lambda_k x_{ik} + u_i^K, \ i = 1, 2, \ldots, m, \\
&\qquad y_{jK} = \sum_{k=1}^{n} \lambda_k y_{jk} - v_j^K, j = 1, 2, \ldots, s, \\
&\qquad \theta_K \geq 0, \ \lambda_k \geq 0, \ k = 1, 2, \ldots, n, \\
&\qquad u_i^K \geq 0, \ i = 1, 2, \ldots, m, \ v_j^K \geq 0, \ j = 1, 2, \ldots, s,
\end{aligned}
\tag{1}
$$

where lex-min stands for lexicographically minimize. $X = (x_{ik})$ is an $m \times n$ matrix of input data and $Y = (y_{jk})$ is an $s \times n$ matrix of output data. We assume that $X > O_m^n$ and $Y > O_s^n$, where O_m^n is an $m \times n$ zero matrix.

The optimal solution to Problem (1) is denoted by $(\hat{\theta}_K(X,Y), \hat{\boldsymbol{\lambda}}(X,Y), \hat{\boldsymbol{u}}^K(X,Y), \hat{\boldsymbol{v}}^K(X,Y))$ since it depends on the input-output data (X,Y). As can be seen easily, $0 < \hat{\theta}_K(X,Y) \leq 1$. The value $\hat{\theta}_K(X,Y)$ is called the *efficiency score* of DMU_K. DMU_K is said to be *weakly efficient* if and only if $\hat{\theta}_K(X,Y) = 1$. The vectors $\hat{\boldsymbol{u}}^K(X,Y)$ and $\hat{\boldsymbol{v}}^K(X,Y)$ show the excess of inputs and the shortage of outputs, respectively. DMU_K is said to be *efficient* if and only if $\hat{\theta}_K(X,Y) = 1$, $\hat{\boldsymbol{u}}^K(X,Y) = \mathbf{0}$ and $\hat{\boldsymbol{v}}^K(X,Y) = \mathbf{0}$.

For the sake of simplicity, the kth columns of X and Y are denoted by $X_{\cdot k}$ and $Y_{\cdot k}$, respectively. $X_{\cdot k}$ and $Y_{\cdot k}$ show the input and output data of DMU_k. A pair of input and output data $(\boldsymbol{x}, \boldsymbol{y})$ is called an activity. The following theorem is well-known (see [2]).

Theorem 1. *When $\hat{\theta}_K(X,Y) < 1$, $\hat{\boldsymbol{u}}^K(X,Y) \neq \mathbf{0}$ or $\hat{\boldsymbol{v}}^K(X,Y) \neq \mathbf{0}$, DMU_K becomes efficient by an improvement which changes $X_{\cdot K}$ and $Y_{\cdot K}$ to $X_{\cdot K}^{\text{new}} = \hat{\theta}_K(X,Y)X_{\cdot K} - \hat{\boldsymbol{u}}^K(X,Y)$ and $Y_{\cdot K}^{\text{new}} = Y_{\cdot K} + \hat{\boldsymbol{v}}^K(X,Y)$, respectively.*

We have the following lemma and theorems.

Lemma 1. *Let $\hat{\boldsymbol{\lambda}}(X,Y) = (\hat{\lambda}_1(X,Y), \ldots, \hat{\lambda}_n(X,Y))$. We have $\hat{\lambda}_K(X,Y) \leq \hat{\theta}_K(X,Y)$. Moreover, $\hat{\theta}_K(X,Y) = 1$ if $\hat{\lambda}_K(X,Y) > 0$.*

Proof. Taking care of the optimality of $\hat{\theta}_K(X,Y)$, it can easily be proven by reduction to absurdity.

Theorem 2. *$\hat{\theta}_K(X,Y)$ is continuous with respect to $X > O_m^n$ and $Y > O_s^n$. Moreover, $\hat{\theta}_K(X,Y)$ is non-increasing with respect to $X_{\cdot K}$ and $Y_{\cdot k}$, $k \neq K$ and non-decreasing with respect to $Y_{\cdot K}$ and $X_{\cdot k}$, $k \neq K$.*

Proof. The last part of the theorem is easily obtained from Lemma 1. Let us prove the first part. To prove the continuity, we can apply Hogan's results [7]. $\hat{\theta}_K(X,Y)$ is also obtained as the optimal value of the following linear programming problem $P(X,Y)$:

$$\begin{aligned}
\text{minimize } & f(\theta_K, \boldsymbol{\lambda} : X, Y) = \theta_K, \\
\text{subject to } & \boldsymbol{g}_1(\theta_K, \boldsymbol{\lambda} : X, Y) = X\boldsymbol{\lambda} - \theta_K X_{\cdot K} \leq \boldsymbol{0}, \\
& \boldsymbol{g}_2(\theta_K, \boldsymbol{\lambda} : X, Y) = Y_{\cdot K} - Y\boldsymbol{\lambda} \leq \boldsymbol{0}, \\
& g_{3k}(\theta_K, \boldsymbol{\lambda} : X, Y) = x_{ik}\lambda_k - x_{iK} \leq 0, \ k = 1, 2, \ldots, n, \\
& g_{4k}(\theta_K, \boldsymbol{\lambda} : X, Y) = -\lambda_k \leq 0, \ k = 1, 2, \ldots, n, \\
& g_5(\theta_K, \boldsymbol{\lambda} : X, Y) = \theta_K - 1 \leq 0, \\
& g_6(\theta_K, \boldsymbol{\lambda} : X, Y) = -\theta_K \leq 0.
\end{aligned} \tag{2}$$

Let $S(X,Y)$ be the feasible set of $P(X,Y)$ and $\boldsymbol{g}_j(\theta_K, \boldsymbol{\lambda} : X, Y) = (g_{j1}(\theta_K, \boldsymbol{\lambda} : X, Y), \ldots, g_{jn}(\theta_K, \boldsymbol{\lambda} : X, Y))$, $j = 3, 4$. Obviously, f, $\boldsymbol{g}_j$, $j = 1, 2, 3, 4$ and g_j, $j = 5, 6$ are continuous with respect to θ_K, $\boldsymbol{\lambda}$, X and Y. For an arbitrary (X,Y) such that $X > O_m^n$ and $Y > O_s^n$, $\boldsymbol{g}_j$, $j = 1, 2, 3, 4$ and g_j, $j = 5, 6$ are convex (linear) with respect to θ_K and $\boldsymbol{\lambda}$. $S(X,Y)$ is uniformly compact at any (X,Y) such that $X > O_m^n$ and $Y > O_s^n$ since $S(X,Y)$ is compact for any (X,Y) such that $X > O_m^n$ and $Y > O_s^n$. Moreover, we have $\{(\theta_K, \boldsymbol{\lambda}) \mid \boldsymbol{g}_j(\theta_K, \boldsymbol{\lambda} : X, Y) < \boldsymbol{0},\ j = 1, 2, 3, 4,\ g_j(\theta_K, \boldsymbol{\lambda} : X, Y) < 0,\ j = 5, 6\} \neq \emptyset$ for any (X,Y) such that $X > O_m^n$ and $Y > O_s^n$. Hence, from Hogan's results (Hogan [7]: Theorems 7, 10 and 12), $\hat{\theta}_K(X,Y)$ is continuous.

Theorem 3. *$-\sum_{i=1}^m \hat{u}_i^K(X,Y) - \sum_{j=1}^s \hat{v}_j^K(X,Y)$ is non-increasing with respect to $X_{\cdot K}$ and $Y_{\cdot k}$, $k \neq K$ and non-decreasing with respect to $Y_{\cdot K}$ and $X_{\cdot k}$, $k \neq K$.*

Proof. It can easily be shown from Lemma 1.

3 Fuzzy Input-Output Data

In this paper, we extend DEA to the case of fuzzy input-output data. Such fuzzy data will be useful in the following cases: the case when data fluctuates and its range is not known exactly, the case when data is not known precisely but roughly, and the case when data is obtained as predicted values. We assume that fuzzy input data $\tilde{x}_{ik}$ and fuzzy output data $\tilde{y}_{jk}$ are represented by L-R fuzzy numbers $(x^{\mathrm{L}}_{ik}, x^{\mathrm{R}}_{ik}, \alpha^{\mathrm{L}}_{ik}, \alpha^{\mathrm{R}}_{ik})_{L^{\mathrm{x}}_{ik}R^{\mathrm{x}}_{ik}}$ and $(y^{\mathrm{L}}_{jk}, y^{\mathrm{R}}_{jk}, \beta^{\mathrm{L}}_{jk}, \beta^{\mathrm{R}}_{jk})_{L^{\mathrm{y}}_{jk}R^{\mathrm{y}}_{jk}}$, respectively. Namely $\tilde{x}_{ik}$ has the following membership function;

$$\mu_{\tilde{x}_{ik}}(r) = \begin{cases} L^{\mathrm{x}}_{ik}\left(\dfrac{x^{\mathrm{L}}_{ik} - r}{\alpha^{\mathrm{L}}_{ik}}\right), & \text{if } r < x^{\mathrm{L}}_{ik}, \\ 1, & \text{if } x^{\mathrm{L}}_{ik} \le r \le x^{\mathrm{R}}_{ik}, \\ R^{\mathrm{x}}_{ik}\left(\dfrac{r - x^{\mathrm{R}}_{ik}}{\alpha^{\mathrm{R}}_{ik}}\right), & \text{if } r > x^{\mathrm{R}}_{ik}, \end{cases} \tag{3}$$

and $\tilde{y}_{jk}$ has a similar membership function. Here, L^{x}_{ik}, R^{x}_{ik}, L^{y}_{jk}, R^{y}_{jk} are reference functions defined as follows: a reference function $L : [0, +\infty) \to [0, 1]$ is an upper semi-continuous and non-increasing function such that $L(0) = 1$ and $\lim_{r \to +\infty} L(r) = 0$. x^{L}_{ik} and y^{L}_{jk} are lower bounds of the range where membership values are unity and x^{R}_{ik} and y^{R}_{jk} are upper bounds of that range. α^{L}_{ik} and β^{L}_{jk} show left spreads and α^{R}_{ik} and β^{R}_{jk} show right spreads. We assume that $\tilde{x}_{ik}$ and $\tilde{y}_{jk}$ are strictly positive, i.e.,

$$\exists \varepsilon_{ik} > 0,\ \forall r \le \varepsilon_{ik},\ \mu_{\tilde{x}_{ik}}(r) = 0 \text{ and } \exists \delta_{jk} > 0,\ \forall r \le \delta_{jk},\ \mu_{\tilde{y}_{jk}}(r) = 0. \tag{4}$$

For convenience, matrices whose (i,k) components are $\tilde{x}_{ik}$ and ε_{ik} are denoted by $\tilde{X} = (\tilde{x}_{ik})$ and $E_X = (\varepsilon_{ik})$, respectively. Similarly, we define $\tilde{Y} = (\tilde{y}_{jk})$ and $\Delta_Y = (\delta_{jk})$. Matrices $\tilde{X}$ and $\tilde{Y}$ can be regarded as fuzzy sets of matrices $X = (x_{ik})$ and $Y = (y_{jk})$ whose membership functions are defined by

$$\mu_{\tilde{X}}(X) = \min_{\substack{i=1,2,\ldots,m \\ k=1,2,\ldots,n}} \mu_{\tilde{x}_{ik}}(x_{ik}), \qquad \mu_{\tilde{Y}}(Y) = \min_{\substack{j=1,2,\ldots,s \\ k=1,2,\ldots,n}} \mu_{\tilde{y}_{jk}}(y_{jk}). \tag{5}$$

4 Extension and Resolution Principles

We regard the optimal value $\hat{\theta}_K(X,Y)$ to Problem (1) as an image of an input-output data (X,Y) by a function. From this point of view, we can extend this function to a function from a pair of matrices with fuzzy components to a fuzzy quantity. To this end, we utilize the extension principle [5]. In this section, a few well-known results are reviewed.

Extension principle is written as follows.

Definition 1. Given a mapping $f : \Omega_1 \times \Omega_2 \times \cdots \times \Omega_q \to \Xi$, the image of fuzzy sets, $\tilde{A}_i \subseteq \Omega_i$, $i = 1, 2, \ldots, q$ by f, is a fuzzy set $f(\tilde{A}_1, \tilde{A}_2, \ldots, \tilde{A}_q) \subseteq \Xi$ defined by the following membership function:

$$\mu_{f(\tilde{A}_1,\tilde{A}_2,\ldots,\tilde{A}_q)}(y) = \begin{cases} \sup\limits_{\boldsymbol{r} \in f^{-1}(y)} \min\left(\mu_{\tilde{A}_1}(r_1), \ldots, \mu_{\tilde{A}_q}(r_q)\right), & \text{if } f^{-1}(y) \ne \emptyset, \\ 0, & \text{if } f^{-1}(y) = \emptyset, \end{cases} \tag{6}$$

where $\boldsymbol{r} = (r_1, r_2, \ldots, r_q)$ and $f^{-1}(y) = \{\boldsymbol{r} \mid f(\boldsymbol{r}) = y\}$.

When $\Omega_1 = \cdots = \Omega_q = \Xi = \mathbf{R}$, we have the following theorem (see [5]).

Theorem 4. *If f is continuous and an h-level set of $\tilde{A}_i$, i.e., $[\tilde{A}_i]_h = \{\omega \mid \mu_{\tilde{A}_i}(\omega) \geq h\}$ is a closed interval $[a_i^{\mathrm{L}}(h), a_i^{\mathrm{R}}(h)]$ for all $h \in (0,1]$ and for $i = 1, 2, \ldots, q$, and f is non-decreasing with respect to $x_1, \ldots, x_{q'}$ and non-increasing with respect to $x_{q'+1}, \ldots, x_q$, then we have*

$$\left[f(\tilde{A}_1, \tilde{A}_2, \ldots, \tilde{A}_q)\right]_h = [f^{\mathrm{L}}(h), f^{\mathrm{R}}(h)], \tag{7}$$

where f^{L} and f^{R} are defined as follows:

$$f^{\mathrm{L}}(h) = f(a_1^{\mathrm{L}}(h), \ldots, a_{q'}^{\mathrm{L}}(h), a_{q'+1}^{\mathrm{R}}(h), \ldots, a_q^{\mathrm{R}}(h)), \tag{8}$$

$$f^{\mathrm{R}}(h) = f(a_1^{\mathrm{R}}(h), \ldots, a_{q'}^{\mathrm{R}}(h), a_{q'+1}^{\mathrm{L}}(h), \ldots, a_q^{\mathrm{L}}(h)). \tag{9}$$

We have the following theorem called the *resolution principle* (see [5]).

Theorem 5. *For any fuzzy set $\tilde{A}$, we have*

$$\mu_{\tilde{A}}(r) = \sup\{h \mid r \in [\tilde{A}]_h\}. \tag{10}$$

5 Fuzzy Efficiency Score

From Theorem 2, $\hat{\theta}_K(X, Y)$ is continuous with respect to $X > O_m^n$ and $Y > O_s^n$. The h-level sets $[\tilde{x}_{ik}]_h$ and $[\tilde{y}_{jk}]_h$ of fuzzy data $\tilde{x}_{ik}$ and $\tilde{y}_{jk}$ are closed intervals for all $h \in (0, 1]$. Thus, by Theorems 2 and 4, we have

$$[\hat{\theta}_K(\tilde{X}, \tilde{Y})]_h = [\hat{\theta}_K(X^{\mathrm{L}}(h:K), Y^{\mathrm{R}}(h:K)), \hat{\theta}_K(X^{\mathrm{R}}(h:K), Y^{\mathrm{L}}(h:K))], \tag{11}$$

where $X^{\mathrm{L}}(h:K) = (X_{\cdot 1}^{\mathrm{L}}(h), \ldots, X_{\cdot K-1}^{\mathrm{L}}(h),\ X_{\cdot K}^{\mathrm{R}}(h), X_{\cdot K+1}^{\mathrm{L}}(h), \ldots, X_{\cdot n}^{\mathrm{L}}(h))$ and $X^{\mathrm{R}}(h:K) = (X_{\cdot 1}^{\mathrm{R}}(h), \ldots, X_{\cdot K-1}^{\mathrm{R}}(h),\ X_{\cdot K}^{\mathrm{L}}(h), X_{\cdot K+1}^{\mathrm{R}}(h), \ldots, X_{\cdot n}^{\mathrm{R}}(h))$. $X_{\cdot k}^{\mathrm{L}}(h)$ and $X_{\cdot k}^{\mathrm{R}}(h)$ are the kth columns of matrices $X^{\mathrm{L}}(h) = (X_{ik}^{\mathrm{L}}(h))$ and $X^{\mathrm{R}}(h) = (X_{ik}^{\mathrm{R}}(h))$ whose (i,k)-components $X_{ik}^{\mathrm{L}}(h)$, $X_{ik}^{\mathrm{R}}(h)$ are defined by

$$X_{ik}^{\mathrm{L}}(h) = x_{ik}^{\mathrm{L}} - \alpha_{ik}^{\mathrm{L}}(L_{ik}^{\mathrm{x}})^{(-1)}(h), \quad X_{ik}^{\mathrm{R}}(h) = x_{ik}^{\mathrm{R}} + \alpha_{ik}^{\mathrm{R}}(R_{ik}^{\mathrm{x}})^{(-1)}(h). \tag{12}$$

Moreover, the pseudo-inverse $(L_{ik}^{\mathrm{x}})^{(-1)}$ is defined by

$$(L_{ik}^{\mathrm{x}})^{(-1)}(h) = \sup\{r \mid L_{ik}^{\mathrm{x}}(r) \geq h\}, \tag{13}$$

and $(R_{ik}^{\mathrm{x}})^{(-1)}$ is defined similarly. $Y_{\cdot k}^{\mathrm{L}}(h)$ and $Y_{\cdot k}^{\mathrm{R}}(h)$ are defined in the same way.

Equation (11) shows that the lower bound of a closed interval $[\hat{\theta}_K(\tilde{X}, \tilde{Y})]_h$ is determined by the lower bounds of closed intervals $[\tilde{x}_{ik}]_h$, $k \neq K$, the upper bound of a closed interval $[\tilde{x}_{iK}]_h$, the upper bounds of closed intervals $[\tilde{y}_{jk}]_h$,

$k \neq K$ and the lower bound of a closed interval $[\tilde{y}_{jK}]_h$. On the other hand, the upper bound of a closed interval $[\hat{\theta}_K(\tilde{X},\tilde{Y})]_h$ is determined by the upper bounds of closed intervals $[\tilde{x}_{ik}]_h$, $k \neq K$, the lower bound of a closed interval $[\tilde{x}_{iK}]_h$, the lower bounds of closed intervals $[\tilde{y}_{jk}]_h$, $k \neq K$ and the upper bound of a closed interval $[\tilde{y}_{jK}]_h$. Those are well-matched with our intuition that the lower bound of $[\hat{\theta}_K(\tilde{X},\tilde{Y})]_h$ is obtained by the worst values among h-level sets of fuzzy input-output data for DMU_K and the upper bounds is obtained by the best values.

Based on resolution principle (10), the fuzzy set of efficiency scores, $\hat{\theta}_K(\tilde{X}, \tilde{Y})$ can be obtained as follows:

$$\mu_{\hat{\theta}_K(\tilde{X},\tilde{Y})}(r) = \sup\{h \mid r \in [\hat{\theta}_K(\tilde{X},\tilde{Y})]_h\}, \tag{14}$$

In practice, the fuzzy set $\hat{\theta}_K(\tilde{X},\tilde{Y})$ is approximately obtained by sufficiently many h-level sets $[\hat{\theta}_K(\tilde{X},\tilde{Y})]_h$. The upper and lower bounds of $[\hat{\theta}_K(\tilde{X},\tilde{Y})]_h$ can easily be calculated by solving lexicographical optimization problems (1).

As shown above, a fuzzy set of efficiency scores is approximately obtained by solving many linear programming problems in fuzzy DEA while a probability distribution on efficiency score is usually estimated by numerical simulations together with solving quite a lot of linear programming problems in stochastic DEA [12].

6 Efficiency Analysis Based on Possibility Theory

Given crisp input-output data, the efficiency of each DMU is judged dichotomously in the traditional DEA. Under fuzzy input-output data, we cannot decide the efficiency of each DMU in a dichotomous way because of the presence of uncertainty. However, we can evaluate the possibility and certainty (necessity) degrees of each DMU's efficiency. Those degrees are defined by possibility and necessity measures.

The possibility Π and necessity measures N are defined as follows: given a fuzzy set $\tilde{A} \subseteq \Omega$ of possible values of an uncertain variable, the possibility and certainty degrees that the realization of the variable is in a crisp set $B \subseteq \Omega$ are respectively defined by

$$\Pi(B|\tilde{A}) = \begin{cases} \sup\limits_{r\in B} \mu_{\tilde{A}}(r), & \text{if } B \neq \emptyset, \\ 0, & \text{if } B = \emptyset, \end{cases} \quad N(B|\tilde{A}) = \begin{cases} \inf\limits_{r\notin B} (1 - \mu_{\tilde{A}}(r)), & \text{if } B \neq \Omega, \\ 1, & \text{if } B = \Omega, \end{cases} \tag{15}$$

where $\mu_{\tilde{A}}$ is a membership function of $\tilde{A}$ and Ω is the universal set.

Possibility and necessity measures satisfy

$$N(B|\tilde{A}) > 0 \Rightarrow \Pi(B|\tilde{A}) = 1, \tag{16}$$

$$\Pi(B|\tilde{A}) > h \Leftrightarrow (\tilde{A})_h \cap B \neq \emptyset, \quad N(B|\tilde{A}) \geq h \Leftrightarrow (\tilde{A})_{1-h} \subseteq B, \tag{17}$$

where $(\tilde{A})_h$ is a strong h-level set of $\tilde{A}$ and defined by $(\tilde{A})_h = \{r \mid \mu_{\tilde{A}}(r) > h\}$. Moreover, when B is closed and $[\tilde{A}]_h$ is compact for all $h \in (0,1]$, we have

$$\Pi(B|\tilde{A}) \geq h \Leftrightarrow [\tilde{A}]_h \cap B \neq \emptyset, \quad N(B|\tilde{A}) \geq h \Leftrightarrow \mathrm{cl}(\tilde{A})_{1-h} \subseteq B, \qquad (18)$$

where cl D is a closure of a crisp set D.

The set of crisp input-output data (X,Y) where DMU_K is efficient can be defined by

$$EF_K = \{(X,Y) \geq (E_X, \Delta_Y) \mid \hat{\theta}_K(X,Y) = 1, \\ \hat{\boldsymbol{u}}^K(X,Y) = \boldsymbol{0},\ \hat{\boldsymbol{v}}^K(X,Y) = \boldsymbol{0}\}. \qquad (19)$$

Thus, under a fuzzy input-output data $(\tilde{X},\tilde{Y})$, the possibility and certainty degrees of the event that DMU_K is efficient are respectively defined by

$$\Pi(EF_K|\tilde{X},\tilde{Y}) \text{ and } N(EF_K|\tilde{X},\tilde{Y}). \qquad (20)$$

When $\Pi(EF_K|\tilde{X},\tilde{Y}) = 0$ which is equivalent to $((\tilde{X})_0 \times (\tilde{Y})_0) \cap EF_K = \emptyset$ by (17), DMU_K is not efficient for any (X,Y) such that $\mu_{\tilde{X}}(X) > 0$ and $\mu_{\tilde{Y}}(Y) > 0$. In other words, it is impossible that DMU_K is efficient under the given fuzzy input-output data. On the other hand, when $N(EF_K|\tilde{X},\tilde{Y}) = 1$ which is equivalent to $((\tilde{X})_0 \times (\tilde{Y})_0) \subseteq EF_K$ by (17), DMU_K is efficient for any (X,Y) such that $\mu_{\tilde{X}}(X) > 0$ and $\mu_{\tilde{Y}}(Y) > 0$. In other words, it is certain that DMU_K is efficient under the given fuzzy input-output data. Moreover, when $\Pi(EF_K|\tilde{X},\tilde{Y}) > 0$, it is possible to some extent (to a degree $\Pi(EF_K|\tilde{X},\tilde{Y})$) that DMU_K is efficient and when $N(EF_K|\tilde{X},\tilde{Y}) > 0$, it is certain to some extent (to a degree $N(EF_K|\tilde{X},\tilde{Y})$) that DMU_K is efficient.

It is significant to discuss methods to check $\Pi(EF_K|\tilde{X},\tilde{Y}) \geq h$ and $N(EF_K|\tilde{X},\tilde{Y}) \geq h$ for an arbitrarily given $h \in (0,1]$ and methods to obtain the improvement guides for DMU_K to be $\Pi(EF_K|\tilde{X},\tilde{Y}) \geq h$ and $N(EF_K|\tilde{X},\tilde{Y}) \geq h$ when $\Pi(EF_K|\tilde{X},\tilde{Y}) < h$ and $N(EF_K|\tilde{X},\tilde{Y}) < h$, respectively. However, from (17), we only have

$$([\tilde{X}]_h \times [\tilde{Y}]_h) \cap EF_K \neq \emptyset \Rightarrow \Pi(EF_K|\tilde{X},\tilde{Y}) \geq h, \qquad (21)$$

$$\mathrm{cl}((\tilde{X})_{1-h} \times (\tilde{Y})_{1-h}) \subseteq EF_K \Rightarrow N(EF_K|\tilde{X},\tilde{Y}) \geq h. \qquad (22)$$

but not the reverse implications because EF_K is not closed. Instead, for any $\varepsilon > 0$, we have

$$\Pi(EF_K|\tilde{X},\tilde{Y}) \geq h \Rightarrow ([\tilde{X}]_{h-\varepsilon} \times [\tilde{Y}]_{h-\varepsilon}) \cap EF_K \neq \emptyset, \qquad (23)$$

$$N(EF_K|\tilde{X},\tilde{Y}) \geq h \Rightarrow ((\tilde{X})_{1-h} \times (\tilde{Y})_{1-h}) \subseteq EF_K. \qquad (24)$$

Therefore, it would be meaningful enough to discuss methods for the sufficient conditions of $\Pi(EF_K|\tilde{X},\tilde{Y}) \geq h$ and $N(EF_K|\tilde{X},\tilde{Y}) \geq h$, i.e., $([\tilde{X}]_h \times [\tilde{Y}]_h) \cap EF_K \neq \emptyset$ and $\mathrm{cl}((\tilde{X})_{1-h} \times (\tilde{Y})_{1-h}) \subseteq EF_K$.

Since $[\tilde{x}_{ik}]_h$ and $[\tilde{y}_{jk}]_h$ are closed intervals, we have

$$([\tilde{X}]_h \times [\tilde{Y}]_h) = \{(X,Y) \mid (X^{\mathrm{L}}(h), Y^{\mathrm{L}}(h)) \leq (X,Y) \leq (X^{\mathrm{R}}(h), Y^{\mathrm{R}}(h))\}. \tag{25}$$

Since $(X,Y) \in EF_K$ is equivalent to $\hat{\theta}(X,Y) = 1$, $\hat{\mathbf{u}}^K(X,Y) = \mathbf{0}$ and $\hat{\mathbf{v}}^K(X,Y) = \mathbf{0}$ in the optimal solution to Problem (1), from Theorem 2,

$$\left\{\begin{array}{l} \hat{\theta}_K(X^{\mathrm{R}}(h:K), Y^{\mathrm{L}}(h:K)) = 1, \\ \hat{\mathbf{u}}^K(X^{\mathrm{R}}(h:K), Y^{\mathrm{L}}(h:K)) = \mathbf{0}, \\ \hat{\mathbf{v}}^K(X^{\mathrm{R}}(h:K), Y^{\mathrm{L}}(h:K)) = \mathbf{0} \end{array}\right\} \Rightarrow \Pi(EF_K|\tilde{X},\tilde{Y}) \geq h. \tag{26}$$

Thus, the sufficient condition for $\Pi(EF_K|\tilde{X},\tilde{Y}) \geq h$ can be checked easily by solving Problem (1) with input-output data $(X^{\mathrm{R}}(h:K), Y^{\mathrm{L}}(h:K))$.

Assuming that the sufficient condition does not satisfied, i.e., $\hat{\theta}_K(X^{\mathrm{R}}(h:K), Y^{\mathrm{L}}(h:K)) < 1$, $\hat{\mathbf{u}}^K(X^{\mathrm{R}}(h:K), Y^{\mathrm{L}}(h:K)) \neq \mathbf{0}$, or $\hat{\mathbf{v}}^K(X^{\mathrm{R}}(h:K), Y^{\mathrm{L}}(h:K)) \neq \mathbf{0}$, let us discuss an improvement guide for DMU_K to satisfy the sufficient condition for $\Pi(EF_K|\tilde{X},\tilde{Y}) \geq h$.

Consider a new fuzzy input-output data $(\tilde{X}^{\mathrm{new}}, \tilde{Y}^{\mathrm{new}})$ obtained by a modification of the Kth column of fuzzy input-output data $(\tilde{X},\tilde{Y})$, i.e.,

$$\tilde{X}^{\mathrm{new}}_{\cdot K} = \hat{\theta}(X^{\mathrm{R}}(h:K), Y^{\mathrm{L}}(h:K))\tilde{X}_{\cdot K} - \hat{\mathbf{u}}(X^{\mathrm{R}}(h:K), Y^{\mathrm{L}}(h:K)), \tag{27}$$

$$\tilde{Y}^{\mathrm{new}}_{\cdot K} = \tilde{Y}_{\cdot K} + \hat{\mathbf{v}}(X^{\mathrm{R}}(h:K), Y^{\mathrm{L}}(h:K)). \tag{28}$$

By the extension principle, $X^{\mathrm{L\,new}}_{\cdot K}(h)$ and $Y^{\mathrm{R\,new}}_{\cdot K}(h)$ are obtained by $X^{\mathrm{L\,new}}_{\cdot K}(h) = \hat{\theta}(X^{\mathrm{R}}(h:K), Y^{\mathrm{L}}(h:K))X^{\mathrm{L}}_{\cdot K}(h) - \hat{\mathbf{u}}(X^{\mathrm{R}}(h:K), Y^{\mathrm{L}}(h:K))$ and $Y^{\mathrm{R\,new}}_{\cdot K}(h) = Y^{\mathrm{R}}_{\cdot K}(h) + \hat{\mathbf{v}}(X^{\mathrm{R}}(h:K), Y^{\mathrm{L}}(h:K))$. By Theorem 1, we have $\hat{\theta}_K(X^{\mathrm{R\,new}}(h:K), Y^{\mathrm{L\,new}}(h:K)) = 1$, $\hat{\mathbf{u}}^K(X^{\mathrm{R\,new}}(h:K), Y^{\mathrm{L\,new}}(h:K)) = \mathbf{0}$ and $\hat{\mathbf{v}}^K(X^{\mathrm{R\,new}}(h:K), Y^{\mathrm{L\,new}}(h:K)) = \mathbf{0}$. From (26), we have $\Pi(EF_K|\tilde{X}^{\mathrm{new}}, \tilde{Y}^{\mathrm{new}}) \geq h$. Hence, DMU_K can be possibly efficient to the degree h by an improvement suggested by (27) and (28).

Now let us discuss the sufficient condition for $N(EF_K|\tilde{X},\tilde{Y}) \geq h$. Using

$$X^{\bar{\mathrm{L}}}_{ik}(h) = x^{\mathrm{L}}_{ik} - \alpha^{\mathrm{L}}_{ik}(L^{\mathrm{x}}_{ik})^{\langle -1\rangle}(h), \quad X^{\bar{\mathrm{R}}}_{ik}(h) = x^{\mathrm{R}}_{ik} + \alpha^{\mathrm{R}}_{ik}(R^{\mathrm{x}}_{ik})^{\langle -1\rangle}(h), \tag{29}$$

we define $X^{\bar{\mathrm{L}}}(h) = (X^{\bar{\mathrm{L}}}_{ik}(h))$ and $X^{\bar{\mathrm{R}}}(h) = (X^{\bar{\mathrm{R}}}_{ik}(h))$, where

$$(L^{\mathrm{x}}_{ik})^{\langle -1\rangle}(h) = \sup\{r \mid L^{\mathrm{x}}_{ik}(r) > h\}, \tag{30}$$

and $(R^{\mathrm{x}}_{ik})^{\langle -1\rangle}$ is defined similarly. We define $Y^{\bar{\mathrm{L}}}(h)$ and $Y^{\bar{\mathrm{R}}}(h)$ in the same manner. Moreover we define $X^{\bar{\mathrm{L}}}(h:K)$, $X^{\bar{\mathrm{R}}}(h:K)$, $Y^{\bar{\mathrm{L}}}(h:K)$ and $Y^{\bar{\mathrm{R}}}(h:K)$ in the same manner as (12) replacing L and R with $\bar{\mathrm{L}}$ and $\bar{\mathrm{R}}$, respectively. Under those definitions, from Theorem 2, we have

$$\left\{\begin{array}{l} \hat{\theta}_K(X^{\bar{\mathrm{L}}}(1-h:K), Y^{\bar{\mathrm{R}}}(1-h:K)) = 1, \\ \hat{\mathbf{u}}^K(X^{\bar{\mathrm{L}}}(1-h:K), Y^{\bar{\mathrm{R}}}(1-h:K)) = \mathbf{0}, \\ \hat{\mathbf{v}}^K(X^{\bar{\mathrm{L}}}(1-h:K), Y^{\bar{\mathrm{R}}}(1-h:K)) = \mathbf{0} \end{array}\right\} \Rightarrow N(EF_K|\tilde{X},\tilde{Y}) \geq h. \tag{31}$$

Hence, the sufficient condition for $N(EF_K|\tilde{X},\tilde{Y}) \geq h$ can also be examined by solving Problem (1) with input-output data $(X^{\text{L}}(1-h:K), Y^{\text{R}}(1-h:K))$.

Assuming that the sufficient condition does not fulfilled, i.e., $\hat{\theta}_K(X^{\text{L}}(1-h:K), Y^{\text{R}}(1-h:K)) < 1$, $\hat{\boldsymbol{u}}^K(X^{\text{L}}(1-h:K), Y^{\text{R}}(1-h:K)) \neq \mathbf{0}$ or $\hat{\boldsymbol{v}}^K(X^{\text{L}}(1-h:K), Y^{\text{R}}(1-h:K)) \neq \mathbf{0}$, let us discuss an improvement guide for DMU_K to satisfy the sufficient condition for $N(EF_K|\tilde{X},\tilde{Y}) \geq h$.

Consider a new fuzzy input-output data $(\tilde{X}^{\text{new}}, \tilde{Y}^{\text{new}})$ obtained by a modification of the Kth column of fuzzy input-output data $(\tilde{X},\tilde{Y})$, i.e.,

$$\tilde{X}_{\cdot K}^{\text{new}} = \hat{\theta}(X^{\text{L}}(1-h:K), Y^{\text{R}}(1-h:K))\tilde{X}_{\cdot K} - \hat{\boldsymbol{u}}(X^{\text{L}}(1-h:K), Y^{\text{R}}(1-h:K)), \quad (32)$$

$$\tilde{Y}_{\cdot K}^{\text{new}} = \tilde{Y}_{\cdot K} + \hat{\boldsymbol{v}}(X^{\text{L}}(1-h:K), Y^{\text{R}}(1-h:K)). \quad (33)$$

By the extension principle, $X_{\cdot K}^{\text{R}\,\text{new}}(h)$ and $Y_{\cdot K}^{\text{L}\,\text{new}}(h)$ are obtained by $X_{\cdot K}^{\text{R}\,\text{new}}(h) = \hat{\theta}(X^{\text{L}}(1-h:K), Y^{\text{R}}(1-h:K))X_{\cdot K}^{\text{R}}(h) - \hat{\boldsymbol{u}}(X^{\text{L}}(1-h:K), Y^{\text{R}}(1-h:K))$ and $Y_{\cdot K}^{\text{L}\,\text{new}}(h) = Y_{\cdot K}^{\text{L}}(h) + \hat{\boldsymbol{v}}(X^{\text{L}}(1-h:K), Y^{\text{R}}(1-h:K))$. By Theorem 1, we have $\hat{\theta}_K(X^{\text{L}\,\text{new}}(1-h:K), Y^{\text{R}\,\text{new}}(1-h:K)) = 1$, $\hat{\boldsymbol{u}}^K(X^{\text{L}\,\text{new}}(1-h:K), Y^{\text{R}\,\text{new}}(1-h:K)) = \mathbf{0}$ and $\hat{\boldsymbol{v}}^K(X^{\text{L}\,\text{new}}(1-h:K), Y^{\text{R}\,\text{new}}(1-h:K)) = \mathbf{0}$. From (31), we have $N(EF_K|\tilde{X}^{\text{new}}, \tilde{Y}^{\text{new}}) \geq h$. Hence, DMU_K can be necessarily efficient to the degree h by an improvement suggested by (32) and (33).

As discussed above, by solving a linear programming problem we can examine wheather the possibility/necessity measure of the event that a specific DMU is efficient is not less than a given degree. On the other hand, we should solve a nonconvex programming problem in order to examine wheather the probability of the same event is not less than a given degree (see [9][14]). The proposed fuzzy DEA is more tractable than the stochastic DEA.

7 Relationship between Two Approaches

Let WEF_K be a set of crisp input-output data (X, Y) where DMU_K is weakly efficient, i.e.,

$$WEF_K = \{(X,Y) \geq (E_X, \Delta_Y) \mid \hat{\theta}(X,Y) = 1\}. \quad (34)$$

Since $\hat{\theta}(X,Y)$ is continuous, WEF_K is closed. Consider possibility and certainty degrees, $\Pi(WEF_K|\tilde{X},\tilde{Y})$ and $N(WEF_K|\tilde{X},\tilde{Y})$, that DMU_K is weakly efficient. By definition, we have $\Pi(WEF_K|\tilde{X},\tilde{Y}) = \mu_{\hat{\theta}(\tilde{X},\tilde{Y})}(1)$ and $N(WEF_K|\tilde{X},\tilde{Y}) = 1 - \sup_{r<1} \mu_{\hat{\theta}(\tilde{X},\tilde{Y})}(r)$. Since $EF_K \subseteq WEF_K$, the following inequalities are valid:

$$\Pi(EF_K|\tilde{X},\tilde{Y}) \leq \Pi(WEF_K|\tilde{X},\tilde{Y}) = \mu_{\hat{\theta}(\tilde{X},\tilde{Y})}(1), \quad (35)$$

$$N(EF_K|\tilde{X},\tilde{Y}) \leq N(WEF_K|\tilde{X},\tilde{Y}) = 1 - \sup_{r<1} \mu_{\hat{\theta}(\tilde{X},\tilde{Y})}(r). \quad (36)$$

When L_{ik}^{x} and R_{ik}^{x}, $i = 1,2,\ldots,m$, $k = 1,2,\ldots,n$ are continuous and when L_{jk}^{y} and R_{jk}^{y}, $j = 1,2,\ldots,s$, $k = 1,2,\ldots,n$ are continuous, (35) and (36) hold with equalities as shown in the following theorem.

Theorem 6. *When reference functions L_{ik}^{x} and R_{ik}^{x}, $i = 1,2,\ldots,m$, $k = 1,2,\ldots,n$ are continuous, we have*

$$\Pi(EF_K|\tilde{X},\tilde{Y}) = \mu_{\hat{\theta}(\tilde{X},\tilde{Y})}(1), \quad N(EF_K|\tilde{X},\tilde{Y}) = 1 - \sup_{r<1} \mu_{\hat{\theta}(\tilde{X},\tilde{Y})}(r). \quad (37)$$

Similarly, the same equalities hold when reference functions L_{jk}^{y} and R_{jk}^{y}, $j = 1,2,\ldots,s$, $k = 1,2,\ldots,n$ are continuous.

Proof. When L_{ik}^{x} is continuous, $(L_{ik}^{x})^{(-1)}$ and $(L_{ik}^{x})^{\langle-1\rangle}$ are strictly decreasing. This is same for R_{ik}^{x}, L_{jk}^{y} and R_{jk}^{y}. On the other hand, let us define $(X(E), Y(\Delta))$ by

$$X_{\cdot k}(E) = \begin{cases} X_{\cdot k} - E_{\cdot k}, \text{ if } k \neq K, \\ X_{\cdot K} + E_{\cdot K}, \text{ if } k = K, \end{cases} \quad Y_{\cdot k}(\Delta) = \begin{cases} Y_{\cdot k} + \Delta_{\cdot k}, \text{ if } k \neq K, \\ Y_{\cdot K} - \Delta_{\cdot K}, \text{ if } k = K. \end{cases} \quad (38)$$

When $\hat{\theta}_K(X,Y) = 1$ and $\sum_{i=1}^{m} \hat{u}_i^K(X,Y) + \sum_{j=1}^{s} \hat{v}_j^K(X,Y) > 0$, as can be shown easily, $\hat{\theta}_K(X(E),Y(\Delta)) = 1$ and $\sum_{i=1}^{m} \hat{u}_i^K(X(E),Y(\Delta)) + \sum_{j=1}^{s} \hat{v}_j^K(X(E),Y(\Delta)) = 0$ hold for any $E > O_m^n$ and $\Delta \geq O_s^n$ and for any $E \geq O_m^n$ and $\Delta > O_s^n$. Using those properties, for any $h \in (0,1]$ and for any $\varepsilon > 0$, $\hat{\theta}_K(X^{\mathrm{R}}(h-\varepsilon:K), Y^{\mathrm{L}}(h-\varepsilon:K)) = 1$, $\hat{\boldsymbol{u}}^K(X^{\mathrm{R}}(h-\varepsilon:K), Y^{\mathrm{L}}(h-\varepsilon:K)) = \mathbf{0}$ and $\hat{\boldsymbol{v}}^K(X^{\mathrm{R}}(h-\varepsilon:K), Y^{\mathrm{L}}(h-\varepsilon:K)) = \mathbf{0}$ hold when $\hat{\theta}_K(X^{\mathrm{R}}(h:K), Y^{\mathrm{L}}(h:K)) = 1$. Thus, we have $\Pi(EF_K|\tilde{X},\tilde{Y}) = \Pi(WEF_K|\tilde{X},\tilde{Y})$. Similarly, we can prove $N(EF_K|\tilde{X},\tilde{Y}) = N(WEF_K|\tilde{X},\tilde{Y})$.

8 Conclusions

In this paper, we have developed two approaches to DEA with fuzzy input-output data: one is a method for obtaining efficiency score of a DMU in terms of a fuzzy set and the other is a method for verifying the possible and necessary efficiency of a DMU to a given degree. We have also discussed an improvement guide for the DMU to be possibly and necessarily efficient to the given degree. Both approaches are natural extensions of the traditional DEA and related each other. Especially, when the fuzzy input-output data is defined by continuous membership functions, possible and necessary efficiency degrees can be obtained from the fuzzy efficiency score.

Though kinds of uncertainty treated in fuzzy DEA and stochastic DEA are different, both deal with the ambiguity in input-output data. Our results show that the fuzzy DEA techniques are more tractable than the corresponding stochastic DEA techniques. The proposed fuzzy DEA will be useful to evaluate the efficiency of DMUs based on subjective information about data fluctuation.

References

1. Andersen, P. and Petersen, N. C. (1993): A Procedure for Ranking Efficient Units in Data Envelopment Analysis. *Management Science* 39(10), 1261–1264.
2. Charnes, A., Cooper, W., Lewin, A. Y. and Seiford, L. M. (Eds.)(1994): *Data Envelopment Analysis: Theory, Methodology and Applications.* Kluwer Academic Publishers, Boston.
3. Charnes, A., Cooper, W. W. and Rhodes, E. (1978): Measuring Efficiency of Decision Making Units. *European Journal of Operational Research* 2, 429–444.
4. Charnes, A. and Neralić, L. (1990): Sensitivity Analysis of the Additive Model in Data Envelopment Analysis. *European Journal of Operational Research* 48, 332–341.
5. Dubois, D. and Prade, H. (1987): Fuzzy Numbers: An Overview. In: Bezdek, J. C. (Ed): *Analysis of Fuzzy Information, Vol.I: Mathematics and Logic.* CRC Press, Boca Raton, FL, 3–39.
6. Guo, P. and Tanaka, H. (1997): Fuzzy DEA with Fuzzy Data. *Proceedings of the 13th Fuzzy Systems Symposium* 685–686.
7. Hogan, W. W. (1973): Point-to-Set Maps in Mathematical Programming. *SIAM Review* 15(3), 591–603.
8. Hougaard, J. L. (1999): Fuzzy Scores of Technical Efficiency. *European Journal of Operational Research* 115 529–541.
9. Huang, Z. and Li, S. X. (1996): Dominance Stochastic Models in Data Envelopment Analysis. *European Journal of Operational Research* 95, 390–403.
10. Inuiguchi, M. (1992): Stochastic Programming Problems versus Fuzzy Mathematical Programming Problems. *Japanese Journal of Fuzzy Theory and Systems* 4(1), 97–109.
11. Morita, H. (1996): Fuzzy Categorical Inputs in Data Envelopment Analysis. *Transactions of the Institute of Systems, Control and Information Engineers* 8(4), 149–156 (in Japanese).
12. Morita, H. and Seiford, L. M. (1999): Characteristics on Stochastic DEA Efficiency: Reliability and Probability Being Efficient. (submitted for publication)
13. Nagano, F., Yamaguchi, T. and Fukukawa, T. (1995): DEA Using Fuzzy Numbers as Output Values. *Communications of the Operations Research Society of Japan* 40(8) 425–429 (in Japanese).
14. Olesen, O. B. and Petersen, N. C. (1995): Chance Constrained Efficiency Evaluation. *Management Science* 41(3), 442–457.
15. Seiford, L. M. and Zhu, J. (1998): Stability Regions for Maintaining Efficiency in Data Envelopment Analysis. *European Journal of Operational Research* 108, 127–139.
16. Thompson, R., Dharmapala, P. S. and Thrall, R. M. (1994): Sensitivity Analysis of Efficiency Measures with Applications to Kansas Farming and Illinois Coal Mining. In: [2], 393–422.
17. Ueda, T. and Kamimura, T. (1997): Data Envelopment Analysis Based on Triangular Fuzzy Numbers. *Proceedings of the 13th Fuzzy Systems Symposium,* 681–682 (in Japanese).

Target Mix Approach for Measuring Efficiency in Data Envelopment Analysis

Tarja Joro
Helsinki School of Economics and Business Administration, Finland

Abstract

Data Envelopment Analysis (DEA) measures the relative efficiency of Comparable entities called Decision Making Units (DMUs). In this study the aim is to measure the efficiency of DMUs with respect to a target output or target input mix provided by a Decision Maker (DM). We assume that the DM wants DMUs to adapt their outputs or inputs to this mix. Proposed approach produces upper and lower bounds to the efficiency score that the DMU would receive if it reorganizes its input and outputs to meet the target mix.

Keywords: Data Envelopment Analysis, Multiple Objective Linear Programming, Efficiency Analysis

1. Introduction

Data Envelopment Analysis (DEA), originally proposed by Charnes, Cooper and Rhodes [1978], has become one of the most widely used methods in management science. DEA measures the relative efficiency of comparable entities called Decision Making Units (DMUs) essentially performing the same task using similar multiple inputs to produce similar multiple outputs. The purpose of DEA is to empirically estimate the so-called efficient frontier based on the set of available DMUs. A DMU is efficient if there is no other unit – existing or virtual – that can either produce more outputs by consuming the same amount or less inputs or produce the same amount or more outputs by consuming less inputs as the DMU under consideration. The former approach is referred to as the output oriented and the latter as the input oriented DEA. DEA provides the user with information about the efficient and inefficient units, as well as the efficiency scores and reference sets for inefficient units. The results of the DEA analysis, especially the efficiency scores, are used in practical applications as performance indicators of DMUs.

When Decision Making Units are evaluated in practice, there is always a reason for this. It might be the allocation of existing or additional resources to units, need to make the operations more profitable by improving the performance of inefficient units, or the desire to reward the most efficient units. The results of the analysis provide a basis for such decisions. Generally there exists a Decision Maker (DM) who has preferences over outputs and inputs. The underlying assumption of the original DEA, however, is that no output or input is *a priori* more important than another. In such a situation, for example, a DMU which is a superior producer of a marginally important output is diagnosed as efficient even if it performs poorly with respect to all other outputs.

The traditional way to incorporate preference information in DEA is to restrict the flexibility of weights. (See Allen, Athanassopoulos, Dyson and Thanassoulis [1997] for a survey.) Another approach is to use some *ideal unit or mix*. In Halme, Joro, Korhonen, Salo and Wallenius [1998] the efficiency of DMUs is defined in the spirit of Data Envelopment Analysis (DEA), complemented with DM's preference information in the form of his/her most preferred input-output vector (i.e. the Most Preferred Solution, MPS).

The aim of this study is to develop an approach for measuring the efficiency of DMUs with respect to a *target mix* provided by a DM. The mix can be an output or an input mix. We assume that the DM wants DMUs to *alter* their outputs and/or inputs to this mix. The DM specifies the mix by articulating a target mix vector. After the mix is articulated, upper and lower bounds for the efficiency score that the DMU would receive, if it reorganizes its inputs or outputs to meet the target mix, are produced. One specific application area for the approach is selection problems. In these problems it is possible to have some 'good' alternatives that are inefficient in DEA sense, but not dominated by any existing DMU. The proposed approach is capable of recognizing such alternatives.

The rest of this paper is organized as follows. Section 2 sets the stage by discussing preliminary considerations. In Section 3 we develop the procedure for our approach. Section 4 illustrates the use of the approach, and section 5 concludes the paper.

2. Basic DEA Models

Assume there are n DMUs each consuming m inputs and producing p outputs. Let $\mathbf{X} \in \Re_+^{m \times n}$ and $\mathbf{Y} \in \Re_+^{p \times n}$ be the matrices, consisting of nonnegative elements, containing the observed input and output measures for the DMUs. It is further assumed that there are no duplicated units in the data set. We denote by x_j (the jth column of $\mathbf{X}$) the vector of inputs consumed by DMUj, and by x_{ij} the quantity of input i consumed by DMUj. A similar notation is used for outputs. Furthermore, we denote $1 = [1, ..., 1]^T$. The basic DEA formulations: CCR model with constant returns to scale (Charnes et al. [1978 and 1979]) and BCC model with variable returns to scale (Banker, Charnes and Cooper [1984]) are reproduced in (2.1a-d) as so-called primal forms.

Output-Oriented CCR Primal (CCR_P - O)	**Input-Oriented CCR Primal (CCR_P - I)**
$max\ Z_{CO} = \theta + \varepsilon(1^T s^+ + 1^T s^-)$ s.t. (2.1a) $\mathbf{Y}\lambda - \theta y_0 - s^+ = 0$ $\mathbf{X}\lambda + s^- = x_0$ $\lambda, s^-, s^+ \geq 0$ $\varepsilon > 0$	$max\ Z_{CI} = \theta + \varepsilon(1^T s^+ + 1^T s^-)$ s.t. (2.1b) $\mathbf{Y}\lambda - s^+ = y_0$ $\mathbf{X}\lambda - \theta x_0 + s^- = 0$ $\lambda, s^-, s^+ \geq 0$ $\varepsilon > 0$
Output-Oriented BCC Primal (BCC_P - O)	**Input-Oriented BCC Primal (BCC_P - I)**
$max\ Z_{BO} = \theta + \varepsilon(1^T s^+ + 1^T s^-)$ s.t. (2.1c) $\mathbf{Y}\lambda - \theta y_0 - s^+ = 0$ $\mathbf{X}\lambda + s^- = x_0$ $\lambda, s^-, s^+ \geq 0$ $1^T\lambda = 1$ $\varepsilon > 0$	$max\ Z_{BI} = \theta + \varepsilon(1^T s^+ + 1^T s^-)$ s.t. (2.1d) $\mathbf{Y}\lambda - s^+ = y_0$ $\mathbf{X}\lambda - \theta x_0 + s^- = 0$ $\lambda, s^-, s^+ \geq 0$ $1^T\lambda = 1$ $\varepsilon > 0$

A DMU is efficient iff $Z_{CO}^* = Z_{CI}^* = Z_{BO}^* = Z_{BI}^* = 1$ (and thus all slack variables s^-, s^+ thus equal zero); otherwise it is inefficient (Charnes et al. 1994).

Define the set

$$\Lambda = \begin{cases} \{\lambda \mid \lambda \in \Re_+^n\}, & \text{for the CCR model} \\ \{\lambda \mid \lambda \in \Re_+^n \text{ and } 1^T\lambda = 1\}, & \text{for the BCC model} \end{cases}$$

and $T = \{ (x, y) \mid x = X\lambda, y = Y\lambda, \lambda \in \Lambda\}$. All efficient DMUs lie on the efficient frontier, which is defined as a subset of points of set T satisfying the efficiency conditions below.

Definition 1. A solution $(Y\lambda^*, X\lambda^*) = (y^*, x^*)$ is *efficient* iff there does not exist another $(y, x) \in T$ such that $y \geq y^*$, $x \leq x^*$, and $(y, x) \neq (y^*, x^*)$.

Definition 2. A point $(y^*, x^*) \in T$ is *weakly efficient* iff there does not exist another $(y, x) \in T$ such that $y > y^*$ and $x < x^*$.

3. Development of Target Mix Efficiency

The aim of the approach developed in this paper is to measure the efficiency of a DMU with respect to *target input or output mix*. The DM provides the mix by identifying a target mix vector. The proposed approach provides upper and lower bounds for the efficiency score the DMUs would receive if it adopts the target mix in outputs (inputs) maintaining its current input (output) levels. The only assumption is that marginal rates of substitution between outputs (inputs) are nonnegative. In the following the approach is formulated with assumption of constant returns to scale (CCR model), but the assumption on variable returns to scale (BCC model) can be used as well.

3.1. Target Input and Target Output Mixes

Figure (3.1a) illustrates the target output mix and Figure (3.1b) the target input mix (referred as $\boldsymbol{y}_t$ and $\boldsymbol{x}_t$ the figures). Both figures represent five DMUs, in Figure (3.1a) consuming all the same amount of one input to produce two outputs and in Figure (3.1b) producing same amount of one output by consuming two inputs.

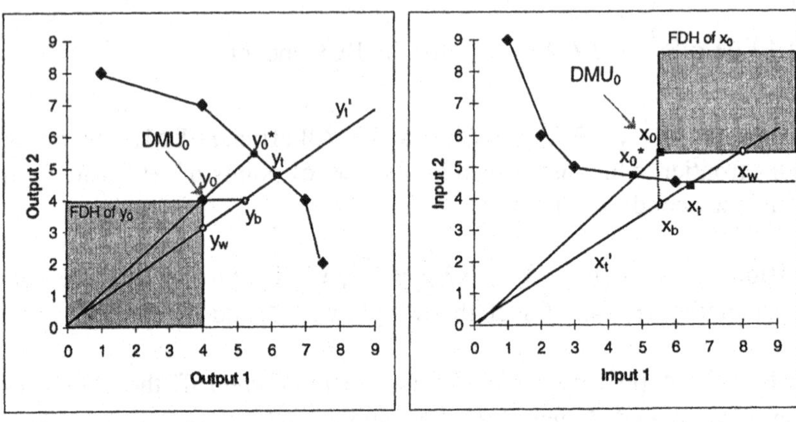

Figures 3.1a and b.
Upper and lower bounds for DMU's position on Target Mix line

DMU$_0$ is inefficient, whereas other DMUs are efficient in DEA sense. The shaded area is the Free Disposable Hull (FDH) of DMU$_0$ (see e.g. Deprins, Simar and Tulkens [1984], Thiry and Tulkens [1992]). The output and input vectors of DMU$_0$ are $\boldsymbol{y}_0$ and $\boldsymbol{x}_0$, $\boldsymbol{y}^*$ and $\boldsymbol{x}^*$ represent the projections provided by standard DEA and $\boldsymbol{y}_t$ and $\boldsymbol{x}_t$ the efficient targets indicated by the target mix.

Let τ_t^y, τ_w^y, τ_b^y, (τ_t^x, τ_w^x, τ_b^x) be nonnegative scalars, $\boldsymbol{y}_0$ ($\boldsymbol{x}_0$) be the output (input) vector of DMU$_0$, and $\boldsymbol{y}_t^{'}$ ($\boldsymbol{x}_t^{'}$) an arbitrary vector identifying the target output (input) mix. Both vectors consist of nonnegative elements, at least one of them being strictly positive.

Lemma 1. A ray from origin through $\boldsymbol{y}_t'$ intersects the weakly efficient surface at the target output $\boldsymbol{y}_t = \tau_t^y \boldsymbol{y}_t^{'}$ where $\tau_t^y = \max \{\tau_t^y \mid \tau_t^y \boldsymbol{y}_t^{'} \leq \mathbf{Y}\boldsymbol{\lambda}, \boldsymbol{\lambda} \geq 0\}$. A ray from origin through $\boldsymbol{x}_t'$ intersects the weakly efficient surface at the target input $\boldsymbol{x}_t = \tau_t^x \boldsymbol{x}_t^{'}$ where $\tau_t^x = \min \{\tau_t^x \mid \tau_t^x \boldsymbol{x}_t^{'} \geq \mathbf{X}\boldsymbol{\lambda}, \boldsymbol{\lambda} \geq 0\}$).

We first identify the possible output (input) vectors of the DMU after it has adopted the target output or input mix. According to the principle of Strong Free Disposability, DMU_0 can achieve any point on its FDH by simply discarding some outputs or by consuming more inputs. Thus when DMU_0 adopts the target mix, and moves itself to the target mix line, its position on the line will be at least y_w (x_w).

Lemma 2. A ray from origin through y_t (x_t) intersects necessarily the FDH of DMU_0. The intersection $\tau_w^y y_t$ ($\tau_w^x x_t$), referred as y_w (x_w) is the point from the ray that is weakly dominated by y_0 (x_0).

Lemma 3. On the ray from origin through y_t (x_t) it is possible to find a point $y_b = \tau_b^y y_t$ ($x_b = \tau_b^x x_t$) that weakly dominates y_0 (x_0).

Proofs are omitted as trivial. Lemma 2 defines the worst possible position. It assumes that the marginal rate of substitution between the outputs (inputs) is zero. In Lemma 3 the best possible position after adoption of the target mix is obtained by assuming that the marginal rate of substitution is ∞. The worst position might be realistic in some cases, but the best position is an open limit for the true score.

Thus it holds for the actual output vector y_a (input vector x_a) that the unit would receive after reorganizing its outputs (inputs):

$$\begin{bmatrix} y_w \\ x_0 \end{bmatrix} \le \begin{bmatrix} y_a \\ x_0 \end{bmatrix} \le \begin{bmatrix} y_b \\ x_0 \end{bmatrix} \text{and} \begin{bmatrix} y_0 \\ x_b \end{bmatrix} \le \begin{bmatrix} y_0 \\ x_a \end{bmatrix} \le \begin{bmatrix} y_0 \\ x_w \end{bmatrix} \tag{3.1}$$

Note that best position can be superefficient (see Figure 3.1b). This means that with the assumption of the marginal rate of substitution between the outputs (inputs) being ∞ the unit would shift the efficient frontier when adopting the desired mix.

3.2. The Best and Worst Positions

The best and worst positions for the unit after adapting the target mix can be found by projecting the target vector on the output (input) vector of the unit evaluated with an *achievement (scalarizing) function* (see Wierzbicki [1986]).

In output oriented case the worst position y_w is found by optimizing the following scalarizing function over the points dominated by unit's output vector y_0 (see Fig. 3.2a):

$$\min \left\{ \max_r \left(\frac{y_{tr} - y_{wr}}{y_{tr}} \right) \right\} \tag{3.2}$$

$$\text{st} \quad y_{wr} \le y_{0r}, \forall r$$

This is equal to

$$\begin{aligned} &\min \tau \\ &\text{st} \quad \tau \ge \left(\frac{y_{tr} - y_{wr}}{y_{tr}} \right), \forall r \\ &\quad y_{wr} \le y_{0r}, \forall r \end{aligned} \quad \Leftrightarrow \quad \begin{aligned} &\min \tau \\ &\text{st} \quad y_{wr} + \tau y_{tr} \ge y_{tr}, \forall r \\ &\quad y_{wr} \le y_{0r}, \forall r \end{aligned} \tag{3.3}$$

Respectively, the best position can be found by optimizing the following scalarizing function over the points dominating the current outputs (see Fig. 3.2b):

$$\max \left\{ \min_r \left(\frac{y_{tr} - y_{br}}{y_{tr}} \right) \right\} \tag{3.4}$$

$$\text{st} \quad y_{br} \ge y_{0r}, \forall r$$

equal to

$$\begin{aligned} &\max \tau \\ &\text{st} \quad \tau \le \left(\frac{y_{tr} - y_{br}}{y_{tr}} \right), \forall r \\ &\quad y_{br} \ge y_{0r}, \forall r \end{aligned} \quad \Leftrightarrow \quad \begin{aligned} &\max \tau \\ &\text{st} \quad y_{br} + \tau y_{tr} \le y_{tr}, \forall r \\ &\quad y_{br} \ge y_{0r}, \forall r \end{aligned} \tag{3.5}$$

Thus with target output and input mixes the best and worst positions y_b, y_w, x_b and x_w can be found with following LPs:

Best position with Target Output Mix	Worst position with Target Output Mix
$\max \tau$ s.t. $y_b + \tau y_t \leq y_t$ (3.6a) $y_b \geq y_0$	$\min \tau$ s.t. $y_w + \tau y_t \geq y_t$ (3.6b) $y_w \leq y_0$
Best position with Target Input Mix	**Worst position with Target Input Mix**
$\min \tau$ st $x_b + \tau x_t \geq x_t$ (3.6c) $x_b \leq x_0$	$\max \tau$ st $x_w + \tau x_t \leq x_t$ (3.6d) $x_w \geq x_0$

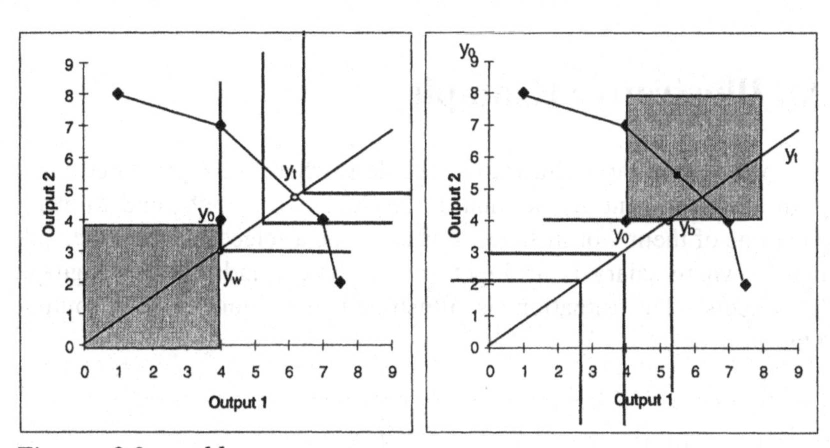

Figures 3.2a and b.
Achievement Scalarizing Functions

3.3. Calculation of the Efficiency Scores

The next step of the analysis is to use the obtained best and worst possible output (input) vectors instead of original ones in a standard DEA analysis with selected scale and orientation assumptions. The obtained range is $[\theta^y_w, \theta^y_b)$ or $[\theta^x_w, \theta^x_b)$ for the target mix efficiency score. With target output mix the range would be:

$[\theta^y_w = \max\{\theta^y_w \mid \mathbf{Y}\lambda \geq \theta^y_w y_w, \mathbf{X}\lambda \leq x_0\}, \theta^y_b = \min\{\theta^y_b \mid \mathbf{Y}\lambda \geq \theta^y_b y_b, \mathbf{X}\lambda \leq x_0\})$

and with target input mix

$[\theta^x_w = \max\{\theta^x_w \mid \mathbf{Y}\lambda \geq y_0, \mathbf{X}\lambda \leq \theta^w_x x_w\}, \theta^x_b = \min\{\theta^x_b \mid \mathbf{Y}\lambda \geq y_0, \mathbf{X}\lambda \leq \theta^x_b x_b\})$.

As mentioned, the upper bound for DMU's position may be superefficient, and thus the corresponding efficiency score may exceed one. This means that after the DMU has adapted the target mix it may become superefficient, and shift the efficient frontier. Some information can also be deducted from the length of the interval – the longer it is, the further the DMU lies from the target mix line.

4. An Illustrative Example

Let us look into a naïve illustrative problem where there are three units using the same amount of one input to produce two outputs, and we want to select one of them. For instance, this could be a selection from three job applicants where salary is an input and analytical and foreign language skills outputs. The situation is illustrated in Figure 4.1 in output orientation.

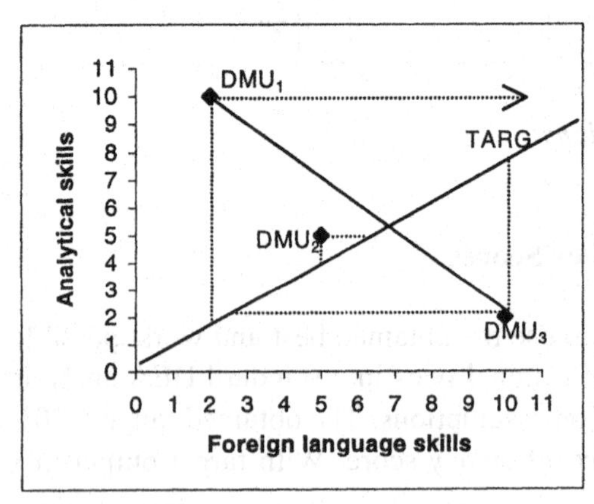

Figure 4.1

In classical DEA DMUs 1 and 3 are efficient and DMU_2 is inefficient. Now let us assume, that the Decision Maker wants to select a unit that performs relatively well in both aspects, i.e. has reasonable analytical and foreign language skills. DMU_2, although inefficient, would be a reasonable compromise.

Let Decision Maker's target output mix be identified with the line TARG. The dotted lines in Figure 4.1 indicate the upper and lower positions on the target mix line for each DMU, and the corresponding efficiency scores are based on them. Since here it is plausible that there is no substitution, the lower bounds for the efficiency scores are used. Thus the DMU who is inefficient would receive the best score.

5. Conclusion

In this paper we have presented a novel way to incorporate preference information in DEA. The information is gathered from the Decision Maker in the form of the target output or target input mix. The approach produces a possible range for efficiency scores if the DMU under consideration would alter its input and/or output structure according to the target mix. The approach is capable of finding 'good' compromise solutions that are inefficient, but have a input and/or output mix preferred by the DM and are not dominated by any existing DMU.

References

Allen, R., A. Athanassopoulos, R.G. Dyson and E. Thanassoulis (1997), "Weights Restrictions and Value Judgements in Data Envelopment Analysis: Evolution, Development and Future Directions", Annals Of Operations Research 73, 13-34

Banker, R.D., A. Charnes and W.W. Cooper (1984), "Some Models for Estimating Technical and Scale Inefficiencies in Data Envelopment Analysis", Management Science 30, 1078-1092.

Charnes, A., W.W. Cooper and E. Rhodes (1978), "Measuring Efficiency of Decision Making Units", European Journal of Operational Research 2, 429-444.

Charnes A., W.W. Cooper, A.Y. Lewin, and L.M. Seiford (1994), Data Envelopment Analysis: Theory, Methodology and

Deprins, D., L. Simar and H. Tulkens (1984), "Measuring Labour Efficiency in Post Offices", chapter 10 (pp. 243-267) in M. Marchand, P. Pestieau and H. Tulkens (eds.), The Performance of Public Enterprises: Concepts and Measurement, North-Holland, Amsterdam.

Halme, M., T. Joro, P Korhonen, S. Salo and J. Wallenius (1998), "A Value Efficiency Approach to Incorporating Preference Information in Data Envelopment Analysis", Management Science 45, 103-115

Thiry, B. and H. Tulkens (1992), "Allowing for Technical Inefficiency in Parametric Estimates of Production Functions" Journal of Productivity Analysis 3 (1/2), 45-66.

Wierzbicki, A. (1980), "The Use of Reference Objectives in Multiobjective Optimization", in G. Fandel and T. Gal (Eds.), Multiple Objective Decision Making, Theory and Application, Springer-Verlag, New York.

The TEAM Model for Evaluating Alternative Adaptation Strategies

Susan Herrod Julius and Joel D. Scheraga
Global Change Research Program, Office of Research and Development
USEPA

Abstract[1]

Advances in the scientific literature have focused attention on the need to develop adaptation strategies to reduce the risks, and take advantage of the opportunities, posed by climate change and climate variability. Adaptation needs to be considered as part of any response plan. But appropriate adaptive responses will vary across different geographic regions since the potential consequences of climate change and variability for human and natural systems will vary regionally in scope and severity. The assessment of consequences and selection of appropriate adaptation strategies is a complex challenge for regional and local decision makers. To aid in these assessments, the U.S. EPA developed a decision support software system called the Tool for Environmental Assessment and Management (TEAM) that employs a multi-criteria approach for evaluating actions to address climate change impacts. Applications of TEAM have revealed some strengths of this tool: (1) transparency of the methodology used in TEAM is important, particularly to international audiences; (2) the structure encourages users to consider strategies and attributes from an array of disciplines, which leads to more effective outcomes; (3) the ability to consider and understand tradeoffs of noncomparable attributes is essential to good decision making; and (4) TEAM fosters communication and consensus among participants in the decision making process, in particular through the use of visual display features and sensitivity analyses.

Keywords: Multi-criteria decision making, climate change, adaptation, agriculture, sea level rise, water resources, Egypt

1 Introduction

Management problems are often complex because of the need to consider multiple objectives, the need to formulate and consider different aspects of a problem, and the need to assess the importance and relevance of these aspects in some consistent way. As noted by Zeleny (1982):

---

[1] The authors acknowledge valuable contributions made by Dr. Helmy M. Eid in conducting the case study using TEAM that is featured in this paper. The authors also thank Decision Focus Incorporated for their assistance in developing TEAM for the Climate and Policy Assessment Division of the U.S. Environmental Protection Agency. Any remaining errors are the sole responsibility of the authors. The views expressed in this paper are the authors' own and do not represent official EPA policy.

> "decision making is ultimately the most difficult (and potentially the most rewarding) activity because a 'model' of any reasonable richness will return multiple criteria, forcing us to choose not only among the possible courses of action but also among the means of evaluating such actions."

Supporting the decision making process calls for a systematic approach to making choices and providing useful insights in complex situations.

Climate change is an example of a particularly difficult and complex problem facing decision makers today. This environmental problem provides an excellent illustration of the potential complexity of a decision-making process because of the multi-disciplinary nature of the issue, uncertainties about the potential magnitude, timing and effects of climate change, uncertainties about the effectiveness of different courses of action to adequately address the potential impacts, and the existence of many alternative societal problems competing for scarce resources that could be used to address concerns about climate change (Smith and Chu 1994, Scheraga and Julius 1995).

The U.S. Environmental Protection Agency has developed a decision support software system called the Tool for Environmental Assessment and Management (TEAM), to assist decision makers trying to assess risks posed by climate change and to select adaptive responses. TEAM employs a multi-criteria approach for evaluating actions to address climate change impacts to water resources, coastal zones, and agriculture.

2 Evolution of TEAM

The historic Framework Convention on Climate Change (FCCC) was signed by 154 countries in June 1992. The ultimate objective of this Convention, as proclaimed in Article 2, is to prevent dangerous anthropogenic interference in the climate system, to achieve this within a time frame sufficient to allow ecosystems to adapt naturally to climate change, to ensure that food production is not threatened, and to enable economic development to proceed in a sustainable manner.

Two different but complementary mechanisms exist for fulfilling the goals of the Convention: *mitigation* of greenhouse gas emissions and *adaptation* to the potential effects of climate change. Mitigation refers to policies intended to reduce anthropogenic emissions of greenhouse gases which contribute to global climate change. Adaptive actions are those responses taken to reduce damages to human and natural systems resulting from climate change.

Anticipatory adaptation refers specifically to those actions taken before the effects of climate change are apparent. These measures contrast with actions taken in *reaction* to the impacts of climate change. Often, anticipatory investments can be completed at lower costs than comparable actions that are taken when impacts are observable. Also, delaying action may render some strategies ineffective or impossible to implement if long lead times are required. Failure to anticipate particular climate impacts may cause irreversible damages to natural systems and the loss of unique cultural resources. The development of TEAM was undertaken to help decision makers consider alternative anticipatory adaptation strategies for reducing the risks posed by climate change.

3 Features of TEAM Oriented to Decision Makers

3.1 Steps of TEAM Software

Each step in TEAM is designed to provide easy management of information and the ability to manipulate presentation of information while investigating tradeoffs. In the first step of TEAM, the geographic region is identified, and the resource (e.g., water basin, agricultural system, coastal zone) is characterized according to its vulnerabilities to climate change and other stressors. The appropriate geographic scale will depend on the type of resource being evaluated and the classification of the resource or site. TEAM helps establish the appropriate scale by guiding the user through a series of questions to define the analysis and evaluate the facts.

The second step is to specify the actions to be considered for addressing the identified vulnerable system or resource. The types of actions that may be specified range in nature from action oriented (e.g., building sea walls), to planning oriented (e.g., legal or regulatory changes such as zoning rules and building codes). The system provides a list of suggested actions within each of these categories, and allows for additions or modifications to the list.

The third step, selection of evaluation criteria, is central to this tool. TEAM provides suggested criteria, and allows for additions to be made to the list. This structure encourages consideration of a broad spectrum of factors that may be important to a decision, ensuring comprehensiveness to the decision making process.

Prior to scoring the strategies (step four), the user is encouraged to make an explicit decision about the time horizon over which the assessment will be made and enters that time horizon into TEAM. The user chooses the time horizon that is appropriate for the particular planning decision and the choice is displayed throughout the scoring process to ensure consistency in evaluation.

The fourth step is evaluating and scoring the selected actions. This step entails comparing options against the criteria selected to assess the performance of each. This step elicits judgements about the performance of options and promotes articulation of the reasoning behind judgements. The system encourages qualitative assessments of performance, but allows for quantitative comparisons to be made where that data exist. Scoring is done based on the user's evaluation of the effectiveness of a given strategy in meeting goals as expressed through the criteria selected. The scores entered are based on a relative comparison of specified strategies, and the scoring categories of "excellent," "good," "fair," and "poor" may vary from criterion to criterion.

The final step, assessing results, promotes consideration of the consequences of each action under review. TEAM provides alternative presentations of the information through different visual displays to aid the investigation of consequences and tradeoffs implied by choosing particular actions. Each action may be compared against the other for the criteria selected. The criteria are not aggregated to provide a single index of performance for each action. In this way, clear or subtle differences in alternative courses of actions for each criterion may be understood.

TEAM has six different types of visual displays from which to choose to foster further exploration of the data, which may lead to insights that a single display would fail to reveal (see Smith et al. 1995 for a description of each visual display). Sensitivity analyses complement this exploration and understanding of priorities and tradeoffs. These sensitivity analyses may be performed by going back to previous stages of the process and changing inputs (criteria, strategies, scores, absolute data such as costs), or selecting different assumptions under which to conduct the analyses and modifying the inputs accordingly, or simply modifying weights on the criteria to reflect judgements of their relative importance. Uncertainty is addressed through the ability to conduct sensitivity analyses on the uncertain variables and assess the effect on the overall results.

3.2 Limitations of TEAM

One limitation of TEAM is that the user can manipulate the data to produce a desirable outcome. There are some checks within TEAM to prevent this (e.g., displaying weights when a user has applied them to criteria), but they may not be adequate to prevent such manipulation. Also, the qualitative scoring is only as good as the user's knowledge of the performance of different strategies. Group participation may help the quality of the analysis and this approach to the use of TEAM is recommended. Finally, TEAM has no check on the internal consistency of the user's inputs with respect to data or scoring.

4 Applications of TEAM and Insights Gained

In March 1995, seven case studies were conducted in collaboration with colleagues at the University of Cairo's Faculty of Agriculture, and at the University of Alexandria's Department of Environmental Studies. These case studies were funded as part of the United States Country Studies Program to assist developing countries in assessing their vulnerability to climate change and possible adaptation options. In 1997, several researchers from the University of Cairo were able to visit the United States to conduct further analysis that built on one of the case studies performed in 1995. Because the two phases of this case study reveal the usefulness of TEAM in a decision process, from the formative stage of the problem to making preliminary recommendations to the government of Egypt, it is the focus of discussion below. The first phase of the case study is an assessment of the vulnerability of wheat production in the Nile Delta area. The second phase is an assessment of a broader range of alternative crops and rotation practices using historical data and information gathered from crop models.

4.1 Phase I - Wheat Production in the Nile Delta Area

The site for the study was the Nile Delta area which encompasses 22,000 square kilometers on the Mediterranean coast. Although this area accounts for only 3% of the country's land area, it provides 45% of the nation's cultivated land. In 1984 and 1987, agriculture accounted for about 20% of Egypt's gross domestic product (Hansen 1991, Strzepek 1995). Egypt's agricultural water supply comes entirely from irrigation and the only source for water is the Nile River. Agricultural uses consume 80% of the water budget (Shahin 1985, Strzepek et al. 1995).

Climate change is likely to have a significant effect on the supply of water available for irrigation and other uses. Runoff may decrease as a result of higher evaporation and changes in precipitation. These changes, along with increased evapotranspiration under a warmer climate, will increase water losses in the fields and during storage and transport processes. Increased climate variability (drought, heat waves) could pose an additional threat to agricultural production (Strzepek et al. 1995).

The Egyptian agricultural year is composed of three crop seasons. These seasons and their crops are:

- *winter season:* wheat and barley, berseem and lentils, winter onions, and vegetables, planted between October and December and harvested between April and June;
- *summer season:* cotton, rice, maize, sorghum, sesame, groundnuts, summer onions, and vegetables, with the growing season starting in March and ending in November;
- *late summer season:* rice, sorghum, berseem, and some vegetables, with planting times overlapping the summer growing season, necessitating different years for plantings of summer and late summer crops.

The crop selected as the focus of this initial study was wheat. Wheat is widely grown in the Nile Delta area and is important as a component of the Egyptian diet. Production is viewed as critical to maintaining Egypt's food security. The combination of projected increases in the demand for wheat and the sensitivity of wheat productivity to changes in climate are a cause for concern. If, as expected, this is the trend for most agricultural commodities, the implication is that the agricultural trade balance will be affected and net imports of all agricultural commodities will increase (Fischer et al. 1988).

The three purposes of the first phase of this case study were to: provide hands-on training in the use of the methodology employed in TEAM; prepare specific applications to address climate change vulnerabilities in Egypt, and; introduce the users to a way of structuring information and results from a variety of sources and models to enable a systematic consideration and assessment of adaptation options to address identified vulnerabilities.

Three types of vulnerabilities of wheat production were identified: drought, weed, and heat stress. Among these vulnerabilities, drought was the most serious consideration in the context of potential future climate change. The categories of strategies to choose from included no anticipatory action, changing agricultural practices, switching to a new type of cultivar, switching to a new type of crop, or abandonment of agriculture in favor of an alternative use of the land. After extensive discussion, five candidate strategies were developed: two strategies consisted of changing agricultural practices, two strategies were to switch to a new cultivar, and one strategy was to take no action. They are defined more completely below:

1. *Tillage* – implement minimum tillage and change planting dates; increase government import of herbicides and conduct education programs about timing and cultivating methods.
2. *Mech* – provide government funding to import small-scale machines suitable for Egypt's geography and increase mechanization in farming.
3. *Culti1* – provide government funding for development of drought resistant wheat variety; change planting dates and use more fertilizer; modify existing seed distribution systems to aid in widespread adoption of new cultivar.
4. *Culti2* – provide government funding for development of heat- and drought-resistant wheat variety; change planting dates and use more fertilizer; modify existing seed distribution systems to aid in widespread adoption of new cultivar.
5. *No Policy Action* – no anticipatory action will be taken.

After reviewing the candidate list of attributes in TEAM, seven attributes were selected to evaluate the candidate strategies: up-front costs for implementing strategy; long-term expected net benefits; effectiveness in addressing vulnerability; farm income (yield, price, and cost of production); technical/financing feasibility; distributional impact; and food security.

Once the selection of strategies and evaluation criteria were chosen, each strategy was compared based on the above criteria. The relative performance of the strategies are represented by the scores shown in the following table:

Table 1

	No Policy Action	Tillage	Culti1	Mech	Culti2
short-term cost	Excellent	good	fair	poor	fair
long-term cost	Poor	fair	fair	good	fair
effectiveness	Poor	fair	good	good	excellent
farm income	Poor	fair	fair	good	excellent
feasibility	Excellent	good	poor	fair	good
distrib impacts	Good	excellent	fair	poor	fair
food security	Poor	fair	excellent	good	excellent

The choice of seven attributes added to the complexity of the case because of the need to consider each one in light of the others, for all of the selected strategies. Using TEAM's bubble chart display, the two strategies "Tillage" and "Culti2" appeared to perform relatively better than the others (see Figure 1). The "No Policy Action" should only be pursued if the country had no resources available to allocate to action in the short term. However, switching to the column chart display to view the results made the judgement of the most desirable strategy much more difficult to

Figure 1

determine (see Figure 2). Because of this difficulty, a set of weights were selected by the participants for each of the criteria in order to reflect more strongly their priorities in decision making (see Figure 3). For example, the more important criteria received higher weights than the less important criteria, causing their performance to be emphasized visually over the less important criteria. When these weights were applied, it became more evident that "Culti2" was the best overall strategy, while "Tillage" dropped off in strength of performance.

Despite the good performance of the "Culti2" strategy, participants in the case study raised some of the difficulties involved with implementing this strategy. These difficulties included the government's unwillingness to fund cultivar research and the possible uneven distribution of the strategy's benefits among farmers. The strategy

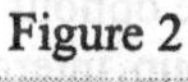

Figure 2

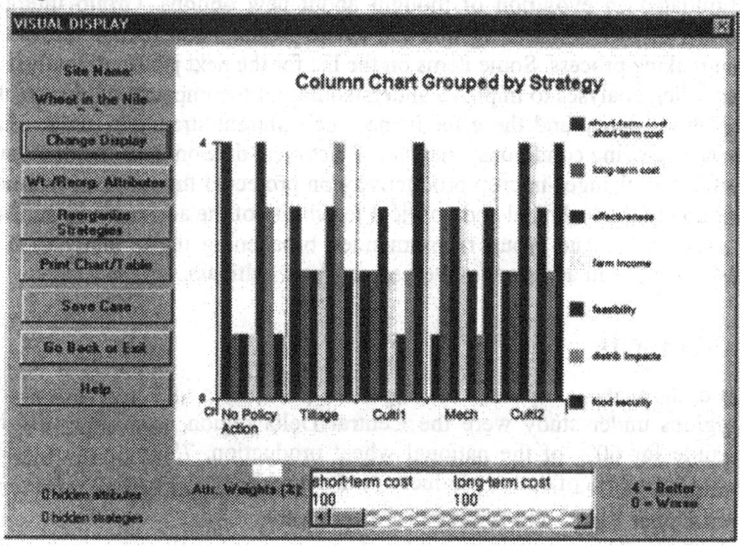

had limitations and required further analysis to improve the probability of its success.

As illustrated above, TEAM's visual displays provided a means of summarizing for the participants the fairly complex evaluation information and allowed them to easily interpret the results. Sensitivity analyses in the form of application of different value functions through weighting schemes were also easy to perform and evaluate.

Figure 3

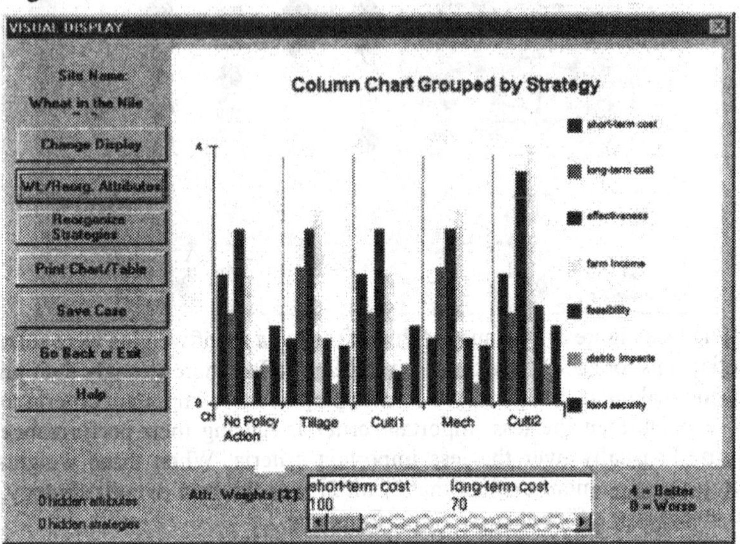

Using TEAM in this initial phase of analysis was critical for developing a decision-oriented mindset among the researchers. Such issues as financing barriers for cultivar development were previously not the focus of concern for most members of the analytic team. The process of considering and evaluating the options above also stimulated an evolution of thought about new options. Group interactions resulted in a focused research agenda that would produce information relevant to the decision making process. Some items on the list for the next phase of analysis were: crop modeling analyses to improve understanding of the impacts of climate change on crop productivity and the effectiveness of different strategies under changed climate and growing conditions; analysis of historical data on farm income; analysis of the effect of changes in crop productivity on projected future farm income; and examination of the technical and political feasibility of the alternatives enumerated in the case study. The group recommended broadening these analyses to other possible strategies in addition to alternative wheat cultivars.

4.2 Phase II

Phase II of the analysis expanded on adaptation alternatives and crops examined. The three regions under study were the Central Delta region represented by Sakha (responsible for 60% of the national wheat production, 75% of the total maize production, and 75% of cotton production), the Middle Egypt region represented by Giza, and Upper Egypt represented by Shandaweel.

Prior to beginning Phase II of the case study, Dr. Helmy Eid conducted quantitative crop modeling which generated evidence about the direction and magnitude of productivity responses of various crops. Information on crop vulnerabilities were from model runs of COTTAM and DSSAT3[2]. Crops considered in this study were wheat, maize, cotton, rice and sugar cane. Dr. Eid also gathered data on historical patterns of crop prices and crop productivity. These sources of data were combined to construct farm incomes under different scenarios[3]. These quantitative results were input into TEAM to examine key trade-offs and interactions among alternative agricultural strategies that were feasible for the Egyptian agricultural economy under changed climatic conditions. The results provided quantitative evidence of crop productivity and vulnerability that differed from expert judgement used in the first phase of the case study.

The decision criteria chosen for the analysis were agricultural income, food security, industry/employment, food culture, water demand and chemical usage. The final list of strategies examined were combinations of different crops and alterations in cropping patterns, alterations in types of cultivars and amounts and varieties of crops to be grown, and changes in other management strategies such as sowing dates and irrigation methods. Summer crops (e.g., maize and cotton) and winter crops (e.g., wheat) were analyzed separately and in a unified annual strategy comparison.

Results from the case study conducted in Phase II are represented in Figure 4. Analysis of price data for cotton showed that over the last 30 years, it has significantly outperformed alternative crops such as maize, sorghum, soybeans, sunflowers and wheat on a revenue per feddan basis. If wheat were to be phased out because of the conflicting soil needs (wheat cannot be grown directly following cotton), current annual farm revenues would be increased by displacing the usual pattern of growing maize followed by growing wheat with growing cotton only.

If farm income were the primary consideration in the selection of strategies, the recommendation coming out of the TEAM analyses would be to shift to growing cotton (see figure 4). All of the information indicates that under climate change, growing incrementally more cotton could have strong economic benefits: cotton productivity may not only increase relative to other major crops, but it may increase absolutely; cotton prices are very strong and it appears possible that they will remain strong in a hotter future climate since cotton is one of the best textiles for use in hot environments; and cotton production also involves more jobs, both in the agricultural sector, and in the textiles industry that can expand with cotton growth. However, other criteria confounded the decision process and it was the role of TEAM to aid in elucidation of research members' values and to build consensus around those values. Minimizing water usage while maximizing farm income was determined to be

---

2 Tsuji, et al., 1995.

3 General Circulation Models (GCMs) used were Goddard Institute for Space Studies (GISS), General Fluid Dynamics Laboratory (GFDL), United Kingdom Meteorological Organization (UKMO) and Canadian Climate Center Model (CCCM). For descriptions of these and other GCMs, see IPCC(1996) and IPCC (1998).

important. Food security, important in the Phase I analysis, remained important. To a lesser extent, chemical usage was also determined to be important. If these other criteria were emphasized over farm income, then the most attractive strategy would be to grow a combination of the cotton and the super SC10 (maize) crops over the soy/sunflower and sorghum.

These results are reflected in the first recommendation: improve wheat and maize cultivars and continue as normal. The second and third recommendations were

Figure 4

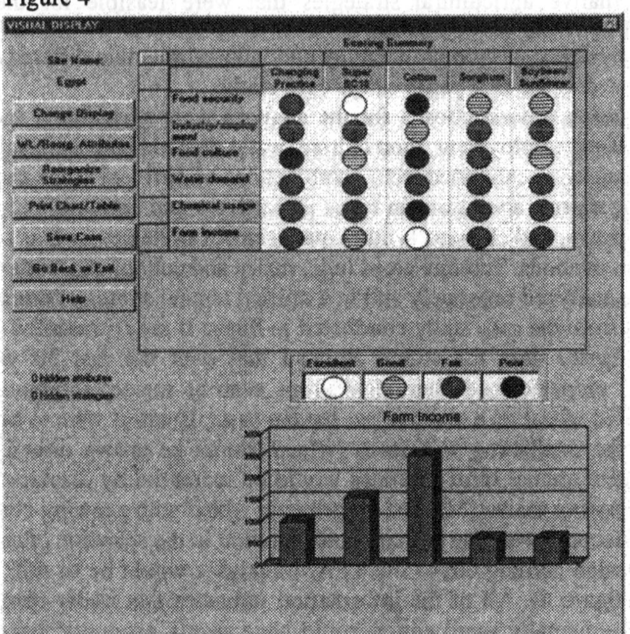

to: shift from maize to cotton and use more winter crops to replace some wheat; or shift to a mix of cotton/sunflowers in the summer and continue to grow an improved wheat cultivar in the winter to the extent feasible, while using winter crops in those areas that can no longer support wheat due to an increase in cotton production. These last two recommendations are based on emphasizing farm income over the other criteria.

One lesson learned from this analysis is that Egypt may be paying heavily for food security, given that it appears to be holding back on cotton production so that more wheat can be grown domestically. Looking at a hotter future, the arguments for more cotton may be enhanced: wheat prices may decline on a global basis whereas income from cotton may be enhanced due to 30% productivity increases projected by the crop models and possible growth in demand with climate change.

More work needs to be done before recommending a major shift toward cotton. The participants in Phase II of the analysis recommended that a third phase be conducted with TEAM that would incorporate research on expanding the cotton supply and the effect this action would have on cotton prices. Any shift toward cotton would need to be phased in carefully and should involve active assessment of potential price effects.

Participants also recommended that analyses be conducted to estimate the water demand requirements for different strategies, as well as any associated changes in chemical usage. Differences of opinion about the importance of each criterion need to be resolved in the next phase, and further research may help to resolve some of these differences. Although cotton performs well economically, it is a controversial recommendation.

A further refinement recommended by participants was to examine annual and/or bi-annual cycles of the best summer and winter options that are mutually compatible with principles of crop rotation. Results of such analyses would be useful and timely -- the policy environment within Egypt has made recommendations arising from the TEAM analysis more politically feasible, particularly recommendations that call for switching crops. A recent policy of crop liberalization allows farmers the possibility of adapting to more suitable and profitable crops in each area.

4.3 Comments on the Contributions of TEAM

The results of these analyses show the important role TEAM can play in focusing research throughout the decision process to develop more efficient, effective and robust recommendations. Because of the use of TEAM in the initial stages of this case study, further research using crop models and historical analysis provided quantitative results that were sometimes not intuitive and were needed for more informed decision making.

Throughout this process, individuals participating in the case study emphasized the importance of the transparency of the results. The importance of transparency became apparent in a number of ways: Phase I analyses were examined by those who participated in Phase II and the transparency of the underlying assumptions were critical to understanding how and why particular strategies outperformed others; communication about assumptions was important among various members of the analytic teams in both phases of the case study, especially where expertise might provide differing viewpoints about performance of strategies; and communication of results and recommendations to government officials would require transparency in order to convince the government of the validity of the recommendations, especially where results may be contradictory to current national policy.

Multiple disciplines were required to participate in the decision making process from the beginning. Because of the involvement of people from different backgrounds, the research agenda for the intervening period between phases of the case study addressed the variety of information needs discovered in the first phase. The involvement of representatives of the Egyptian government were critical to understanding the feasibility of different options and for providing an overall sense of the importance of different criteria in the policy making process.

This case study revealed that the criterion considered of primary importance in developing agricultural policy -- national security -- may have led to lower economic returns. This may be an acceptable tradeoff providing the government made its decision with full knowledge of the priorities implicit in their decision. The TEAM framework makes this tradeoff clear. In the future, the Egyptian government may decide to change its agricultural policy, or it may decide to continue with current practices, depending on their determination of the importance of domestic production of wheat. However, if climate change threatens the ability of Egypt to continue producing wheat, this tradeoff is all the more important to understand.

Finally, group processes proved critical to enumerating options for the agriculture sector and eliciting values. Differing values led to different recommendations for crop selections in the second phase of analysis, especially with respect to economic performance, pointing to the need for further work to assess performance of strategies along other criteria. With TEAM, not only was research focused to provide the most useful results for the decision making process, but a foundation was built on which values were understood and consensus could be reached to make final recommendations.

References

Hansen B. (1991) The Political Economy of Poverty, Equity, and Growth: Egypt and Turkey. Oxford University Press, New York Oxford.

(IPCC) Intergovernmental Panel on Climate Change (1996) Climate Change 1995: The Science of Climate Change, Contribution of Working Group I to the Second Assessment Report of the Intergovernmental Panel on Climate Change. Houghton J.T., Meira Filho L.G., Callander B.A., Harris N., Kattenberg A., Maskell K. (Eds.). Cambridge University Press, Cambridge.

(IPCC) Intergovernmental Panel on Climate Change (1998) The Regional Impacts of Climate Change: An Assessment of Vulnerability. Watson R.T., Zinyowera M.C., Moss R.H. (Eds.). Cambridge University Press, New York.

Janis I.L., Mann L. (1977) Decision Making. The Free Press, New York.

Quade E.S. (1982) Analysis for Public Decisions. North Holland, New York.

Scheraga J.D., Julius S.H. (1995) Decision Support Systems for Evaluating Alternative Adaptation Strategies. Draft paper presented at the U.S. Country Studies Vulnerability and Adaptation Workshop, St. Petersburg, Russia, May 22-25, 1995.

Shahin M. (1985) Hydrology of the Nile Basin. Elsevier, Amsterdam.

Smith A.E., Chu H.Q. (1994) A Multi-Criteria Approach for Assessing Strategies for Anticipatory Adaptation to Climate Change. Paper prepared for the Climate and Policy Assessment Division, Office of Policy, Planning and Evaluation, of the U.S. Environmental Protection Agency under EPA contract number 68-W2-0018, Subcontract No. 353-1, submitted to American Water Works Association Journal.

Smith A.E., Chan N., Chu H.Q., Helman C.J., Kim J.B. (1995) Documentation of Adaptation Strategy Evaluator Systems, vol. 1. Prepared for the U.S. Environmental Protection Agency under EPA Contract No. 68-W2-0018.

Strzepek K.M., Onyeji C., Saleh M., Yates D. (1995) An Assessment of Integrated Climate Change Impacts on Egypt. In: Strzepek K.M., Smith J.B. (Eds.) As Climate Changes: International Impacts and Implications. Cambridge University Press, Cambridge.

Tsuji G.Y., Jones J.W., Uhera G., Balas S. (1995) Decision Support System for Agrotechnology Transfer, v3.0. (DSSAT3). Three Volumes. IBSNAT, University of Hawaii, Honolulu.

Zeleny M. (1982) Multiple Criteria Decision Making. McGraw-Hill, New York.

Fuzzy Multi-Criteria DSS: Their Applicability and Their Practical Value

Anastasia Koulouri and Valerie Belton
Management Science, University of Strathclyde,
40 George Street, Glasgow G1 1QE, Scotland, UK.

Abstract. Most forms of Multiple Criteria Analysis / Decision Aid require the decision maker to make many preference judgements. We have found, from our experience with using a simple Multiple Attribute Value Function (MAVF) approach in both experimental and organisational settings, that in practice those involved in the decision process are often reluctant or have difficulty to specify such judgements in a precise way, preferring to retain some degree of imprecision or ambiguity.

Whilst there is much theoretical work on the use of concepts from Fuzzy Set Theory in MCDA and a number of outranking methods, in particular, incorporate these ideas, little attention has been paid to the potential for integrating these ideas within the framework of the MAVF approach. This is the focus for the work presented in this paper.

We begin by briefly describing a fuzzy version of MAVF approach and its implementation as an extension of the V•I•S•A software. Our concern is both with the theoretical validity and the practical usefulness of this new approach. We also describe a programme of experimental work comparing the fuzzy and the non-fuzzy MAVF approaches. Finally, we discuss the issues arising from this work together with some theoretical concerns.

Keywords. Multi-Attribute Value Function Theory; Fuzzy Preference Judgements; Decision Support Systems (DSS).

1 Introduction

Uncertainty, imprecision, ambiguity and equivocality are all terms associated with complexity in decision making. Multiple Criteria Decision Analysis (MCDA), as a collection of methods, seeks to help decision makers (DMs) clarify their values and uncertainties. It does this by proposing a process which, through careful elicitation, structure and synthesis of values and information, helps the DMs learn about and come to understand a problem, thereby enabling them to arrive at a decision or to resolve the issue. However, there may be a number of reasons, contextual, personal or political, which cause the DM to be reluctant to be too explicit or precise.

As a means of addressing some of these issues, by incorporating imprecision in the analysis rather then resolving it, notions of Fuzzy Set Theory (FST) were introduced in decision analysis methodologies. The following quote from Terano et al. [1, pp.1] is a justification for the use of FST: "Above all else, the outstanding feature of fuzzy sets is the ability to express the amount of ambiguity in human thinking and subjectivity (including natural language) in a comparatively undistorted manner."

Bellman & Zadeh were the pioneers in this field and, in 1970, were the first to introduce fuzzy notions in decision analysis [2]. Claiming that: "Much of the decision making in the real world takes place in an environment in which the goals, the constraints and the consequences of possible actions are not known precisely" [2, pp. B141], they developed a fuzzy MCDA model in correspondence to the classical maxmin optimisation model. Since then, several fuzzy MCDA methods have been developed in "direct" or "indirect" correspondence to the classical ones. For a thorough and systematic review of the existing methodologies the reader is referred to Chen & Hwang [3], where a classification of the methods is proposed; the theoretical background and the algorithm of each method are presented together with illustrative examples. Strengths and weaknesses of the methods are also discussed.

Whilst a lot of effort has been paid to the development of models integrating FST notions into MCDA, less attention has been paid to the behavioural implications of applying these approaches in practice. Moreover, although there is a substantial literature highlighting some of the difficulties faced by DMs in response to the elicitation questions posed by, for example, Multi-Attribute Utility Theory [4]; we are not aware of similar work exploring the use of fuzzy methods.

Our investigation aims at the exploration of DMs' evaluations of a fuzzy model for MCDA and in comparison with a parallel classical model. To do this we modified the "fuzzified" Multi-Attribute Value Function (MAVF) model introduced by Bonissone [5 & 6]. Our model has been implemented as part of V•I•S•A, a MAVF based decision support system. Currently, we are in the process of conducting a series of experiments in order to elicit DMs' views, comparing the classical and fuzzy approaches and their implementation.

In this paper, we begin with a brief review of the classical MAVF additive model and its fuzzy extension. We then go on to describe the software implementation of the fuzzy model, and report on our experience from using both -the classical and the fuzzy- models in an experimental setting. Finally, we reflect on our observations to date and discuss the main issues that emerged during the course of the experiments and from the data collected in particular those which we feel are fundamental to the theoretically sound and practically feasible integration of FST notions into MCDA.

2 The "Classical" Model

Let $A=\{ a_j ; j=1,...,m \}$ be a set of possible alternatives (courses of action, options, decisions, strategies, etc.) and $C=\{ c_i ; i=1,...,n \}$ a set of criteria with respect to which the alternatives are to be assessed. One of the simplest aggregation models for the selection of the most preferred alternative is the ***Multi-Attribute Value***

Function Additive Model [7, Chapter 4], according to which the overall attractiveness of an alternative in *A* is given by the equation:

$$U(a_j) = \Sigma_i\, w(c_i) \bullet u_i(a_j)$$

where $w(c_i)$ is a scaling factor which reflects the relative importance of the ith criterion with $0<w(c_i)\leq 1$ and $\Sigma_i w(c_i)=1$; and, $u_i(a_j)$ is the partial attractiveness of the jth alternative with respect to the ith criterion with $0\leq u_i(a_j)\leq 100$. Note that it is necessary to define two reference points in order to specify the partial attractiveness scale. A common approach is to let 0 describe the worst plausible performance and 100 the best plausible performance.

The most preferred alternative a_p is the one with $U(a_p)= \max\{U(a_j)\ ;\ j=1,...,m\}$. Having said that, we would like to underline that "... there is no such thing as a "right answer" even within the context of the model ..." [8, pp.54] . Moreover, we would like to emphasise that decision analysis "... is an aid to decision making - to help structure a problem, to integrate objective measurement with value judgment, to make explicit and manage subjectivity, to provide a focus for discussion - and above all to facilitate learning and understanding, about the problem, about one's own priorities and values and about those of others involved in the process ..." [9].

Finally, it should be noted that an essential condition for the justification of additivity is that of preference independence of the criteria [7, pp.120].

3 The Fuzzy Model

The first to develop a fuzzy MAVF additive model were Baas and Kwakernaak [10], who in 1977, reformulated a probabilistic MAVF model described by Kahne in [11 & 12] (as cited by Baas & Kwakernaak [10]). Claiming that: "... the sort of uncertainty that comes into play here (in decision making) is better represented by the notion of fuzziness than that of chance" [10, pp.48], they formulated a method which allows the representation of linguistically expressed preference judgments and criteria weights as fuzzy sets, whereas the overall attractiveness of an alternative is obtained using the Extension Principle [13, Chapter 5; 14, pp.36-40]. Since then, several new and modifications of existing methods have been developed. For example, Jain [15]; Cheng & McInnis [16] (as cited by Cheng & Hwang [3, pp.315]); Dubois & Prade [17]; Dong et al. [18] and Dong & Wong [19]; Tseng & Klein [20]. However, a review of different approaches is beyond the scope of this paper and, for more details, the reader is referred to Cheng & Hwang [3, pp. 292-329], Dubois & Prade [17] and Ribeiro[21].

In a direct "fuzzification" of the classical model briefly described in Section 2, Bonnissone [5 & 6], in the early 80's, suggested an approach where the preference judgements and the criteria weights are still given in the form of linguistic variables, which subsequently are "interpreted" into triangular or trapezoidal fuzzy numbers (see Figure 1), denoted by (a,b,c) and (k,l,m,n), respectively. The aggregation is performed by means of fuzzy arithmetics (for more details on fuzzy arithmetics the reader is referred to Cheng & Hwang [3, pp.66-99] and Kaufmann & Gupta [22, Chapter 3]) and the concept of a dominance relation is applied in order to obtain "... the overall degree to which each of the alternatives dominates the others..." [6].

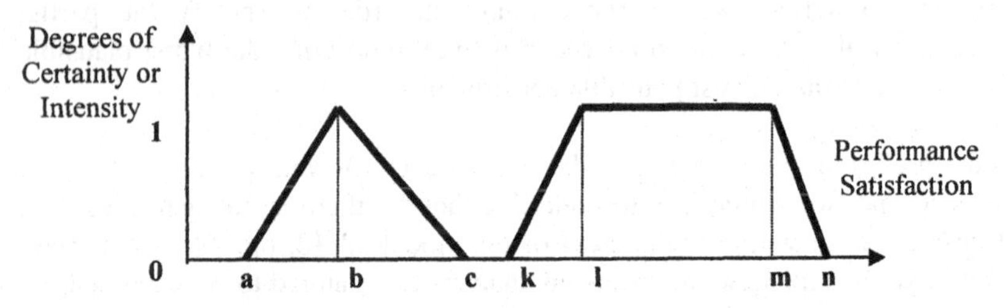

Fig.1 Triangular and Trapezoidal Fuzzy Scores

However, we feel that the use of linguistic variables for the assessment of the alternatives' performance is not completely justifiable: their "interpretation" into triangular or trapezoidal fuzzy numbers appears to be arbitrary [5, pp.104 & 6, pp. 334] and the DM does not seem to have any control over the process.

In the model we use for our investigation the preference judgements are assigned by the DM directly in the form of triangular or trapezoidal fuzzy numbers, whereas, the weights remain crisp. The aggregation is performed using simple fuzzy arithmetics: that is the multiplication of a fuzzy number by a constant and the addition of two fuzzy numbers.

It should be noted that we decided on the partial fuzzification of the model, that is the use of crisp weights, due to both computational and conceptual complications involved in the use of fuzzy weights. Concerning the computational difficulties: the product of two triangular or trapezoidal fuzzy numbers is a non-linear operation (that is: it is not necessarily a triangular or a trapezoidal fuzzy number, respectively); and, therefore, for practical applications various approximations have been developed [13, pp.65; 14, pp.55; 22, Chapter 6 & 23]. For a review of different approximation methods see Negi [24], as cited by Chang & Lee [23].

Moreover, the cognitive complexities of combining two fuzzy entities would not contribute to the conceptual simplicity and transparency of the approach, which is one of the fundamental requirements in any sort of decision analysis [8, pp.55]. Hence, we felt that it was more appropriate to start our exploration of the potential for integrating fuzzy numbers into the classical MAVF additive model, by introducing them initially only into the scoring procedure.

It is also worth mentioning that the introduction of fuzzy sets for the expression of preference judgements and weights in the classical MAVF additive model has two practical implications [13, pp. 272; 25; 26, pp.136 & 27]. First of all, the aggregation procedure for the calculation of the overall attractiveness of the alternatives has to be modified, since manipulation of fuzzy data is involved. Secondly, the alternatives' overall attractiveness is in the form of a fuzzy set as well. Consequently, the question of the comparison and ranking of fuzzy preference judgements or -as we will refer to it- the "defuzzification procedure" has to be addressed.

Considering the question of the defuzzification: since the early '70s, many authors have addressed the issue of the comparison and ranking of fuzzy numbers (for a comprehensive review of defuzzification procedures see Cheng & Hwang [3, Chapter 4]). However, we have fundamental objections to the rationale of using such processes in this context. First of all, we feel that leaving some ambiguity in the final results gives to the DM a sense of "being in control" over the decision that needs to be made. Secondly, it seems inconsistent to use fuzzy notions in order to grasp and express the ambiguities involved in human decision making processes and surrounding "real-world" decision making problems and, subsequently, to "defuzzify" the results at the end of the analysis. Thirdly, it could be rather annoying for the DMs to be submitted to the process of the elicitation of fuzzy scores in order to incorporate into the model a richer representation of the situation at hand and, then be presented with an answer that suppresses all the information into a single value. Finally, it opposes the nature of decision analysis itself, as a dynamic learning process [7, pp.342; 8, pp.54 & 28, pp.1].

In other words, is it possible for a crisp value to be representative of a situation characterised by ambiguity? Are the defuzzified results reflecting the values of the DM? Is it not contradictory to the aims of decision analysis to look for a specific answer to the problem? And, most importantly, would the use of such a procedure essentially exclude the DM from the analysis of a problem that he/she owns and absolve him/her of any responsibilities for the decision made?

4 The Implementation: Fuzzy V •I •S •A

V•I•S•A stands for Visual Interactive Sensitivity Analysis and it is a Multi-Criteria DSS based on the MAVF additive model (see Section 2). For a discussion on the motivation for the development of V•I•S•A, its use and its strengths and weaknesses the reader is referred to Belton & Vickers [29].

The fuzzy MAVF additive model described on Section 3 was implemented in V•I•S•A by incorporating triangular and trapezoidal fuzzy numbers into the existing classical scoring procedure of the software. Two graphic displays were designed for the interactive scoring of the alternatives' performance.

The first is the ***Fuzzy Bar Chart*** (see Figure 2). On this graphic display, the classical Bar Chart has been modified and fuzzy scores (for all the alternatives on one of the criteria) are represented by "fuzzily" coloured bars. The solid coloured part of the bar represents the modal interval of a trapezoidal fuzzy score or the modal value of a triangular fuzzy score. The non-solid part represents the support of the fuzzy scores, excluding their modal values. The fuzzy scores can be entered or modified by pointing with the cursor on one of the four point-values and dragging it up or down.

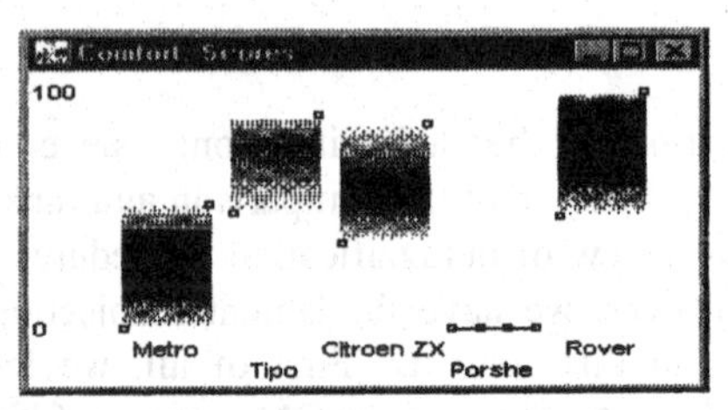

Fig.2 Fuzzy Bar Chart

The second is the ***Individual Fuzzy Score Graph*** (see Figures 3 & 4). On this display, the graph of the membership function of the fuzzy score (for one of the alternatives on one of the criteria) can be "drawn" from the beginning or modified, by pointing with the cursor to the point-values and dragging them left or right to the desired level.

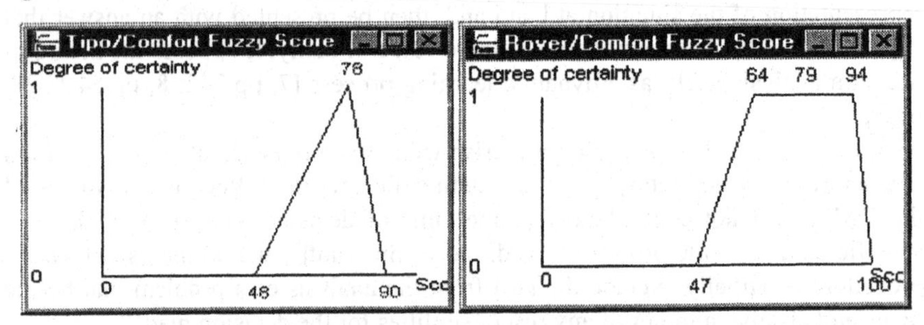

Fig.3 Individual Fuzzy Score Graph (triangular)

Fig.4 Individual Fuzzy Score Graph (trapezoidal)

Finally, the ***Overall Fuzzy Score Graph*** (see Figure 5) is an alternative visualisation of the information presented on the Fuzzy Bar Chart (Figure 2), where the Individual Fuzzy Score Graphs for all the alternatives on one of the criteria are synthesised in a single display. Note that in V•I•S•A each alternative is assigned a different colour and, hence, the display shown on Figure 5 is easier to see and understand.

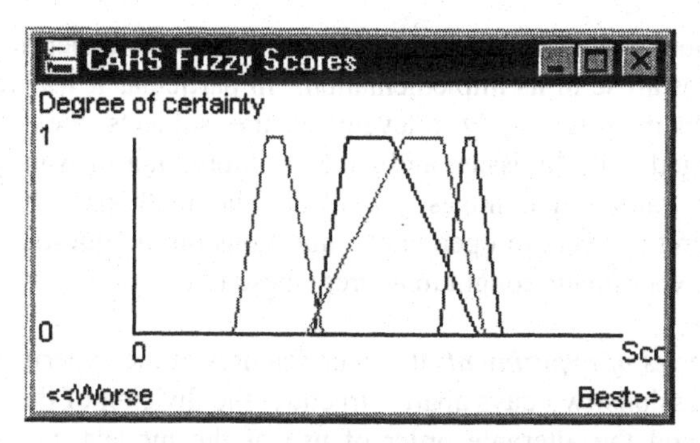

Fig.5 Overall Fuzzy Score Graph

It should also be mentioned that the software allows the exploration of the effect of different weighting strategies and, in a way, compensates for the partial fuzzification of the model, by permitting the interactive modification of the weights [30].

5 The Experimental Work

In order to compare the classical and the fuzzy models (Sections 2 &3) and elicit DMs views on their use, two series of experiments, using V•I•S•A and F-V•I•S•A, were performed .
More specifically, these experiments were aimed at:

- The assessment of the closeness of the results of the classical and the fuzzy analyses to the DMs' gut-feelings, together with the DMs satisfaction with the outcome of the analysis and their confidence on it.
- The assessment and comparison of the models' value for reflecting on MCDM problems; exploring and clarifying muddled issues, perplexing thoughts and conflicting feelings; and, balancing preferences and beliefs.
- The evaluation of the models' "user-friendliness" and practicality and their overall value as decision making aiding tools.

The ***first series of experiments*** involved volunteer graduate and MSc students from the Department of Management Science of the University of Strathclyde. The experiments were structured in two sessions that were two days apart. The group of subjects was equally divided into two subgroups; one subgroup using V•I•S•A in the first session and F-V•I•S•A in the second, and vice versa for the other subgroup. The experimental task was the selection of a job from a collection of four alternatives (different for each session). A list of relevant criteria from which the DMs could choose or to which they could add was provided. For the collection of the feedback, a questionnaire -complemented with comparative questions for the second session- was used.

However, although careful attention had been paid to the experimental design, a number of issues arose in its implementation. In particular it was felt that even though the problem was highly relevant to the subjects they did not get sufficiently engaged with the issue, and gave an impression of wanting to please the experimenter rather than honestly evaluate the methods. It also proved difficult to interpret answers to open ended questions on the questionnaires. The second series of experiments sought to address these issues.

For the ***second series of experiments*** the main features of the experimental design, that is the "two sessions-two days apart" structure, the division of the subjects into two sub-groups and the alternate order of use of the models, remained intact. However, the sessions, involving PhD students from the Department of Management Science of the University of Strathclyde, were one-to-one and included the structuring of the problem that the subjects chose to work on from a selection of MCDM problems. For the collection of the feedback, the Thinking-Aloud procedure was used (for more details on the technique see Svenson [31]; for a review on process-tracing techniques in studies of the decision making processes, see Svenson [32] as cited by Dahlstrand & Montgomery [33]). The use of the Thinking-Aloud procedure was complemented by a short interview focused on interesting issues that emerged during the session.

The new experimental design proved to be more effective. The subjects were more engaged in the process and showed genuine interest. Additionally, the use of the Thinking-Aloud procedure -although intimidating for some of the subjects and quite demanding for the analyst- gave a richness of data that would have been difficult -if not impossible- to acquire through the use of questionnaires.

6 The Results

We begin with some general observations from the second series of experiments. First of all, the subjects assigned the scores interactively using the visual displays provided by the softwares (see [30] & Section 4), rather than "directly" giving numerical values. In the case of F-V•I•S•A, they made use of the Fuzzy Bar Charts rather than the Individual Fuzzy Number Graphs (Section 4). Most importantly, they seemed to have difficulties in grasping the meaning of fuzzy scores and understanding the rationale (principle, logic) behind them.

In general the subjects considered the use of ***fuzzy scores*** to be valuable for the portrayal of reality and useful for the assessment of alternatives surrounded by ambiguity, by allowing the expression of these ambiguities and their incorporation into the model. On the other hand, some subjects found the fuzzy scoring procedure more time consuming than the classical one. They also expressed concern that the incorporation of vagueness in the final outcome could be counterproductive, possibly prompting further analysis, possibly indermining the credibility of the approach in the eyes of the DM. Another concern was that allowing the DM to express his/her views in a vague manner might nurse indecisiveness. Finally, they found the use of fuzzy scores more difficult than that of the crisp ones in that it involves the evaluation of alternatives with respect to two dimensions (instead of one) and requires the introduction of three or four values (instead of one) for the determination of a single score.

Concerning the use of ***crisp scores***, the subjects felt that it is limiting in that it forces the representation of the DM's thoughts and feelings with a single value. This was viewed as a rather simplistic way of describing reality, leading to assumptions and over-simplifications.

A very interesting observation was that those subjects who were familiar with MCDA notions before the experiments, recognising the advantages of the approach for the analysis of ambiguous situations, were in favour of a selective application of fuzziness into the model, whereas the novices in the area were wholeheartedly in favour of it, feeling that it will better capture reality.

7 Reflections

As we have already mentioned in the previous section, one of our first observations during the course of the experiments was the adversity the subjects exhibited towards the use of numerical values, together with a feeling of comfort from the use of fuzzy scores for the articulation of their preference judgments. However, although it is widely accepted that the quantification of preferences is a very difficult and painful process, the belief that the use of fuzzy scores instead of crisp ones can relieve this discomfort is ill-founded and potentially misleading, given that fuzzy scores are still precisely defined by numbers. As Simon French wrote [7, pp.369 & 34, pp. 41]: "There is no panacea to be found here (in fuzzy decision analysis) for analysts who dislike setting numerical values".

It emerged from the experiments that fuzzy scores were considered a useful tool for the description of reality and for the representation of one's thoughts and feelings. Nevertheless, placing the emphasis on the modelling rather than the resolution of the ambiguities involved in any decision making process, contradicts the nature of decision analysis as a "facilitative" analysis [7, pp. 369; 35, pp. 8 & 34, pp.41]. Moreover, it assumes that the modelling of reality using fuzzy scores is useful for the moulding of a better informed decision and that the DM's thoughts and feelings are well represented by means of membership functions [7, pp.363; 34, pp.33 & 35, pp.3]. However, there is no indication that the first of these assumptions is true [34, pp.38]. Similarly, there is no indication that the human perception of ambiguity is adequately captured by the notion of fuzzy sets [34, pp.33]. Furthermore, the question of whether or not the human way of classifying and reasoning and the human decision making processes are fuzzy, is still open to debate.

Considering the meaning of membership functions as expressions of preference judgments, we, as our subjects did, have difficulty to comprehend what fuzzy scores mean and what they describe. Do they describe the DM's certainty about or confidence in his/her preferences, the performance of the alternatives under evaluation, or, both? The fact that our subjects considered fuzzy scores useful for the description of reality whereas, at the same time, they found it difficult to understand their meaning, raises our doubts concerning their proper use and emphasises our conviction that the feeling of ease that fuzzy scores nurse is ill-founded and misleading. Furthermore, it is unclear to us how fuzzy scores can be elicited or constructed [36, pp.131 & 35, pp.7]. That is, what questions we –as analysts- have to ask? For example, should the fuzzy scores be formulated first by defining their support and then their modal values or vice versa? Finally, it is rather contradictory for an approach that claims to give freedom and flexibility of expression to the DMs to require a larger volume of input data and be considered by the DMs more difficult and more time consuming than the classical one.

As for the results of the analysis using fuzzy scores, our subjects claimed that they reflected their feelings, but does this mean that the process of expressing their preference judgments in the form of membership functions did help them or would have helped them in a "real-world" situation to make a better informed decision and commit themselves to a specific course of action? [7, pp.367]. Or, as some of our subjects pointed out, might it have nursed their indecisiveness or impaired their trust in the approach?

The questions concerning both computational and conceptual issues related to the introduction of fuzzy numbers into the classical MAVF additive model are still open. Despite that, we feel that there are prospects for a synergy between the two theories, MAVF Theory and Fuzzy Set Theory. We believe that the use of fuzzy numbers for the expression of ambiguous preference judgments or the assessment of alternatives' performance with respect to qualitative criteria or alternatives' performance surrounded by uncertainty might add to the potential of MAVF theory as a decision aiding tool. Towards that direction, we think that it is worthwhile investigating the possibilities of a selective introduction of fuzziness for the performance of sensitivity analysis of results obtained using the classical model, along the lines of Lootsma's suggestion [37, pp.99]. We are also considering the exploration of the development of a conditional "defuzzification" procedure. We certainly feel that this work is only the beginning rather than the end of our investigation into the potential and the implications of integrating Fuzzy Set Theory notions into MAVF Theory.

References

[1] Terano, T.; Asai, K. & Sugeno, M. (1992): "Fuzzy Systems Theory and its Applications". *Academic Press, Inc.*

[2] Bellman, R.E. & Zadeh, L.A. (1970): "Decision – Making in a Fuzzy Environment". In *Management Science, Vol. 17, No. 4, pp. B141-B164.*

[3] Chen, S.-J. & Hwang, C.-L. (1992): "Fuzzy Multiple Attribute Decision Making: Methods and Applications". *In collaboration with Hwang, F..P. Springer-Verlag; Berlin Heidelberg.*

[4] Von Winterfeldt, D. & Edwards, W. (1986): " Decision Analysis and Behavioral Research". *Cambridge University Press.*
[5] Bonissone, P.P. (1980): "A Fuzzy Sets based Linguistic Approach: Theory and Applications". In *Proceedings of the 1980 Winter Simulation Conference; Orlando, Florida. Ören, T.I.; Shub, C.M. & Roth, P.F. (eds.).*
[6] Bonissone, P.P. (1982): "A Fuzzy Sets based Linguistic Approach: Theory and Applications". In *Approximate Reasoning in Decision Analysis; Gupta, M.M. & Sanchez, E. (eds.). North-Holland Publishing Company.*
[7] French, S. (1986): "Decision theory: An Introduction to the Mathematics of Rationality". *Ellis Horwood Limited; Chichester, UK.*
[8] Belton, V. (1990): "Multiple Criteria Decision Analysis - practically the only way to choose". In *Operational Research Tutorial Papers: 1990; Hendry, L.C. & Eglese, R.W. Operational Research Society, Birminghman.*
[9] Belton, V. (1997): "Implementing Multicriteria Decision Analysis". In *Management Science: Theory, Method and Practice Series; working paper no. 97/10.*
[10] Baas, S.M. & Kwakernaak, H. (1977): "Rating and Ranking of Multiple-Aspect Alternatives using Fuzzy Sets". In *Automatica, Vol. 13, pp. 47-58.*
[11] Kahne, S. (1975): "A Procedure for Optimising Development Decisions". In *Automatica, Vol. 11, pp. 261-269.*
[12] Kahne, S. (1975): "A Contribution to Decision Making in Environmental Design". In *Proceedings of IEEE, Vol. 63, pp. 518-528.*
[13] Zimmermann, H.-J. (1991): "Fuzzy Set Theory and its Applications". *Kluwer Academic Publishers, Boston.*
[14] Dubois, D. & Prade, H. (1980): "Fuzzy Sets and Systems: Theory and Applications". *Academic Press, Inc.*
[15] Jain, R. (1977): "A Procedure for Multiple Aspect Decision Making". In *International Journal of Systems Sciences, Vol. 8, pp.1-7.*
[16] Cheng, Y.M. & McInnis, B. (1980): "An Algorithm for Multiple Attribute, Multiple Alternative Decision Problem based on Fuzzy Sets with Application to Medical Diagnosis". In *IEEE Transactions on Systems, Man and Cybernetics, Vol. SMC-10, pp. 645-650.*
[17] Dubois, D. & Prade, H. (1982): "The Use of Fuzzy Numbers in Decision Analysis". In *Fuzzy Information and Decision Processes. Gupta, M.M. & Sanchez, E. (eds.). North Holland Publishing Company.*
[18] Dong, W.M.; Shah, H.C. & Wong, F.S. (1985): "Fuzzy Computations in Risk and Decision Analysis". In *Civil Engineering Systems, Vol. 2, pp. 201-208.*
[19] Dong, W.M. & Wong, F.S. (1987): "Fuzzy Weighted Averages and Implementation of the Extension Principle". In *Fuzzy Sets and Systems, Vol. 21, pp.183-199.*
[20] Tseng, T.Y. & Klein, C.M. (1992): "A New Algorithm for Fuzzy Multicriteria Decision Making". In *International Journal of Approximate Reasoning, Vol. 6, pp. 45-66.*
[21] Ribeiro, R.A. (1996): "Fuzzy Multiple Attribute Decision Making: a review and new preference elicitation techniques". In *Fuzzy Sets and Systems, Vol. 78, pp. 155-181.*
[22] Kaufmann, A. & Gupta, M.M. (1988): "Fuzzy Mathematical Models in Engineering and Management Science". *Elsevier Science Publishers B.V. (North Holland).*
[23] Chang, P.-T., & Lee, E.S. (1994): "Fuzzy Arithmetics and Comparison of Fuzzy Numbers". In *Studies in Fuzziness; Fuzzy Optimization: Recent Advances. Delgado, M.; Kacprzyk, J.; Verdegay, J.-L. & Vila, M.A. (eds.). Physica-Verlag.*

[24] Negi, D.S. (1989): "Fuzzy Analysis and Optimisation". *PhD Dissertation; Kansas State University, Department of Industrial Engineering, Manhattan, Kansas.*

[25] Zimmermann, H.-J. (1986): "Multi Criteria Decision Making in Crisp and Fuzzy Environments". In *Fuzzy Sets: Theory and Applications; Jones, A, et al. (eds.). D. Reidel Publishing Company.*

[26] Zimmermann, H.-J. (1987): Fuzzy Sets, Decision Making and Expert Systems. *Kluwer Academic Publishers, Boston.*

[27] Werners B. & Zimmermann H.-J. (1989): "Evaluation and Selection of Alternatives considering Multiple Criteria". In *Lecture Notes in Engineering: Expert Systems in Structural Safety Assessment; Jovanovic A.S., Kussmaul, K.F., Lucia A.C., Bonissone P.P. (eds.). Proceedings of an International Course, Stuttgart, FRG. Springer-Verlag.*

[28] French, S. (1991): "Recent Mathematical Developments in Decision Analysis". In *IMA Journal of Mathematics Applied in Business and Industry, Vol. 3, pp. 1-12.*

[29] Belton, V. & Vickers, S. (1990): "Use of a Simple Multi-Attribute Value Function incorporating Visual Interactive Sensitivity Analysis for Multiple Criteria Decision Making". In *Readings in Multiple Criteria Decision Making; Bana e Costa, C. A. (ed.). Springer-Verlag, Berlin.*

[30] Visual Thinking International Ltd. (1996): "V•I•S•A for Windows User Manual".

[31] Svenson, O. (1989): "Eliciting and Analysing Verbal Protocols in Process Studies of Judgment and Decision Making". In *Process and Structure in human Decision Making; Montgomery, H. & Svenson, O. (eds.). John Wiley & Sons Ltd.*

[32] Svenson, O. (1979): "Process Descriptions of Decision Making". In *Organizational Behavior and Human Performance, 23, pp. 86-112.*

[33] Dahlstrand, U. & Montgomery, H. (1989): "Information Search and Evaluative Processes in Decision Making: A Computer-based Process-tracing Study". In *Process and Structure in human Decision Making; Montgomery, H. & Svenson, O. (eds.). John Wiley & Sons Ltd.*

[34] French, S. (1984): "Fuzzy Decision Analysis: Some Criticisms". In *TIMS / Studies in the Management Sciences, Vol. 20, pp. 29-44. Elsevier Science Publishers B.V. (North Holland).*

[35] French, S. (1989): "Fuzzy Sets: the Unanswered Questions". *Research Report Series: Report No 89.6. University of Leeds, School of Computer Studies.*

[36] French, S. (1989): "On being Numerate". *Inaugural lecture on 4 December 1989, when the author was appointed to the Chair of Operational Research and Information Systems.*

[37] Lootsma, F.A. (1997): "Fuzzy Logic for Planning and Decision Making". *Lecture notes a197; Delft University of Technology, Faculty of Mathematicaand Informatics; The Netherlands.*

Problems of Measurement in Decision Analysis

Oleg I. Larichev

Institute for Systems Analysis, Russian Academy of Sciences, 9, Moscow, pr. 60 let Octjabrja, Russia, 117312.

Abstract. The problem of measurement is one from most important in the decision analysis. The human being is indeed the central figure in the process of decision making because the information elicited from the human beings (decision makers, experts) underlies comparison of the alternatives. Therefore, the procedures of measurement (information elicitation) in decision methods must correspond to the possibilities and limitations of human information processing system. The crucial question is: to what extent is the human being as a measurement device accurate and reliable? Psychological experiments prove that these "devices" have specific characteristics that must be taken into account. The evidence is those "devices" have specific features that are to be taken into account. The qualitative verbal measurements as a line of research is proposed.

Keywords. Preference elicitation, verbal measurements, comparison of decision support systems.

1 Introduction

For typical problems of operations research, the information required to solve them is given in the formulation of the problem To solve the decision making problems, on the contrary, one need information provided by the Decision Makers (DM) and experts.. That is why, any decision making method has built-in procedures for elicitation of information from DMs and experts. For example, the experts could measure the quality of alternatives on criteria; the DMs could measure the relative importance of criteria and so on.

In various procedures of information elicitation human beings play the role of measurement devices. By drawing on a technical analogy, these measurement devices can be said to provide an additional information required for elimination of an uncertainty and making a decision.

In this connection, one could formulate the following questions:

-"What are the characteristics of the human being as a measurement device?"

-"To what extent one could rely upon the information provided by such measurement devices?"

-"Is it possible to develop new decision making methods that take into account the characteristics of such measurement device?"

These questions are of great importance. The studies of many psychologists which stem from brilliant papers of A.Tversky, P.Slovic, B. Fischhoff and others (see the summary in Kaheman et al., 1982) demonstrated that the human information processing system is of limited capacity. We have currently a lot of evidence that in some operations of information processing people make errors and run into contradictions. That is why, the decisions based on unreliable information are unreliable as well.

Do we have the way to elicit reliable information from people?

2 Verbal or numerical measurement?

People made decisions many centuries before the appearance of the decision theory. They made measurement of uncertain factors influencing the decisions .The method of such measurement was quite different from that required by the majority of decision making methods. In the typical situation of a choice people expressed their evaluations in verbal form. I would like to stress that in many everyday situations they do it in the same way.

On the contrary, the great majority of decision methods and decision support systems are based on quantitative evaluations. In a popular textbook for businesspersons I met the following sentence : " Mr. X predicted the possibility of bankruptcy for his company as 50% ." I could not imagine that in real life a businessperson could give the estimation of the situation in such way. One could say: "The situation for my company worsens." or: " I really lost a big part of my capital.", or even express the unhappiness in dirty words. Fifty percents, what does it mean? Who can communicate in this way in such a situation?

It is quite clear why numerical information is needed. The majority of decision methods are of mathematical origin. The numbers are very convenient for any transformation. That is why, the presumption that people could feed numbers in decision methods and in computers is also very convenient for researchers.

Verbal information is less convenient for such purpose. However, computer can use verbal variables as the symbols (Simon, 1996). Logical transformation of symbols is possible, as well.

The question is: "To which extent human numerical and verbal measurements are reliable and do we have scientific confirmation of the intuitive guess of bigger reliability for verbal evaluations?"

3 Verbal and numerical probabilities

The famous book of J.von Neuman and O. Morgenstern (1953) provides the fundamentals of the decision theory. According to this book, two main parameters in the scientific representation of a decision are the probability and utility. The subjective probabilities are now very important parameters for a good many decision methods.

That is why the great attention was paid to comparative studies of numerical and verbal probabilities.

In which forms prefer people give and receive the evaluations?

In the experiments with 442 respondents it was found (Wallsten et al., 1993) that 65% of respondents prefer to give the evaluations of uncertain factors in verbal form but 70% prefer to receive this information in quantitative form. This phenomenon has received the name of communication mode preference paradox - CMP. Let us give an illustrative example (Erev, Cohen, 1990).

Let us suppose that there are three officers of different levels from intelligence service who are to give the most informative data to the decision maker. In the accordance with CMP paradox, the officer of middle level prefers to get information from his subordinate officer in quantitative form but gives his information to the officer of upper level in verbal form.

There is the example taken from real life. US national Security Council prepared in 1981 for President R. Reagan the list of five possible operations against Libia. The chances for the success were evaluated in verbal form (Hamm, 1991).

Comparison of pro and contra

There are several reasons for people to give an information in verbal form:

1. People use verbal way of communication much easier than numerical one. The probability theory appeared in many thousand years after development of languages.
2. Words are perceived as more flexible and less precise, with various communicative functions (Teigen, 1988), and therefore seem better suited to describe vague opinions and characterize imprecise beliefs. In the work of Erev and Cohen (1990) there is the statement that " forcing people to give numerical expressions for vague situations where they can only distinguish between a few levels of probability may result in misleading assessments".

However, there are positive factors in utilization of quantitative form of information:

1. People attaches the degree of precision, authority and confidence to numerical statements that they do not ordinarily associate with verbal ones.
2. It is possible to use quantitative methods of information processing (for example, Bayes theorem).

Experimental comparison

The group of researchers headed by Prof. T.Wallsten was undertake many experiments to compare different forms of expressing uncertainty: verbal and quantitative ones.

Usually, an experiment had consisted in some comparisons between lotteries with outputs having verbal of quantitative probabilities. Let us stress that verbal probabilities had the form of labels consisting from one or two words.

Then verbal probabilities have been transformed into quantitative ones by different ways (the scale of the correspondence between verbal expressions and numbers given by a subject or by membership functions). The comparison usually consisted in finding differences between gains in a game achieved by a subject in the dependence from utilization of verbal or numerical probabilities.

A subject was usually a given scale of verbal definitions of probability or he /she could construct such scale selecting some verbal definitions of probability.

In some experiments one subject constructed a scale, made a measurement and second one received from the first an information and made a decision (dyadic decisions; Budescu, Wallsten, 1995).

The experiments demonstrated no essential difference in the profit received by subjects or in the accuracy of evaluations. The conclusion was usually made: the convenience of utilization of verbal expressions is more attractive than a difference in 1-4% of a profit. (Hamm, 1991).

The only strong and evident difference consisted in the number of preference reversals.

Preference reversal

Well known prominence effect consists in the following: people choose one alternative over another according only to the primary dimension (or first criterion) when neither alternative has a clear advantage on secondary dimension (Tversky et al., 1988). Let us recall that in the experiments of A.Tversky et al. subjects were asked to choose one from two pairs of approximately equal expected -value bets. One bet had a higher probability of winning and other bet promised more to win. Subjects were also asked to bid for the bets offered separately. The probabilities were given in numbers.

The results demonstrated that in choice situation bits with higher probability were chosen and bets with bigger profit received higher bids. It is the violation of expected utility theory.

G.Gonzalez-Vallojo and S. Wallsten (1992) repeated the same experiments but verbal labels (more or less probable and so on) expressed the probabilities in bets. The frequency of predicted reversal was significantly less in the verbal as opposed to the numerical display mode.

Comparative verbal probabilities

Some experiments focused on the relationship between the form of communication and degree of indefiniteness of events (Erev, Cohen, 1990). For example, the subjects were asked to estimate the chances of basketball teams to win in games between them. The experimenters noticed that in case of an unknown team (higher indefiniteness) the experts were able to discriminate only two levels of verbal probabilities in comparative forms — for example, 'it is believed that the host always play better than the guests'.

It were stated (Erev, Cohen, 1990) that compelling people to quantitative probability estimates in situations where only a few levels of indefiniteness can be discriminated can result in erroneous estimates. This example shows that some measurements can be carried out only in verbal form with the use of `more probable than' relationships.

Methodical studies of comparative probabilistic estimates (Huber and Huber, 1987) demonstrated that comparative probabilities are much more frequently used by the common people (both adults and children) than quantitative estimates of probabilities of events. The experiments used tasks such as estimation of the probabilities of hitting the sectors of a rotating disk and estimating the winners in competitions and games. The authors of these studies formulated six mathematical principles for comparative probabilities in form of axioms representing the mathematical concept of qualitative probabilities (for example, the transitivity of relations).

The main experimental result obtained with adults and children (above five years old) is as follows: the human comparisons follow completely the principles of the mathematical theory of qualitative probabilities. The authors of these studies concluded that the six principles provided the more reliable foundation for describing the human behavior than the laws of quantitative probability.

4 Human being as measurement device

Methodical studies of probability elicitation give some information about the human being as measurement device. There are other experiments demonstrating the limited accuracy of quantitative measurements for human beings.

The famous experiment of A.Tversky (1969) demonstrated that people neglect small differences in their evaluations which accounts for intransitive behavior in some problems of choice. The inability to take into account the small differences between evaluations leads to elimination of dominating alternatives and retention of the dominated ones (Korhonen et al., 1997).

As was established by psychologists, the human being is not accurate measurement device producing quantitative measurements. It is well known that the accuracy of physical measurements depends on the precision of instrumentation. The same holds for human measurements.

The question is: "To what extent such "inaccuracy" influences the output of decision methods?"

5 Verbal or numerical measurement in decision making methods

To make valid the comparison of verbal and numerical measurements, one must use each kind of measurements at all stages of the decision making method.

The measurements in the decision tasks are closely related with the construction of a decision rule. For various kinds of information, one must provide different methods of decision rules construction, which means that for verbal information given one must provide a special qualitative form of a decision rule construction. Only then the comparisons of the efficiency (verbal vs. numerical) could be valid.

The decision problem

The experimental study was done to compare verbal and numerical methods of decision making (Larichev et al., 1995). The subjects were college students of Texas A&M University nearing graduation, who were in job search process, facing opportunities similar to those given in the study.

Let us suppose that a college graduate has several offers (after interviews) and he (or she) is to make a decision. Every variant is acceptable, but of course, one variant is better upon one aspect and the other - upon the other. So, the student has the multicriteria problem. The student was asked to solve it with the help of the appropriate multicriteria method.

Let there be Q criteria, upon which N alternatives are evaluated. Each alternative a_i (i=1,2, ..., N) corresponds to the vector $a_i = (a_{i1}, a_{i2}, .. a_{iQ})$.

Four criteria are used as the focus for the study: *salary, job location, job position* (type of work involved), and *prospects* (career development and promotion opportunities). The following alternatives were used:

FIRM	SALARY	JOB LOCATION	POSITION	PROSPECTS
a1	\$30 000	Very attractive	Good enough	Moderate
a2	\$35 000	Unattractive	Almost ideal	Moderate
a3	\$40 000	Adequate	Good enough	Almost none
a4	\$35 000	Adequate	Not appropriate	Good
a5	\$40 000	Unattractive	Good enough	Moderate

It is easy to note that in this case there are three possible values upon each criterion. The greater the salary, the more attractive it would be to a rational subject. Thus, we have four criteria with three possible values each and the values upon each criterion are rank-ordered from the most to the least preferable one.

It is evident, that there are no dominated alternatives. Therefore, comparison of these alternatives requires some value function, which would take into account the advantages and disadvantages of each alternative upon each criterion.

Two decision support systems based on numerical measurements

Two decision support systems based on Multiattribute Utility Theory (MAUT- Keeney and Raiffa, 1976; Keeney,1992) were used for the solution of the problem given above. These systems are LOGICAL DECISION (Smith and Speiser, 1991) and DECAID (Pitz,1987). The third DSS was one based on Verbal Decision Analysis (see below).

Both decision support systems LOGICAL DECISION and DECAID were used to solve this task. Both systems implement ideas of multiattribute utility theory, providing possibilities for construction of an additive utility function for the case of risky decisions, and additive value function for decision making under certainty. In our study, we used only additive value functions.

The value function obtained from both systems would therefore have the linear form,

$$v(a) = \sum_{i=1}^{Q} k_i v_i(a_i)$$

where: a is an alternative ,estimated over each of the Q criteria , k_i is the coefficient of importance for the i-th criterion , a_i is the value of alternative a on criterion i , and v_i is the value function for the i-th criterion.

Both systems are easy to use, have flexible dialogue and graphical tools to elicit decision maker's preferences.

The main difference in the systems (besides interface) is the way of determination of numerical values upon separate criteria and criteria weights. In DECAID pure graphical (direct) estimation is used (a point on the line of the size 1). In LOGICAL DECISION there is a possibility to use special function for criterion values. To determine the parameters of this function it is enough to mark the "middle" value for the criterion (sure thing for a lottery with 50% possibility for the best and the worst variant).

Criteria weights are also defined in a different manner in these two systems. In LOGICAL DECISION criteria weights are defined on the basis of trade-offs in a rather traditional way (Keeney, 1992). In DECAID weights are elicited directly (in a graphical way - point on a line), though the system provides also the possibility to make trade-offs, but after that the result is presented as points on lines. Thus, it is possible to consider it as direct elicitation of criteria weights.

Taking into account the commonness of the approach implemented in both systems and also the similarity of information, received from a DM in the process of task solution, the attempt to solve the above described task with the help of these systems must lead to the same result.

Decision Support System ZAPROS

The third DSS is one from the family of Verbal Decision Analysis (Larichev and Moshkovich, 1997). Only verbal measurements are used on all stages of this method. ZAPROS uses ranking rather than rating information, but the additive overall value rule is correct if there is an additive value function. In ZAPROS the additive rule does not provide the summation of values, but rather the means of obtaining pair-wise compensation between components of two alternatives.

The following procedure was used for the preference elicitation from subjects.

Subjects were asked to compare several specially formed alternatives by pairs. For each pair two alternatives were different only on two criteria evaluations (one evaluation was best for each alternative) and had equal evaluations (best or worst) on other criteria.

For the task presented above it was necessary for subjects to compare the pairs of alternatives different on each pair of four criteria. The example of typical question is:

«What do you prefer: the firm giving salary $ 40 000 with adequate location or the firm giving salary $35 000 with very attractive location? Please, take into account that on the criteria «Position» and «Prospects» both firms are good.».

While comparing these alternatives, subjects were to choose one of the following responses:

alternative 1 is more preferable than alternative 2;

alternative 2 is more preferable than alternative 1;

alternative 1 and 2 are equally preferable.

Implementation of such simple system for comparison of pairs of alternatives gives us a possibility for simple check of the received comparisons on the basis of transitivity:

if $a > b$ and $b > c$, then $a > c$;

if $a > b$ and $b = c$, then $a > c$;

if $a = b$ and $b = c$, then $a = c$;

where a, b, c -alternatives, symbol > means more preferable and symbol = - equally preferable.

The method also provides verification of the received comparisons for transitivity and allows change some of the responses on the request of the user to eliminate intransitivity. It also guarantees that comparison of each pair of alternatives from this set is supported by at least two responses of the user.

Let us note that such way of preference elicitation is psychologically valid (Larichev,1992). The received information allows one to build the joint ordinal scale combining all evaluations on separate criteria scales. The joint ordinal scale provides the possibility for the construction of partial ranking for every given set of alternatives.

Thus, this ranking may be used for comparison of initial 5 alternatives because in our task additive value function is supposed to be the right one and criteria were formed to be preferentially independent. This algorithm does not guarantee comparison of all alternatives because for some pairs of alternatives ZAPROS gives only incomparability relation.

The comparison of three decision support systems

Each subject from the group used all three DSS for the solution of the problem presented above. The difference in the outputs of methods consisted in following: some pairs of alternatives had not been compared with ZAPROS method. Simple method of preferences elicitation used by ZAPROS gave no possibility (in general case) to compare all given alternatives. ZAPROS gave only partial ranking of alternatives.

In the difference from it, two other methods gave the complete ranking for given alternatives. Also, LOGICAL DECISION and DECAID gave numerical values of the utility for all alternatives.

The results of the experiment were analysed in different form: the ranking of given alternatives, the ranking of special alternatives used in ZAPROS, the ranking of criteria weights and so on.

First of all, it was found very low correlation between the outputs of LOGICAL DECISION and DECAID. The ANOVA test demonstrated that for the group of subjects the outputs of LOGICAL DECISION and DECAID have not been statistically significant in measurements of criteria weights and ranking of alternatives.

The following results were very interesting: the outputs of pairs LOGICAL DECISION-ZAPROS and DECAID- ZAPROS were correlated and it was statistically significant. It means that only for alternatives compared by ZAPROS the relations were essentially the same.

It is possible to give the following explanation to the results.

The alternatives that could be ordered by ZAPROS are in the relations closed to ordinal dominance. Such relations are more stable. More, they were constructed in very reliable way: verbal measurements, psychologically correct way of preference elicitation, possibility to check information and eliminate contradictions.

Two complete orders constructed by LOGICAL DECISION and DECAID were based on numerical measurements and weighted sum of alternatives estimations by criteria. The difference in the utility (even small) defined the final order of alternatives. The errors (even small) made by people while performing numerical measurements resulted in quite different orders of alternatives.

6 Instrumental analogy

Imagine that you have N unreliable balances weighing with great and poorly predictable errors. For example, you put the same object on balance several times and each time get different readings, that is, you cannot know which reading is true.

Let there is another, (N+1)-t balance also measuring inaccurately. We weigh on it the N balances as objects. Now, let us assume that we have K objects and want to 'evaluate' them by weighing each of them on the balances, multiplying each measurement by the weight of the *I*-th balances, and summing the results. The sum of N weights for us is the 'value' of each object. By comparing these 'values' we determine the object which is most valuable to us. Obviously, for balances with substantial errors the results of comparisons can be valid only if one object weighs on each balance much more than another. In all other cases the results of such weighing are questionable.

Unfortunately, we cannot improve our balances, but can 'calibrate' them before weighing by means of precise balance weights, which are weighed on the imperfect balances, the results being laid off on a scale. Thus, the balances are 'marked'. One can readily see that the results of measurements based on this scale are incontrovertible only if one object is much heavier than another. In all other cases the results are questionable.

Now, we can give the explanation of the above example with balances. The inaccurate balances (measurers) are person(s) evaluating objects by N criteria and establishing the importance indices ('weights') of these criteria. It will be recalled once more that it is measurement of purely subjective factors that is concerned.

There exist numerous experimental proofs of the inaccuracy of human measurements of subjective factors on continuous scales. We mention again the inaccuracies in measuring subjective probabilities and in assigning weights of criteria (Borcherding et al., 1994).

The above measurements with 'calibration,' in our view, correspond to the transition from qualitative scales to quantitative ones by means of establishing a direct correspondence between the qualitative notions and numbers.

7 Sensitivity check

Some researchers believe that accuracy of measurements is inessential in decision making — inaccurate measurements are admissible because checking the result for sensitivity to variations of parameters completes analysis.

Sensitivity analysis is useful if there is a good scenario for it. However, the analysis of differences in several parameters simultaneously is complicated and does not help to DM. Effective, comprehensive and useful sensitivity analysis is quite difficult. It is much more the skill than the science.

Many questions about sensitivity arise when isolating the best alternative in a group. If there are only two criteria, then the problem is still simple enough; but if there are about five to seven criteria, which is the case in many applications, then the sensitivity check for weights and criteria-based estimates becomes a complicated problem.

Let us ask: "What is the aim of sensitivity analysis?" If several alternatives by turns become the best with variations in weights and estimates, what can the analyst tell to the DM? It is difficult to convince the DM that the Pareto-optimal alternatives are close in utilities. The need for reliable measurements brings us to the need for of qualitative measurements in a natural language.

8 Qualitative or verbal measurements

We regard decision making in unstructured problems as the domain of human activity where quantitative (the more so, objective) means of measurement are not developed, and it is unlikely that they will appear in future. Therefore, it is required to estimate the possibility of doing reliable qualitative measurements. Following R. Carnap, we turn to the methods for measuring physical magnitudes that were used before the advent of reliable quantitative measurements. Before the invention of balances, for example, objects were compared in weight using two relationships — equivalence (E) and superiority (L), that is, people determined whether the objects are equal in weight or one is heavier than the other. There are four conditions to be satisfied by E and L (Carnap, 1969):

1. E is the equivalence relationship,
2. E and L must be mutually exclusive,
3. L is transitive, and

For two objects a and b either (i) a E b, or (ii) a L b, or (iii) b L a.

One can easily see that the above scheme enables one to carry out relatively simple comparisons of objects in one quality (weight). It is required here that all objects be accessible to the measurement maker (expert).

Let us consider this example, measurement of weight. By lifting objects, people performed relative measurements based on the binary relationships E and L. At the next step, however, it was required to compare the measurements made by different persons (experts, so to speak) at different times or made by one person for different sets of objects. This became possible when people agreed about common points on the scales of measurement. For weight, for instance, the following points could be defined:

(1) So heavy, that one can hardly lift an object.
(2) Not heavy, that everybody can lift it without difficulty.
(3) So light that everybody can fling it.

We see that these definitions are not precise, but anyway they provide a ground for agreement. Using such definition, we have an absolutely ordinal scale with discrete estimates. Measurement is reduced to classifying objects as belonging to one of the estimates or to an interval between estimates.

Two more remarks are due. It is obvious that the thus-constructed absolute ordinal scale cannot have many values; otherwise, they will be poorly distinguishable by the measurement makers. To come to terms easier, it is required to identify commonly understandable and identically perceived points on the scale and explain their meaning in detail. Therefore, these scales must have detailed verbal definitions of estimates (grades of quality). Moreover, these definitions focus on those estimates on the measurement scale that were emphasized by the persons constructing the scale (for example, they could be interested only in very heavy and very light objects). Thus, the estimates on the ordinal scale are defined both by the persons interested in one or another kind of measurements (in our case, it is the DM) and by the distinguishability of estimates, that is, the possibility of describing them verbally in a form understandable to experts and DMs.

There is no reason to question the fact that before the coming of reliable methods of quantitative measurement of physical magnitudes, they were already measured qualitatively. These methods today could seem primitive because we have much more reliable quantitative methods. Yet, there is no doubt that the pre-quantitative (qualitative) methods of measuring physical magnitudes did exist. When they were superseded by the quantitative methods, they were treated with negligence as something 'unscientific' and obsolete. The progress of physics gave rise to the well-known statement that the science appears wherever the number (quantity) occurs. To our mind, these declarations refer mostly to the natural sciences, but in the sciences dealing with human behavior qualitative measurements were and will be most reliable.

9 Conclusion

We conclude by formulating the following requirements to human measurements in decision processes.

The measurements must be made in a language that is natural to the DMs and their environment.

In the case of quantitative variables (criteria) it is advisable to use discrete scales with the evaluations representing some intervals meaningful for "measurement makers".

The ordinal scales with verbal evaluations are best for measurements.

For the cases with great uncertainty comparative verbal measurements (better, worse, and so on) are the best for eliciting information from human beings.

In the general case, one could take the discrete evaluations on criteria scales as the output of measurement process. Such evaluations frequently have verbal labels or verbal descriptions.

The accuracy (reliability) of measurements in decision making is extremely important because it defines the choice of the best alternative. It is only natural that the decision makers on whom human lives depend want to put questions in a language understandable to them and get unambiguous answers.

The research is partly supported by Russian Foundation for Basic Research, grant 98-01-00086.

References

Borcherding K., Schmeer S., Weber M. (1993). "Biases In Multiattribute Weight Elicitation." *Contributions To Decision Research.*,(Edited by : J.-P. Caverni, M. Bar-Hillel, F. N. Barron, H. Jungermann), North-Holland.

Budescu D., Wallsten T. (1995) "Processing Linguistic Probabilities: General Principles And Empirical Evidence." *The Psychology Of Learning And Motivation,* V. 32, Academic Press.

Carnap R. (1969) "*Philisophical Foundation Of Physics.*", London, Basic Books Inc. Publishers.

Erev I., Cohen B. (1990) "Verbal Versus Numerical Probabilities: Efficiency, Biases, And The Preference Paradox.", *Organizational Behavior And Human Decision Processes,* 45, 1-18.

Hamm R.(1991) "Selection Of Verbal Probabilities Solution For Some Problems Of Verbal Probability Expression" , *Organizational Behavior And Human Decision Processes,*48, 193-223.

Huber B., Huber O. (1987) "Development Of The Concept Of Comparative Subjective Probability", *Journal Of Experimental Child Psychology,* 44, 304-316.

Gonzalez-Vallejo C., Wallsten T. (1992) "The Effects Of Communication Mode On Preference Reversal And Decision Quality", *Journal Of Experimental Psychology: Learning, Memory And Cognition,* 18, 855-864.

Kahneman D., Slovic P., Tversky A. (*Eds.*) (1982*)* *"Judgement Under Uncertainty: Heuristics And Biases.*" Cambridge University Press, Cambridge.

Keeney R. L. (1980) "*Siting energy facilities.*", New York, Academic Press.

Keeney R. L., Raiffa H. (1976) *"Decisions With Multiple Objectives: Preferences And Value Tradeoffs"*, New York, Wiley.

Korhonen P., Larichev O. ,Moshkovich H., Mechitov A., Wallenius J. (1997): "Choice behaviour in a Computer-Aided Multiattribute Decision Task", *Journal of Multi-Criteria Decision Analysis,* v 6., 233-246.

Larichev O. I., Olson D. L., Moshkovich H. M., Mechitov A. I. (1995) "Numerical Vs. Cardinal Measurements In Multiattribute Decision Making: How Exact Is

Exact Enough?" *Organizational Behavior And Human Decision Processes*, 64, 1, 9-21.

Larichev O. I., Moshkovich H. M., (1995) "ZAPROS-LM-A Method And System For Rank-Ordering Of Multiattribute Alternatives". *European Journal Of Operations Research*, V. 82, 503-521.

Larichev O. I. (1992) "Cognitive Validity In Design Of Decision-Aiding Techniques", *Journal Of Multicriteria Decision Analysis*, V. 1, N 3, 127-138.

Larichev O.I., Moshkovich H.M. (1997) *"Verbal Decision Analysis For Unstructured Problems."*, Boston, Kluwer Academic Publishers.

Pitz G.F.(1987) "*DECAID Computer Program.*" Carbondale, IL, University of Southern Illinois.

Smith, G.R., Speiser F.(1991) *"Logical Decision: Multi-Measure Decision Analysis Software."* Golden, CO: PDQ Printing.

Simon H.A.(1996) "Computational Theories Of Cognition", *The Philosophy Of Psychology*.(Edited by W.O' Donohue, R.F. Kitchener) London, Sage Publications.

Teigen K. (1988) "The Language Of Uncertainty", *Acta Psychologica*, 68, 27-38.

Tversky A., Sattach S., Slovic P.(1988). "Contingent Weighting In Judgment And Choice", *Psychological Review*, 95, 371-384.

Tversky A. (1969) "Intransitivity of preferences", *Psychological Review*, 76, 31-48.

Von Neumann J., Morgenstern O. (1947) "*Theory Of Games And Economic Behavior.*" Priceton, NJ, Princeton University Press.

Wallsten T., Budescu D., Zwick R. (1993) "Comparing The Calibration And Coherence Numerical And Verbal Probability Judgments", *Management Science*, 39, 176-190.

MULTICRITERIA DECISION AID TECHNIQUES: SOME EXPERIMENTAL CONCLUSIONS

D.L.Olson

College of Business, Texas A&M University, College Station, TX 77843

A.I.Mechitov, H.M.Moshkovich

College of Business, University of West Alabama, Livingston, AL 35470

Abstract

Decision aids are computer systems intended to assist users in analyzing tradeoffs in their decisions. Some of the methods implemented in decision aids have been tested over a series of past experiments conducted by the authors. This paper presents the authors' evaluation of the methods with respect to cognitive complexity. Decision aids must be easy to understand and use if they are to be implemented by nonspecialists. The need for improved sensitivity and recognition of cultural differences were cited as needed areas of development. The importance of methods to support learning was emphasized as an area in need of research activity.

Keywords: Decision aids, multiattribute analysis, selection decisions

1 Introduction

Many real-life decisions involve selection of one or more alternatives from a given set, and selection decisions are usually complicated by the existence of multiple, often conflicting objectives, criteria, or influencing factors [9],[11],[28],[29]. Over the past few decades, a number of interesting methods and systems to support selection decision making have been presented (see overviews [4],[25]). These tools have been developed throughout the world, incorporating different ideas. A popular European term for this class of decision support tool is decision aid.

As more and more decision aids appear in the market, the investigation of how users (not only specialists in decision analysis) are able to use these systems and interpret their results on a broad scale is necessary. In recent years in cooperation with our colleagues we carried out a series of experiments [13], [18], [19], [[23], [24], [26], verifying the results in different settings, trying to understand the comparative effectiveness of their implementation, the role of decision models used, and finally to reveal possible cultural differences in using the systems. The overview of those results and related conclusions are discussed in this paper.

2 Decision Aid Methods

One of the most popular approach in multicriteria decision making is that of multiattribute utility theory, which is based on the concept of a decision maker preference function. A preference function is assumed of the form:

$$value_i = \sum_{j=1}^{k} w_j score_{ij}$$

where each alternative *i* is evaluated by some score *ij* on each criterion *j*. The simplest form is the linear form given above, although nonlinear forms exist when preferential dependencies exist. The scores *ij* can reflect nonlinear returns to scale as well. If there is a value function, the alternative with the greatest value is the choice that the decision maker should prefer. There have been a number of techniques that have been developed to implement this idea. The best known are multiattribute utility theory [6] and SMART (or, when swing weighting is added, SMARTS [7]).

The analytic hierarchy process [28] is also based on this preference function form, although it uses a different approach to estimate relative value of criteria as well as scores of alternatives over these criteria. The method operates by developing a hierarchy of criteria and obtaining relative values through ratio pairwise comparisons of relative value. When AHP is used with two levels (criteria at the upper level, alternatives at the lower level), values obtained for the upper levels appear very similar to the multiattribute value function, while scores for alternatives at the lower level are similar to attribute importance scores. Ratio pairwise comparisons of attribute importance and alternative values are combined using an additive value function to calculate overall scores of alternatives.

AHP has received heavy criticism, including concern over rank reversal [2] as well as concern that ratios obtained through pairwise comparison do not consider differences in scale [5], [30]. In response to some of these criticisms, a number of methods have branched from AHP, including the geometric mean approach [1], and REMBRANDT, the basic ideas of which were outlined in Lootsma [21].

The paradigm of an underlying value function also serves as the basis for two methods that operate through the logic of elimination. The preference cone method [14] uses decision maker selection between pairs of alternatives from the choice set, and compares measures of the selected and rejected alternatives over criteria as a basis for a preference cone. A rational decision maker is assumed to be consistent, and therefore transitivity is used as the basis to logically eliminate other alternatives in the choice set.

A concept similar to the preference cone applies in the case of the Russian method ZAPROS [16], [17]. In ZAPROS, the choices presented to the decision maker are strictly controlled to ensure that the decision maker can grasp the tradeoffs involved. The performances of alternatives are measured over criteria in categories, thus disregarding minor differences in criteria attainment levels across alternatives. Only two criteria vary in the pairwise comparison. The results of decision maker selections from a set of theoretical alternatives is used as the basis of a joint ordinal scale, a preference mapping that can be used to generate a partial order for all possible alternatives.

The outranking methods coming from France and Belgium use a different paradigm. Alternatives are compared on the basis of two measures: concordance and discordance. Concordance reflects those cases where one of a pair of alternatives is superior to the other. Discordance reflects the reverse case, where the first alternative is inferior to the other. Minor differences can be disregarded, with a number of alternative functions available to measure relative value. Outranking relationships are developed based on weighted concordance and discordance indices. There have been a number of versions developed. PROMETHEE [3] is a popular implementation of these ideas.

Preemptive selection is an old concept that appeared in the form of preemptive goal programming in continuous selection decisions [10], [20]. The decision rule is to have the decision maker identify criteria of importance, set target attainment levels (with more than one target per criterion allowed), prioritize (rank order) the importance of attaining each target, and apply the method to the alternatives available. The most important target is considered. All alternatives not satisfying that target are eliminated. If no alternatives attain the target, those alternatives closest to the target are selected. If there is more than one, further priority targets are considered to break the tie. The method proceeds until the desired number of alternatives is left. There are two theoretical complaints about preemptive decision rules. First, it is contended that preemptive decision rules can lead to selection of dominated alternatives. This is easy to remedy, by making sure that at least one target is ambitious enough not to be satisfied, and breaking ties with less important priority goals. Wierzbicki in [31] demonstrated another way to guarantee nondominated solutions by minimizing distance from an ideal point determined by simultaneous attainment of all targets. Another theoretical limitation of the preemptive approach is that the tradeoff between target levels is not evaluated. When moving from one preemptive priority level to the next, there is in effect an infinite weight given to the more important target relative to the less important target.

3 Cognitive Efforts Consideration

Comparison across multiattribute decision aids is not easy. One of the most important complications is that there is not a clear, objectively best, decision in multiple attribute environments [8], [27], [32]. The choice of a decision maker for a particular alternative depends on that individual's preference system, and this preference structure is implicit without a precise, accurate description. Therefore, a lot of attention should be given to the ways in which preferences have been elicited, the models used, and the stability of the results obtained as well as to changes of the values of various parameters used (sensitivity analysis).

In [15] Larichev proposed that validity of decision processes depends on complexity of elementary operations required of decision makers by various multicriteria decision aids. An operation was classified as complex if psychological research indicates that in performing such operations the decision maker displays many inconsistencies and makes use of simplifying strategies. An operation was classified as admissible if psychological research indicated that people were capable of performing these operations with minor inconsistencies, and if they could employ complex strategies. Operations that are admissible but for small dimension are those that research indicates can be performed with minor inconsistencies given that the number of criteria, alternatives, or multiattribute estimates are small enough that they can be dealt with without major inconsistencies. Those operations classified as uncertain were those where insufficient psychological research had been conducted in order to evaluate

4 Review of Experimental Studies and Discussion

In all our studies we have applied the mentioned decision aid systems in a classroom setting on decision problems that the students cared about and were knowledgeable about. Therefore, the subjects were naïve users, using decision aids directly. The subjects, whether in Russia or the U.S., were above average in computer expertise, but had received only rudimentary training in multiattribute methods. In these studies, subjects used several different decision aids for solution of a multiple criteria decision problem in their decision making domain. The focus of the investigation was consistency of results among system, including rank ordering of alternatives, criteria, and alternative scores. Additionally, subjects answered questionnaires evaluating features of the system, such as ease of use, satisfaction with the result, satisfaction with the process, etc. Details such as sample size are given in Table 1 and in referenced works. The aim of those experiments was to measure the results of using systems, as well as a subjective questionnaire about user opinions of systems. The systems tested yielded similar results when the choice set involved distinct differences [18]. ZAPROS and preference cones forced consistency. Subjects sometimes inconsistent when using AHP, despite being informed by the consistency index. When subjects were tested with alternatives designed to be quite similar (thus involving more difficult decisions), those methods involving more robust input from the decision maker were found to yield more stable results [19], [26]. Most research effort in the field has focused on accurate estimation of weights in the preference function. A consistent finding across our studies is that not only are accurate weight estimates important, but the alternative scores on the criteria are equally important [24], [26].

Table 1: Study Characteristics

Study	Systems	Subjects	Alternatives	Criteria
LMMO	AHP, PC, ZAPROS	28	30	5
LOMM	DECAID, LD, ZAPROS	16	8	4
OMSM	DECAID, LD, ZAPROS	21	8	4
MSO	SMART, ZAPROS	2,2	48,45	6,5
KLMMW	VIMDA, SCP, SCPPAR	43,42	98	5
MMO	AHP, LD, PROM, ZAPROS	17,17	5	4

Studies are coded:
LMMO - Larichev, et al., 1993
LOMM - Larichev, et al., 1995
OMSM - Olson, et al., 1996
MSO - Moshkovich, et al., 1997
KLMMW - Korhonen, et al., 1997
MMO - Mechitov, et al., 1997
Methods are coded:
AHP – analytic hierarchy process
PC – preference cones
LD – Logical Decision (MAUT)
MSO involved replications in two different job selections.
KLMMW involved replications in Helsinki, Moscow
MMO involve replications in the US, Russia

Subjects typically make errors, in that they have inconsistent ratings of scores across systems, and will occasionally have reversal of relative importance of criteria across systems [26]. Furthermore, systems based on the same model (LOGICAL DECISION and DECAID, an implementation of SMART) have been found to yield different results for some subjects [26]. These two systems are different in the way in which they elicit input about both utility scoring and weighting of criteria. LOGICAL DECISION uses lottery tradeoffs for weights and uses curve fitting for scoring. DECAID simply uses graphical scales where the user inputs data with a cursor. Both of these systems yielded results that were closer to the AHP and ZAPROS system results than they were with each other. The DECAID system involved the most dominated solutions selected. This indicates that LOGICAL DECISION is more accurate (in the sense that the decision aid matched the post-analysis holistic choice of the subject) than was DECAID. Larichev, et al. [18] found that the preference cone method forces consistency, but when subjects are faced with a difficult choice, the method can be drastic and discard solutions that might be quite attractive. In this sense, it is an unforgiving system.

Our series of experiments also yielded subject response about relative cognitive complexity. SMART and AHP were both found to be quite easy to use. [18]. In [19], Logical Decision was found to be quite complex by subjects. In [26], AHP was criticized by subjects for involving many questions, while Logical Decision was criticized for asking abstract questions (in lottery tradeoffs). ZAPROS was found to involve both abstraction as well as redundancy. Supporters of each system could argue that their results were best, but subjective measures indicated that subjects were not confident with either the easy entry of data available with DECAID nor with the complex data entry available in LOGICAL DECISION. Our research finds that decision aids must be easy to understand and use, but also must provide feedback about consistency of preference selections.

In the Mechitov, et al. study [23], exposing both US and Russian students to MAUT (both in LOGICAL DECISION and DECAID), AHP, ZAPROS, and PROMETHEE, prior results were confirmed. The greatest confidence in results, both for US and Russian students, was placed on the outcome of AHP, perhaps because AHP provides a clear and straightforward decision process (due to the small dimensionality of tasks the number of pairwise comparisons in this experiment was quite small). There was a difference between the two groups of students. US students seemed to prefer quantitative results, as with LOGICAL DECISION, DECAID, and AHP, that were used as a basis for full rank ordering. Russian students placed less confidence on numeric scores, stating that they were less confident on precise input values. EXPERT CHOICE was rated highly, possibly because it requires inexact input, but returns seemingly exact numerical output.

While not tested formally, we have also exposed students to AIM [22], a learning method. This approach allows users to change aspiration levels and explore the set of available alternatives. While based on a very small sample size, the AIM system was very popular with specific students, who appreciated the ability to examine tradeoffs among alternatives.

Korhonen, et al. in [13] compared the learning method VIMDA with two forms of preemptive systems. Subjects were students in Finland (43) and Russia (42). The pre-experiment view was that compensatory rules (MAUT) would be used by humans when faced with a fairly small set of alternatives, but preemptive elimination rules would more often be used when a large alternative set was present. It was hypothesized that this use of preemptive selection would result in selection of more dominated alternatives. The test set included five criteria and 98 alternatives.

Measures were ease of use, satisfaction with chosen alternative, speed of convergence, percentage of dominated alternatives, and consistency across systems. There were no significant differences found in ease of use across systems. The preemptive method was rated significantly better than VIMDA in satisfaction of final outcome in Moscow, but not in Helsinki. In part this was because VIMDA selected one preferred solution [12], while the preemptive method generated a set of alternatives. The preemptive method was significantly faster than VIMDA. VIMDA eliminates dominated alternatives, but when subjects were allowed to identify a small set of other attractive alternatives, dominated alternatives were included by many subjects. There were usually dominated alternatives in the final set of alternatives obtained with the preemptive methods. However, subjects were quite satisfied with their choices anyway. There was low consistency of subject choices across methods.

These studies indicate that cognitive load depends not only on the type of elementary decision operations involved, but also on the type of decision problems, the model used, time and effort, and other factors. Thus AHP was rated as very easy to use by subjects in [18] and [33], while SMART (implemented either through DECAID or by spreadsheet) was rated as very easy to use in [26] and ZAPROS and LOGICAL DECISION were consistently rated as very hard to use.

5 Conclusions

A number of interesting and useful decision aids have been developed over the past few decades. They differ in what they intend to provide decision makers. Logical Decision and ZAPROS both seek formal and rigorous measurement of preference. Both ask users to provide difficult input. (ZAPROS asks many questions, while Logical Decision asks hard questions.) However, while the other methods may be easier to use, they may well sacrifice accuracy (if matching post-analysis holistic choice is accurate).

Analysis needs to focus on decision maker learning about tradeoffs. Learning methods, such as VIMDA, AIM, and VISA provide decisiion makers with views of such tradeoffs. However, decision makers might consider a variety of criteria. PROMETHEE and VIMDA both allow the user to block out specific criteria or alternatives as desired, a useful feature in this dynamic environment.

We feel that there is a need to focus research on decision aids to the impact of dynamic decision maker environments. There are a number of interesting questions that could be studied. Cognitive effort and aspects of learning are important. Group aspects also are important. There is much interesting work that needs to be conducted in the application of decision aids to decision making.

References

[1] Barzilai, J., Cook, W., and Golanyi, B. (1987): Consistent Weights for Judgements Matrices of the Relative Importance for Alternatives. *Operations Research Letters* 6:3, 131-134.

[2] Belton, V. and Gear, T. (1983): On a Short-Coming of Saaty's Method of Analytic Hierarchies. *Omega* 11:3, 228-230.

[3] Brans, J.P. and Vincke, P. (1985): A Preference Ranking Organization Method: The PROMETHEE Method. *Management Science* 31, 647-656.

[4] Buede, D.M. (1992): Software review: Overview of the MCDA software market. *Journal of Multi-Criteria Decision Analysis* 1(1), 59-61.

[5] Dyer, J.S. (1990): Remarks on the Analytic Hierarchy Process. *Management Science* 36:3, 249-258.

[6] Dyer, J.S. and Sarin, R.K. (1979): Measurable Value Functions. *Operations Research* 27, 810-822.

[7] Edwards, W., and Barron, F.H. (1994): SMARTS and SMARTER: Improved Simple Methods for Multiattribute Utility Measurement. *Organizational Behavior and Human Decision Processes* 60, 306-325.

[8] Edwards, W., Kiss, I., Majone, G. and Toda, M. (1984): What constitutes a good decision? *Acta Psychologica* 56, 5-27.

[9] Goodwin, P., & Wright, G. (1991): *Decision Analysis for Management Judgments*, Wiley, New York.

[10] Ignizio, J.P. (1976): *Goal Programming and Extensions*, Heath, Lexington, MA.

[11] Keeney, R.L., and Raiffa, H. (1976): *Decisions with Multiple Objectives: Preferences and Value Tradeoffs*, John Wiley & Sons, New York.

[12] Korhonen, P. (1988): A Visual Reference Direction Approach to Solving Discrete Multiple Criteria Problems. *European Journal of Operational Research* 34:2, 152-159,

[13] Korhonen, P., Larichev, O., Mechitov, A., Moshkovich, H., and Wallenius, J. (1997): Choice Behavior in a Computer-Aided Multiattribute Decision Task. *Journal of Multi-Criteria Decision Analysis* 6, 233-246.

[14] Korhonen, P., Wallenius, J., and Zionts, S. (1984): Solving the Discrete Multiple Criteria Problem Using Convex Cones. *Management Science* 30:11, 1336-1345.

[15] Larichev, O.I. (1992): Cognitive validity in design of decision-aiding techniques. *Journal of Multi-Criteria Decision Analysis* 1, 127-138.

[16] Larichev, O.I., and Moshkovich, H.M. (1991): *ZAPROS: A Method and System for Ordering Multiattribute Alternatives on the Base of a Decision-Maker's Preferences*, All-Union Research Institute for Systems Studies, Moscow.

[17] Larichev, O.I., and Moshkovich, H.M. (1995): ZAPROS-LM: A Method and System for Rank-Ordering of Multiattribute Alternatives. *European Journal of Operational Research* 82, 503-521.

[18] Larichev, O.I., Moshkovich, H.M., Mechitov, A.I. and Olson, D.L. (1993): Experiments comparing qualitative approaches to rank ordering of multiattribute alternatives. *Journal of Multi-Criteria Decision Analysis* 2:1, 5-26.

[19] Larichev, O.I., Olson, D.L., Moshkovich, H.M., and Mechitov, A.I. (1995): Numeric vs. Cardinal Measurements in Multiattribute Decision Making: (How Exact is Enough?). *Organizational Behavior and Human Decision Processes* 64, 9-21.

[20] Lee, S.M. (1972): *Goal Programming for Decision Analysis*, Auerbach Publishers, Philadelphia.

[21] Lootsma, F.A. (1993): Scale Sensitivity in a Multiplicative Variant of the AHP and SMART. *Journal of Multi-Criteria Decision Analysis*, 2, 87-110.

[22] Lotfi, V., Stewart, T.J., and Zionts, S. (1992): An Aspiration-Level Interactive Model for Multiple Criteria Decision Making. *Computers and Operations Research* 19:7, 671-681.

[23] Mechitov, A.I., Moshkovich, H.M. and Olson, D.L. (1997): Computer Systems Illustration of Decision Aid Techniques in CIS Courses. *Journal of Computer Information Systems* XXXVII:4, 1-7.

[24] Moshkovich, H.M., Schellenberger, R., and Olson, D.L. (1998): Data Influences the Result More than Preferences. *Decision Support Systems* 22(1), 73-84.

[25] Olson, D.L. (1996): *Decision Aids for Selection Problems*, Springer, New York.

[26] Olson, D.L., Moshkovich, H.M., Schellenberger, R., and Mechitov, A.I., (1996): Consistency and Accuracy in Decision Aids: Experiments with Four Multiattribute Systems. *Decision Sciences* 26, 723-748.

[27] Rohrmann, B. (1986): Evaluating the usefulness of decision aids: A methodological perspective. In *New Directions in Research on Decision Making*, Brehmer, B., Jungermann, H., Lourens, P. and Sevon, Eds., North-Holland, Amsterdam.

[28] Saaty, T.L. (1980): *The Analytic Hierarchy Process*. McGraw-Hill International, New York.

[29] Watson, S.R. and Buede, D.M. (1987): *Decision Synthesis: The Principles and Practice of Decision Analysis*, Cambridge University Press, New York.

[30] Watson, S.R. and Freeling, A.N.S. (1982): Assessing Attribute Weights. *Omega* 10(6), 582-583.

[31] Wierzbicki, A.P. (1982): A Mathematical Basis for Satisfying Decision Making. *Mathematical Modelling* 3, 391-405.

[32] Wierzbicki, A.J. (1997): On the Role of Intuition in Decision Making and Some Ways of Multicriteria Aid of Intuition. *Journal of Multi-Criteria Decision Analysis* 6, 65-76.

[33] Zapatero, E.G., Smith, C.H. and Weistroffer, H.R. (1997): Evaluating Multiple-Attribute Decision Support Systems. *Journal of Multi-Criteria Decision Analysis* 6, 201-214.

Cone Decomposition for the Solution of Efficient Extreme Points in Parallel

Ralph E. Steuer and Craig Piercy
University of Georgia and Towson University, USA

Abstract

The solution of a multiple objective linear program (MOLP) is often characterized by the enumeration of all efficient extreme points. Unfortunately, the larger the problem, the larger the number of efficient extreme points and the longer the computation time required. Using vector maximum algorithms, the set of efficient extreme points is effectively enumerated by maximizing LPs whose objective function gradients point into the relative interior of the criterion cone (cone generated by strictly positive linear combinations of the gradients of the MOLP objective functions). In this way, each efficient extreme point can be associated with a conal subset of the criterion cone. In this paper, we present a divide and conquer routine for computing efficient extreme points by decomposing the criterion cone into subsets and then solving the sub-problems associated with these subset criterion cones in parallel. Computational results are also reported.

Keywords: Efficient points, multiple objective linear programming, vector maximum algorithms, parallel computing

1 Introduction

Many authors including Yu and Zeleny (1975), Gal (1977), Isermann (1977), and Ecker and Kouada (1978) have worked on algorithms for computing all efficient extreme points of a multiple objective linear program. However, drawbacks have been the number of efficient extreme points and the computation time involved. Other authors have attempted to address these drawbacks. Recently, Mavrotas, Diakoulaki and Assimacopoulos (1998) experimented with bounding the objective functions and Wiecek and Zhang (1995) investigated parallelization possibilities in computer codes. In this paper, we investigate a parallelization of a different sort based upon decompositions of the criterion cone.

In Section 2 we draw relevant observations from vector maximum theory. In Section 3 we describe the cone decomposition approach for computing all efficient extreme points using parallel processors. Sections 4 and 5 present computational results, and Sections 6 and 7 comment on future directions.

2 Observations from Vector Maximum Theory

An MOLP can be written in regular form

$$\begin{array}{ll} \max\{c^1 x = z_1\} \\ \vdots \\ \max\{c^k x = z_k\} \\ s.t. \quad x \in R \end{array}$$

or in vector maximum form

$$\text{"max"}\{Cx = z \mid x \in S\}$$

where $S = \{x \in R^n \mid Ax = b, x \geq 0, \ b \in R^m\}$ and C is the $k \times n$ criterion matrix whose rows are the c^i gradients of the MOLP objective functions.

Let weighting vector space be the set of all strictly positive weighting vectors

$$\Lambda = \{\lambda \in R^k \mid \lambda_i > 0, \sum_{i=1}^{k} \lambda = 1\}$$

It is known that in an MOLP, $\overline{x} \in S$ is an efficient extreme point if and only if there exists a $\lambda \in \Lambda$ such that $\overline{x}$ corresponds to an optimal simplex tableau of the weighted-sums LP

$$\max\{\lambda^T Cx \mid x \in S\}$$

Consequently, if we solve the family of weighted-sums LPs

$$\{\max\{\lambda^T Cx \mid x \in S\} \mid \lambda \in \Lambda\}$$

for all maximizing extreme points we will obtain precisely the set (Steuer (1986, Section 9.10)) of all efficient extreme points of the MOLP. With $\lambda \in \Lambda$, we know that $\lambda^T C \in R^n$ points into the relative interior of the criterion cone. Hence it is observed that each efficient extreme point is associated with a convex conal subset of the relative interior of the criterion cone.

Therefore, the following three-step idea comes to mind: (a) decompose the criterion cone into subset criterion cones, (b) solve the sub-problem MOLPs associated with the different subset criterion cones, and (c) take the union of all extreme points generated. As long as the union of the relative interiors of the subset criterion cones equals the relative interior of the original criterion cone, this will produce exactly the set of all efficient extreme points. Justification for this is based upon the fact that the conal subset of each efficient extreme point will have a nonempty intersection with at least one of the subset criterion cones causing each efficient extreme point to be generated at least once.

3 Decomposition of the Criterion Cone

With c^1 and c^2 the objective function gradients, consider the following MOLP whose set of efficient extreme points is $\{x^1, x^2, x^3, x^4\}$ in Figure 1.

$$\begin{array}{lrcl} \max & \{x_1 + 6x_2 & = & z_1\} \\ \max & \{7x_1 + x_2 & = & z_2\} \\ s.t. & x_1 & \leq & 8 \\ & x_2 & \leq & 8 \\ & 3x_1 + x_2 & \leq & 26 \\ & x_1 + x_2 & \leq & 12 \\ & x_1 + 3x_2 & \leq & 26 \\ & x_1, x_2 & \geq & 0 \end{array}$$

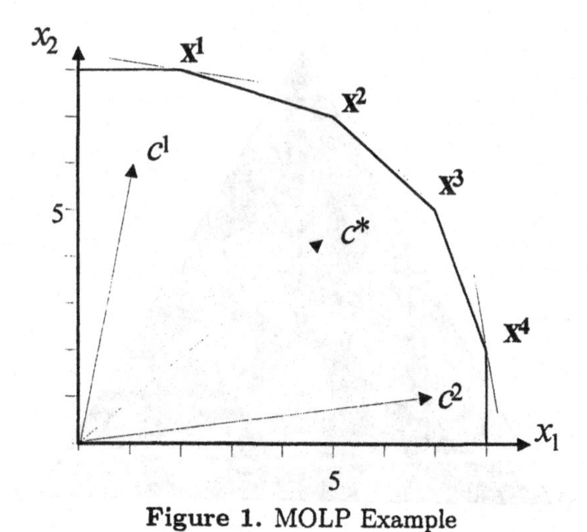

Figure 1. MOLP Example

To decompose the criterion cone, let us introduce c^* as a "central" gradient where

$$c^* = \tfrac{1}{2}c^1 + \tfrac{1}{2}c^2$$

to divide the original criterion cone into subset criterion cones. Visualizing weighted-sums LP gradients pointing into the relative interiors of the subset criterion cones, we see $\{x^1, x^2, x^3\}$ to be the set of extreme points efficient with respect to the subset criterion cone defined by c^1 and c^*, and $\{x^3, x^4\}$ to be the set of extreme points efficient with respect to the subset criterion cone defined by c^* and c^2.

Two things are seen here. The first is that the set of all efficient extreme points can indeed be broken into subsets by cone decomposition. The second is that the resulting subsets of efficient extreme points need not be mutually exclusive. That is, there can be overlap. This is illustrated by efficient extreme point x^3 as its conal subset has a nonempty intersection with each of the two subset criterion cones.

Schemes for decomposing the criterion cone of 3-objective MOLPs can be portrayed using weight-space graphs from the software TRIMAP (Climaco and Antunes (1989)). The weight-space graphs are helpful because they represent cross-sections of the criterion cone. In the weight-space graph of Figure 2 we see the cross-section of the criterion cone of a 3-objective MOLP. In the graph, the weighting vectors $\lambda^1 = (1,0,0)$, $\lambda^2 = (0,1,0)$ and $\lambda^3 = (0,0,1)$ correspond to the three objective function gradients c^1, c^2 and c^3. The 48 differently shaded regions correspond to the cross-sections of the conal subsets of the MOLP's 48 efficient extreme points.

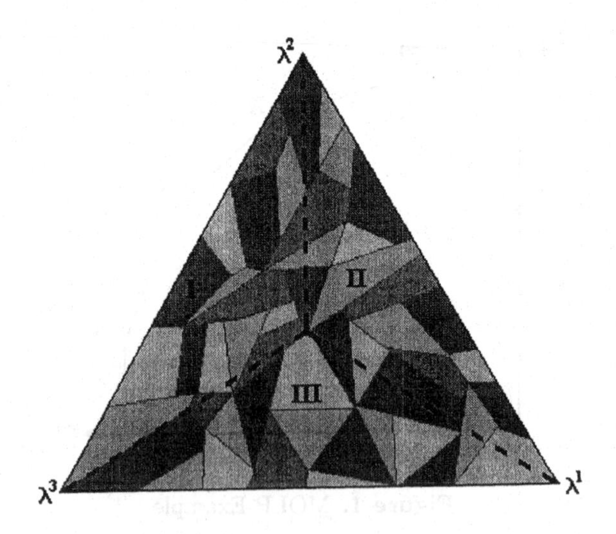

Figure 2. TRIMAP Style Representation of 3-Objective MOLP Weight Space

The strategy for decomposing the criterion cone into subset criterion cones employed in this paper is in accordance with the dashed lines. This type of decomposition would result by forming a c^* central gradient by taking an equally weighted convex combination of the original c^i gradients.

From the differently shaded regions, we see that 19 of the MOLP's 48 efficient extreme points are efficient with respect to the subset criterion cone whose cross-section is Section I. Similarly, 20 of the MOLP's 48 efficient extreme points are efficient with respect to the subset criterion cone of Section II, and a different group of 20 are efficient with respect to the subset criterion cone of Section III. In this example, we have 11 overlapping efficient extreme points. These are the extreme points whose conal subsets straddle more than one subset criterion cone. Although in this example the number of extreme points efficient with respect to the different subset criterion cones is relatively balanced, this is not always the case.

Based on these ideas, several questions arise. How much of a gain in solution time can we expect for various problem sizes? What is the extent of overlap in larger problems? How balanced are the numbers of efficient extreme points over the subset criterion cones? Might there be other strategies for subdividing the criterion cone?

4 Experimental Design

To conduct computational experiments, ADBASE (Steuer (1998)) was used. Its random problem generator (described in Steuer (1994)) was called upon to create MOLPs of the sizes in Table 1.

Table 1. Problem Size Parameters

Parameter	Values
Number of Objectives k	3,4,5
Number of Constraints m	20,30,40,50,60
Number of Variables n	10,25,50,75,100

For each combination of the problem size parameters, randomly generated MOLPs were solved for all efficient extreme points, for a grand total of 998 problems. After decomposing the criterion cone of each of the 998 MOLPs into k subset criterion cones, each of the sub-problems associated with the different subset criterion cones was solved.

The decomposition of each criterion cone was carried out as in Figure 2. That is, a c^* central gradient where

$$c^* = \frac{1}{k}c^1 + \frac{1}{k}c^2 + \ldots + \frac{1}{k}c^k$$

was formed by taking an equally weighted convex combination of the original criterion cone gradients c^i. Using c^* and $k-1$ of the c^i, k different subset criterion cones were formed thus defining the criterion cones of the sub-problem MOLPs.

5 Computational Results

Data on the following were collected to measure the effectiveness of the cone decomposition approach:

EXT_i = number of efficient extreme points of the original MOLP i
$EXT_{j,i}$ = number of efficient extreme points of sub-problem j of MOLP i
CPU_i = solution time for the original MOLP i
$CPU_{j,i}$ = solution time for sub-problem j of MOLP i

A measure of the gain in time by solving the sub-problems in parallel is as follows:

$$\textit{Percent Time Gain}_i = \frac{CPU_i - \max_{j=1,\dots,k} \{CPU_{j,i}\}}{CPU_i} \times 100$$

This formula determines the gain in time due to parallellizing the MOLP by first subtracting the longest sub-problem time from the original MOLP time. A unitless measure is then obtained by dividing the gain by the original MOLP time. In Figure 3, this measure is charted against the original MOLP number of efficient extreme points. The black diamonds signify 3-objective MOLPs, the gray boxes signify 4-objective MOLPs, and the triangles signify 5-objective MOLPs. The percentage time gain quickly increases to greater than 30% for modestly sized problems and for large problems the gain ranges up to 70%.

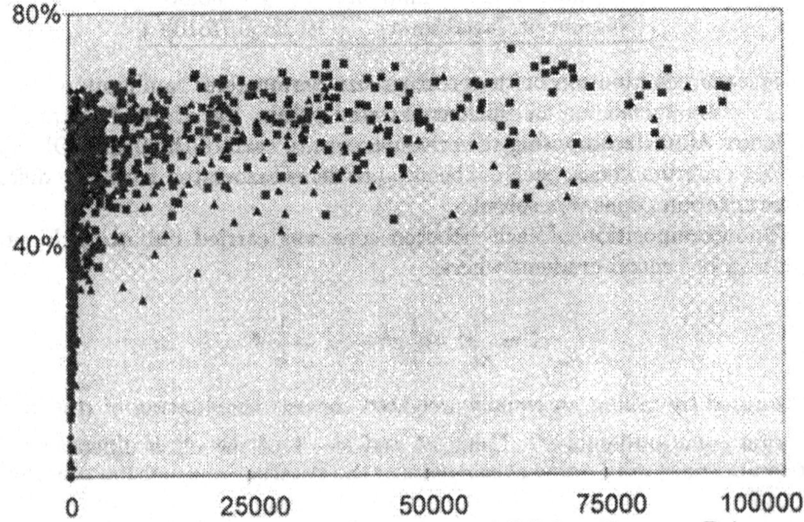

Figure 3. Percent Time Gain vs. Original Efficient Extreme Points

A desirable condition for solving a MOLP in parallel would be if the original MOLP efficient extreme points were spread out in a balanced fashion over the subset criterion cones. For example, for a MOLP that is divided into 3 sub-problems, it would be desirable for about one-third of the efficient extreme points to be generated by each sub-problem. When this is not the case, the problem is said to be in imbalance. A measure of imbalance is defined in a manner that is similar to the statistical concept of variance. We calculate a difference between the actual proportion of efficient extreme points in a subset criterion cone and the proportion that would be there under perfect balance. We then square these differences and find an average of these squared differences. The calculation is as follows:

$$Imbalance_i = \frac{\sum_{j=1}^{k} \left(\frac{EXT_{j,i}}{EXT_i} - \frac{1}{k} \right)^2}{k}$$

In Figure 4, we see that imbalance over the sub-problems tends to become less pronounced as the total number of original MOLP efficient extreme points increases.

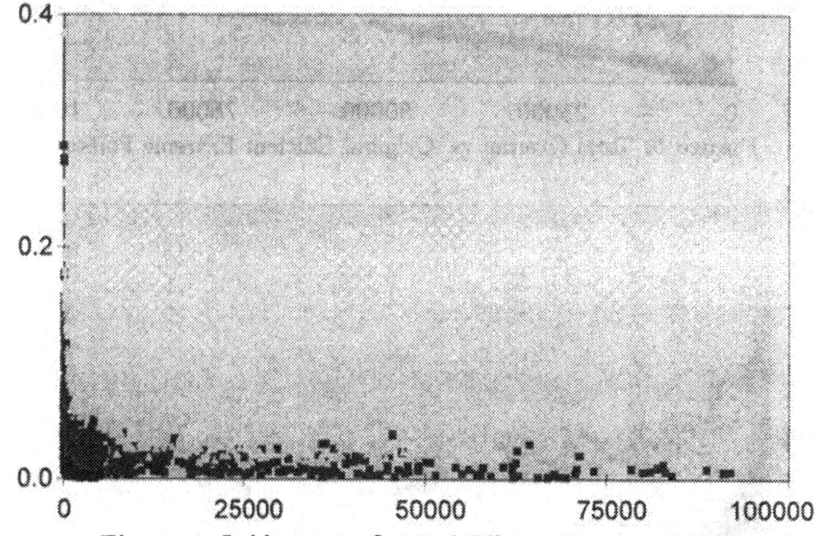

Figure 4. Imblance vs. Original Efficient Extreme Points

When there is overlap, the processors solve for the overlapping points multiple times. We measure the number of overlapping points as follows:

$$Total\ Overlap_i = \left(\sum_{j=1}^{k} EXT_{j,i} \right) - EXT_i$$

$$Percent\ Overlap_i = \frac{Total\ Overlap_i}{\sum_{j=1}^{k} EXT_{j,i}} \times 100$$

Here we are measuring the total overlap of each MOLP and converting it to percentages for comparisons between problem sizes.

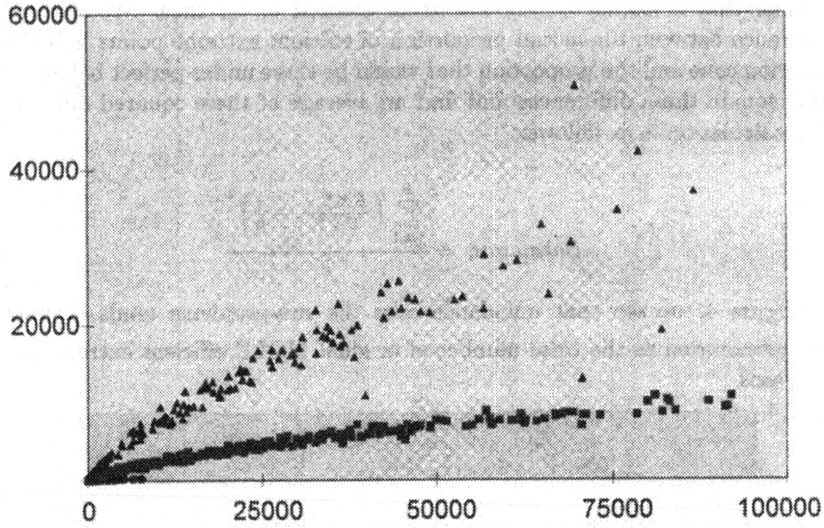

Figure 5. Total Overlap vs. Original Efficient Extreme Points

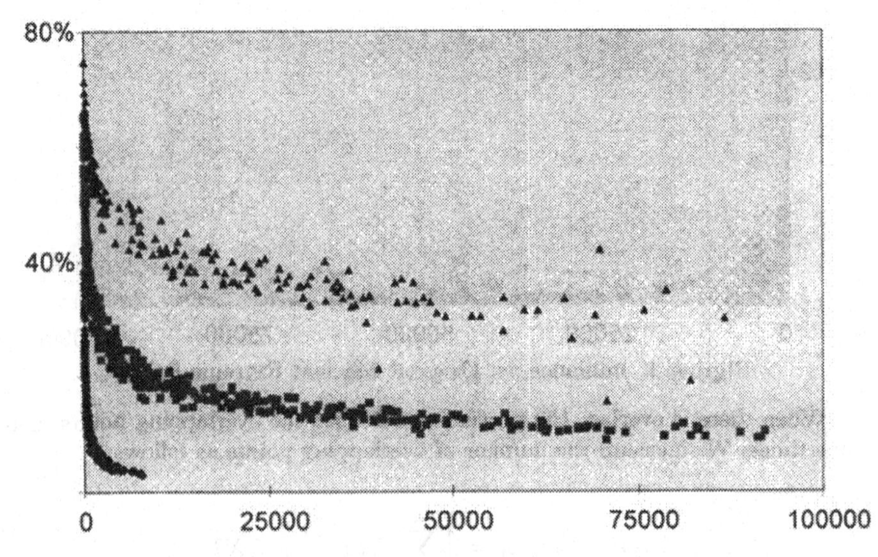

Figure 6. Percent Overlap vs. Original Efficient Extreme Points

Figures 5 and 6 show the relationship of these two measures to the number of original MOLP efficient extreme points. The total number of overlapping points increases with the number of original efficient extreme points. On the other hand, these numbers as a percentage of the total number of points generated by the sub-problems show a decrease as problem size increases. In addition, these figures illustrate that overlap increases at a greater rate with the number of objectives.

As another way of describing overlap, Table 2 provides frequency distributions on the average number of times efficient extreme points were computed over the sub-problem MOLPs for 3-objective and 4-objective MOLPs.

Table 2. Average Number of Overlap Counts

MOLP Size	Once	Twice	Three	Four
3 Objectives	974.98	75.03	1.49	-
4 Objectives	6727.19	1414.41	110.31	2.81

6 Other Decomposition Strategies

Other decomposition strategies are possible such as the one indicated by the dashed lines in Figure 7. In this strategy, three new gradient vectors are formed so as to divide the criterion cone of a 3-objective MOLP into four subset criterion cones. Then four sub-problem MOLPs are formed each using three of the now six gradients in correspondence with Figure 7.

Using this new strategy, data for 372 three-objective MOLPs were obtained representing all combinations of the 3-objective problem size parameters. Figures 8, 9 and 10 provide a comparison between the new decomposition strategy and the previous strategy. While a small percentage gain in solution time is seen, overlap and imbalance show no significant differences.

7 Concluding Remarks

The computational results reported in this paper provide evidence of a potential for cone decomposition strategies for parallelizing large MOLPs. The results show that gains in solution time greater than 50% are possible, and in particular, as problem size increases:

- percentage gain in solution time increases.
- percentage of sub-problem overlap decreases.
- numbers of efficient extreme points become more balanced across the sub-problems.

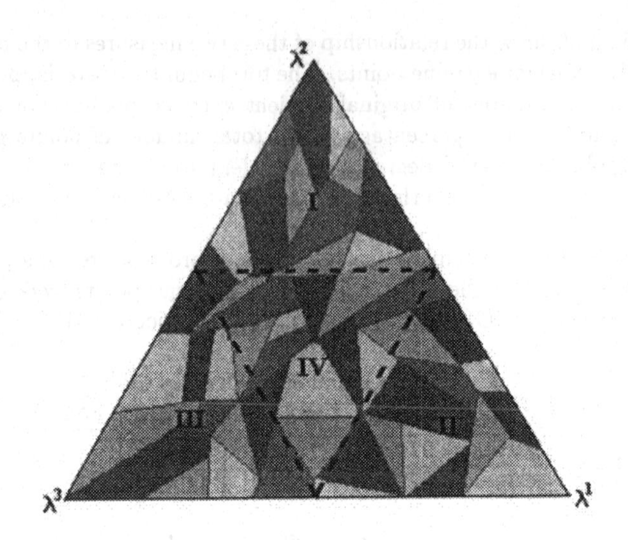

Figure 7. Alternative 3-Objective Decomposition Strategy

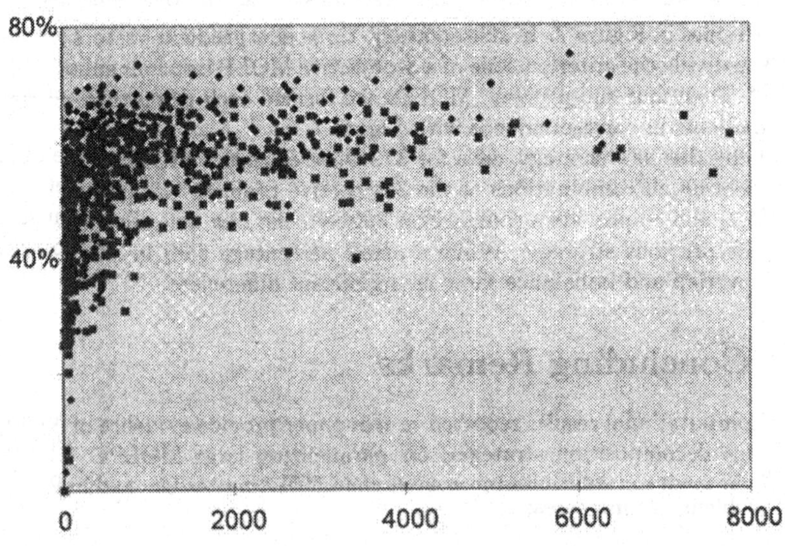

Figure 8. Percent Time Gain vs. Original Efficient Extreme Points

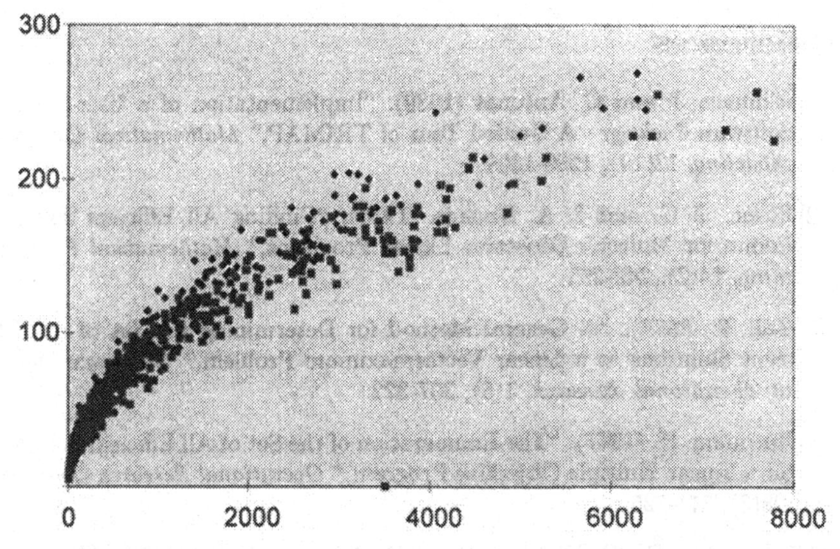

Figure 9. Imbalance vs. Original Efficient Extreme Points

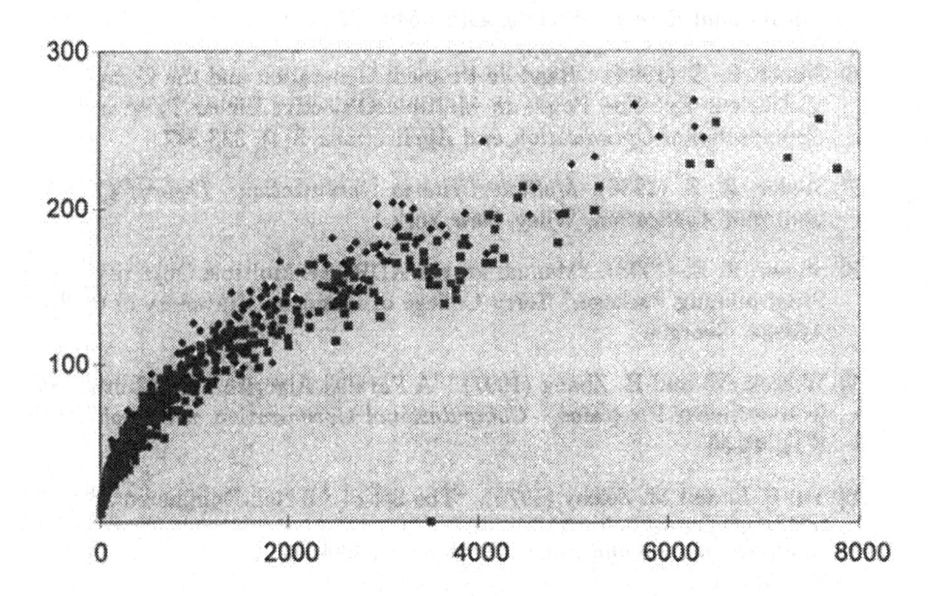

Figure 10. Total Overlap vs. Original Efficient Extreme Points

References

[1] Climaco, J. and C. Antunes (1989). "Implementation of a User-Friendly Software Package - A Guided Tour of TRIMAP," *Mathematical Computer Modeling*, 12(10), 1299-1309.

[2] Ecker, J. G. and I. A. Kouada (1978). "Finding All Efficient Extreme Points for Multiple Objective Linear Programs," *Mathematical Programming*, 14(2), 249-261.

[3] Gal, T. (1977). "A General Method for Determining the Set of All Efficient Solutions to a Linear Vectormaximum Problem," *European Journal of Operational Research*, 1(5), 307-322.

[4] Isermann, H. (1977). "The Enumeration of the Set of All Efficient Solutions for a Linear Multiple Objective Program," *Operational Research Quarterly*, 28(3), 711-725.

[5] Mavrotas, G., D. Diakoulaki and D. Assimacopoulos (1998). "Bounding MOLP Objective Functions: Effect on Efficient Set Size," *Journal of the Operational Research Society*, 49(5), 549-577.

[6] Steuer, R. E. (1994). "Random Problem Generation and the Computation of Efficient Extreme Points in Multiple Objective Linear Programming," *Computational Optimization and Applications*, 3(4), 333-347.

[7] Steuer, R. E. (1986). *Multiple Criteria Optimization: Theory, Computation, and Application*, Wiley, New York.

[8] Steuer, R. E. (1998). "Manual for the ADBASE Multiple Objective Linear Programming Package," Terry College of Business, University of Georgia, Athens, Georgia.

[9] Wiecek, M. and H. Zhang (1997). "A Parallel Algorithm for Multiple Objective Linear Programs," *Computational Optimization and Applications*, 8(1), 41-56.

[10] Yu, P. L. and M. Zeleny (1975). "The Set of All Non-Dominated Solutions in Linear Cases and a Multicriteria Simplex Method," *Journal of Mathematical Analysis and Applications*, 49(2), 430-468.

Scheduling to Minimize Maximum Earliness and Number of Tardy Jobs Where Machine Idle Time is Allowed

Meral Azizoğlu, Murat Köksalan and Suna Kondak
Middle East Technical University, Department of Industrial Engineering, Ankara 06531, Turkey

Abstract

We address the bicriteria problem of minimizing number of tardy jobs and maximum earliness on a single machine. The combination of these two criteria captures the internal and external concerns of the manufacturer in a balanced way. We assume that the machine can be kept idle at times in order to reduce maximum earliness.

We show that the problem of minimizing the number of tardy jobs subject to the constraint that maximum earliness is kept at its minimum value is solvable in polynomial time. We present a polynomial time algorithm for the problem of minimizing maximum earliness subject to no tardy jobs. The algorithm has led to the development of another polynomial time algorithm for the problem of minimizing maximum earliness for a given set of tardy jobs.

Keywords: Scheduling, Maximum Earliness, Number of Tardy Jobs

1 Introduction

Scheduling activities in practice often require balancing the internal and external considerations of a manufacturer. The internal efficiency is usually measured by flow time-related measures to represent work in process inventories, and earliness-related measures to represent finished goods inventories. On the other hand, the external efficiency is usually measured by the customer satisfaction levels. Customer satisfaction can be represented by tardiness-related measures such as total tardiness, maximum tardiness and number of tardy jobs. In this study we consider a bicriteria problem where the internal efficiency is measured by maximum earliness and the external efficiency is measured by number of tardy jobs.

There are a number of models in the literature that deal with various combinations of earliness and tardiness penalties. Lee and Vairaktarakis (1993) and Hoogeveen (1992) review various algorithms and complexity results. Garey et al. (1988) and Hoogoveen (1996) proposed polynomial time algorithms for the problems involving maximum earliness and maximum tardiness penalties. Hoogeveen and van de Velde (1997) and Koksalan et al. (1998) study the NP-hard problem of minimizing maximum earliness and total completion time. Both studies consider no machine idle time allowed and machine idle time allowed cases. The work most closely related to this study is that of Azizoglu et al. (1997) who consider the same criteria where machine idle time is not allowed.

In the following section we define our problem environment. In Section 3 we present polynomial time algorithms for three cases of maximum earliness and number of tardy jobs problem, namely minimizing number of tardy jobs while keeping maximum earliness at its minimum, minimizing maximum earliness subject to no tardy jobs, and minimizing maximum earliness for a given set of tardy jobs. We conclude in Section 4 with a summary and suggestions for future research.

2 Problem Environment

We consider a single machine problem of scheduling a set of n independent jobs. Each job i has a processing requirement of p_i units and a customer specified due date d_i. For a given schedule S, $C_i(S)$ is the completion time of job i. The earliness of job i in S, $E_i(S)$, can be computed as $Max\{d_i-C_i(S),0\}$ and the maximum earliness of S, $E_{max}(S)$, is $Max_i\{E_i(S)\}$. The tardiness of job i in S, $T_i(S)$, can be computed as $Max\{C_i(S)-d_i, 0\}$. Letting $U_i(S)$ be a function that takes a value of 1 if $T_i(S)>0$ and 0 otherwise, the number of tardy jobs in S, $n_T(S)$, is $\Sigma_i U_i(S)$.

A schedule S is said to be efficient with respect to n_T and E_{max} if there exists no schedule S' with either $n_T(S') < n_T(S)$ and $E_{max}(S') \leq E_{max}(S)$ or $n_T(S') \leq n_T(S)$ and $E_{max}(S') < E_{max}(S)$.

Maximum earliness and number of tardy jobs are two conflicting criteria in the sense that the former is a non-increasing function of completion times whereas the latter is a non-decreasing function of completion times. In the literature there are two approaches to deal with two conflicting criteria: hierarchical and simultaneous. In the hierarchical approach, the secondary (i.e. the less important) criterion is minimized subject to the constraint that the value of the primary (i.e. the more important) criterion is kept at its optimum. In the simultaneous approach there are two methods: generation of all efficient schedules and optimization of a composite function of the two criteria.

It is well known that the number of tardy of jobs is minimized by Moore's Algorithm (MA) (Moore 1968) with no machine idle time between the jobs. The maximum earliness value of any schedule S can be forced to its minimum value of zero by inserting $E_{max}(S)$ units of idle time before the commencement of the first job. Hereafter we denote minimum values of number tardy and maximum earliness by n_T^* and E_{max}^*, respectively.

We extend the three field classification scheme of Graham et al. (1979) and denote the single machine problems of our concern as:
$1| E_{max}=E_{max}^*| n_T$: minimizing the number of tardy jobs subject to the constraint that the maximum earliness is at its minimum value,
$1| n_T = n_T^*| E_{max}$: minimizing the maximum earliness subject to the constraint that the number of tardy jobs is at its minimum value,
$1| | E_{max} | T$: minimizing the maximum earliness for a given set of tardy jobs.

3 The Hierarchical Problems

In this section we present polynomial time solutions for the hierarchical problems of $1| E_{max} = E_{max}^*| n_T$ and for the special case of $1|\, n_T = n_T^*|\, E_{max}$ when $n_T^* = 0$. The existence of the latter algorithm has led to the development of a polynomial time algorithm for $1|\,|\, E_{max}|\, T$.

3.1 A Polynomial Time Algorithm for $1|\, E_{max} = E_{max}^*|\, n_T$

An efficient schedule can be found by solving the $1|\, E_{max} = E_{max}^*|\, n_T$ problem. The corresponding n_T value, $\overline{n_T}$, gives an upper bound on the n_T values of all efficient schedules. The range of n_T values of efficient schedules is thus $[n_T(MA), \overline{n_T}]$. When $E_{max} = E_{max}^* = 0$, there is no job that completes before its due date. Minimizing the number of tardy jobs becomes equivalent to maximizing the number of jobs that complete exactly on their due dates. We refer to those jobs as on-time jobs. Algorithm 1 maximizes the number of on-time jobs thereby minimizing the number of tardy jobs in O(nlogn) time (see Azizoglu et al. 1998).

Algorithm 1

Let n_{OT} denote the number of on-time jobs.

Step 0. $n_{OT} \leftarrow 0$, $t \leftarrow 0$, $U = \{1,2,...,n\}$

Order the jobs such that $d_1 \le d_2 \le ... \le d_n$.

Step 1. Let $\pi = \{i \mid i \in U, d_i - p_i \ge t\}$.

If $\pi = \phi$ then Stop, $\overline{n_T} = n - n_{OT}$.

Let job k be the job in π having smallest index.

Schedule job k between $[d_k - p_k, d_k]$.

Let $n_{OT} \leftarrow n_{OT} + 1$, $t \leftarrow d_k$ and $U \leftarrow U \setminus \{k\}$.

Repeat this step.

3.2 A Polynomial Time Algorithm for $1|\, n_T = n_T^*|\, E_{max}$ when $n_T^* = 0$

An efficient schedule can be found through the solution of the $1|\, n_T = n_T^*|\, E_{max}$ problem. The problem is NP-hard in the strong sense (See Lee and Vairaktarakis, 1993). We show that if $n_T^* = 0$, i.e. if Moore's algorithm produces no tardy jobs, then Algorithm 2 solves the problem in O(nlogn) time (see Azizoglu et al. 1998).

Algorithm 2

Step 0. Order the jobs such that $d_1 \leq d_2 \leq ... \leq d_n$.

Step 1. Shift the jobs, ordered in Step 0, to the right as much as possible, without making any one of them tardy.
Separate the jobs into blocks where each block contains a set of jobs that are processed consecutively without any idle time.

Step 2. Treat each block as a separate problem. In each block, starting from the last position, assign the job with the largest slack time, i.e. $d_i - p_i$, among the jobs that incur zero tardiness when assigned to that position.

3.3 A Polynomial Time Algorithm for $1 \,||\, E_{max} \,|\, T$

The algorithm we proposed in Subsection 3.2 has been the basis for the development of a polynomial time algorithm to minimize maximum earliness for a given tardy set T. Before introducing our algorithm, we need to present the following result.

Theorem 1. There exists an optimal schedule for the $1 \,||\, E_{max} \,|\, T$ problem, in which all early jobs are processed before all tardy jobs.
Proof. Suppose S is an optimal schedule in which some tardy jobs are processed before early jobs. Put all tardy jobs to the end without changing the start times of early jobs and get Schedule S'. Clearly, E_{max} (S) = E_{max} (S') and n_T(S) = n_T (S'), so S' is also optimal. But now the early jobs are processed before the tardy jobs.

#

The theorem directly leads to the following result: Once Set T is defined, in search for an optimal solution it is sufficient to schedule only the jobs in Set E. The tardy jobs can be scheduled at any time after the last early job. Algorithm 3 states the solution formally.

Algorithm 3

Step 1. Schedule the jobs in Set E by Algorithm 2.

Step 2. Append the jobs in Set T to the schedule found in Step 1.

Theorem 2. Algorithm 3 solves the $1 \,||\, E_{max} \,|\, T$ in O(nlogn) time.

Proof. From Theorem 1 we know that there exists an optimal schedule in which the jobs in Set E appear before all tardy jobs. Hence the problem reduces to solve the $1| n_T = n_T^*| E_{max}$ problem where $n_T^*=0$, for Set E which can be solved by Algorithm 2.
The complexity of the algorithm is dominated by Step 1 that is implemented in O(nlogn) steps.

#

4. Conclusions

We have analyzed the bicriteria problem of minimizing maximum earliness and number of tardy jobs on a single machine. We have proposed polynomial time algorithms for several versions of the problem. A branch and bound algorithm can be developed for the NP-hard problem of generating all efficient schedules.

A challenging area for future research is to extend the ideas given in this paper to three criteria problems. Makespan is a criterion that keeps the schedule length short. Such a third criterion could be useful to control machine idle time.

References

[1] Azizoglu, M. , Koksalan, M., and Kondakci, S., "Scheduling to Minimize Maximum Earliness and Number of Tardy Jobs where Machine Idle Time is Allowed," *Technical Report No: 98-10*, Dept. of Industrial Engineering, METU, (1998).

[2] Azizoglu, M. , Kondakci, S. and Koksalan, M., "Single Machine Scheduling with Two Criteria: Maximum Earliness and Number Tardy," *Technical Report No: 97-02*, Dept. of Industrial Engineering, METU, (1997).

[3] Garey, M.R., Tarjan, R.E., and Wilfong, G.T., " One processor Scheduling with Symmetric Earliness and Tardiness Penalties," *Mathematics of Operations Research,* **13**, 330-348, (1988).

[4] Graham, R.L., Lawler, E.L., Lenstra, J.K., and Rinnooy Kan, A.H.G., " Optimization and Approximation in Deterministic

Sequencing and Scheduling: A Survey," *Annals of Discrete Mathematics,* **5**, 287-326, (1979).

[5] Hoogeveen, J.A., "Single Machine Bicriteria Scheduling," *PhD Thesis,* CWI, Amsterdam, (1992).

[6] Hoogeveen, J.A., "Minimizing Maximum Promptness and Maximum Lateness on a Single Machine," *Mathematics of Operations Research,* **21**, 100-114, (1996).

[7] Hoogeveen, J.A. , and van de Velde, S., "A Single Machine Scheduling Model for Coordinating Logistics Activities in a Simple Supply Chain," *Management Report No: 30(13)*, Rotterdam School of Management, Erasmus University, (1997).

[8] Koksalan, M., Azizoglu, M. and Kondakci, S., "Minimizing Flow Time and Maximum Earliness on a Single Machine," *IIE Transactions,* **30**, 192-200, (1998).

[9] Lee, C.Y. and Vairaktarakis, G.L., "Complexity of Single Machine Hierarchical Scheduling: A Survey," *Complexity in Numerical Optimization,* **19**, 269-298, (1993).

[10] Moore, J.M., "An n Job, One Machine Sequencing Algorithm for Minimizing the Number of Late Jobs," *Management Science,*" **15**, 102-109, (1968).

Using Compromise Programming in a Stock Market Pricing Model

Enrique Ballestero
Technical University of Madrid[1]

[1]ETS Ingenieros Agrónomos, Ciudad Universitaria 28040 Madrid (Spain)

Abstract

Institutional and private investors in stock markets insistently demand pricing models to decide their strategies. We proposed an operational research (OR) model based on few assumptions, the main of them on diversification. This model is solved by resorting to recent results in compromise programming (CP). Thus, we obtain the expected risk premiums of securities.

Keywords: Compromise Programming Shadow Prices, Pricing Models, Stock Market Analysis.

1 Introduction

Besides portfolio selection analysis which was initiated by Markovitz (1952, 1987), the field of Capital Asset Pricing Models (CAPM) is relevant in stock market approaches with a wide range of literature [see e.g., Elton and Gruber (1984), Reilly (1985), Copeland and Weston (1988)]. Indeed, CAPM is an equilibrium model describing the cross-sectional relationship between securities' rates of returns. In this paper, an OR pricing model leading to the expected risk premiums of securities (and therefore to their respective expected prices) is proposed. Our assumptions and methodology broadly differ from those obtained by other authors in the field [Fama (1976), Brealey and Myers (1996)]. From the definition of certain well-diversified portfolios (termed the p-portfolios) we introduce three specific assumptions on arbitrage, risk and diversification. Under these assumptions we formulate a system of ∞ linear programming models where the unknown variables are the expected risk premiums of securities. This system is proven to have a single solution by resorting to recent results in CP literature. The expected risk premiums obtained are proportional to the expected risk premium of the market portfolio (defined in Section 2). The factors of proportionality for the different securities are called OR betas in this paper. Such a terminology is justified as the OR betas and the standard CAPM betas are tested to be close. The test in this paper is performed through an example referred to an illustrative stock market with 10 securities (see Section 5). This result seems to be appealing as underpinning the standard CAPM betas, which have been commented in the literature [see, e.g., Arnott (1983)].

Henceforward the following notation is used: $\hat{R}$ = random returns in one period T; E = expected returns; V = variance of $\hat{R}$; σ = standard deviation; D = dividends to pay during the period T; M = market portfolio; p = diversified portfolio with the same E and V than the market portfolio M; W_{Mt} and W_{pt} = beginning values of M and p respectively; W_{Mt+1} and W_{pt+1} = ending (expected) values of M and p respectively; π_M and π_p = expected risk premium for M and p respectively; (N_1, N_2 ... N_j ... N_n) = p-portfolio composition (in percentage units, that is, $N_1 + N_2 + ... + N_n = 1$); N_{jM} (j = 1, 2, ...n) = market portfolio composition; (W_{1t}, W_{2t}, ...W_{jt}, ... W_{nt})= beginning prices of securities; (W_{1t+1}, W_{2t+1}, ...W_{jt+1}, ... W_{nt+1})= ending (expected) prices of securities; Cov(j,h)= covariance of (j, h) securities; (π_1, π_2, ... π_j, ... π_n) = expected risk premiums of securities; i = interest rate for riskless bonds; X_j^* = ideal or anchor value for the jth security; X_{j*} = anti-ideal or nadir value for the jth security (both anchor and nadir defined in Section 4).

2 Common Definitions and The Family of p-portfolios

Remember that returns on an investment are defined as the investment ending value minus the beginning value plus the cash flows (or dividends).

Definition 1. *The Market portfolio* is a largely diversified portfolio M including all securities in proportion to their market value.

Definition 2. The p-portfolios are well-diversified portfolios satisfying $E_p = E_M$ and $V_p = V_M$. In Section 3, the conditions guaranteeing the diversification of a p-portfolio will be specified.

Definition 3. Given a portfolio p its expected risk premium π_p is the difference between the p-expected return E_p and the risk free return. That is:

$$\pi_p = E_p - W_{pt} i$$

where W_{pt} is the beginning value of p and $W_{pt}i$ is the risk free return. Equivalently, the expected risk premium can be defined as the value π_p derived from:

$$W_{pt} = \frac{W_{pt+1} + D_p - \pi_p}{(1+i)}$$

Where D_p = dividends. The variables on the right-hand side of the equation are formulated as expectations.

The right-hand side of the above equation is the present value of (Ending value + Dividends)$_p$ corrected by the expected risk premium π_p. From the above equation we get:

$$W_{pt} = (E_p - \pi_p)/i \qquad (1)$$

Since $E_p = W_{pt+1} - W_{pt} + D$, that is, capital gains plus dividends. Let us highlight equation (1) such as obtained from the above relationships.

In equation (1) the variables can be specified either in percentage changes (as usual) or in dollars. Indeed, the relationships above equation (1) reflect two different types of trade-off: (a) money at the beginning of the investment period versus money at the end of this period, this trade-off being established by the present value; and (b) riskless money versus risking money, this trade-off being established through the expected risk premium. That is, according to the normal behaviour of the markets we assume risk aversion (not risk neutrality). Suppose, for example, that expectations over the ending value plus dividends are equal to 1.008. Consider the trade-off (b). Due to risk aversion, the investor is not indifferent between the expected (ending value plus dividends) = 1.008 (under risk) and a riskless value equal to 1.008. On the contrary, he will be indifferent between [1.008 (under risk) plus π_p] and a riskless value equal to 1.008. Therefore, we should establish both trade-offs by taking the present value of the equivalent riskless value, that is, $(1.008 - \pi_P)/(1 + i)$.

From equation (1) we get:

$$\pi_p = E_p - W_{pt}\, i \qquad (2)$$

Note that W_{pt} (as well as π_p) can differ from a portfolio p to another portfolio p. In other words, we do not assume equal expected value nor equal expected risk premium for the p-portfolios.

Definition 4. Analogously, the expected risk premium π_j of the jth security derives from:

$$W_{jt} = (E_j - \pi_j)/\, i \qquad (3)$$

According to (1) the expected risk premium π_M of the market portfolio derives from:

$$W_{Mt} = (E_M - \pi_M)/i \qquad (4)$$

3 Diversification Constraints of the p-portfolios

In defining the p-portfolios (see Section 2, Definition 2), we have pointed out that the p-portfolios should satisfy a diversification constraint. This constraint is formulated as follows: $N_j \geq \lambda\, N_{jM}$ for all j

where λ is a positive parameter ($0 < \lambda < 1$). The greater is the λ value, the more diversified are the p-portfolios. For example, if the greatest N_{jM} value is 0.01 and $\lambda = 0.95$, then the greatest N_j component cannot be higher than 0.0595. Generally, we have the following property: If N_{hM} is the greatest component of the market portfolio M and the above constraints are satisfied, the maximum component of any portfolio p cannot be greater than $(1-\lambda) + \lambda\, N_{hM}$. Hence, in the particular case $\lambda = 0.95$ and $N_{hM} = 0.01$ we get $(1-\lambda) + \lambda\, N_{hM} = 0.0595$.

Proof. The greatest component attainable N_{jp} is the N_{hp} component given by the following structure:

$N_{1p} = \lambda\, N_{1M}$

$N_{2p} = \lambda\, N_{2M}$

....

$N_{hp} = (1 - \lambda) + \lambda N_{hM} \geq \lambda N_{hM}$

....

$N_{np} = \lambda N_{nM}$

since $\sum N_{jp} = \sum N_{jM} = 1$. Therefore, the greatest N_{jp} is equal to $(1 - \lambda) + \lambda N_{hM}$ and the property is demonstrated.

Check the above result for $\lambda = 0.95$ and $N_{hM} = 0.01$ by the following linear programming:

Max N_{hp}

subject to

$N_{jp} \geq 0.95 N_{jM}$ for all j

$$\sum_{j=1}^{n} N_{jp} = 1$$

with the non-negativity constraints, for any numerical values of N_{jM} satisfying

$N_{jM} \leq 0.01$ and $\sum_{j=1}^{n} N_{jM} = 1$.

4. Expected Risk Premiums and Pricing Equation

As noted, the model herein does not belong to the CAPM class. Neither its assumptions nor its methodology are CAPM [see a list of the CAPM assumptions in Elton and Gruber (1984), p. 275-276 and Reilly, (1985), p. 239-240]. Particularly, we do not need the hypothesis: "all investments are mean-variance efficient" (or another assumption leading to this assertion). In other words, the assumption of a single efficient frontier for all investors is not required. The market portfolio M is not assumed to be mean-variance efficient. Although the assumption of quadratic preferences is commented by Arrow (1965, pp 96-97) and Pratt (1964, p. 132), this assumption (or alternatively, Normally distributed returns) is a necessary hypothesis for the CAPM results [see Ross (1978), Chamberlain (1983) and Berk (1997)], but it is not required for the model herein. The assumptions in the proposed model are the following:

-Assumption 1. Competitive markets.

-Assumption 2. Two well-diversified portfolios of equal total risk also have the same expected risk premium. Then, a well-diversified p-portfolio and the market portfolio M have the same risk premium $(\pi_p = \pi_M)$ as they are of equal variance.

Justification. According to CAPM, the p-portfolio total risk σ_p^2 = systematic risk plus unsystematic risk = $\sigma_p^2 \rho^2(p, M) + \sigma_\varepsilon^2$ where the unsystematic risk σ_ε^2 tends to zero due to diversification [see, e.g. Copeland and Weston (1988) pp. 198-199]. Then, the covariance term and the total risk are very close, or tantamount, correlation $\rho(p,M)$ becomes high. Thus, Assumption 2 is consistent with these CAPM results although this assumption has been introduced regardless of them.

- Assumption 3. We have:

$$\sum_{j=1}^{n} \pi_j \; N_j \; \geq \; \pi_M \tag{5}$$

that is, given a portfolio p the weighted sum of the (j = 1, 2,... n) expected risk premiums (separately considered) is greater than (or equal to) the expected risk premium π_M of the market portfolio.

Justification. This assumption seems to be plausible due to $\Sigma \pi_j N_j \geq \pi_p$ (because p-portfolio diversification effect) while, on the other hand, π_p is equal to π_M according to Assumption 2.

Consider the n securities within the market. Furthermore, consider the p-portfolio set defined as the set of diversified portfolios p satisfying $E_p = E_M$ and $V_p = V_M$. Portfolios p can be mean-variance inefficient. Therefore if (N_1, N_2, ..., N_j, ..., N_n) is the composition of a portfolio p we have:

$$E_p = E_1 N_1 + E_2 N_2 + \ldots + E_j N_j + \ldots + E_n N_n = E_M \tag{6}$$

$$V_p = \sum_j V_j N_j^2 + \sum_{hj} Cov\,(j,h)\, N_h N_j = V_M \tag{7}$$

$$N_1 + N_2 + \ldots + N_j + \ldots + N_n = 1 \tag{8}$$

Moreover, portfolios p satisfy the diversification constraints (see Section 3), that is:

$$N_j \geq \lambda\, N_{jM} \; (0 < \lambda < 1) \text{ for all } j \tag{9}$$

where λ is a positive parameter close to 1 (e.g., $\lambda = 0.95$). For diversified portfolios, the λ value can be ranged over 0.90-1.00

In competitive markets (Assumption 1) the arbitrage will minimise the gap between W_{Mt} and [$W_{1t} N_1 + \ldots + W_{jt} N_j + \ldots + W_{nt} N_n$]. From (3) - (4) the minimisation of this gap can be written as follows:

$$Min\left[W_{Mt} - \sum_{j=1}^{n} W_{jt} N_j\right] = Min\left[\frac{E_M - \pi_M}{i} - \sum_{j=1}^{n} \frac{(E_j - \pi_j)}{i} N_j\right]$$

according to (3) and (4). On the other hand, since $E_M = Ep = \sum E_j N_j$ [according to (6)] we have:

$$Min\left[W_{Mt} - \sum_{j=1}^{n} W_{jt} N_j\right] = Min\left[\frac{\sum_{j=1}^{n} \pi_j N_j - \pi_M}{i}\right] = Min\left[\sum_{j=1}^{n} \pi_j N_j - \pi_M\right] \geq 0$$

(See Assumption 3.) Then, the system of market prices (W_{1t}, W_{2t} ... W_{jt} ... W_{nt}) will derive from:

$$\text{Min}\left[\sum_{j=1}^{n} \pi_j N_j - \pi_M\right] \textit{subject to} \sum_{j=1}^{n} \pi_j N_j \geq \pi_M \qquad \textbf{(10)}$$

for every portfolio (N_1, N_2, ... N_j ... N_n) belonging to the p-portfolio set. To formulate one of the model (10), take a finite number of p-portfolios, e.g., take 200 p-portfolios. With their corresponding weights (N_1, N_2, ... N_j ... N_n) establish 200 constraints in (10). Choose the left-hand side of any constraint as the objective function to minimise. According to a recent theorem in the CP literature [Ballestero and Romero (1993). Corollary 2], the above system (10) of ∞ linear programming models has a single solution which is independent of the objective function chosen in this way and also independent of the linear model formulated. This single solution is given by:

$$\pi_j = \frac{\pi_M}{(X_j^* - X_{j*})\left[1 + \sum_{j=1}^{n} X_{j*}/(X_j^* - X_{j*})\right]} \qquad \textbf{(11)}$$

where the coefficients π_j / π_M will be called the OR betas. In equation (11) the parameters X_{j*} and X_j^* are the nadir or anti-ideal and the anchor or ideal values respectively [Zeleny (1982), Yu (1985)], defined as follows:

$X_{j*} = \lambda N_{jM}$ [see constraint (9)].

X_j^* is a value derived as follows. First step, in equation (7) we make the following substitution:
$N_1 = X_{1*}$; $N_2 = X_{2*}$ ... $N_{j-1} = X_{j-1*}$; $N_{j+1} = X_{j+1*}$ ... $N_n = X_{n*}$

Second step, we take out N_j from this equation, this result being the X_j^* anchor value.

Equation (11) depends on λ. However for the λ values corresponding to diversified portfolios (namely, λ ranged over 0.90-0.98) we obtain close solutions (see Table 3). Now we introduce (11) into equation (3). Thus, we get:

$$W_{jt}i = E_j - \frac{\pi_M}{(X_j^* - X_{j^*})\left[1 + \sum_{j=1}^{n} X_{j^*} / (X_j^* - X_{j^*})\right]} \tag{12}$$

On the other hand we have:

E_j = ending price minus beginning price plus dividends =

$$= W_{jt+1} - W_{jt} + D_j \tag{13}$$

From (12) and (13) we straightforwardly obtain:

$$W_{jt+1} = W_{jt}(1+i) - D_j + \frac{\pi_M}{(X_j^* - X_{j^*})\left[1 + \sum_{j=1}^{n} X_{j^*} / (X_j^* - X_{j^*})\right]} \tag{14}$$

From equation (4) we obtain:

$$\begin{aligned} \pi_M &= E_M - W_{Mt}i = W_{Mt+1} - W_{Mt} + D_M - W_{Mt}i = \\ &= W_{Mt+1} - W_{Mt}(1+i) + D_M \end{aligned} \tag{15}$$

Finally, by introducing π_M from (15) into equation (14), this equation yields the expected price of the jth security, that is:

$$W_{jt+1} = W_{jt}(1+i) - D_j + \frac{W_{Mt+1} - W_{Mt}(1+i) + D_M}{(X_j^* - X_{j^*})\left[1 + \sum_{j=1}^{n} X_{j^*} / (X_j^* - X_{j^*})\right]} \tag{16}$$

Equation (16) explains the expected price of every security as depending upon (a) the riskless interest rate i; (b) the market values W_{Mt} and W_{jt} at the beginning of the investment period; (c) the components of the market portfolio through $X_{j^*} = \lambda N_{jM}$ as well as the variances and covariances through the parameters X_j^*; and (d) the expected value W_{Mt+1} of the market portfolio at the end of the investment period, as well as the expected dividends D_M and D_j on the market portfolio and the jth security, respectively for the investment period. Therefore, the application of equation (16) to predictions only requires predicting the variables W_{Mt+1} (expected market value) and D_j (dividends) for the short period of investment. As the dividends are usually announced, their prediction is obviously possible. Regarding the expected W_{Mt+1} value, it can be surrogated by a proxy such as S&P500 and predicted from market information on S&P500 futures (see Section 5).

5 Using the Valuation Formula for Predictions

There is huge amount of literature on whether or not futures prices are unbiased predictor of the future price. Remember that a capital market is defined as efficient if the current stock prices adjust immediately to continuous information changes. As a consequence, security prices would fully reflect all available information, and the analyst could not employ prediction methods worthily. However, the different hypotheses on efficiency in capital markets (namely, the so-called weak, semi-strong and strong hypotheses) lead to uncertain conclusions in the light of the empirical tests [see Elton and Gruber (1984) Chapter 15, Reilly (1985) Chapter 7]. Here, we assume imperfect efficiency, so that some investors can resort to financial analysis to achieve superior returns or to predict prices in the short term. Whether or not this is a reasonable assumption is an empirical question.

Once derived the valuation formula (16) the following question raises. How is this formula to be used?

As well-known, in stock markets there are futures markets for government bonds as well as for market indexes (e.g., S&P500 futures). Unfortunately there are not futures markets for every security. Therefore, the market price of every security cannot be directly predicted from the market information.

In this situation, we can resort to the valuation formula (16) to predict the market prices of every security under the above commented assumption. Let

Table 1. Information from the Stock Market (10 securities)

Mean Values (60 monthly observations)										
	0.015	0.012	0.011	0.009	0.015	0.01	0.012	0.009	0.008	0.004
Variance and Covariance Matrix (60 monthly observations)										
	1	2	3	4	5	6	7	8	9	10
1	4.63966E-08	5.8357E-09	3.9017E-09	3.3152E-09	5.6646E-09	4.9092E-09	1.196E-08	4.2655E-09	3.8147E-09	5.5222E-10
2	5.8357E-09	9.087E-09	1.753E-09	1.8569E-09	1.456E-10	1.4668E-09	2.1877E-09	2.586E-09	1.0898E-09	1.1535E-10
3	3.9017E-09	1.753E-09	1.9032E-08	2.007E-09	2.61E-09	2.9496E-09	5.0085E-09	1.8105E-09	1.018E-09	1.5942E-10
4	3.3152E-09	1.8569E-09	2.007E-09	2.5482E-09	1.5749E-09	9.152E-10	2.5327E-09	1.1723E-09	4.4139E-10	1.1997E-10
5	5.6646E-09	1.456E-10	2.61E-09	1.5749E-09	2.287E-08	2.8385E-09	5.0896E-09	2.6363E-09	1.601E-09	4.351E-10
6	4.9092E-09	1.4668E-09	2.9496E-09	9.152E-10	2.8385E-09	4.3465E-09	2.4394E-09	5.4194E-10	8.7741E-10	1.4979E-10
7	1.196E-08	2.1877E-09	5.0085E-09	2.5327E-09	5.0896E-09	2.4394E-09	1.4902E-08	3.6233E-09	1.7611E-09	3.3492E-10
8	4.2655E-09	2.586E-09	1.8105E-09	1.1723E-09	2.6363E-09	5.4194E-10	3.6233E-09	4.2629E-09	9.0639E-10	1.1673E-10
9	3.8147E-09	1.0898E-09	1.018E-09	4.4139E-10	1.601E-09	8.7741E-10	1.7611E-09	9.0639E-10	1.7537E-09	7.2208E-11
10	5.5222E-10	1.1535E-10	1.5942E-10	1.1997E-10	4.351E-10	1.4979E-10	3.3492E-10	1.1673E-10	7.2208E-11	8.2532E-11
Information on the market portfolio M and beta coefficients.										
Weights, N_{jM}	0.0845	0.0963	0.1045	0.1093	0.1068	0.0985	0.0867	0.0976	0.1059	0.1079
Mean value of M =	0.0103649			Variance of M= (60 monthly obervations)				3.00067E-09		
Cov(j,M)	8.27132E-09	2.50035E-09	4.01546E-09	1.60295E-09	4.57503E-09	2.07203E-09	4.6797E-09	2.10569E-09	1.2742E-09	2.05979E-10
CAPM Betas	2.756489907	0.833264429	1.338186344	0.534198524	1.524667954	0.690522395	1.559551463	0.701738991	0.424636234	0.06864435

us highlight the use of this formula by the following example.

In May, 1, 1998, an analyst wants to predict the security prices in the short run (e.g. on June, 1998). For this intent, he imagines the following investment: To buy a S&P500 portfolio in the spot market on May, 1, 1998; and selling it in the spot market on June, 1, 1998. Let, for example, W_{Mt}= 1.016 (May, 1) be the beginning price and W_{Mt+1} (June, 1) be the ending price corresponding to this investment. To predict the ending price, the analyst utilises the futures market information. The steps are as follows:

(a) To observe the price of S&P500 index in the futures market corresponding to June, 1, 1998. Suppose this price is 1.03. As noted, it is a good surrogate or proxy for the market portfolio value. In this way, we have:

W_{Mt+1} (June, 1) = 1.03

(b) The riskless interest rate i on May, 1, is computed as follows.

Consider a riskless bond (or a portfolio of riskless government bonds) denoted by F. Also consider its price W_F (May, 1) in the spot market. The rate i (May, 1) is derived from the following equation:

W_F (May, 1) = F (i)

where F(i) is the present value of all riskless payments (interest cash-flow and principal at the maturity of the bond) from May, 1, 1998, this present value being computed at the unknown rate i.

Thus, we obtain, e.g., i = 0.003.

(c) For simplicity, in our example, we suppose a market with only ten securities instead of the real world market with 500 securities or more. In Table 1, information from an illustrative market with ten securities is shown. This information includes the mean values, variances and covariances using 60 monthly observations. Also in this table, the market portfolio M percentage composition (N_{1M}, N_{2M}, ..., N_{jM}, ... N_{nM}) (n = 10) is recorded, as well as the mean value and variance of M as derived from the indicated 60 monthly observations.

The variance and covariance as well as the market portfolio weights will be employed in the computing process [see paragraph (f) below].

(d) On May, 1, the dividends D_j to be paid by the companies in the near months are known [see Table 2 column (7)]. As the composition of the market index (surrogate or proxy for the market portfolio), is also known, a perfect foresight for dividends D_M can be straightforwardly obtained on May, 1. In Table 2 this aggregated dividend D_M (corresponding to May) is also shown at the bottom of column (7).

Table 2. Computing process for $\lambda = 0.95$

(1)	(2)	(3)	(4)	(5)	(6)	(7)	(8)	(9)	(10)
0.080275	0.0979667	0.0176917	4.53744102	0.36909080	2.70936041	0.03915568	0.008	1.074	1.10837768
0.093385	0.1490401	0.0556551	1.67792350	1.16109797	0.86125377	0.01244684	0.003	1.275	1.28827184
0.099275	0.1345005	0.0352255	2.81827085	0.73488785	1.36075185	0.01966559	0.004	0.995	1.01365059
0.103835	0.1932115	0.0893765	1.16177071	1.86460671	0.53630613	0.00775070	0.002	0.984	0.99270270
0.101460	0.1325702	0.0311102	3.26130980	0.64903288	1.54075399	0.02226698	0.003	1.447	1.47060798
0.093575	0.1626246	0.0690496	1.35518526	1.44053915	0.69418454	0.01003235	0.001	1.064	1.07622435
0.082365	0.1136304	0.0312654	2.63438178	0.65227073	1.53310574	0.02215644	0.004	1.055	1.07632144
0.092720	0.1608936	0.0681736	1.36005727	1.42226370	0.70310449	0.01016127	0.005	0.933	0.94096027
0.100605	0.2123966	0.1117916	0.89993345	2.33223909	0.42877251	0.00619662	0.003	0.890	0.89586662
0.102505	0.7591311	0.6566261	0.15610863	13.6987847	0.07299918	0.00105498	0.003	0.489	0.48852198
			19.86238230				0.0035	1.016	1.03

(1) Nadir values $X_{j*} = 0.95N_{jM}$; (2) Anchor values X_j^* (derived as indicated in Section 5, f) ; (3) = (2)-(1) = $X_j^* - X_{j*}$; (4) = (1)/(3) = $X_{j*}/(X_j^*-X_{j*})$; (5) = (3) [1+Σ(4)] = $(X_j^*-X_{j*})[1 + \Sigma X_{j*}/(X_j^* - X_{j*})]$; (6) = 1/(5), that is, the OR betas ; (7) = π_M /(5) = 0.014452/ (5), that is the expected risk premiums of securities; (8) = D_j (dividends of the securities) as well as D_M (bottom); (9) = W_{jt} (beginning prices of the securities) as well as W_{Mt} (bottom) ; (10) = W_{jt+1} (ending prices of the securities, that is, predictions) as well as W_{Mt+1} (bottom).

On the other hand the beginning values W_{jt} of securities on May 1 are shown in Table 2, column (9).

(e) The expected value of M as derived from the futures market is:

E_M (May) = W_{Mt+1} (June,1) $- W_{Mt}$ (May,1) $+ D_M$ (May) = 1.03 - 1.016 + 0.0035 = 0.0175

From this prediction, the risk premium of the market portfolio becomes:

π_M (May)= E_M (May) - W_{Mt} (May,1) × i = 0.0175 - 1.016 × 0.003 = 0.014452

(f) The anti-ideal (or nadir) X_{j*} values are given by (9), that is, $X_{j*} = \lambda N_{jM}$ (see the N_{jM} values in Table 1). For the computing process in Table 2 we have taken λ =0.95. However, the final results by taking other λ values, such as λ =0.90 and λ =0.98 are shown in Table 3. Check that the different λ values lead to close solutions.

To obtain the X_j^* anchor value in equation (7) we substitute the (X_{1*}, X_{2*}, ... X_{j-1*}, X_{j+1*} ... X_{n*}) nadir values for the (N_1, N_2, ...N_{j-1}, N_{j+1}...N_n) and take out $N_j = X_j^*$. The variance and covariance in equation (7) are derived from time series of returns [see the above paragraph (c) and Table 1].

(g) The computing process to introduce those anchor and nadir values into equation (16) is developed in Table 2.

From Table 2 [column (5)] taking into account that π_M (May) = 0.014452, we obtain the expected risk premium π_j (May) for every security [see column (7)]. Thus, e.g., for the security j = 2 we get

$$\pi_2 = 0.014452/1.16109797 = 0.0124$$

(h) As a result of the computing process in Table 2, we obtain the predicted prices of securities [see column (10)]. These prices have been computed with λ =0.95. However as noted above, the results for other λ values, corresponding to diversified portfolios (λ =0.90 and λ =0.98) are very close to those obtained with λ =0.95, (see Table 3).

Table 3. Comparing Results: OR betas for λ =0.90-0.95-0.98 and CAPM betas.

(1)	(2)	(3)	(4)	(5)	(6)	(7)
2.68277424	2.70936041	2.73479198	2.75648991	0.97325741	0.98290235	0.99212842
0.88671799	0.86125377	0.84497448	0.83326443	1.06414957	1.03358998	1.01405322
1.38330365	1.36075185	1.346799	1.33818634	1.03371527	1.01686275	1.00643606
0.54047963	0.53630613	0.53479083	0.53419852	1.011758	1.00394535	1.00110877
1.56066144	1.54075399	1.53071384	1.52466795	1.02360742	1.01055052	1.00396538
0.70032836	0.69418454	0.69169138	0.69052239	1.01420079	1.00530343	1.00169289
1.51830352	1.53310574	1.54737859	1.55955146	0.97355141	0.98304274	0.99219463
0.70745684	0.70310449	0.70191034	0.70173899	1.00814811	1.00194589	1.00024417
0.43404493	0.42877251	0.42617159	0.42463823	1.02215226	1.00973599	1.00361098
0.07656609	0.07299918	0.07052936	0.06864435	1.11540262	1.06344041	1.02746051

(1) OR beta, namely, a coefficient proportional to the jth security risk premium, for λ =0.90 ; (2) OR beta for λ =0.95; (3) OR beta for λ =0.98 ; (4) Standard CAPM betas; (5) = (1)/(4); (6) = (2)/(4); (7) = (3)/(4).

Therefore, all information required for predictions through equation (16) are observable from the stock markets. However, the accuracy of price foresight raises a statistical problem on the deviations between actual and expected prices in the stock market. This empirical problem is not undertaken in this paper.

6 OR Betas and Standard CAPM Betas: A Conclusion

Equation (11) can be written as follows:

$$E_j - W_{jt} i = \frac{E_M - W_{Mt} i}{(X_j^* - X_{j*})\left[1 + \sum_{j=1}^{n} X_{j*}/(X_j^* - X_{j*})\right]} \tag{17}$$

according to (3) and (4). Now compare (17) to the standard CAPM equation:

$$E_j - R_F = \beta_j (E_M - R_F) \tag{18}$$

where R_F is the risk free rate and β_j is the standard CAPM beta coefficient of the jth security.

Note that the form of equation (17) is equal to the form of equation (18). However, these equations come from different models derived from very different assumptions and methodology. Then the similarity between (17) and (18) raises a critical question. Will the CAPM beta coefficients β_j be related to the OR beta coefficients

$$\frac{1}{(X_j^* - X_{j*})\left[1 + \sum_{j=1}^{n} X_{j*}/(X_j^* - X_{j*})\right]} \tag{19}$$

in any way? In Table 3 we find an interesting result. The empirical values for the CAPM beta coefficients almost coincide with the empirical values of (19), despite they have been obtained by quite different methodologies and computing processes. This test seems to lead to a conclusion, the OR model is able to explain relevant aspects of the capital market behaviour under a few plausible assumptions. Since the OR betas are virtually equal to the CAPM betas, the predictions from the OR model herein will also coincide with the predictions from the CAPM betas.

References

(1) Arnott, R.D. (1983). "What Hath MPT Wrought: Which Risk Reap Rewards?" in Bersnstein P.L. and Fabozzi F.J. Streetwise. The Best of the Journal of Portfolio Management. Princeton University Press. Princeton. 41-47.

(2) Arrow, K. (1965). *Aspects of the Theory of Risk Bearing*. Academic Book Store, Helsinky.

(3) Ballestero, E. and Romero, C (1993). "Weighting in compromise programming: A theorem on shadow prices", *Operations Research Letters* 13, 325-329.

(4) Berk, J.B. (1997). "Necessary Conditions for the CAPM," *Journal of Economic Theory* 73, 245-257.

(5) Brealey, R.A. and Myers, S.C. (1996). Principles of Corporate Finance (5th Edition). McGraw-Hill, New York.

(6) Chamberlain, G. (1983). "A Characterization of the Distributions that Imply Mean-Variance Utility Functions," *Journal of Economic Theory* 29, 185-201.

(7) Copeland, T. and Weston, J (1988). *Financial theory and corporate policy*. Addison-Wesley Publishing Company, Inc.

(8) Elton, E.J. and Gruber, M.J. (1984). *Modern portfolio theory and investment analysis*, John Wiley & Sons, New York

(9) Fama, E.F. (1976). *Foundations of finance*. Basic Books. New York.

(10) Markovitz , H.M. (1952). "Portfolio Selection", *Journal of Finance* 7, 77-91.

(11) Markowitz, H.M. (1987). Mean-Variance Analysis in portfolio choice and capital markets. Basil Blackwell, New York.

(12) Pratt, J.W. (1964). "Risk Aversion in the Small and in the Large", *Econometrica*, 32, 1-2, 122-136.

(13) Reilly, F.K., (1985). Investment analysis and portfolio management, CBS College Publishing.

(14) Ross, S.A. (1976). "Mutual Fund Separation in Financial Theory - The Separating Distributions," *Journal of Economic Theory* 17, 254-286.

(15) Yu, P.L. (1985) *Multiple Criteria Decision Making. Concepts, Techniques and Extensions*, Plenum Press, New York.

(16) Zeleny, M. (1982). *Multiple Criteria Decision Making*. McGraw Hill. New York.

Redistribution of Funds for Teaching and Research among Universities

G. Fandel and T. Gal[1]

Abstract

The Ministry of Science and Research of the state of North-Rhine Westphalia has initiated a procedure for the performance- and success-based redistribution of funds for teaching and research among universities. The fundamental procedure to determine a solution for this decision situation that is accepted by all universities may be described as a three-level decision process with multiple objectives. The three decision levels are the ministry, the rectors' council and the individual universities. Criteria for a performance- and success-oriented distribution of funds in teaching are the proportions of academic personnel employed, the proportions of students in the first 4 semesters, and the proportions of graduates. Criteria for assessing successful research are the proportions of outside funds and the proportions of PhDs.

Given these criteria, the solution process consists of agreeing on the weights for the criteria. Here, the ministry prefers the proportions of students in the first 4 semesters and the proportions of graduates. On the other hand, at the rectors' council weights are sought that do not cause the re-distribution of the budget for teaching and research to deviate too much from the actual distribution of funds among universities. Each individual university, however, is interested in weights that lead to its receiving the lion's share of the redistribution budget.

Distance minimizing, linear programming with multiple objectives and goal programming are the mathematical approaches with whose help the problem of redistribution among universities can be coped with practically on the basis of

---

1 Prof. Dr. Günter Fandel and Prof. Dr. Dr. Tomas Gal, Department of Economics, FernUniversität, D-58084 Hagen, Germany. We would like to thank Dipl.-Kfm. Thomas Pitz and Dipl.-Kfm. Armin Rudolph, M.A. (WSU) of the Department of Economics, FernUniversität, for their support during calculating and implementing the computer program.

real data. The findings derived with these methods will then be compared with the realized solutions.

Keywords: multiple criteria decision making, MCDM, University, redistribution of funds, allokation of budgets

1. Starting situation

At the instigation of the parliament of the state of North-Rhine Westphalia the Science and Research Ministry has started to redistribute some of the funds for teaching and research among the 15 universities in the state in accordance with defined criteria for performance and success. The budget available for redistribution amounts to DM 148.58 million and has to be provided by the universities themselves from 50 % of the funds from the approved budget for teaching and research.

The criteria on which the redistribution is based differ for the two areas of teaching and research. For teaching these are for each university

(1) the proportion of academic personnel employed,

(2) the proportion of students in the first 4 semesters, and in the case of the distance teaching at the FernUniversität in addition half of the part-time students in the first 8 semesters, because due to their employment these have twice the regular study period with half the study load as against full-time students, and

(3) the proportion of graduates, with the analogously modified conversion of the number of part-time graduates at the FernUniversität as under (2).

To record achievements and successes in research the redistribution takes into account

(4) the proportion of funds from outside sources and

(5) the proportion of successfully finished PhDs

in each university.

The starting situation outlined here can be described by the data gathered in Table 1. Column 1 shows the universities in North-Rhine Westphalia included in the redistribution of funds for teaching and research. These are numbered consecutively in column 2 with i, $i = 1,2,\ldots,15$. Columns 3 to 7 contain the proportions a_{ij} of the universities i in the redistribution criteria j, $i = 1,2,\ldots,5$, in accordance with statements (1) to (5).

Table 1: Initial data in 1996 for redistribution in 1997

		Unweighted parameters (%)					Actual distribution	
		Acad. personnel	Stud.	Grads.	Outside funds	PhDs	DM mill.	%
University	No.	a_{i1}	a_{i2}	a_{i3}	a_{i4}	a_{i5}	B_i	$P_i = B_i/B$
1	2	3	4	5	6	7	8	9
Aachen	1	13.46	10.30	15.33	15.40	20.34	20.44	13.76
Bielefeld	2	5.24	5.58	3.61	10.31	4.70	7.44	5.01
Bochum	3	10.66	8.74	8.84	11.17	10.95	15.16	10.21
Bonn	4	9.89	9.18	11.89	12.96	15.95	17.20	11.58
Dortmund	5	8.61	8.52	7.70	5.00	7.26	10.28	6.92
Düsseldorf	6	4.51	5.42	2.88	4.70	5.94	5.97	4.02
Cologne	7	8.17	11.98	10.61	8.71	10.81	15.82	10.65
Münster	8	9.29	10.86	12.26	9.15	10.47	17.23	11.59
DSH Cologne	9	1.08	1.62	2.10	1.17	0.37	1.80	1.21
Duisburg	10	4.86	3.70	3.92	2.17	2.82	5.80	3.90
Essen	11	5.97	6.56	5.00	3.23	3.12	7.39	4.97
Paderborn	12	6.03	5.49	6.34	3.68	2.40	7.79	5.24
Siegen	13	4.60	4.00	3.72	4.04	2.02	6.04	4.07
Wuppertal	14	5.50	5.50	4.44	4.39	2.17	6.78	4.56
FU Hagen	15	2.13	2.56	1.36	3.92	0.66	3.43	2.31
Total		100.00	100.01	100.00	100.00	99.98	148.58 = B	100.00

Column 8 indicates the amount B_i, rounded off to the nearest DM 10,000, that the university i has currently to provide for the redistribution from its own funds, and column 9 shows the corresponding percentage rate P_i they have of the actual distribution at the moment, with $i = 1,2,...,15$.

The redistribution among universities in North-Rhine Westphalia may be understood as a n-person negotiation problem equivalent to a multiple criteria decision making (MCDM) problem (see, e.g., RATICK et al. 1980, SAKAWA et al. 1994). The task is to determine for the five criteria referred to a weighting vector

$$g^* = (g_1^*, g_2^*, ..., g_5^*)'$$

that leads to a redistribution solution in the sense of a compromise which is accepted by all universities. The negotiation process is formalized in that it takes place at the level of the Landesrektorenkonferenz, the council of the rectors of all 15 universities in North-Rhine Westphalia. At the Landesrektorenkonferenz the universities concerned bring in their own individual ideas for determining the weighting vector, and the Science and Research Ministry of the state of North-Rhine Westphalia acts as moderator with a political mandate from parliament. Of course, interests at the different levels of negotiation differ:

(a) Each university prefers that weighting vector $\hat{g}^i$ which is granting the greatest amount from the redistribution pool. However, the perfect solution, namely to satisfy all maximum individual interests simultaneously at one go, is not feasible because of the zero sum character of the negotiation problem. A compromise solution must therefore be found.

(b) At the level of the Landesrektorenkonferenz a compromise solution with regard to the determination of weighting vector g will be sought that leads to a redistribution which is as close as possible to the current actual distribution. With a zero sum distribution problem, this guarantees that each university which is a loser in the distribution process does not have to take too much losses. Here the distance concept in one or other form might be an appropriate representation (see, e.g., SAWARAGI et al. 1985) of conservative behaviour in the negotiation process. Note that there is no weighting vector for generating the current actual distribution; otherwise there would be no formal distance minimizing problem at this level.

(c) In order to represent the public, to which it is accountable, the ministry strongly prefers weighting vector g with high values of g_2 and g_3 emphasizing the important role of the proportions of students in the first 4 semesters, and in particular the proportions of graduates. It is these two criteria that permit the achievements and successes of the universities as academic training locations to be justified most plausibly to the public, and in particular to students as the consumers of these teaching services.

2. Solutions under consideration

The vector $\hat{z} = (\hat{z}_1, ..., \hat{z}_{15})'$ represents the perfect solution in GEOFFRION'S definition (GEOFFRION 1965; see also, e.g., STEUER 1986, SAWARAGI et al. 1985); its deviations from the actual distribution from col. 3, table 2 (corresponding to col. 8, table 1), are shown in col. 4, table 2. The perfect solution is not feasible because it has only positive deviations which is in contradiction to the constant sum negotiation, i.e. there is no $g \in G$ with $z(g) = \hat{z}$.

In connection with the determination of the perfect solution it is seen that the five weighting vectors g^{ℓ}

$$\begin{aligned} \text{with} \quad & g_j^{\ell} = 1, \text{ if } j = \ell, \\ \text{and} \quad & g_j^{\ell} = 0, \text{ if } j \neq \ell, \\ & \ell, j = 1, \ldots, 5, \end{aligned}$$

in the 15 optimization problems for determining the individual maximal solutions of the 15 universities each occur three times as optimal solutions. This suggests that a compromise solution can be seen in a uniformly distributed vote of the universities for the possible components of the weighting vector, i.e. that the weighting vector

$$\tilde{g} = (0{,}2;\ 0{,}2;\ 0{,}2;\ 0{,}2;\ 0{,}2)'$$

is selected as a compromise solution for the negotiation problem from the universities' point of view. The deviation differences from the actual distribution occurring for solution $\tilde{g}$ are shown in columns 5 and 6 of table 2 both absolutely and as percentages. Regarding the percentage deviations of this proposed solution from the actual distribution, it quickly becomes clear that the universities are hardly likely to accept $\tilde{g}$ unanimously as a compromise, because five universities would lose nearly 10% of their actual budget and two universities would gain about 20%. A margin of deviation of this kind in this frequency appears not to be acceptable as a solution for the universities. For this reason, the Landesrektorenkonferenz has never discussed a compromise solution that lies as close as possible to the perfect solution.

A solution of this kind for the distance norm $q = 2$ was calculated here for the purposes of a comparative analysis. It is shown in columns 7 and 8 in table 2 in its absolute and percentage deviation differences as against the actual distribution. A glance at these two columns quite obviously supports the presumption referred to above that this solution is not acceptable as a compromise solution because of the margins of deviation and their frequencies. This is reflected in the terms $\|z\|_1$ and $\|z\|_\infty$ in column 7 in table 2, which indicate the sum of the absolute deviations and the maximum absolute deviation from the initial distribution and have poorer values than the solution for $\tilde{g}$, which has already been subjected to criticism. In addition, the corresponding weighting vector $\hat{g}^{(2)}$ is biased towards research, and this is not appropriate for universities as institutes of higher education.

Instead of this, the Landesrektorenkonferenz tended to look for compromise solutions the redistribution results of which were "as close as possible" to the actual distribution.

This substitute programme would have an absolute minimum for all distance norms $q \geq 1$, if a weighting vector g^0 existed for which the deviations from the actual distribution would be equal to zero, but there is not such a g^0. The equation system with the values in columns 3 to 7 in table 1 as coefficients and the values in column 9 of table 1 as the right-hand side of the equations quite obviously does not contain any solution, because it is redundant.

In concrete terms, the redistribution problem has now been solved for the distance norms $q = 1$, $q = 2$ and $q = \infty$. For $q = 1$ and $q = 2$ equivalent goal programming approaches in accordance with, e.g., IJIRI (1965), IGNIZIO (1978), ROMERO (1991) have been applied for calculating the solutions. This can also be used for $q = \infty$ (Tschebyscheff's approximation; see, e.g., STEUER 1986).

The results of these approaches are shown in columns 9 to 14 in table 2 in their absolute and relative deviation differences to the actual distribution. A comparison of the percentage figures in columns 10, 12 and 14 with those in column 6 of table 2 shows directly that with the deviation differences for the universities as against the actual distribution there are on the whole smaller percentage rates, if we overlook the Deutsche Sporthochschule (DSH Cologne; $i = 9$) as the runaway with the redistribution profits because of the smallest budget, and the fact that the University of Bochum $(i = 3)$ is the biggest loser, and the University of Paderborn $(i = 12)$ is suddenly a winner.

The sum of the absolute deviations $\|z\|_1$ and the maximum absolute deviation $\|z\|_\infty$ for the weighting vectors $g^{(1)}$, $g^{(2)}$ and $g^{(\infty)}$ as solutions for the optimization problem variant to come as close as possible to the actual distribution with the distance norms $q = 1$, $q = 2$ and $q = \infty$ are particularly smaller than those for the equal distribution vector $\tilde{g}$. This can be seen in the last two lines of table 2 in the comparison.

The additional modification, namely restricting the loss differences for the universities to a fixed percentage rate α by means of additional constraints, leads to hardly any solution changes for $0.1 \leq \alpha \leq 1$; on the other hand the solution space for $\alpha < 0.058$ is empty, i.e., if redistributions are required which bring less than 5% loss for all participating universities with regard to the initial distribution, there is no solution $g \in G$.

Columns 15 and 16 in table 2 show the real negotiated target distribution in its absolute and percentage deviation differences. It is surprising that the sum of the

Approach		Actual distribution	Perfect solution	Equal distribution		Closest to perfect solution q=2		Closest to the actual distribution						Negotiated target distribution	
								q=1		q=2		q=			
		$\nexists g \in G$	$\nexists g \in G$	$\tilde{g} = \begin{pmatrix} 0{,}2 \\ 0{,}2 \\ 0{,}2 \\ 0{,}2 \\ 0{,}2 \end{pmatrix}$		$\hat{g}^{(2)} = \begin{pmatrix} 0{,}0000 \\ 0{,}0000 \\ 0{,}1942 \\ 0{,}4305 \\ 0{,}3752 \end{pmatrix}$		$g^{(1)} = \begin{pmatrix} 0{,}0000 \\ 0{,}2614 \\ 0{,}5120 \\ 0{,}1746 \\ 0{,}0520 \end{pmatrix}$		$g^{(2)} = \begin{pmatrix} 0{,}0503 \\ 0{,}2647 \\ 0{,}4648 \\ 0{,}1720 \\ 0{,}0482 \end{pmatrix}$		$g^{(\infty)} = \begin{pmatrix} 0{,}1546 \\ 0{,}1348 \\ 0{,}5051 \\ 0{,}2045 \\ 0{,}0000 \end{pmatrix}$		$g^{S} = \begin{pmatrix} 0{,}20 \\ 0{,}20 \\ 0{,}35 \\ 0{,}20 \\ 0{,}05 \end{pmatrix}$	
			Dev. from col. 3	Dev. from col. 3		Dev. from col. 3		Dev. from col. 3		Dev. from col. 3		Dev. from col. 3		Dev. from col. 3	
University	No.	DM mill.	DM mill.	DM mill.	%	DM mill.	%	DM mill.	%	DM mill.	%	DM mill.	%	DM mill.	%
1	2	3	4	5	6	7	8	9	10	11	12	13	14	15	16
Aachen	1	20,44	9,78	1,80	8,80	5,18	25,32	0,79	3,88	0,60	2,92	0,92	4,51	0,68	3,31
Bielefeld	2	7,44	7,88	1,31	17,63	2,82	37,91	0,51	6,90	0,61	8,23	0,73	9,86	1,11	14,87
Bochum	3	15,16	1,43	-0,20	-1,32	0,64	4,19	-1,30	-8,58	-1,19	-7,84	-0,92	-6,08	-0,66	-4,37
Bonn	4	17,20	6,49	0,59	3,40	3,41	19,80	0,00	0,00	-0,19	-1,12	-2,19	-1,27	-0,31	-1,81
Dortmund	5	10,28	2,52	0,74	7,25	-0,81	-7,88	0,75	7,28	0,83	8,09	0,72	6,99	0,84	8,14
Düsseldorf	6	5,97	2,85	0,99	16,64	1,17	19,66	0,00	0,00	0,11	1,83	-0,26	-4,30	0,31	5,20
Cologne	7	15,82	1,98	-0,88	-5,55	-1,16	-7,33	0,00	0,00	-0,17	-1,08	-0,92	-5,83	-0,92	-5,79
Münster	8	17,23	0,99	-1,76	-10,24	-2,00	-11,60	-0,50	-2,89	-0,71	-4,10	-0,92	-5,83	-1,37	-7,97
DSH Col.	9	1,80	1,32	0,08	4,47	-0,24	-13,47	0,76	41,89	0,69	38,27	0,70	38,95	0,43	23,59
Duisburg	10	5,80	1,42	-0,61	-10,48	-1,71	-29,46	-0,60	-10,34	-0,52	-8,93	-0,33	-5,75	-0,36	-6,23
Essen	11	7,39	2,36	-0,29	-3,97	-2,14	-28,98	0,04	0,55	0,14	1,86	0,04	0,52	0,13	1,74
Paderborn	12	7,79	1,63	-0,68	-8,71	-2,27	-29,15	0,30	3,89	0,31	3,95	0,58	7,41	0,20	2,58
Siegen	13	6,04	0,79	-0,58	-9,61	-1,26	-20,83	-0,45	-7,53	-0,38	-6,28	-1,58	-2,62	-0,20	-3,36
Wuppertal	14	6,78	1,39	-0,24	-3,56	-1,48	-21,83	0,04	0,61	0,14	2,04	0,26	3,84	0,26	3,89
FU Hagen	15	3,43	2,39	-0,28	-8,02	-0,17	-4,85	-0,34	-9,83	-0,28	-8,16	-0,22	-6,33	-0,12	-3,52
$\lVert z \rVert_1$		0	45,23	11,03	-	45,25	-	6,39	-	6,86	-	7,90	-	7,90	-
$\lVert z \rVert_\infty$		0	9,78	1,80	-	5,18	-	1,30	-	1,19	-	0,92	-	1,37	-

Table 2: Solution variants for the negotiation problem

absolute deviations is worse or the same, but that the maximum absolute deviations are greater so that

$$\|z(g^{(1)})\|_1 < \|z(g^{(\infty)})\|_1 = \|z(g^{(S)})\|_1$$

$$\|z(g^{(\infty)})\|_\infty < \|z(g^{(1)})\|_\infty = \|z(g^{(S)})\|_\infty .$$

At the same time, the weighting components g_2^S and g_3^S have significantly smaller values than $g_2^{(1)}$ or $g_3^{(1)}$. From the universities' and the ministry's point of view it might seem more rational on the basis of the distance concept or of equivalent goal programming approaches to prefer solution $g^{(1)}$ to solution $g^{(2)}$. On the other hand, applying the Tschebyscheff approximation, $g^{(\infty)}$ would be clearly preferred in a comparison with g^S because it leads to much smaller maximum deviations showing simultaneously the same deviation sum.

3. Some critical remarks on the suitability of the redistribution concept

Although we must in principle welcome the attempt to distribute budgets among universities according to criteria of loads and success in order to achieve an efficient allocation of resources among and within universities, this control concept does have some weak points which are discussed briefly below and can lead to reservations with regard to a redistribution approach.

Concentrating a redistribution of funds for teaching and research basically on the proportion of students in the first 4 semesters and the proportion of graduates is, from the point of view of research in universities, certainly problematic because research in general is weighted much less in comparison with teaching, and the criterion of the proportion of successfully finished PhDs in particular is treated laughably with a weighting of 5%. Universities which today place little emphasis on stimulating their own young academics will experience a considerable setback in their research potential in the medium to long term, and this will also affect the quality of their teaching because of the unity of teaching and research. Generally, it is questionable whether this control approach of a load- and success-oriented redistribution of funds in teaching and research between universities will be successful and lead to a perceptible effect on an economic redistribution of resources in universities, if we remind ourselves that the redistribution is installed as a constant sum concept, and that universities have been hopelessly underfinanced in the area of personnel for ten years with simultaneous increases in the number of students. So the impression is created that this is rather a redistribution of considerable deficiencies than a selective reward for successful activities in teaching and research.

But even within the intended redistribution concept reservations can be seen as to whether the implied means-effects relationships are actually correct. In a period in which, in the Federal Republic of Germany, the retention of the status of a student is socially advantageous, in which students are tending to study even longer in various disciplines, and in which even students of traditional universities have called for the possibility of part-time studying, it is questionable that with the reward for their proportion of graduates universities will also be able to have an effective influence on the motivation of their students with regard to finishing their studies quickly and successfully. What is even more aggravating in this evaluation is that rewarding the proportion of students in the first 4 semesters makes little sense when we take into consideration the present situation for graduates on the employment market in Germany. For example, it can already be seen now that graduates from mechanical engineering and engineering science degree courses are enrolling in the distance teaching FernUniversität for a second degree course in economics to avoid unemployment after they graduate with their first degree.

4. References

Fandel, G.: Optimale Entscheidung bei mehrfacher Zielsetzung, Berlin et al. 1972.

Geoffrion, A.M.: A Parametric Programming Solution to the Vector Maximum Problem, with Applications to Decisions under Uncertainty, Stanford/California 1965.

Ignizio, J.P.: A Review of Goal Programming: A Tool for Multiobjective Analysis, J. Oper. Res. Soc. 29, 11, 1109-1119, 1978.

Ijiri, Y.: Management Goals and Accounting for Control, Amsterdam 1965.

Ratick, S.J., J. Cohon, C. ReVelle: Multiobjective Programming Solutions to n-person Bargaining Games, J.N. Morse (ed.): Organizations: Multiple Agents with Multiple Criteria, Lecture Notes in Econ. and Math. Syst., Berlin et al. 1981, 296-320.

Romero, C.: Handbook of Critical Issues in Goal Programming, Oxford 1991.

Sakawa, M., I. Nishizaki: N-Person Cooperative Games with Multiple Scenarios, paper presented at the XIth Int. Conf. on MCDM, Coimbra, Portugal 1994.

Sawaragi, Y., H. Nakayama, T. Tanino: Theory of Multiobjective Optimization, Orlando 1985.

Steuer, R.: Multiple Criteria Optimization: Theory, Computation and Application, New York et al. 1986.

Multicriteria Optimization Accounting for Uncertainty in Dynamic Problem of Power Generation Expansion Planning

Vladimir I. Kalika[1] and Shimon Frant[2]

[1] University of Haifa, Natural Resources and Environmental Research Center, Mount Carmel, Haifa 31905, Israel

[2] Israel Electric Corporation, Haifa, Israel

Abstract. The new approach (including software developed) enables selecting a predetermined number of "reasonable" power generation system (PGS) expansion alternatives from their initial set (ISA) in accordance with multiple criteria. The dynamic problem is considered, since each alternative from ISA is a possible dynamic strategy of PGS expansion during the planning period (30-50 years). The approach is interpreted as a multi-stage process including: (1) Constructing the ISA as a dynamic tree of possible combinations of generating units to be installed in each planning year. The ISA may include a vast number of alternatives and a "reduction-recovery" procedure has been developed to enable solving the problem in a desired computer time. (2) Determining a criteria assessment vector (CAV) for each initial alternative according to the developed calculation models. (3) Multicriteria optimization of the CAV set, interrelated to the ISA, using various techniques (TOPSIS, Monte Carlo simulation, Pareto optimization), modified to account for uncertainty and decision makers' estimations. The main method to account for uncertainty is by obtaining "stable" solutions resulting from multi-variant computations.

Keywords: Multicriteria decision making; dynamic aspects; uncertainty; scenarios, multi-variant computations

1. Introduction

An approach is proposed to solve a specific problem of medium/long term power generation system (PGS) expansion planning, interpreted as follows.

A predetermined number of the "best", called "reasonable", alternatives of the PGS expansion is selected from their initial set according to a given (a priori) multiple criteria (economical, environmental, etc.). Each such alternative is a possible dynamic strategy of PGS expansion during a given planning period (up to 30-50 years). This strategy has to reflect a successive "movement" of PGS through years of the planning period, from a state of PGS in each preceding planning year to the appropriate state of PGS in the next year. The initial set of alternatives (ISA) reflects all possible alternatives or all possible combinations of such states of PGS, taken one at a time for each planning year.

The purpose of proposed calculation process to solve the problem is to find a set of "reasonable" alternatives (RAS), including a predetermined number of the "best" (in a certain sense) alternatives, selected in accordance with all given objectives (criteria) jointly. The RAS is treated as a basis for a further finding of a final solution of the whole PGS expansion planning problem. This final solution may be considered either as a single alternative, chosen from the RAS, or as some combination of RAS alternatives, taken in a form preferable by decision makers. Thus, the final choice of the expansion plan is to be left to the decision makers' judgment on a basis of qualitative analysis of the RAS derived.

The specific features of the considered problem may be presented as follows:

1. Inherent dynamics since a selection is to be only made among dynamic strategies of PGS expansion through prolonged planning period.

2. The initial set of alternatives (ISA) in the problem may include a vast number of alternatives. This is resulted from both the a. m. dynamic aspect and the method of forming an alternative as a sequence of PGS states for all planning years, when each such state, named configuration, is taken one at a time in each year. If there is a considerable number of configurations for each planning year (in our case up to 10000), the number of possible alternatives may be a vast.

3. Multicriteria optimization is a very essential aspect to solve this planning problem. Many leading countries (including Israel) perform a similar planning task using long range planning models, such as EGEAS, PROSCREEN, WASP, taking into consideration only a single economical criterion. However, in fact, this criterion incorporates different criteria jointly (in particular, minimization of general cost, of rates, of investments, etc.). Each of them is not so compatible with other, that makes sense to reflect the economical objective by a set of such different criteria. Besides, environmental criteria are growing in importance at present time regarding the problem of PGS expansion planning. These criteria have to reflect minimization of damage caused by pollution emissions of electricity production. As seen from a lot of publications (Chattopadhyay 1994 and others), basic pollutants caused by electricity production are the sulfur dioxide (SO_2), the nitrogen oxides (NO_x), the solid particulate matters (SPM) and the carbon dioxide (CO_2). Taking into account the existing opinions (see, e.g., Chattopadhyay 1994, p. 426: "While the impact of fly ash is limited to local areas, that of CO_2 is a global concern, and SO_2 and NO_x have both local and regional impacts"), and also the state of initial data, etc., three criteria, expressing the SO_2 , NO_x and SPM emissions, were chosen as independent environmental criteria for our local problem. And finally, it would be advisable to consider some criteria reflecting the problems of PGS functioning, e.g., reliability. We should underline that all three groups of the above criteria are independent and it is difficult to accumulate them in a single general criterion.

4. As for any multicriteria optimization (MCO) problem (e.g., Steuer 1986), our problem to be solved has to take into account decision makers' (DM) estimations, expressing either relations between criteria (mainly, their priorities) or other factors. The decision makers may express these estimations in different

forms, including a situation of full uncertainty when they can not give such estimations at all.

5. Uncertainty is inherent to the problem in consequence of different reasons: (a) uncertainty especially inherent to the process of medium/long term planning itself, owing to the long planning period; (b) the DM estimations, taken to make the criteria scalar sum for some used MCO techniques, may be either subjective and not quite reliable or inappropriate at all, if a situation of full uncertainty occurs; (c) many parameters of the problem are not reliable in their nature.

A lot of publications are devoted to multicriteria analysis of PGS expansion (Kok 1987; Amogai & Leung 1989, 1991; Lootsma et al. 1990; Chattopadhyay 1994; Balestieri & Correia, 1997; and others). They reflect the following basic aspects of such an analysis: construction of criteria; techniques and procedures of multicriteria decision making, including performing multicriteria optimization and accounting for uncertainty factors, etc. However, these publications does not contain an approach to solve our problem with accounting for all the above mentioned specific features jointly. Such an approach has been developed to solve the problem of medium/long term PGS expansion planning for the conditions prevailing in Israel (Kalika & Frant 1996).

According to this proposed approach, the considered solution process, intended to reach a set of "reasonable" alternatives (RAS), may be interpreted as a multi-stage decision-making (MSDM) process including the following stages:

1. Creation of an initial set of alternatives (ISA). This stage is dedicated to account for the a. m. features of the problem: dynamic aspect and possibility of a vast number of initial alternatives. A combinatorial algorithm has been developed to generate the ISA with taking into account conditions of the PGS expansion. This algorithm includes also a “reduction-recovery” procedure, developed to overcome obstacles when too many alternatives emerge (Sections 3, 4).

2.Definition of an initial set of criteria assessment vectors (ISCAV), interrelated to the ISA, in accordance with the multiple criteria (in fact, their number may be without limitation) and criteria calculation models. Such model has been developed for each criterion (original development was needed for environmental criteria) to create a criteria assessment vector for each ISA alternative that is performed simultaneously with creation of this alternative (Sections 3, 4).

3. Performance of a multi-step man-machine procedure (MSMMP) to reduce the ISA to the RAS, from step to step of MSMMP. Various MCO and other techniques, including TOPSIS (Massam 1988), Pareto optimization (Steuer 1986), Monte Carlo simulation (Hammersley and Handscomb 1964), may be used in a framework of MSMMP. This third stage of MSDM process takes into account three the above last-mentioned features of the problem. So, the MCO techniques, used by MSMMP, were adapted to take into account uncertainty factors and DM estimations. Using techniques on various MSMMP steps corresponds, mainly, to the manner of considering DM estimations. Various versions of MSMMP, called scenarios, may be carried out to meet purposes of customers and to reflect situations regarding data, DM opinions, etc. (Section 5).

According to the proposed approach, the main way to account for uncertainty and DM opinions in MSDM process is to perform multi-variant computations to obtain "stable" solutions for all performed variants of computations (Section 6).

The proposed approach to solve the problem for the conditions prevailing in Israel was implemented in the computer model "MCALTPES-96", which was tested using the real data of Israeli electric system expansion. This testing allowed to make the conclusion that the proposed approach is viable (Section 7).

2. Mathematical Model of the Problem

The considered problem formally may be presented as a common multicriteria optimization model (see, e.g., Steuer 1986):

$$\min z_k (x_j), \ k = 1,..., K \quad (1)$$

$$\text{s.t.} \quad x_j \in A \quad (2)$$

where k (or j) is a number of objective (or alternative); x_j is a vector of variables or, that the same, j-point corresponding to j-alternative; z_k $[z_k (x_j)]$ is k- objective function or, that the same, k- criterion [its value for j-alternative]; $A = \{x_j\}$ is the ISA or, that the same, a set of all j-points x_j..

We would like to note that in our problem the vector x_j. has a complicated structure where each component of vector x_j. may be interpreted as a dynamic variable to be used in such form only for the first and second stages of MSDM process. For the third stage of MSDM when multicriteria optimization in a framework of MSMMP is performed, each j-vector is considered as a single general variable or j-point. In this case we consider the A as a discrete set with a finite number of j-points (general variables) x_j, for each of which a vector of numerical values $Z(x_j) = \{z_1 (x_j), ..., z_K (x_j)\}$ will be calculated, and multi-criteria optimization process will be performed on a space

$$Z = \{Z(x_j) \,/\, x_j \in A\} \quad (3)$$

3. Algorithm to Construct the ISA (ISCAV) as a Dynamic Tree

Both the first and second stages of MSDM process are performed jointly to construct the ISA and the ISCAV simultaneously. If an initial alternative from ISA is constructed, then a criteria assessment vector from ISCAV corresponding to this alternative is formed, too. According to the algorithm proposed, the ISA is constructed by way of successive forming of subsets ISA(t) for all planning t-years, by going from the ISA(t-1) for the preceding (t-1)-year to the ISA(t) for the next t-year. When the ISA(t) is formed for the current t-year, then the ISCAV(t), interrelated to ISA(t), is calculated, too. Thus, the ISA (as also ISA(t) for each t-year, t=1,..,T) is constructed as a dynamic tree (a graph) of PGS states (configurations) for all considered planning years. Such PGS state for each t-year, presented as a node of this constructed dynamic graph and named t-configuration (or, simply, configuration), is expressed by a set of power units to

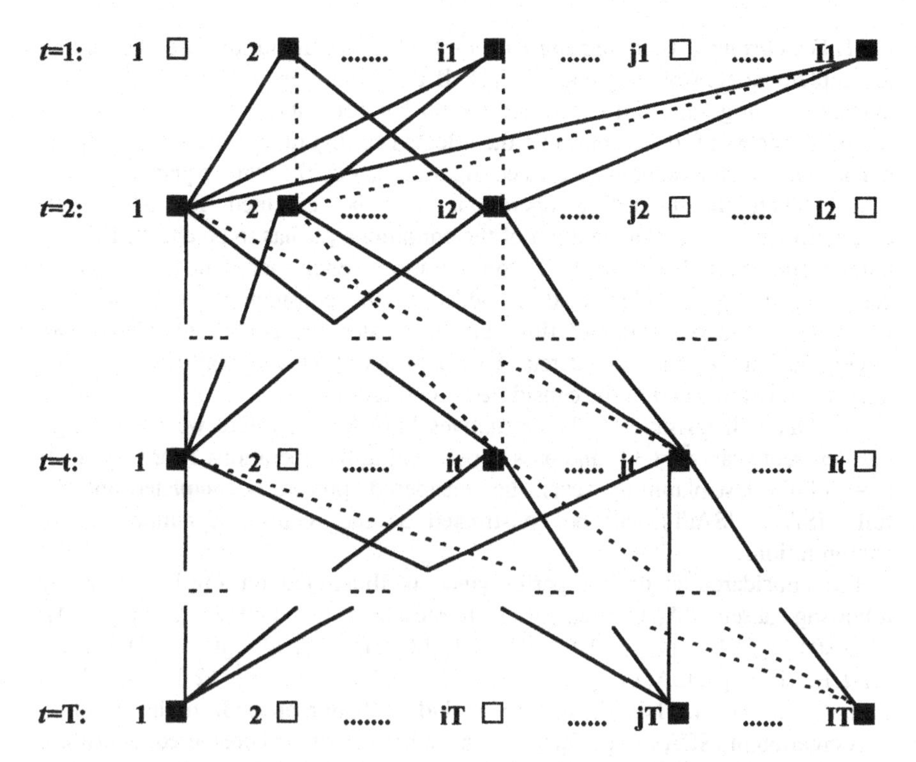

Figure 1. Illustration of the dynamic tree of configurations.

to be installed in specified locations in this *t*-year. For instance, in our computations each configuration has included three types of power units: coal power station with capacity of 550 MW (CPS-550) which could be installed in four locations (A, B, C, D); gas turbines (GT-115) corresponding to three locations, suitable for their erection; the same for combined cycles (CC-330). Thus each configuration is presented by a set of 10 integers, indicating how many units of a planned candidate (such candidates reflect 10 pairs: CPS-550/A, CPS-550/B, ...) are presented in this configuration.

The algorithm is reflected on Figure 1 through consideration of the subsets: ISA(1)={1,2,...,i1,...,j1,..I1};ISA(2)={1,2,...,i2,..,j2,...I2);...ISA(t)={1,2,...,it_t,..,jt, ...It}; ... ISA(T) = {1,2,...,iT,...,jT,...IT} for the successive years t^*=1,2,...,t,...,T. Configurations of these subsets are presented by the black or white squares with their numbers 1, 2,..., i_1, etc. This algorithm to construct the dynamic tree (the ISA) is based on checking two types of conditions, characterizing: [1] interior ones for each configuration (e.g., can t-configuration provide an installation of new gas turbines' units only by pairs in this t-year?); [2] connections between pair configurations taken from two neighboring years, i.e., for a pair of t- and (t+1)-configurations (e.g., a pair of CPS-550 may be installed only for two neighboring years). Thus, the algorithm to construct the ISA is as follows:

1. For current t^*-year, starting with $t^*=1$, all t^*-configurations are checked on an implementation of the condition [1]. If this condition is implemented for a current t^*-configuration, it is selected for further computations, otherwise it will not be considered further (on Fig.1 the selected configurations are marked by the black squares, and the non-selected configurations - by the white squares).

2. When the preceding operations will be executed for all (t^*+1)-configurations, an implementation of the conditions [2] has to be checked for all pairs of the selected t^*- and (t^*+1)-configurations. Each such pair, for which the conditions [2] are implemented, is added to the considered t^*-configuration, which is in the beginning of this pair. It creates the new (t^*+1)-alternative, having its end in the considered (t^*+1)-configuration and continuing the t^*-alternative with its end in the considered t^*-configuration.

3. Once all possible (t^*+1)-alternatives have been created, we must assign new current year $t^{**}=t^*+1$ and pass to performing the operations 1 for t^{**}-year. If $t^{**}=T$ (T-the last planning year), the considered process is completed and the full ISA = ISA(T) will be constructed as the required dynamic tree of configurations.

The considered algorithm performance is illustrated on Fig.1, where the following subsets ISA(t),t=1,2,..,t ,..,T are presented: ISA(1)=[{2},..,{i1},..,{I1}]; ... ISA(t) = [{2,1,...,1}, ... {2,1,...,it}, ... {2,I2,...,1},... {2,i2,...,jt}, ... {I1,1,...,1}, ...{i1,1,...,it},...];...ISA(T)= [{2,1,...,1,...,1}, ... {2,1,...,1,...,jT}, {2,1,...,it,...,1},... {2,1,...,it,...,jT},...]. Thus, the full ISA includes all alternatives from the ISA(T).

A creation of ISCAV is performed in a framework of the considered algorithm. If a configuration is selected according to the operation 1 of this algorithm, then a criteria assessment vector corresponding to this configuration is determined on a basis of calculation models, developed for each criterion separately. In some cases development of such calculation models can lead to certain methodological obstacles. We have such a case for the environmental criteria, where it was needed to develop original calculation models (Kalika & Frant 1996).

4. The "Reduction - Recovery" Procedure

Since the given number of configurations may be very large for each planning year, the number of alternatives of ISA, constructing in accordance with the a.m. dynamic algorithm, may grow rapidly in correspondence with the long-run planning period and become enormously large for a such period. It requires a vast computer time to construct the ISA. For this case, the special "reduction - recovery" procedure has been developed to obtain the reduced ISA, if we have no the required computer time. Otherwise the full ISA may be constructed, since this procedure is performed without loss of ISA information.

The considered procedure is performed in a framework of the a.m. basic algorithm of constructing the sets {ISA(t), t=1,..,T}. According to this algorithm, for some t-year a number of t-alternatives in ISA(t) exceeds a given (by man-machine dialogue) limit at the first time, the "reduction" procedure should be started to reach a reduced ISA(t+1)^ for the next (t+1)-year. In this case such t-year is considered as if it was a new initial year of the planning period and only the conditions [1] are checked for all t-configurations from ISA(t). For all pairs of the selected (providing the condition [1]) t-configurations from ISA(t) and of the given (a priori) (t+1)-configurations, performance of the conditions [2] are checked. All selected pairs form the new reduced ISA(t+1)^, including the strategies only for two years [t, t+1]. In this case the full ISA(t), corresponding to years [1, t], is memorized. Further, in a framework of the "recovery" procedure it is possible to construct a full set ISA(t+1), linking both the full ISA(t) and the reduced ISA(t+1)^. Once the ISA(t+1)^ has been constructed, the basic algorithm should be used again to construct ISA(t+2) on a basis of consideration of the ISA(t+1)^ as a usual ISA(t+1). The process is to be continued up to t=T-1.

As is evident from the foregoing, the above "reduction" process is performed without loss of information, regarding a construction of the full ISA. However, the "recovery" process, successfully linking the obtained full ISA(t) or the reduced ISA(t)^ for all t=1,..,T, may be performed with reduction of the full number of alternatives, if there is computer time limitation. It can be carried out in various ways (specified by dialogue) according to available conditions.

Thus, if the MSDM first and second stages are performed without using the "reduction - recovery" procedure, then only the full ISA and ISCAV, presented as the sets I = ISA(T) and Z = ISCAV(T), corresponding to the mathematical conditions (2) and (3), may be reached at the close of calculation process using the full required computer time. Otherwise, if the "reduction - recovery" procedure is used (at least for one t-year), the reduced sets I^=ISA(T)^ ($I^\wedge \subseteq I$) and Z^= ISCAV(T)^ ($Z^\wedge \subseteq Z$) may be obtained for the reduced computer time.

5. Multicriteria Optimization in a Framework of the MSMMP

On the third stage of MSDM process the model (1)-(3) has to be solved, i.e., multicriteria optimization in the space (3) is performed to reach the RAS:

$$R = \{x_j \ / \ Z(x_j) \in R[Z]\} \qquad (4)$$

where R is the RAS, $R[Z]$ is a set of a predetermined number of criteria assessment vectors $Z(x_j)$ from ISCAV, where the sets R and $R[Z]$ are interrelated. The MSMMP, presented on Fig.2, corresponds to the model "MCALTPES-96" including three steps, which ensure performance of MCO techniques, modified to account for uncertainty and DM estimations, and of the procedures to perform Monte Carlo simulation on a basis of consideration of criteria weights as random variables. Among these techniques (Kalika 1991, Kalika & Frant 1996):

- Pareto optimization with variable uncertain ranges (technique 1.1 on Fig.2).
- TOPSIS techniques, based on scalar summing over all criteria using their weights. These techniques, used at all three MSMMP steps (Fig.2), differ by the following methods of creating the criteria weights, determined to be used

for Monte Carlo simulation performance: (1) without use of criteria weights, when all criteria have equal weights (technique 1.2); (2) using the given criteria

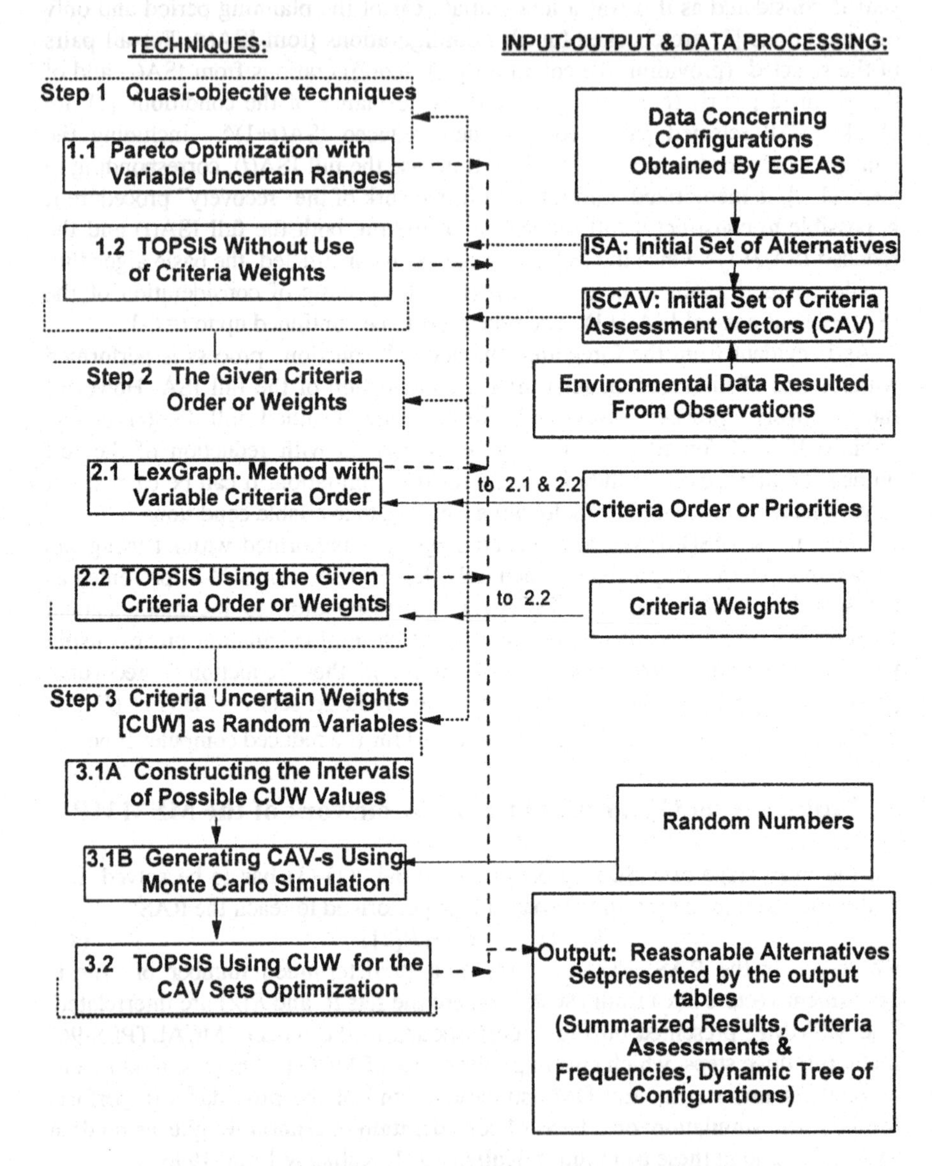

Figure 2. Multi-Step Man-Machine Procedure (MSMMP)

order or weights, when the weights values either are assigned (a priori) by decision makers or are found on a basis of given criteria priorities using special heuristic procedures developed (techniques 2.2); (3) using criteria uncertain weights (CUW) to optimize the ISCAV (technique 3.2), based on Monte Carlo simulation (technique 3.1B), for which the required intervals of possible CUV values are constructed using heuristic procedures (techniques 3.1A) often without the DM estimations.

- Lexicographic method with variable criteria orders (technique 2.1).

As seen from Fig. 2, the MSMMP steps reflect different situations regarding extent of uncertainty and a manner of considering DM estimations. The more advanced is the step, the more DM estimations should be taken into account. So, the first MSMMP step uses "quasi-objective" techniques (without using the DM estimations) and corresponds to a situation of determinancy. At the second MSMMP step DM estimations are used to give values of criteria weights (or criteria orders to construct the weights using heuristic procedures) for TOPSIS technique, or only of criteria orders for Lexicographic method. The appropriate situations may be treated as partially uncertain ones. The third MSMMP step considers the criteria weights as random variables, that allows to treat a situation either as partially uncertain, when the possible intervals of CUV values are determined on a basis of DM estimations (technique 3.1A), or as one closed to full uncertainty, if the required data are given, mainly, hypothetically.

The MSMMP performance is a "movement" from the ISA to the RAS, i.e., from a large (maybe vast) number of alternatives to their few predetermined number. This "movement" is performed in accordance with a fixed scenario of MSMMP. Since all MSMMP steps and techniques do not need to be used in each scenario, a fixed scenario should reflect a certain combination of such steps and techniques. Some of the proper scenarios may be characterized as follows (Fig.2):

1. The successive performance of techniques 1.1 and from step 3 reflects a most full and successive scheme of computations without changing ISA (ISCAV) data.
2. The successive performance of techniques 2.2 and from step 3 corresponds to a partially uncertain case of using a Monte Carlo simulation.
3. Only step 3 is performed and Monte Carlo simulation applies to the full ISA.
4. Only technique 2.2 is used with variation of criteria weights values, based on DM assignments, that allows to estimate "roughly" a possible content of RAS.

The very important aspect of MSMMP is connected with accounting for uncertainty and DM estimations on a basis of multi-variant computations (MVC) to obtain their "stable" solutions. This accounting is carried out in various ways:

1. Variation of MSMMP scenarios may be used in a framework of one MVC series to obtain RAS, reflecting this series "stable" solutions. This variation as well as changing initial data or criteria calculation models allow to account for general uncertainty, caused by initial data state or the long planning period, etc.
2. Use of Monte Carlo simulation allows to reflect different possible values of criteria weights, considered as random variables. The required intervals of such values to make MVC performance are constructed either using DM estimations (case of partial uncertainty) or hypothetically (case closed to full uncertainty).
3. All the a.m. MCO techniques, used in MSMMP, were modified in order to give possibility to perform MVC. According to these techniques, MVC may be

dictated by: (a) varying the values of uncertain ranges (technique 1.1) as well as criteria orders (techniques 2.1, 2.2) or weights (technique 2.2); (b) using Monte Carlo simulation (technique 3.2).

6. Finding "Stable" Solutions in Multi-Variant Computations

In our approach multi-variant computations (MVC), used to obtain "stable" solutions, is the main method to account for uncertainty. We may classify MVC regarding a level of parameters changed for going to a next variant of MVC as:

1. *Interior MVC*, when the changes are carried out in a framework of performed technique or procedure itself without a variation of parameters, assigning limits for these computations. An example of such MVC is performance of Monte Carlo simulation using technique 3.2 without changing the limiting parameters.
2. *Intermediate MVC*, reflecting a repeated performance of interior MVC based on changing their limiting parameters, fixing new conditions for this repetition. A series of the a.m. Monte Carlo simulations caused by changing the given number of simulations may present the proper example of such MVC.
3. *Exterior MVC*, caused by a variation of either scenarios of MSMMP without a change of the ISCAV as a space for multicriteria optimization (case MVC-3.1) or this space itself (case MVC-3.2). A successive performance of both the scenarios 1 and 2 is an example of case MVC-3.1, a change of parameters of criteria calculation models that involves a new ISCAV - of case MVC-3.2.

Performance of each l-variant of computations from the MVC (l=1,..,L) of every of the a.m. levels is completed by obtaining a subset S_l of sub-optimal alternatives from the ISA (ISCAV). Each set S_l should include a predetermined number N_l of alternatives, where the number N_l is assigned either a priori or by dialogue. The resulting sets $\{S_l, l=1,..,L\}$, completing these MVC performance, are analyzed to find a set $S^\wedge$ of "stable" solutions for this MVC series. The set $S^\wedge$ should include all j-alternatives, having maximum of entering into $\{S_l, l=1,..,L\}$. A full set $S^{\wedge\wedge}$ of j-alternatives, entering into sets $\{S_l, l=1,..,L\}$, and a frequency P_j of entering of such j-alternative into the sets $\{S_l, l=1,..,L\}$ are defined as:

$$P_j = n_j / L, \ j \in S^{\wedge\wedge}; \qquad S^{\wedge\wedge} = \bigcup S_l \ (l=1,..,L) \tag{5}$$

where n_j is a number of sets $\{S_l, l=1,..,L\}$, including the considered j-alternative.

Thus the proposed methodology to form the required "stable" set $S^\wedge$ is as:

1. According to (5), the value P_j is determined for each j-alternative from the set $S^{\wedge\wedge}$ and this value P_j is compared with the given limit $P^\wedge$ (at first $P^\wedge=1$).
2. If $P_j \geq P^\wedge$, then the considered j-alternative is included into the set $S^\wedge$, otherwise this j-alternative excluded from further consideration, and according to p.1, the value P_j is determined for a next j-alternative (from $S^{\wedge\wedge}$), and so on .
3. Once all j-alternatives from the set $S^{\wedge\wedge}$ have been checked, then a number ($N[S^\wedge]$) of j-alternatives, entering into the set $S^\wedge$, is compared with a limit $N^\wedge$, given a priori or by dialogue for the considered MVC. If $N[S^\wedge] \geq N^\wedge$, then the set

$S^\wedge$ is already formed. Otherwise we have either to decrease the limit $P^\wedge$ or to fix the incomplete set $S^\wedge$.

7. Testing the Proposed Approach and Conclusions

The approach, regarding scenario 3 of MSMMP, has been tested using the developed model "MCALTPES-96" and its software on FORTRAN (for DEC workstations on UNIX OS) for conditions prevailing in the Israeli PGS. Ten of planned candidates (CPS-550, GT-115, CC-330 with their possible locations, see section 3) were considered to be installed during the planning period 1997-2010. Five criteria (economic, reliability and three environmental, regarding SO_2, NO_x and SPM pollution emissions) has been considered.

More then thirty test-variants (in the most cases the RAS included 50 alternatives) were performed with variation of parameters values regarding: (a) the "reduction-recovery" procedure, (b) calculation models for the environmental criteria, (c) DM estimations to be used for determining the criteria weights; (d) a Monte Carlo simulation, and others. An analysis of obtained results allows to underline the following aspects.

A vast number of initial alternatives in a full ISA was expected, since a huge number (up to 10000) of configurations for each planning year has been reached with help of the EGEAS[1] model. Thus it was impossible (from computer time limitation) to obtain the full ISA, and using the "reduction - recovery" procedure allowed to construct the reduced ISA (ISCAV), including from 2500 to 40000 alternatives. The analysis of influence of ISA (ISCAV) size on the resulting RAS showed: if the ISA had smaller size, the RAS had worse indicators. The variation "(c)" regarding the criteria weights had, as expected, serious effects on the RAS contents, while the variations "(b)" and "(d)" provided more stable results.

Thus, the approach , presented by the model MCALTPES-96 developed to solve the problem of medium/long term planning of Israeli PGS expansion, takes into account the following features of the problem:

1. Dynamic aspect, due to the necessity to consider dynamic strategies of PGS expansion through a long planning period (up to 30-50 years) ,which was taken into account by constructing an initial set of alternatives (ISA) as a dynamic tree of PGS states (named configurations) for all years of the planning period.
2. Emergence of a vast number of alternatives in the ISA, due to a considerable number of configurations in each planning year, that may be overcame using the algorithm to construct ISA jointly with the "reduction - recovery" procedure.
3. Multicriteria aspect, due to the necessity to consider separately different criteria, that involved need to perform multicriteria optimization (MCO) of an initial set of criteria assessment vectors (ISCAV), interrelated to the ISA, in a

[1] EGEAS is Electric Generation Expansion Analysis developed by EPRI and supported by Stone & Webster Management Consultants.

framework of a multi-step man-machine procedure (MSMMP) developed, using various MCO techniques (including also a Monte Carlo simulation).

4. Necessity to consider different forms of decision makers' estimations, expressing, mainly, relations among criteria, that may be achieved by adaptation of MSMMP steps and techniques to the forms of these estimations.

5. Uncertainty, inherent in the problem due to the long planning period, e criteria incomparability, unreliability of many initial parameters values, etc. These are taken into account using: (a) multi-variant computations to obtain "stable" solutions, (b) special techniques including MCO methods, modified to account for uncertainty (all in a framework of MSMMP), (c) various scenarios for MSMMP.

Testing one of the scenarios of MCALTPES-96, using the software and data of the Israeli electric system, allowes one to conclude that the proposed approach is viable. The developed methodology (the MSMMP to make multicriteria optimization accounting for uncertainty) may be implemented for other proper problems, including spheres differing from the problem of PGS expansion.

References

Amagai, H. and Leung, P., Multiple criteria analysis for Japan's electric power generation mix. *Energy Systems and Policy* **13**, 219-236 (1989).

Amagai, H. and Leung, P., The trade-off between economic and environmental objectives in Japan's power sector. *The Energy Journal* **12**, 95-104 (1991).

Balestieri, J.A.P. and Correia, P.B., Multiobjective linear model for pre-feasibility design of cogeneration systems. *Energy - The International Journal* **22**, 537-548 (1997).

Chattopadhyay, D., Systems approach to emissions reduction from a power system in India. *Energy Sources* **16**, 425-438 (1994).

Hammersley, J.M. and Handscomb, D.C., *MonteCarlo Methods*, Wiley, New York (1964).

Kalika, V.I., *Regional Fuel and Energy Complex*, Nauka, Moscow (1991).

Kalika, V. and Frant, S., Multicriteria analysis accounting for uncertainty factors in electricity generation expansion planning. *Proc. 12th Power Systems Computation Conference* **1**, 145-151 (1996).

Kok, M., Energy modeling with multiple objectives and multiple actors. *Energy Systems and Policy* **11**, 21-40 (1987).

Lootsma, F.A., et al., Choice of a long-term strategy for the national electricity supply via scenario analysis and multicriteria analysis. *European Journal of Operational Research* **25**, 216-284 (1990).

Massam, B.N., Multi-criteria decision making (MCDM): Techniques in planning, *Progress in Planning* **30**, 1-84 (1988).

Steuer, R.E, *Multiple Criteria Optimization: Theory, Computation and Application*, Wiley, New York (1986).

Time-Dependent Capital Budgeting with Multiple Criteria

Kathrin Klamroth* and Margaret M. Wiecek
Clemson University, USA

Abstract

In this paper we introduce a time-dependent multiple criteria model of capital budgeting and propose a dynamic-programming-based solution approach to finding all the efficient solutions defined as sequences of projects that are consecutively performed and bring benefit to a company. An illustrative example is enclosed.

Keywords: Multiple criteria capital budgeting, multiple criteria knapsack problem, time-dependent objectives

1 Introduction

Capital budgeting is a well known problem in managerial economics. The problem concerns a company confronted with a variety of possible investment projects and a fixed capital budget independent of the investment decisions. The cost and the revenue associated with every project are assumed to be known. The objective is to select from among the projects the particular projects that lead to the highest earnings for the company.

The problem has been given a lot of attention by economists, management scientists, industrial engineers, operations researchers, and mathematicians, and the related literature is very rich. Among many others, Weingartner (1963) studied capital budgeting in the context of mathematical programming.

The traditional theory of capital budgeting uses a single objective, usually in the form of a maximization of company's revenues. In the late nineteen sixties and early nineteen seventies researchers proposed to extend the traditional model with multiple objectives, as the particular projects can be selected with respect to more than a single objective, see Ansoff (1968), Carsberg (1974), and Bromwich (1976). The other objectives could include appreciation, operational effectiveness, uniqueness, dual use potential etc., all yielded by the selected investment projects.

Hawkins and Adams (1974) proposed a goal programming model of capital budgeting. Bhaskar (1979) more formally recognized multiple objective

*On leave from the Department of Mathematics, University of Kaiserslautern, Kaiserslautern, Germany.

This work was partially supported by ONR Grant N00014-97-1-0784 functions and proposed a generalized goal programming approach. Capital budgeting with multiple objectives was also studied by Lee and Lerro (1974), Ignizio (1976), Thanassoulis (1985), Corner et al. (1993), and others. An interactive procedure using a multiple criteria linear integer model was proposed by Gonzalez et al. (1987). More recent papers proposed models extended with additional features such as time preferences (see Vetschera (1985)), stochastic and dynamic elements (see Turney (1990)), risk (see Lin (1993)), and multiple decision makers (see Kwak et al. (1996)).

In the framework of mathemtical programming, a typical capital budgeting model with multiple criteria is based on the multiple criteria knapsack problem (MCKP), a known combinatorial optimization problem with applications in many other areas such as transportation planning, conservation biology, packaging and loading.

The bi-criteria knapsack problem (BCKP) was studied by Rosenblatt and Sinuany-Stern (1989) whose work was continued later by Eben-Chaime (1996). Several recent papers by Ulungu and Teghem (1994), Ulungu and Teghem (1997) and Visée et al. (1996) dealt with the BCKP or the MCKP. Villarreal and Karwan (1981) were perhaps the only ones who proposed dynamic programming (DP) approaches to the MCKP with multiple constraints.

Being motivated by applications of capital budgeting in affordability analysis in which the time is an important decision parameter, we propose a time dependent capital budgeting model with multiple criteria. Including the time makes this model distinct in comparison to those in the literature and requires a new approach specially tailored for dynamic problems. In particular, we assume that the vector of objective functions is composed of time-dependent functions. The resulting model applies to the decision situation when the projects have to be selected consecutively in time subject to a known and fixed budgetary constraint so that time-dependent criteria are maximized while the time needed to perform the selected projects is minimized. As a result, the time-dependent multiple criteria knapsack problem (TDMCKP) yields efficient portfolios being sequences rather than groups of the projects.

In the proposed approach we follow upon a DP formulation presented by Villarreal and Karwan (1981). While adapting their formulation to the time-dependent case, we modify the forward approach of the time-dependent multiple criteria dynamic programming introduced by Kostreva and Wiecek (1993).

In Section 2 we present the TDMCKP and the DP-based solution approach is developed in Section 3. In Section 7 we include a tri-criteria example that is solved by AMADEuS, a decision support tool based on the proposed model and methodology. The paper is concluded in Section 5.

2 The time-dependent model

Given a set of n projects of interest to a company, let $\{x_1, \ldots, x_n\}$ be a set of elements representing the projects and let $S := \{1, \ldots, n\}$ be the related index set.

We assume that only one project can be performed at a time and that during a decision process (that starts at time zero) some projects will be selected at consecutive times in order to be performed. Every sequence of projects to be performed corresponds to a sequence $x := \{x_{j(r)}\}_{r=1}^{p}$ of elements $x_{j(r)}, r = 1, \ldots, p$, where $j(r) \in S := \{1, \ldots, n\}$.

Given a fixed available budget b, we model the budgetary constraint as $a(x) \leq b$, where $a(x)$ is the function defined as

$$a(x) = a(\{x_{j(r)}\}_{r=1}^{p}) := \sum_{r=1}^{p} a_{j(r)} \tag{1}$$

and $a_{j(r)}$ is the cost coefficient of the project $x_{j(r)}, j(r) \in S$. We additionally assume that the cost coefficients $a_i, i \in S$, and the budget b are positive integers.

Consequently, the set X of all the feasible sequences of projects of the TDMCKP is defined as $X := \{x : a(x) \leq b\}$, where each sequence x satisfies

$$x \in \{\{x_{j(r)}\}_{r=1}^{p} \; : \; p \in \mathbb{N}, \;\; j(r) \in S, r = 1, \ldots, p\}. \tag{2}$$

Note that due to the fact that all the cost coefficients are positive integers, all the feasible solutions in X are finite. Namely we get that $p \leq b$ for all $\{x_{j(r)}\}_{r=1}^{p} \in X$.

Given m objective functions $f_i(x), i = 1, \ldots, m$, of interest to the company, we assume that f_1 measures the time needed to accomplish the projects and therefore it should be minimized while the other functions $f_i, i = 2, \ldots, m$, represent other criteria that should be maximized. Each of the functions is a real-valued and time dependent function of x. We define the vector objective function as

$$f(x) := [f_1(x), f_2(x), \ldots, f_m(x)]^T := \sum_{r=1}^{p} c_{j(r)}(t^r(x)), \tag{3}$$

where (3) involves a vector sum and t is a continuous variable, $t \geq 0$, representing the time, that can be calculated as

$$\begin{aligned} t^1(x) &= 0, \\ t^{s+1}(x) &= t^s(x) + c^1_{j(s)}(t^s(x)), \qquad s = 1, \ldots, p. \end{aligned} \tag{4}$$

For every $j \in S$, $c_j(t) = [c_j^1(t), \ldots, c_j^m(t)]^T$ is a vector objective related to choosing the project x_j at time t. Elements $c_j^i(t)$, $i = 1, \ldots, m$, $j = 1, \ldots, n$, are defined to be real-valued functions of time t and are not assumed to be continuous. In particular, for every $j \in S$, $c_j^1(t)$, $j = 1, \ldots, n$, is a positive function measuring the time needed to accomplish the project x_j if its realization has started at time t, and the other components $c_j^i(t)$, $i = 2, \ldots, m$, represent the earnings, revenue, appreciation, etc., generated by selecting the project x_j if its realization has started at time t.

We also assume that there is no waiting time between choosing and performing two consecutive projects in a sequence of projects, i.e. once a project $x_{j(s)}$ has been selected, another project $x_{j(s+1)}$ is being selected right after.

We formulate the time-dependent multiple criteria knapsack problem (TDMCKP) as:

$$\begin{aligned} \text{vmax}^* \quad & f(x) = [f_1(x), f_2(x), \ldots, f_m(x)]^T \\ \text{s.t.} \quad & a(x) \leq b. \end{aligned} \tag{5}$$

As we are interested in maximizing the objective functions $f_i(x)$, $i = 2, \ldots, m$ and in minimizing the time simultaneously, the operator $vmax^*$ in (5) denotes the maximization of $[-f_1(x), f_2(x), \ldots, f_m(x)]^T$, i.e.

$$\begin{aligned} & \text{vmax}^* \, [f_1(x), f_2(x), \ldots, f_m(x)]^T \\ & \qquad := \text{vmax} \, [-f_1(x), f_2(x), \ldots, f_m(x)]^T. \end{aligned} \tag{6}$$

We also find it convenient to consider the TDMCKP with the right-hand-side $k = 1, \ldots, b$ of the budgetary constraint and denote this problem by k-TDMCKP. A feasible solution $x := \{x_{j(r)}\}_{r=1}^p$ of the k-TDMCKP is a sequence of projects $x_{j(r)}, r = 1, \ldots, p, p \leq k$, so that $j(r) \in S$ and $\sum_{r=1}^p a_{j(r)} = k$ for $k = 1, \ldots, b$.

Solving (5) is understood as generating its efficient (Pareto) solutions (sequences of projects). A feasible solution $\hat{x} \in X$ is said to be an efficient solution of (5) if there is no other feasible solution $x \in X$ such that $f_1(x) \leq f_1(\hat{x})$ and $\forall i \in \{2, \ldots, m\}$ $f_i(x) \geq f_i(\hat{x})$ with at least one strict inequality.

Let $\mathcal{X}_e$ denote the set of efficient solutions of (5) and let $\mathcal{Y}_e$ denote the image of $\mathcal{X}_e$ in the objective space, that is $\mathcal{Y}_e = f(\mathcal{X}_e)$, where $f = [f_1, \ldots, f_m]^T$. $\mathcal{Y}_e$ is referred to as the set of nondominated criterion vectors of the efficient solutions (sequences of projects) of (5).

3 A dynamic programming (DP) approach

3.1 Partial knapsack problems and the states of DP

The TDMCKP is partitioned into a finite set of subproblems that determine the set of states Q, defined as $Q := \{q(0), q(1), \ldots, q(b)\}$. The initial state is defined to be empty (independently from the cost function a and the objective functions), i.e. $q(0) = \emptyset$, and the state $q(k)$, $k = 1, \ldots, b$, represents all the feasible solutions of the k-TDMCKP, i.e.

$$q(k) := \{\{x_{j(r)}\}_{r=1}^{p} \ : \ p \leq k, \ j(r) \in S, r = 1, \ldots, p, \ \sum_{r=1}^{p} a_{j(r)} = k\}.$$

In other words, a state represents all the feasible sequences of projects such that the total cost of each sequence is equal to a partial budget $k, 1 \leq k \leq b$.

Since there may occur nondominated solutions in all the states, the set of final states Q_F is given by $Q_F := \{q(0), q(1), \ldots, q(b)\}$.

3.2 The DP network

The decision of adding a project x_j to a solution sequence $x \in q(k)$ results in an increase of the right-hand side k by a_j and thus corresponds to a transition of x from the state $q(k)$ to the state $q(k + a_j)$. Observe that with the definition of the states, the original problem is represented as a loop-free sequential decision process, i.e. a process whose states can be indexed from 0 to b, so that a transition from a state $q(k)$ always occurs to a state $q(l)$ such that $k < l$, for any $k, l = 0, 1, \ldots, b$ (see Ibaraki (1987)).

Without loss of generality we assume that the system is in the state $q(0)$ at time $t = 0$.

For all $j = 1, \ldots, n$, $c_j^1(t) > 0$ represents the time needed to make a transition from a state $q(k)$ to a state $q(k + a_j)$ for $k = 0, 1, \ldots, b - a_j$ given that the system is in the state $q(k)$ at time t, that is the time needed to accomplish the project x_j if it has started at time t. The transition to the state $q(k + a_j)$ is completed at time $t + c_j^1(t)$, which corresponds to the fact that the project x_j is accomplished at time $t + c_j^1(t)$.

The states defined above and the possible transitions between them yield a network whose nodes and arcs are defined by these states and transitions, respectively. This network does not have any circuits since the states and the transitions form a loop-free decision process. Associated with every arc of this network is a criterion vector $[c_j^1(t), \ldots, c_j^m(t)]^T$ related to adding the project x_j to a sequence of projects at time t.

3.3 Outline of the DP procedure

Given the network, we are in the position to apply the forward approach of Kostreva and Wiecek (1993). They developed the forward approach to find the set of all nondominated (shortest) paths from a given source node to every other node in the network whose links carried a time-dependent vector cost. They considered a general network whose every node could be connected to every other node. The costs were assumed to be real-valued positive and monotone increasing functions of time. These assumptions were necessary to establish the Principle of Optimality for Dynamic Multiple Objective Networks.

Given the special structure of our network, we may relax and change some of their assumptions due to the fact that the feasibility constraint of the TDMCKP yields a circuit-free network and that our problem involves maximization rather than minimization. In general, we can allow all the objective functions to be positive and/or negative functions of time no matter whether we pose a maximization or a minimization problem. In both cases the optimal objective function (vector) value will be necessarily bounded as we have a finite number of states (nodes), a finite number of transitions (arcs), and no circuits in the network. However, as we have chosen $c_j^1(t), j = 1, \ldots, n$ to represent the time, we require these functions to be positive, while the other functions $c_j^i(t), i = 2, \ldots, m, j = 1, \ldots, n$, representing general criteria of interest may be of any sign. In fact, a transition from a state $q(k)$ at a given time t_1 to a state $q(l)$ may for some criterion $i, i \in \{2, \ldots, m\}$, yield an objective value $c_j^i(t_1) < 0$, which means that adding the project x_j at time t_1 to a sequence of projects is not beneficial at time, however, the resulting solution sequence may still be nondominated. But the same project x_j could be added at a time $t_2, t_2 \neq t_1$, so that the corresponding objective value would be positive, $c_j^i(t_1) > 0$, and make the project x_j competitive.

3.4 Assumptions

The following assumption is necessary for the principle of optimality for the TDMCKP to hold.

Assumption 1 *For all $t_1, t_2 \geq 0$, if $t_1 \leq t_2$, then*

(a) $t_1 + c_j^1(t_1) \leq t_2 + c_j^1(t_2)$ for all $j = 1, \ldots, n$, and

(b) $c_j^i(t_1) \geq c_j^i(t_2)$ for all $i = 2, \ldots, m$, and $j = 1, \ldots, n$.

Assumption 1 (a) requires that if a project x_j is initiated at a time t_1 or at a later time t_2, then with the earlier start time it has to be accomplished earlier than with the later start time. In other words, the earlier a project is started, the earlier it has to be completed. We observe that if $c_j^1(t), j = 1, \ldots, n$, are monotone increasing functions of time, then this assumption holds. Assumption 1 (b) simply requires that the other components of the objective functions $c_j^i(t)$ for all $i = 2, \ldots, m$ and $j = 1, \ldots, n$ be monotone decreasing functions of time. In the context of the model this implies that (for example) the revenue generated by a project x_j decreases in time, or in other words, the later the project is initiated the less revenue it brings. We believe that both assumptions naturally fit into the model as they mathematically interpret the commonly made assertions in capital budgeting.

3.5 Principle of optimality

Let $f(x) := f(\{x_{j(r)}\}_{r=1}^{p})$ be a nondominated criterion vector of the k-TDMCKP accomplished at time $t^{p+1}(x)$ that can be computed using (4).

Let $G(q(k)) = \text{vmax}^*\{f(x) : x \in q(k)\}$ be the set of all the nondominated criterion vectors of the k-TDMCKP.

The principle of optimality for dynamic multiple criteria networks established in Kostreva and Wiecek (1993) adapted to the time-dependent capital budgeting network model yields the following theorem.

Theorem 1 *Principle of Optimality for the TDMCKP.*

Under Assumption 1, an efficient sequence of projects $x^p = \{x_{j(r)}\}_{r=1}^p$ of the k-TDMCKP accomplished at time $t^{p+1}(x^p)$ has the property that each subsequence of projects $x^s = \{x_{j(r)}\}_{r=1}^s, 1 \leq s < p$ accomplished at time $t^{s+1}(x^s), t^{s+1}(x^s) \leq t^{p+1}(x^p)$, is an efficient sequence of projects of the $(\sum_{r=1}^s a_{j(r)})$-TDMCKP.

Theorem 1 results in the following recursive equations for $t^p > 0$:

$$\begin{aligned} G(q(0)) &= \{\underline{0}\} \\ G(q(k)) &= \text{vmax}^*\{f(x^{p+1} = \{x_{j(r)}\}_{r=1}^{p+1}) : x^{p+1} \in q(k)\} \\ &= \text{vmax}^*\{f(x^p = \{x_{j(r)}\}_{r=1}^p) + c_{j(p+1)}(t^{p+1}(x^p)) : \\ &\qquad f(x^p) \in G(q(k - a_{j(p+1)})),\ j(p+1) \in S, \\ &\qquad k - a_{j(p+1)} \geq 0\}, \qquad k = 1, \ldots, b, \end{aligned}$$

where the operation *vmax** computes the nondominated criterion vectors according to (6) in the set whose every element is a vector sum of a nondominated criterion vector of the efficient sequence of projects of the $(k - a_{j(p+1)})$-TDMCKP accomplished at time $t^{p+1}(x^p)$, and the criterion vector $c_{j(p+1)}(t)$ evaluated at time $t = t^{p+1}(x^p)$.

Since all the states are final, the set of all the nondominated criterion vectors $\mathcal{Y}_e$ is obtained as the vector-maximum of the union of the sets $G(q(k))$, $k = 1, \ldots, b$; i.e.

$$\mathcal{Y}_e = \text{vmax}^* \bigcup_{k=1,\ldots,b} G(q(k)).$$

Note that in each step of the recursion two or more nondominated criterion vectors may correspond to different efficient sequences of projects, however composed of the same projects, which shows that different criterion vectors can be achieved while choosing the same projects to a sequence but at different times. We discuss this and similar situations in Section 4.

4 Example

We now present a didactic example of the time-dependent capital budgeting problem with three criteria ($m = 3$).

Assume there are four projects ($n = 4$) of interest to the company. The fixed budget equals 3 and the cost coefficients a_j, $j = 1, \ldots, 4$ of each project are given by $a_1 = 1$, $a_2 = 2$, $a_3 = 1$, $a_4 = 1$. The criteria include the time of performing the projects to be minimized, and the revenue and appreciation yielded by the projects to be maximized. The objective vectors $c_j(t)$, $j = 1, \ldots, 4$ related to each project are defined as

$$c_1(t) = \begin{bmatrix} 1 \\ 10 - t^2 \\ 40 - t \end{bmatrix}, \; c_2(t) = \begin{bmatrix} 2 \\ 70 - 2t^2 \\ 10 - t \end{bmatrix}, \; c_3(t) = \begin{bmatrix} t + 1 \\ 20 - 30t^2 \\ 20 - 2t \end{bmatrix}, \; c_4(t) = \begin{bmatrix} 2t + 1 \\ 30 \\ 10 - 2t \end{bmatrix}.$$

The resulting TDMCKP has the following form:

$$\begin{aligned} \text{vmax}^* \quad & f(x) = [f_1(x), f_2(x), f_3(x)]^T \\ \text{s.t.} \quad & a(x) \leq 3. \end{aligned} \tag{7}$$

The possible transitions between states for this example problem are represented by the arcs in the network given in Figure 1. The objective vector $c_j(t) = [c_j^1(t), c_j^2(t), c_j^3(t)]^T$ of each transition and the corresponding variable x_j are identified for each arc and denoted by the vector $[j, c_j]$.

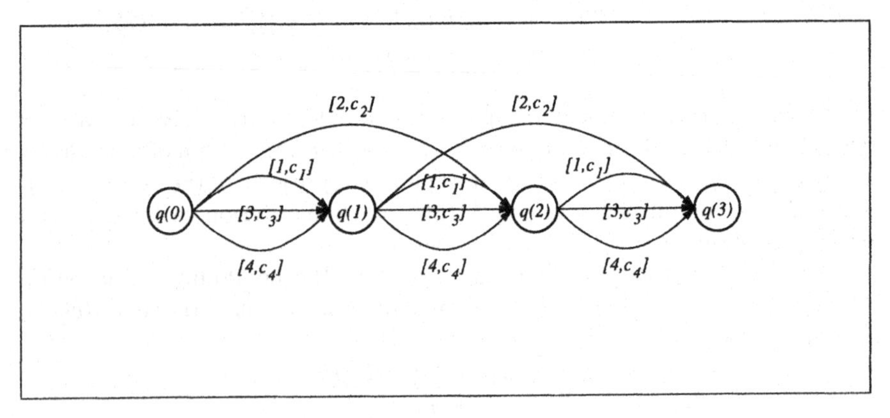

Fig. 1: The DP network for the example problem (7).

Applying the recursive equations developed we obtain the following sets $\mathcal{X}_e$ and $\mathcal{Y}_e$:

$$\begin{aligned} \mathcal{X}_e = \{ & \{\}, \{x_1\}, \{x_3\}, \{x_4\}, \{x_1x_1\}, \{x_3x_1\}, \{x_4x_1\}, \{x_2\}, \{x_1x_1x_1\}, \{x_3x_1x_1\}, \\ & \{x_4x_1x_1\}, \{x_1x_2\}, \{x_3x_2\}, \{x_4x_2\}, \{x_4x_3x_1\}, \{x_4x_4x_1\}, \{x_1x_1x_4\}, \\ & \{x_3x_1x_4\}, \{x_4x_1x_4\}, \{x_2x_4\}, \{x_1x_4x_4\}, \{x_3x_4x_4\}, \{x_4x_4x_4\}\}, \end{aligned}$$

$$\mathcal{Y}_e = \left\{ \begin{bmatrix} 0 \\ 0 \\ 0 \end{bmatrix}, \begin{bmatrix} 1 \\ 10 \\ 40 \end{bmatrix}, \begin{bmatrix} 1 \\ 20 \\ 20 \end{bmatrix}, \begin{bmatrix} 1 \\ 30 \\ 10 \end{bmatrix}, \begin{bmatrix} 2 \\ 19 \\ 79 \end{bmatrix}, \begin{bmatrix} 2 \\ 29 \\ 59 \end{bmatrix}, \begin{bmatrix} 2 \\ 39 \\ 49 \end{bmatrix}, \begin{bmatrix} 2 \\ 70 \\ 10 \end{bmatrix}, \right.$$

$$\begin{bmatrix}3\\25\\117\end{bmatrix}, \begin{bmatrix}3\\35\\97\end{bmatrix}, \begin{bmatrix}3\\45\\87\end{bmatrix}, \begin{bmatrix}3\\78\\49\end{bmatrix}, \begin{bmatrix}3\\88\\29\end{bmatrix}, \begin{bmatrix}3\\98\\19\end{bmatrix}, \begin{bmatrix}4\\48\\65\end{bmatrix}, \begin{bmatrix}5\\54\\54\end{bmatrix},$$
$$\left.\begin{bmatrix}7\\49\\85\end{bmatrix}, \begin{bmatrix}7\\59\\65\end{bmatrix}, \begin{bmatrix}7\\69\\55\end{bmatrix}, \begin{bmatrix}7\\100\\16\end{bmatrix}, \begin{bmatrix}13\\70\\50\end{bmatrix}, \begin{bmatrix}13\\80\\30\end{bmatrix}, \begin{bmatrix}13\\90\\20\end{bmatrix}\right\}.$$

Note that in this example different criterion vectors are achieved by choosing the same projects to a solution sequence at different times. For example, the solution sequences $\{x_4x_2\}$ and $\{x_2x_4\}$ with the criterion vectors $[3, 98, 19]^T$ and $[7, 100, 16]^T$ are both nondominated. Similarly, the solution sequences $\{x_4x_4x_1\}$, $\{x_4x_1\ x_4\}$ and $\{x_1x_4x_4\}$ with the criterion vectors $[5, 54, 54]^T$, $[7, 69, 55]^T$ and $[13, 70, 50]^T$ are all nondominated.

Furthermore we observe that shorter times are achieved when only one or two projects are performed which may not be of high priority to the company. On the other hand, none of the sequences includes all the projects, at most three projects can be selected in any case, and project 2 seems to be the least popular in all the sequences.

In order to make a final decision what projects should be selected, the company would have to specify additional preferences. For example, if the preference was to perform three different projects, two sequences $\{x_4x_3x_1\}$ and $\{x_3x_1x_4\}$ would be the candidates for the final optimal solution. The decision maker would have to choose between the criterion vectors $[4, 48, 65]^T$ and $[7, 59, 65]^T$. The vectors show that performing project 4 at the beginning rather than at the end saves time, yields less revenue, and keeps appreciation at the same level. The final decision would be then between the time and the revenue.

The model and solution approach are part of AMADEuS, a DP-based decision tool developed by Klamroth et al. (1999). AMADEuS generates nondominated solutions for 4 decision scenarios based on different variations of the MCKP. Figure 2 depicts a solution window of AMADEuS with the nondominated solutions in the space of the 2nd and 3rd criterion for example problem (7). The value of the 1st criterion is given next to each solution.

5 Conclusions

We developed the time-dependent multiple criteria knapsack problem (TDMCKP) and used it to model time-dependent capital budgeting with multiple criteria. The novelty of the formulation comes from the fact that the solution set of the problem includes efficient sequences of projects to be performed consecutively over time. We believe that the new model significantly enhances the traditional multiple criteria knapsack model and could be applied in many decision making situations involving capital budgeting where only one project can be selected and performed at a time. This decision situation can be also viewed as scheduling projects subject to a budgetary constraint and so that multiple criteria are optimized.

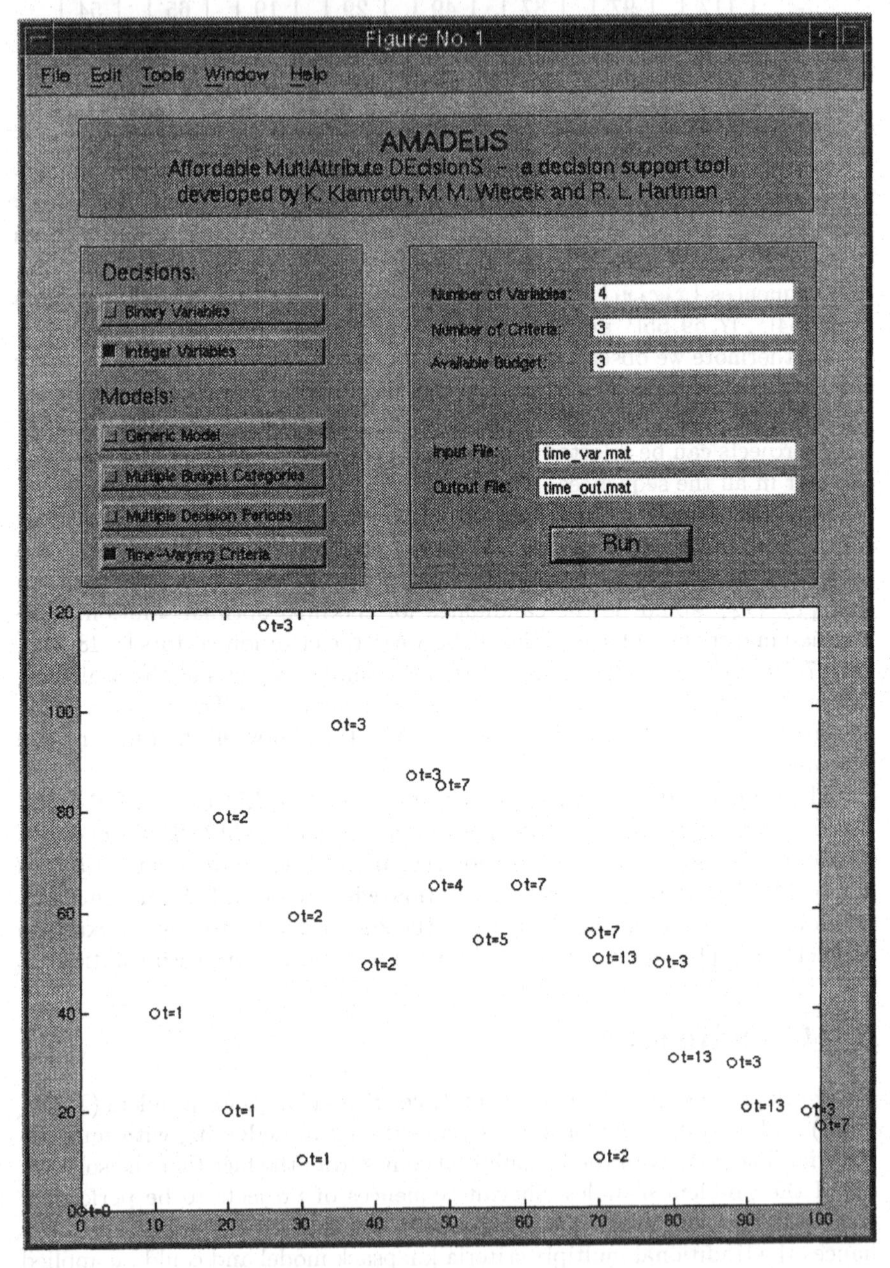

Fig. 2: Solution window of AMADEuS.

References

Ansoff, H. I. (1968). *Corporate Strategy.* Penguin.

Bhaskar, K. (1979). "A multiple objective approach to capital budgeting." *Accounting and Business Research,* 9, 25–46.

Bromwich, M. (1976). *The Economics of Capital Budgeting.* Penguin.

Carsberg, B. V. (1974). *Analysis for Investment Decisions.* Accountancy Age Books. Haymarket Publishing.

Corner, J. L.; Deckro, R. F. and Spahr, R. W. (1993). "Multiple-objective linear programming in capital budgeting." *Advances in Mathematical Programming and Financial Planning,* 3, 241–264.

Eben-Chaime, M. (1996). "Parametric solution for linear bicriteria knapsack models." *Management Science,* 42, 1565–1575.

Gonzalez, J. J.; Reeves, G. R. and Franz, L. S. (1987). "Capital budgeting decision making: An interactive multiple objective linear integer programming search procedure." *Advances in Mathematical Programming and Financial Planning,* 1, 21–44.

Hawkins, C. A. and Adams, R. A. (1974). "A goal programming model for capital budgeting." *Journal of Financial Management,* pages ?–?

Ibaraki, T. (1987). "Enumerative approaches to combinatorial optimization, part ii." In Hammer, P. L., editor, "Annals of Operations Research," volume 11, pages 343–602. Baltzer, Basel.

Ignizio, J. P. (1976). "A approach to capital budgeting with multiple objectives." *The Engineering Economist,* pages 259–272.

Klamroth, K.; Wiecek, M. M. and Hartman, R. L. (1999). "AMADEuS, Affordable MultiAttribute DEcisionS, a decision support tool for multiple criteria capital budgeting." Technical Report 670, Dept. of Math. Sc., Clemson University, Clemson, SC.

Kostreva, M. and Wiecek, M. (1993). "Time dependency in multiple objective dynamic programming." *Journal of Mathematical Analysis and Applications,* 173, 289–307.

Kwak, W.; Shi, Y.; Lee, H. and Lee, C. (1996). "Capital budgeting with multiple criteria and multiple decision makers." *Review of Quantitative Finance and Accounting,* 7, 97–112.

Lee, S. M. and Lerro, A. J. (1974). "Capital budgeting for multiple objectives." *Management Science,* 36, 1106–1119.

Lin, T. W. (1993). "Multiple-criteria capital budgeting under risk." *Advances in Mathematical Programming and Financial Planning*, 3, 231–239.

Rosenblatt, M. and Sinuany-Stern, Z. (1989). "Generating the discrete efficient frontier to the capital budgeting problem." *Operations Research*, 37, 384–394.

Thanassoulis, E. (1985). "Selecting a suitable solution method for a multiple-objective programming capital budgeting problem." *Journal of Business Finance and Accounting*, 12, 453–471.

Turney, S. T. (1990). "Deterministic and stochastic dynamic adjustment of capital investment budgets." *Mathematical Computation and Modelling*, 13, 1–9.

Ulungu, E. and Teghem, J. (1994). "Application of the two phases method to solve the bi-objective knapsack problem." Technical report, Department of Mathematics & Operational Research, Faculté Polytechnique de Mons, Belgium. Submitted.

Ulungu, E. and Teghem, J. (1997). "Solving multi-objective knapsack problem by a branch-and-bound procedure." In Climaco, J., editor, "Multicriteria Analysis," pages 269–278. Springer-Verlag. To appear.

Vetschera, R. (1985). "Time preferences in capital budgeting - an application of interactive multiobjective optimization." *Methods of Operations Research*, 50, 649–660.

Villarreal, B. and Karwan, M. H. (1981). "Multicriteria integer programming: A (hybrid) dynamic programming recursive approach." *Mathematical Programming*, 21, 204–223.

Visée, M.; Teghem, J.; Pirlot, M. and Ulungu, E. (1996). "Two-phases method and branch and bound procedures to solve the bi-objective knapsack problem." Technical report, Department of Mathematics & Operational Research, Faculté Polytechnique de Mons, Belgium. Submitted.

Weingartner, H. M. (1963). *Mathematical Programming and the Analysis of Capital Budgeting Problems.* Prentice-Hall, Englewood Cliffs, N. J.

Sustainability Indicators for Multiple Criteria Decision Making in Water Resources: An Evaluation of Soil Tillage Practices using Web-HIPRE

Jason K. Levy[1], Keith W. Hipel[1,2], and D. Marc Kilgour[3]

[1] University of Waterloo, Department of Systems Design Engineering, Waterloo ON N2L 3G1, Canada
[2] Department of Statistics and Actuarial Science, University of Waterloo
[3] Wilfrid Laurier University, Department of Mathematics, Waterloo ON N2L 3C5, Canada

Abstract. Agricultural nonpoint source pollutant loadings — sediment, pesticides, and nutrients — are characterized by the difficulty and cost of their control, the large areas involved, and their significant impact on water resources systems. This paper considers a multiple-attribute decision making problem in which three soil tillage alternatives — conventional tillage, no-till, and ridge-till — are evaluated based on a finite number of hierarchically arranged attributes. Web-HIPRE, a newly developed Java-applet for multiple criteria decision making by Hämäläinen [10], is used to show that no-till agriculture is the most desirable farm management practice with respect to surface and groundwater objectives in Southwestern Ontario, Canada.

1 Introduction

René Dubos, renowned biologist and essayist, claimed that "when the Egyptians introduced the plow 7,000 years ago they provided mankind with the single technological innovation which has had the most profound and lasting influence on the earth" [7]. Even today, a majority of North American farmers follow the ancient practice of cleanly tilling their fields. This has resulted in large-scale losses of productive soil through surface erosion, particularly during heavy rains. Technologies from systems engineering currently exist to make our agricultural systems more sustainable: the challenge will be in marshalling enough political courage, scientific expertise, and societal commitment to embrace a more holistic and integrated approach to sustainable development [14]. To this end, a comprehensive multiple criteria analysis is *sine qua non* for the identification of sustainable agricultural practices.

Increasing concerns over the integrity of both surface and groundwater systems have led to the development of Best Management Practices (BMPs) for reducing the amount of contaminants generated by nonpoint sources. One important BMP, conservation tillage, is widely believed capable of protecting water quality by reducing runoff from farm fields [6].

2 Study Area and Alternatives

This paper considers three alternative approaches to farming with different impacts on environmental and water resource systems: conventional (CONV) tillage, ridge-till and no-till. The latter two are conservation tillage practices. North American farmers are increasingly turning to conservation tillage systems — capable of "stopping erosion", "building soil", and "fighting pollution" — over traditional methods that rely on plowing or intensive tillage. The objective of this multiple criteria analysis is to determine the best soil tillage strategy to ensure the sustainability of water resources systems in Southwestern Ontario, Canada.

2.1 Case Study Location

Barton and Farmer [2] collected benthic invertebrate data from streams in Southwestern Ontario, located in the Kintore, Pittock, Kettle and Essex basins. The streams are small (less than 2 meters wide) and the predominant crop in all basins is a corn/soybean/wheat rotation, with small livestock operations in the Kintore, Pittock, and Kettle basins. The streams were previously studied for the effects of conservation tillage on erosion, downstream chemical water quality, and crop yields, by the Upper Thames River Conservation Authority and the Soil and Water Environmental Enhancement Program (SWEEP). The final report of the SWEEP study was prepared by Walker and Tossell [20].

2.2 Conventional (CONV) Tillage Practices

Conventional soil tillage practices may contribute to the degradation of water resources systems and to the loss of agricultural land: excessive plowing speeds the breakdown of crop residues and can make production marginal as early as the next planting season. CONV systems involve turning crop residues under at least once a year (usually more) with a mouldboard plow.

2.3 Conservation (CONS) Tillage Practices

Conservation tillage practices leave valuable crop residues (stems, stalks, and leaves) on fields after harvest, thereby increasing organic matter in the top two or three inches of soil (as surface crop residues slowly decompose). The less the soil is tilled, the more stored carbon that is available to improve long-term fertility.

No-till farming, perhaps more accurately described as slot planting, is becoming increasingly common in row-crop farming. Canadian farmers use the term "direct-seeding" since the farmer makes a small groove in the soil to plant the seed. No additional tillage is performed for seedbed preparation: the soil is left undisturbed from harvest to planting except for nutrient injection.

In the *ridge-till* system, planting is completed in a seedbed prepared on ridges with sweeps, disk openers, coulters, or row cleaners. Residue is left on the surface between ridges, which are rebuilt during cultivation. Weed control is accomplished primarily with herbicides. The scope of this paper will be limited to the comparison of conventional tillage systems with no-till and ridge-till farm management practices. Other conservation tillage techniques (mulch-till, reduced- till, zone-till, and strip-till) are often considered modifications of the no-till and ridge-till approaches [1].

3 Hierarchical Arrangement of Criteria

Changing plowing and sowing practices to conservation tillage can improve the health of rivers and streams and mitigate soil loss. But soil tillage systems have other implications, including groundwater quality, herbicide costs, and agricultural yield. Moreover, while conservation tillage can reduce the total *loading* (mass) of pesticide in runoff, pesticide *concentrations* in runoff can increase. In addition, no-till farming will likely increase the fixed costs of planting and herbicide costs. On the other hand, conservation tillage can reduce the time required to prepare and plant a field by up to two-thirds: making fewer trips across the field with decreased horsepower requirements reduces maintenance and fuel costs. Clearly there are many important trade-offs to consider.

3.1 Pesticide Contamination of Surface Waters

Agricultural pesticides are effective tools for increasing crop yields. However, their use has resulted in non-point source contamination of surface waters in the Great Lakes Basin. Considerable research has focused on the differences in non-point source pollutant loadings across soil tillage systems [9]. Conventional tillage techniques pose the highest risk of introducing contaminants into surface runoff.

Conservation tillage can improve surface water quality by retaining or biodegrading pesticides on the field: crop residues hold soil particles and associated nutrients and pesticides, tending to slow water runoff, and allowing the water more time to soak into the soil. Conservation tillage also improves infiltration due to the macropores created by earthworms and old plant roots that are left intact. No-till has sometimes resulted in complete elimination of pesticide runoff from fields.

Weed control may be the most challenging facet of no-tillage systems. Without tillage, weed growth begins before crop growth, sometimes making a chemical burndown necessary (to kill existing vegetation) before or after planting. Perennial weeds are particularly hard to control in no-till systems.

3.2 Nutrient Loading and Soil Erosion

Sediment, pesticides, and nutrients from agricultural surface runoff have greater negative impacts on streams and rivers in North America than pollution from any other source [4]. Conservation tillage systems can significantly reduce soil erosion and nutrient losses (particularly nitrogen and phosphorous) [3]. Depending on the amount of crop residue left on the surface, erosion can be reduced by up to 90% [1]. The no-till system leads to maximum residue preservation and a corresponding reduction in soil erosion by raindrop impact.

3.3 Aquatic Ecosystem Health

Pollution from agricultural activities can significantly stress aquatic communities: the primary effects of farm tillage practices on aquatic ecosystems are sedimentation and habitat degradation due to erosion and eutrophication. There are two types of biological indicators that are useful in measuring aquatic ecosystem quality — ambient biological monitoring and bioassay methods. Ambient biological monitoring samples one or more organisms from a water body while bioassays assess chemical, cellular or genetic changes within an organism.

Of course, there are other environmental considerations. Conventional tillage releases soil carbon into the atmosphere as carbon dioxide, where it can combine with other gases to contribute to global warming. And surface residues from conservation tillage systems provide quality shelter and food for wildlife (such as ducks and small mammals) at critical times.

4 Web-HIPRE and Water Resources Decision Making

Recent years have witnessed a growing interest in developing multistrategy systems that can integrate two or more reasoning types and/or computational paradigms into a single system [13]. Multistrategy systems have the potential for greater competence and versatility than monostrategy systems because they can solve a wider range of problems using the complementarity of individual learning strategies. Since human learning is clearly multistrategy, decision support systems which allow for a variety of MCDM approaches are most likely to be capable of replicating the heuristics of human problem solving.

Web-HIPRE (HIerarchical PREference analysis software) is a Java-applet for multiattribute decision making based on HIPRE 3+ (Hämäläinen [10]). It is the first web-based interactive multicriteria decision analysis tool and, to our knowledge, the only approach to support multiple MCDM approaches: this allows the users themselves to test and compare alternative MDCM methods. As environmental management becomes increasingly inclusive and multidisciplinary, decision makers should have access to a number of MCDM technologies. If multiple MCDM approaches are meaningfully included in the decision making process, then stakeholders will be better prepared to address notions of complexity, uncertainty, and equity — concepts central to sustainability [12].

4.1 Value Tree Construction

The decision problem of choosing the best tillage strategy is illustrated in Fig. 1. The value tree is structured hierarchically from the goal (best agricultural management practice) to first level criteria (surface loss, sediment yield, percolation to groundwater, and ecosystem health) and on to second and third level criteria. Note that the alternatives (conventional tillage, no-till, and ridge-till) are the lowest level (rightmost) elements of the value tree. The alternatives are evaluated with respect to each of the criteria. Table 1 describes the acronyms used in Fig. 1.

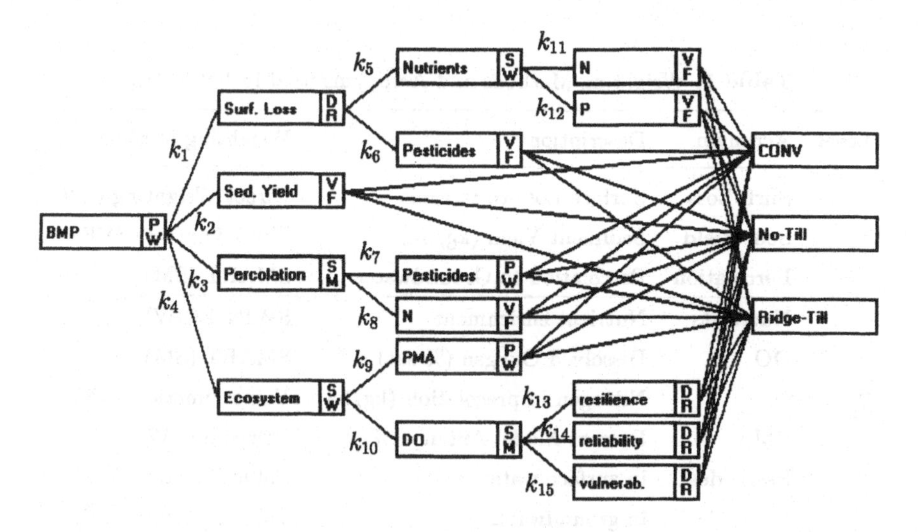

Fig. 1. Hierarchical Arrangement of Value Tree

Fig. 1 illustrates the hierarchical arrangement of attributes in the soil management problem: sediment yield (x_2), pesticides in surface loss (x_6), pesticides in percolation (x_7), Nitrogen in percolation (x_8), Percent Model Affinity (x_9), Nitrogen in surface loss (x_{11}), Phosphorous in surface loss (x_{12}), DO resilience (x_{13}), DO reliability (x_{14}), and DO vulnerability (x_{15}). Web-HIPRE calculates the overall value (V) of the jth alternative using the formula

$$V(x_j) = k_1[k_5(k_{11}x_{11j} + k_{12}x_{12j}) + k_6x_6] + k_2x_{2j} + \\ k_3(k_7x_{7j} + k_8x_{8j}) + k_4[k_9x_{9j} + k_{10}(k_{13}x_{13j} + k_{14}x_{14j} + k_{15}x_{15j})] \quad (1)$$

and provides graphical output for data analysis and results.

The particular weighting method used depends on the nature of the criteria, the amount of information available, and the preferences of the decision maker. For example, the amount of phosphorous and nitrogen in runoff was evaluated with standard value (scoring) functions because such functions are widely used. The value functions were obtained from the work of Yakowitz et al. [21] and Chase and Duffy [5], both of whom evaluated the environmental and economic consequences of alternative farming practices.

Pairwise comparisons were used to evaluate soil tillage alternatives with respect to PMA since, to our knowledge, value functions have yet to be constructed. For PMA, a matrix of pairwise comparisons (importance ratio judgements between pairs of alternatives or attributes) in the framework of AHP [17] was used to elicit the relevant evaluations. The decision maker is asked to compare pairs of attributes; for each pair, which attribute is more important, and by how much?

Table 1. Weighting Methods and Acronyms used in Value Tree

Level	Acronym	Description	Weighting Method
	Surf. Loss	Surface Loss to streams	Direct Weighting (DR)
1	Sed. Yield	Sediment Yield (kg/ha)	Value Function (VF)
	Percolation	Percolation to Groundwater	SMART (SM)
	Nutrients	Nutrient enrichment	SWING (SW)
	DO	Dissolved Oxygen (% sat)	SMART (SM)
2	N	Nitrogen in precolation (kg/ha)	Value Function (VF)
	PMA	Percent Model Affinity (%)	Pairwise (PW)
	Pesticides	in surface water	Value Function (VF)
		in groundwater	Value Function (VF)
	Resilience	System Resilience	Direct Weighting (DR)
	Reliability	System Reliability	Direct Weighting (DR)
3	Vulnerab.	System Vulnerability	Direct Weighting (DR)
	P	Phosphorous in runoff (kg/ha)	Value Function (VF)
	N	Nitrogen in runoff (kg/ha)	Value Function (VF)

In direct weighting (DR), the weights of the alternatives are entered directly. Web-HIPRE provides text fields for weights, and a slider for graphical input; the software is also capable of normalizing the weights. The sustainability indicators of Section 5 will be implemented using DR.

5 Direct Weighting and Sustainability Indicators

In order to evaluate the sustainability of systems using WEB-HIPRE, we consider the notions of reliability and resilience, as defined by Loucks [16]. Let the status of the system at time t, $t = 1, 2, 3...n$, be represented by the variable X_t, where the possible values of X_t are divided into two sets: S, the set of satisfactory values ($X_t \in S$) and F, the set of unsatisfactory values ($X_t \in F$).

The probability that the system output is in S is its reliability, α:

$$\alpha = P(X_t \in S) \tag{2}$$

Resilience, γ, describes how quickly a system returns to a satisfactory state, once it has entered an unsatisfactory state:

$$\gamma = P(X_{t+1} \in S \mid X_t \in F) \tag{3}$$

Next, we propose an original measure of system vulnerability (the expected magnitude of failure). As an estimator, we use σ, the average of the largest failure extent in a failure episode over all failure episodes.

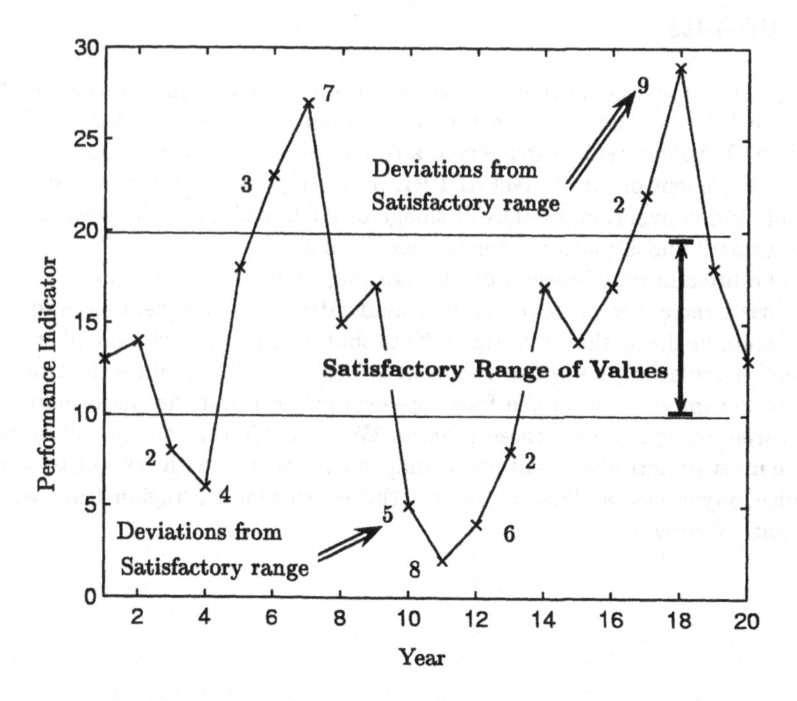

Fig. 2. Simulated time series of a system performance indicator

Figure 2 shows an illustrative time series of values of a typical system performance indicator, along with associated satisfactory and unsatisfactory ranges. Each criterion will have its own range of satisfactory values. The measures of engineering resilience may be computed as follows for the time series in Figure 2:

$$\begin{aligned} \text{Reliability:} \quad & \alpha = \tfrac{10}{20} = 0.5 \\ \text{Resilience:} \quad & \gamma = \tfrac{4}{10} = 0.4 \\ \text{Vulnerability:} \quad & \sigma = \tfrac{4+7+8+9}{4} = 7.0 \end{aligned}$$

To include these statistical measures in the Web-HIPRE framework, it is convenient to convert each statistic into a measure that ranges from 0 to 1 and for which higher values are preferred. The reliability and resilience measures already have the desirable properties. For the vulnerability measure, identify the *largest* value of σ for each criterion (among all alternatives being compared) and divide the average failure extent of the alternative by this maximum value. A second step is necessary to convert the vulnerability measure so that higher values are preferred: subtract each relative vulnerability measure from 1.

6 Results

Using data from the aforementioned reports on stream conditions in Southwestern Ontario and applying the sustainability measures of Section 5, we evaluated the three soil management alternatives on the basis of the hierarchical arrangement of Fig. 1. Web-HIPRE produced overall values for ridge-till, no-till, and conventional (CONV) tillage of 0.52, 0.59, and 0.33, respectively (see Table 2 and Equation 1 for further details).

The breakdown of these values according to the top-level criteria of surface loss to streams, percolation to groundwater, sediment yield, and aquatic ecosystem health is shown in Fig. 3. Note that no-till is superior on all criteria except percolation, on which ridge-till is best. Fig. 3 combines information about the importance of the four top-level criteria and the performance of the three alternatives on these criteria. We conclude that the no-till system is the most preferred agricultural management practice with respect to water quality objectives, at least in the Southwestern Ontario region from which the data is drawn.

Table 2. Evaluations for CONV, No-Till, and Ridge-Till Alternatives

		Indicator Scores: x_{ij}		
x_i : Indicator	Scaling Factor k_i	CONV $j = 1$	No-Till $j = 2$	Ridge-Till $j = 3$
Surface Loss	$\mathbf{k_1 = 0.1}$			
x_5 : Nutrients	$k_5 = 0.7$			
x_{11} : *N*	$k_{11} = 0.5$	0.3	0.8	0.6
x_{12} : *P*	$k_{12} = 0.5$	0.5	0.9	0.7
x_6 : Pesticides	$k_6 = 0.3$	0.2	0.9	0.8
Sediment Yield	$\mathbf{k_2 = 0.3}$			
x_2 : Sed. Yield		0.5	0.74	0.68
Percolation	$\mathbf{k_3 = 0.2}$			
x_7 : Pesticides	$k_7 = 0.3$	0.7	0.2	0.3
x_8 : N	$k_8 = 0.7$	0.2	0.2	0.6
Ecosystem	$\mathbf{k_4 = 0.4}$			
x_9 : PMA	$k_9 = 0.2$	0.2	0.9	0.8
x_{10} : DO	$k_{10} = 0.8$			
x_{13} : *Resilience*	$k_{13} = 0.33$	0.2	0.45	0.35
x_{14} : *Reliability*	$k_{14} = 0.33$	0.3	0.5	0.2
x_{15} : *Vulnerability*	$k_{15} = 0.33$	0.1	0.6	0.3
Overall score $V(\boldsymbol{x}_j)$:		0.33	0.59	0.52

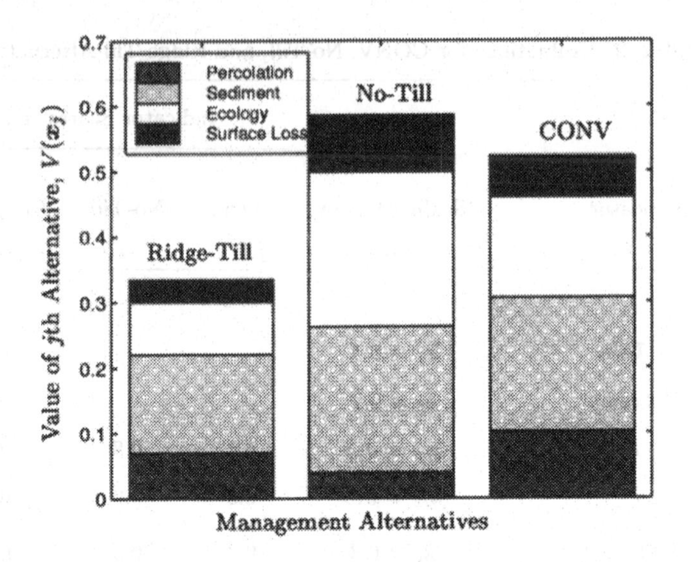

Fig. 3. Overall values of the three soil management alternatives

7 Conclusion

We agree with the philosophy of Hämäläinen and Salo [11] that "the separation of competing [multiple criteria] methodologies into isolated schools of thought is regrettable". Alternative MCDM methods are too often marketed as competing products, rather than as complementary approaches for decision analysis; there is a great need for decision support systems that allow practitioners the choice of tools for the particular problems that they are facing, since different techniques have their strengths and weaknesses [18]. Web-HIPRE is the first decision technology to integrate common techniques from multiattribute value theory, such as AHP [17], SMART [8], and SWING [19]. By integrating pairwise comparisons, direct scoring, and value functions into a holistic system, a decision model can more realistically and comprehensively evaluate alternatives. Using an appropriate combination of multicriteria analysis techniques we have shown that no-till is a desirable alternative in the context of water resources sustainability. Future work will consider additional economic, social, and environmental criteria related to soil management practices, thereby integrating institutional and organizational issues [15].

Bibliography

[1] J. S. Angle, G. McClung, M. S. McIntosh, P. M. Thomas, and D. C. Wolf. Nutrient losses in runoff from conventional and no-till corn watersheds. *Journal of Environmental Quality*, 13:431–435, 1984.

[2] D. R. Barton and M. E. D. Farmer. The effects of conservation tillage practices on benthic invertebrate communities in headwater streams in southwestern Ontario, Canada. *Environmental Pollution*, 94:207–215, 1997.

[3] B. T. Bowman, G. J. Wells, and D. J. King. Transport of herbicides and nutrients in surface runoff from corn cropland in southern Ontario. *Canadian Journal of Soil Science*, 74:59–66, 1993.

[4] J. B. Braden and S. B. Lovejoy. *Agriculture and Water Quality: International Perspectives*. Lynne Rienner Publishers, Boulder, 1990.

[5] C. A. Chase and M. D. Duffy. An economic analysis of the nashua tillage study: 1978-1987. *Journal of Production Agriculture*, pages 91–98, 1991.

[6] C. M. Cooper. Biological effects of agriculturally derived surface water pollutants on aquatic systems – a review. *Journal of Environmental Quality*, pages 402–408, 1993.

[7] R. Dubos, editor. *A God Within*. Charles Scriber's Sons, 1980.

[8] W. Edwards and F. H. Barron. SMARTS and SMARTER: Improved simple methods of Multi-Attribute Utility measurement. *Organizational Behavior and Human Decision Processes*, pages 306–325, 1994.

[9] R. S. Fawcett, B. R. Christensen, and D. P. Tierney. The impact of conservation tillage on pesticide runoff into surface water – a review and analysis. *Journal of Soil and Water Conservation*, pages 126–135, 1994.

[10] R. P. Hämäläinen. *Web-HIPRE Version 1.0 User's Guide*. Systems Analysis Laboratory, Helsinki University of Technology, Helsinki, Finland, 1998.

[11] R. P. Hämäläinen and A. A. Salo. The issue is understanding the weights (Rejoinder). *Journal of Multi-Criteria Decision Analysis*, pages 340–343, 1997.

[12] J. K. Levy, K. W. Hipel, and D. M. Kilgour. A holistic approach to sustainable development: The Graph Model for Conflict Resolution. *Information and Systems Engineering*, 1:159–177, 1995.

[13] J. K. Levy, K. W. Hipel, and D. M. Kilgour. An intelligent approach for resolving multiple- stakeholder groundwater resources conflicts. In *Proceedings of the International Conference on Water Resources and Environmental Research: Towards the 21st Century*, volume 1, pages 183–190, Kyoto, Japan, 1996.

[14] J. K. Levy, K. W. Hipel, and D. M. Kilgour. Systems for sustainable development: challenges and opportunities. *Systems Engineering*, 1:28–35, 1998.

[15] E. T. Loehman and D. M. Kilgour, editors. *Designing Institutions for Environmental and Resource Management.* Edward Elgar, Cheltenham, 1998. 368 pp.

[16] D. P. Loucks. Quantifying trends in system sustainability. *Hydrological Sciences Journal*, 42:513–530, 1997.

[17] T. L. Saaty. *The Analytic Hierarchy Process.* McGraw-Hill, New York, 1980.

[18] A. A. Salo and R. P. Hämäläinen. On the measurement of preferences in the analytic hierarchy process. *Journal of Multi-Criteria Decision Analysis*, 6:309–319, 1997.

[19] D. von Winterfeldt and W. Edwards. *Decision Analysis and Behavioral Research.* Cambridge University Press, London, 1986. 604 pp.

[20] R. R. Walker and R. W. Tossell. Environmental monitoring of sweep pilot watersheds. Final Report 85, Water Resources Branch, Ontario Ministry of the Environment, Toronto, Canada, 1992. Vols 1 and 2.

[21] D. S. Yakowitz, J. J. Stone, L. J. Lane, P. Heilman, J. Masterson, J. Abolt, and B. Imam. A decision support system for evaluating the effects of alternative farm management practices on water quality and economics. *Water Science and Technology*, pages 47–54, 1993.

Soft-OR and Multi-Criteria Decision Analysis for Group Decision Support: A Case Study in Fisheries Management.

Brett E. Malyon[1] and Theodor J. Stewart[2]

[1] Management Science Department, University of Strathclyde, Glasgow G1 1QE, Scotland
[2] Department of Statistical Sciences, University of Cape Town, Rondebosch 7700, South Africa

Abstract. As in most fisheries world-wide, the South African rock lobster fishery is fraught with complexities and conflict. In addition to ensuring the sustainability of fishing resources, decision-makers are increasingly also required to weigh up the political, economic and social aspects of fisheries in a transparent and participative way. This is a particularly difficult task in light of the uncertainty surrounding biomass estimates and projections of future resource productivity. Fisheries management decisions typically impact upon numerous stakeholder groups, and there is a powerful lobby for less bureaucracy, and more stakeholder participation in fisheries management. The South African situation is unique in that fisheries management policy has been significantly influenced by the government's Reconstruction and Development Programme (RDP), aimed at rectifying the wrongs of the apartheid past. This research was carried out at a time of great change and transition, when a new fisheries policy for the country was being formulated and negotiated. As consultants, our task was to structure the messy problems facing the scientific working group upon whose recommendations management decisions are taken. Fishing company directors, government officials, scientists, labour union representatives and informal fishermen were involved in the planning process. A soft-OR method called SODA, as well as multi-criteria decision analysis (MCDA), were used to provide group decision support to this diverse group of stakeholders.

Keywords. problem structuring, SODA, MCDA, fisheries management

1 Introduction

Management science practice is significantly broader and more involved than just modelling techniques. Problem structuring is a critical first phase of any intervention. However, when working with a group of decision-makers, a consensual formulation of the problem seldom emerges. Planning decisions are often clouded by equivocality and uncertainty. Particularly when considering strategic issues, actors often have quite different views about the present and future decision environments. In structuring problems of this nature, consultants are often required to manage the 'group dynamic' in the client group. This role as facilitator is a demanding one, and requires specialist skills and knowledge [Phillips & Phillips (1993), Rosenhead (1996)].

Several members of the MCDA community have also stressed the need for a broader approach to MCDA decision support, to incorporate research from fields such as cognitive science and organisational studies [Bouyssou et al (1993)]. Idea generation techniques, such as the Nominal Group Technique, are commonly used in MCDA workshops to help groups to structure problems [e.g. Belton (1997), Stewart & Scott (1995)]. Recently, several researchers have also suggested using soft-OR together with MCDA for a more "mature" approach to multi-criteria decision aid [Belton et al (1997), Ostanello (1997)].

SODA (short for "**S**trategic **O**bjectives **D**evelopment & **A**nalysis") is one such soft-OR approach. The SODA methodology was designed to aid group problem structuring through a facilitated process. In workshops, computer supported facilitation (using the Decision Explorer (DE) software[1]) is used to capture and structure organizational knowledge into a qualitative model or map. Over the past few years, SODA has been used in a diverse range of organizations, particularly for the exploration and development of strategy [Eden & Ackermann (1998)]. One or more workshops are typically held with the key players in the organization; stakeholders or other interested parties can also be included.

The aim of this research was two-fold. Firstly, to explore the effectiveness of the SODA approach in a highly-political fisheries management context, involving participants from diverse interest groups. We anticipated that our role as facilitator, and the results of the SODA intervention, might be quite different to the experiences of Eden [e.g. Eden (1989)], who had worked largely with private organizations. Secondly, the research aimed to consider how SODA might provide problem structuring for subsequent MCDA modelling in practice.

2 The SODA Approach

SODA relies heavily on the "cognitive mapping" technique developed by Eden (1988). Cognitive mapping is a form of cause mapping, but is distinct from other mapping approaches in that it attempts to model a person's cognition. Eden's technique draws on Kelly's (1970) Personal Construct Theory for the theoretical basis used in constructing cognitive maps. Ideas (more formally 'constructs') are entered as short sentences or phrases. The constructs are connected by arrows, denoting causal or 'means-ends' relationships. In this way, trains of thought or argument can be captured, and a qualitative model or 'map' of a person's perceptions built relating to the problem situation (see Figure 1 under section 3.1.1. for an example DE map). A number of guidelines and conventions should be followed in constructing cognitive maps - the interested reader can see Eden et al (1983) and Ackermann et al (1990) for more details.

Cognitive mapping can be used in one-to-one interviews before a workshop, in order to construct a database of relevant information relating to the problem. Maps

[1] Produced by Banxia, UK. Earlier versions of the software were called COPE.

from a number of interviews are then merged into a group map. In this way, a single all-encompassing 'picture' or model of the problem situation is created, and forms the focus for negotiation and a consensual understanding of the problem issues. Alternatively, 'strategy' maps can be developed 'live' in workshops to structure problems with a group [Eden & Ackermann (1998)]. The group or strategy map is divided into clusters to represent the different issues and themes to be explored in the workshop. In practice, group maps might contain many hundreds of concepts. For this reason, the Decision Explorer software is used to manage this 'rich' information through structuring and analysis. See Eden & Ackermann (1998) for more details of the analyses available in DE to assist in structuring maps.

3 The Intervention

Within South Africa's Department of Sea Fisheries, the West Coast Rock Lobster Working Group (WCRLWG) is responsible for conducting scientific research and providing management advice. The group has important input into decisions relating to Total Allowable Catches (TACs)[2], minimum size restrictions, and other measures for the conservation and sustainable use of the rock lobster stock. Prior to our intervention, the WCRLWG had for some time been developing an Operational Management Procedure (OMP)[3]. The OMP would effectively automate the TAC decision process, so reducing the amount of discussion and negotiation around each annual TAC decision.

Through involvement with certain members of the WCRLWG, the authors were aware of the OMP project, and the progress to date. We saw it as a useful opportunity for action research, and approached the chairman of the WCRLWG to suggest a problem structuring exercise to assist them in exploring the broader issues surrounding the implementation of the OMP. It was decided that a number of interviews would be held to capture the various stakeholders' views, followed by a group workshop in which the issues raised would be discussed further. In this preliminary meeting, ten key representatives were identified, including government officials, academic scientists, independent scientific consultants, industry directors and labour representatives. We emphasised that our role would be that of independent and impartial facilitators in assisting the group to structure the problems at hand.

At the time, the lobster fishery was facing several perplexing problems. Recent years had seen scientists dumbfounded by strangely low somatic growth rates

---

[2] The TAC is the overall total level of fishing allowed in a particular fishing season i.e. restricting the fishing industry to a specific tonnage of fish that can be harvested in that season.

[3] An OMP is defined in the government white paper (1997) 'A Marine Fisheries Policy for South Africa' as: "a scientifically evaluated process which defines the manner in which the available data on a resource are used to determine the level of a control measure such as a TAC, thereby incorporating a harvesting strategy".

measured in lobsters. The low growth had placed considerable pressure on the resource. TACs had declined significantly over the past decade. As a result, the industry had been forced to scale down operations and lay off employees. Poaching also presented a major problem, with illegal and recreational catches on the increase. The poaching problem was exacerbated by a lack of effective policing due to funding shortages. Perhaps the most perplexing problem facing the industry related to the political uncertainty regarding their future, with increasing pressure on politicians to grant commercial fishing rights to communities previously deprived such rights under the apartheid regime.

One of the working group's primary objectives in developing an OMP was to set in motion a systematic and long-term stock-rebuilding strategy. The government white paper nevertheless stressed that any long-term management plans should be developed through a 'co-operative process', taking socio-economic issues into consideration. As in most fisheries world-wide, the greatest challenge to managers is the need for sensible and equitable trade-offs between political, economic, social and conservation concerns; both short and long-term [Parsons (1993)]. Several researchers have proposed MCDA as a means for assisting fisheries managers to explicate and make sense of the trade-offs they face [e.g. Healey (1984)]. However, such proposals have generally met with criticism, mainly on the grounds of the complex dynamics of the decision process, requiring negotiation and bargaining [Leschine (1988)]. Perhaps as a result thereof, very few cases of successful application of MCDA have actually been reported in fisheries management.

3.1 SODA Interviews

In the ensuing months, the ten participants were interviewed. Rather than having a pre-planned agenda, an unstructured interviewing approach was adopted. Participants were asked what they saw as the key issues facing the management of the fishery at the time, and the ideas were actively mapped as they were raised. The recorded concepts and links provided the cues for further exploratory questions, such as "Why is that important?" and "What are the implications of this?". This questioning technique is similar to Keeney's Value Focused Thinking (VFT) approach for identifying 'fundamental' objectives [Keeney (1992)]. As with VFT, cognitive mapping helps to determine whether an identified objective is merely a means to some higher end, or whether it is an 'ultimate objective'. This unstructured interviewing style was adopted at the outset to encourage individuals to talk about the problem as *they* saw it.

3.1.1 An Example

A portion of a map from an interview is shown in Figure 1. When questioned about the problems facing the management of the fishery, the participant said that there was "too much extraction" from the resource. The interviewer proceeded to ask what a satisfactory alternative might be to this situation. The response was: "a limited number of 'industrial' users". This response formed the opposite pole to the initial statement, giving further meaning and clarifying the initial concept. The ellipses denote the words 'rather than'. So, concept number 1 is read as: 'Too much extraction *rather than* a limited number of industrial users'.

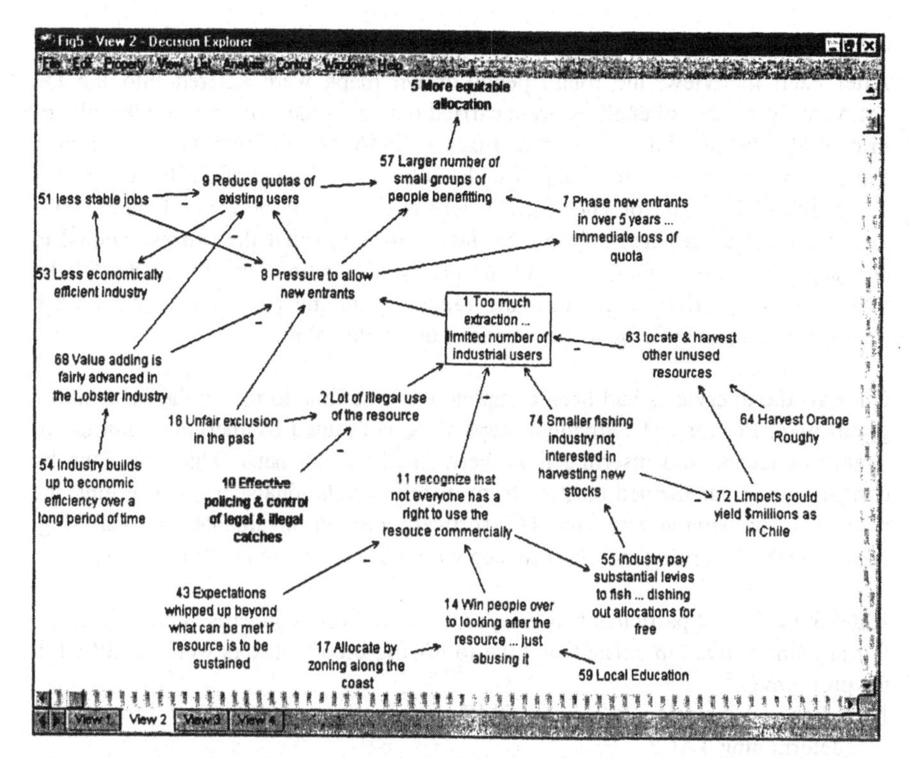

Figure 1. A portion of a map drawn during an interview

From this initial concept, the map was fleshed-out by considering a) the consequences and b) possible explanations for this, and subsequent concepts, as they arose. Arrows are inserted to denote causality. For instance, to the question "What is an *explanation* for there being too many users?", the answer was that there was a lot of illegal use of the resource. This was added as concept no. 2. Similarly, a negative arrow denotes an inverse relationship. For example, the link between concept 11 and concept 1 captures the interviewee's feeling that it should be made clear to the parties concerned that not everyone has the right to fish commercially, and therefore, that the number of industrial users should be limited.

Whilst drawing the cognitive map, the interviewer attempts to order the concepts into a hierarchy, with *goals* at the top, *strategies* leading into the goals, and *options* at the lowest level of the map [Ackermann et al (1990)]. In this way, a meaningful structure is given to the problem. In Figure 1, "5. More equitable allocation" is a goal, "9. reducing the quotas of existing users" is an example of a strategy, and "17. allocate by zoning along the coast" and "64. harvesting Orange Roughy" are two of the options which were identified. By considering possible explanations for concepts, as well as satisfactory alternatives to situations, decision alternatives can be created. Just looking at this portion of one interview map, it is clear that there are several trade-offs to be made in allocating access rights; political, economic and social issues all come into the equation.

3.1.2 Analysing the Maps & Follow-Up

After each interview, the rough pencil-drawn maps were entered into the DE software. A number of analyses were carried out using the software, to identify key or central concepts and so cluster the maps according to different themes or issues. Follow-up interviews were arranged with each interviewee to check the validity of the maps, and expand or alter them where necessary. A number of participants commented that they had enjoyed the interviews, and that the process helped to crystalize their own thoughts about the problem situation. We found the SODA interviews very effective for building rapport with the participants, what Eden (1989) refers to as building a client-consultant relationship.

Once all the interviews had been completed, it was time to merge the maps into a group map. Portions of individual maps were combined by merging common or similar concepts, and inserting links between different maps. This was done by comparing similar-themed clusters from different individuals' maps. Merging the maps was no simple task; the DE software proved invaluable for analysing, organising and merging the ±700 concepts drawn from the ten individual maps.

Although different participants displayed quite different concerns and emphases, a fair amount of overlap existed, and the following broad clusters were identified in the group map:

1. Choosing between a simple OMP versus an annual review process for determining TAC?
2. Including qualitative factors in decision-making
3. Determining a robust indicator of resource status
4. Rebuilding the industry
5. Rebuilding the resource
6. Employing a multi-disciplinary approach to decision-making
7. Measures for more effective policing & control of the fishery
8. Co-management schemes
9. Developing new markets
10. Providing fair & equitable access rights
11. Ensuring transparency & accountability in management

It was found, in analysing and clustering the concepts in the group map, that the various clusters were very inter-related. This suggested that the implementation of the OMP should not be considered in isolation of the other problem areas. For example, a key component of the *OMP* (no. 1) was a systematic and long-term lobster *stock rebuilding* program (no. 5). However, *stock rebuilding* cannot be considered without due consideration to the political pressures for broader *access rights* (no. 10), the economic difficulties facing the industry (no. 4), or the problem of poaching and ineffective *policing* (no. 7). Furthermore, the effectiveness of policing to reduce poaching (no. 7), possibly through *co-management* schemes (no. 8), has a fundamental impact on the uncertainty of biomass estimates (no. 3), and the overall precision of the OMP and TAC decisions (no. 1). In addition, co-management schemes would only be possible through appropriate changes in access rights (no. 10).

3.2 The Workshop

The workshop was divided into two sessions. In the morning, the SODA map was explored, allowing participants to view others' perspectives and negotiate through discussing the validity of the concepts and links in the map. In the afternoon, a brainstorming exercise was held to consolidate the information in the merged map into a value tree and, if time permitted, to identify measurable criteria for each objective in the tree.

3.2.1 The Morning Session

After a brief introduction, an overview of the group map was presented and cycled through [Eden (1989)]. The overview map was created by 'collapsing' the group model onto a number of key goals and themes, and showed the inter-relationships between them. Most of the morning session was dedicated to the first two cluster maps; relating to the OMP, and inclusion of qualitative factors in decision-making. These were considered particularly relevant topics at the time. Due to the magnitude of the map, much time was spent just viewing what others had said in the interviews, exploring how others' views and beliefs related to one's own. As a result, few concepts and links were added or altered in the workshop. Nevertheless, the DE model proved effective as a basis for negotiation, and discussion soon centred on the purpose, benefits and shortcomings of the OMP. Several participants criticised the OMP for being too mathematical, and inadequate for dealing with the economic, political and social complexities inherent in lobster fishery management.

At the end of the morning session, the participants agreed that the DE model was a valid representation of the problem situation, and that it reflected the different points of view. Although discussion of the group map no doubt helped participants to better understand the issues, no consensus was reached as to a way forward. This was perhaps to be expected in light of the significantly different interests and priorities held by the different stakeholder groups. Therefore, we felt an MCDA approach would be useful for further structuring and modelling of different parties' values and preferences. This could provide further insight into the problem, as well as helping the group to identify and make sense of the trade-offs. It was also hoped that MCDA could help to break the deadlock by focusing on those areas of disagreement that really mattered to the outcome of the decision [Phillips (1990)].

3.2.2 The Afternoon Session

In the afternoon session, participants were asked to consider what issues or measures were important for comparing different OMP systems. Participants had a few minutes to write their own ideas on post-its, before sticking them onto five large pieces of paper hung on the walls of the conference room. With the help of the facilitator, the 'Post-its' were grouped according to a number of emerging themes. The participants were able to look at what others had written, and could piggy-back on the ideas, either giving support or counter arguments. In the DE model, most objectives reflected broad statements of policy intent. In contrast, the Post-it exercise approached the identification and structuring of objectives from the 'bottom-up' [Buede (1986)]. Once all the post-its had been grouped under broad headings, it was time to bring the workshop to a close.

3.3 Interpreting the SODA Model in an MCDA Framework

After the workshop, our primary aim was to interpret the SODA map within an MCDA framework. The goals and strategies which had been identified in the group map, together with the criteria raised in the brainstorming session, were used to construct a value tree. The resulting tree is shown in Figure 2.

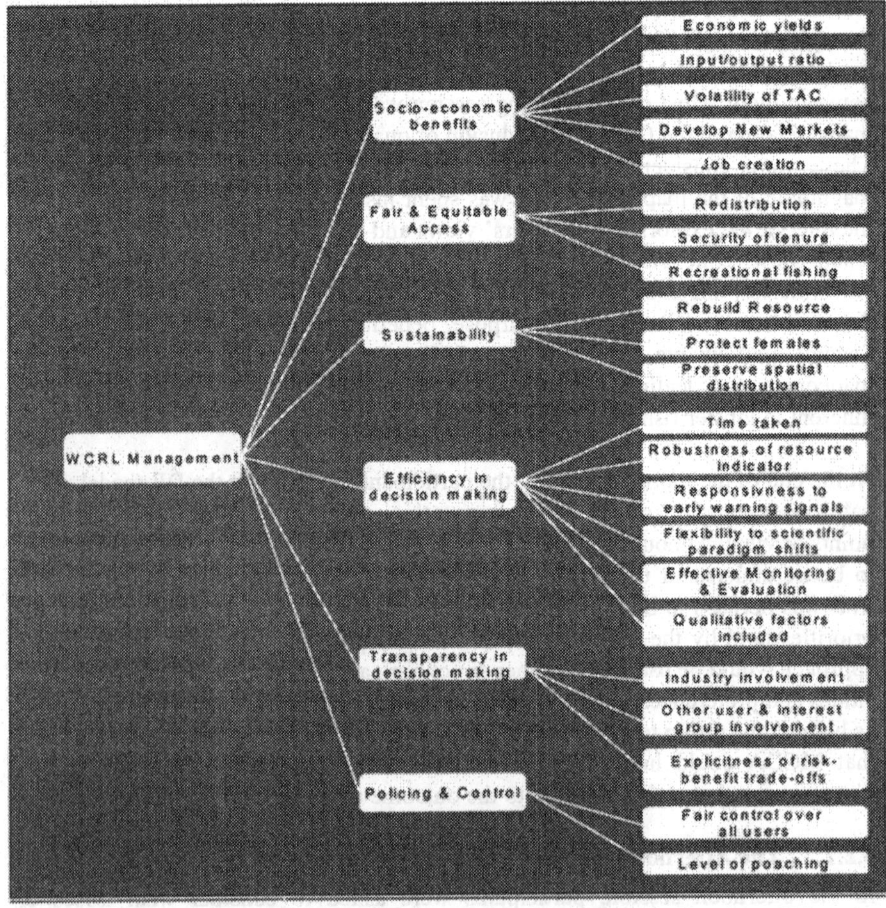

Figure 2. Value Tree for the Future Management of the West Coast Rock Lobster Fishery

Although the participants were generally happy with the value tree, the problem as defined by the tree was too broad. It was impossible, at this level of abstraction, to construct meaningful sets of options (what Stewart & Scott (1995) call 'policy scenarios') for achieving the objectives. This outcome stressed the importance of careful structuring of the problem. It was necessary to determine those objectives which were fundamental to the *particular* decision context, in order to bound the problem and guide the search for alternatives [Keeney (1992)].

Keeney (1992) points out that "a specific decision context is only a part of a larger one, which is itself only a part of a still larger one, until the strategic decision context is reached". Value Focused Thinking (VFT) provided valuable insights as to how the problem could be bounded, and helped us towards a more workable MCDA formulation of the problem. In our case, it was logical to focus on the future operations and management of the WCRLWG as the specific decision context, but still taking the different stakeholders' views into account. This led to the selection of those objectives from Figure 2 which were considered i) essential for considering the development of the OMP and ii) related to issues which fall within the mandate and operations of WCRLWG. Through consultation with the group members, a fundamental objectives hierarchy (FOH) was constructed (Figure 3). The FOH is shown above the dotted line, with a means-ends objectives network (MEN) below. This shows the connection between those objectives which are considered to be fundamental (i.e. ends in themselves), and those which are means to achieving these higher objectives [Keeney (1992)].

Constructing the FOH was no straightforward task, since it entailed differentiating between fundamental and means-ends objectives. Therefore, this exercise was done through considerable consultation with the participants to ensure that the FOH was correct and complete for guiding future decision-making.

3.4 Choosing an OMP

In the following few months, the working group's task was to evaluate a relatively small number of OMP[4] formulae. Computer simulations would be used to compare their performance in terms of robustness, volatility, average yield and other criteria. Based on these findings, the director of Sea Fisheries would make the final choice of OMP algorithm to be implemented.

At this point, MCDA held great potential to assist with evaluating the different OMP alternatives. In consultation with a number of the participants, the following criteria were identified for this purpose, derived largely from the FOH and MEN

[4] The OMP being developed consisted essentially of a set of decision rules, where the TAC for the following year was to be based upon a biological model of the resource. Each year, new data would be fed into the OMP, which then automatically calculated the TAC for the following fishing season.

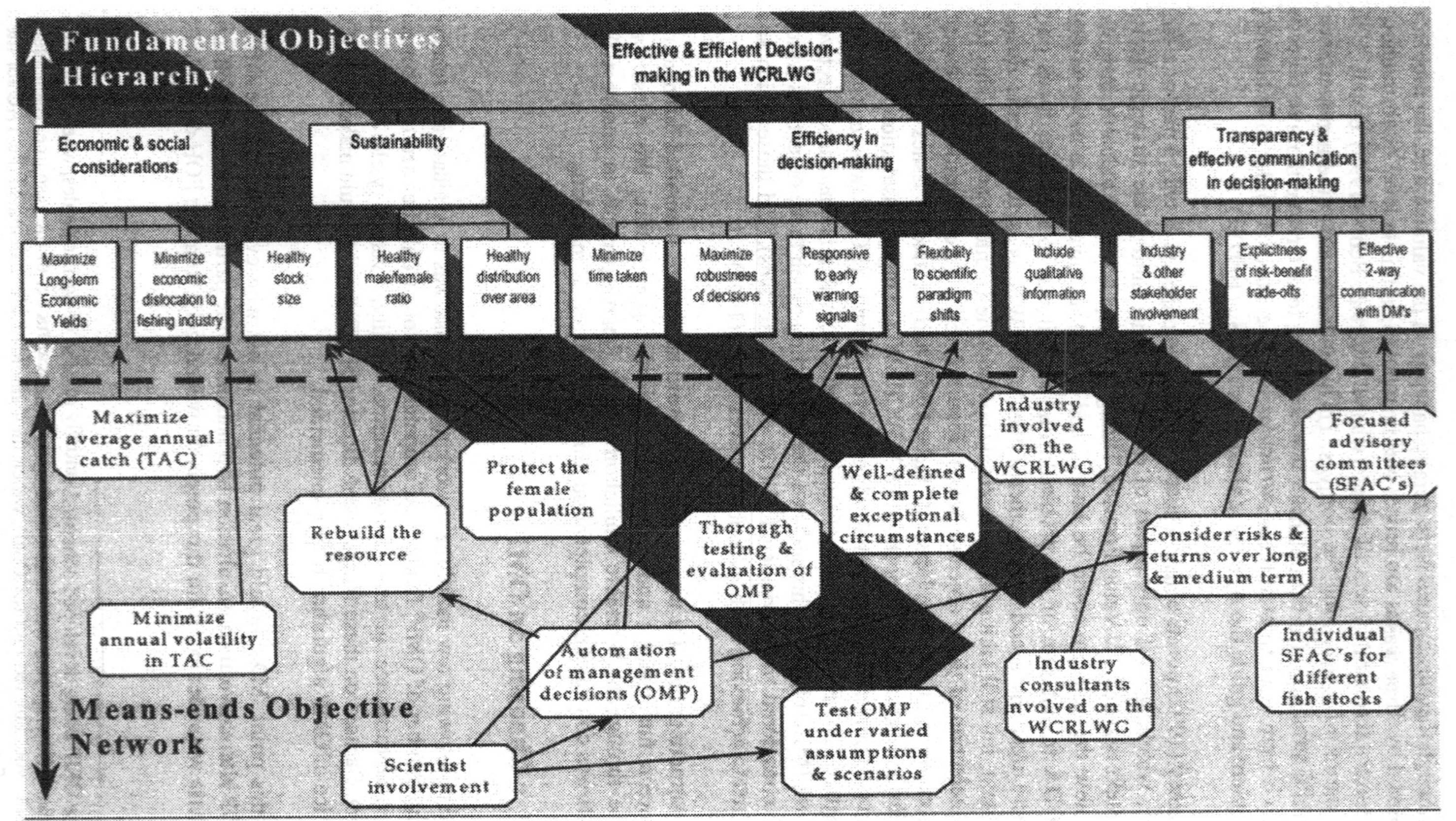

Figure 3. Fundamental Objectives Hierarchy and Means-ends Objective Network for the West Coast Rock Lobster Working Group

in Figure 3:

1. Maximize average TAC.
2. Minimize the volatility of the TAC.
3. Rebuild the resource.
4. Maximize robustness.
5. Responsive to early warning signals.

However, due to time-constraints, our further involvement was not possible. Furthermore, even at this late stage, several participants were not convinced that the an OMP was a good idea. Therefore, our continued involvement was not considered appropriate in light of our previous commitment to unbiased facilitation, and engendering a participative decision process.

4 Conclusions

4.1 Structuring the Problem

The different stages of the intervention highlighted different levels of decision. SODA was useful for modelling the problem at the strategic level. Within this broader context was the narrow focus of the operations and management of the West Coast Rock Lobster Working Group (WCRLWG). Within this framework was the still more focused problem of the OMP, including design, selection and implementation. The criteria eventually identified for evaluating OMP alternatives covered a very small section of the strategic problems addressed by the DE map.

In problem structuring there is always a trade-off between divergence (to ensure that no important issues are overlooked) and convergence (focusing sufficiently in order to reach a decision). SODA was essential for capturing the complexity and richness of the situation, and proved extremely useful for managing group process. Different players had significantly different perspectives, and the interviews were effective for capturing the diverse viewpoints. Quite different problem constructions and emphases arose. In future, the DE model can serve as an Executive Information System (EIS), and might influence how the various parties approach future situations & encounters.

This initial structuring provided the back-drop for further analysis. However, the magnitude and complexity of the DE model made the transition to an MCDA model difficult. Value Focused Thinking proved invaluable for bounding the original problem definition to the level of the WCRLWG, and implementation of the OMP, whilst not losing sight of the broader, strategic context. For instance, several key issues, such as poaching and access rights, were not included in the fundamental objectives hierarchy in Figure 3, since they fall outside the scope of the working group's control. Nevertheless, as was shown in the DE map, it was important not to ignore these issues altogether.

Although the working group consists solely of scientists, it was important to look at the broader strategic issues surrounding the scientific modelling of the resource. It would appear that this is a major dilemma in fisheries world-wide: to decide how to incorporate qualitative factors into decision processes often dominated by biological models of fish stocks.

4.2 Who was our Client?

Eden (1989) stresses the importance of deciding who one's client is early on, in order to manage the political process thereafter. They suggest that the consultant identify a single person or small group of people as one's client, in order to ensure their commitment to the process. Our approach, by necessity, was different. It was crucial that we were seen as impartial and unbiased, and not working for any particular individual or group. As a result, there was perhaps not as much commitment to the process as might otherwise have been the case. Nevertheless, the majority of participants expressed their enjoyment in participating, and some saw it as a particularly useful opportunity to give greater input into the decision process.

4.3 Management Science in Fisheries Management

Management science has an important contribution to make to the field of fisheries management, particularly with the trend toward more strategy-oriented approaches within OR, such as SODA [Ormerod (1997)]. Stephenson & Lane (1995) in their 'plea for conceptual change' urge that empirical research be conducted in different fisheries, incorporating management science with fisheries science to provide management advice. They suggest that such empirical research "will bring about the evolution required for successful management of fisheries systems, rather than of fish, and will promote the framework and tools required to develop effective fisheries co-management." This essentially is what this research set out to do.

References

Ackermann F, Eden C, Cropper S. (1990): Cognitive Mapping - A User's Guide. *Working Paper Series on the Theory, Method & Practice of Management Science, University of Strathclyde* 90/2.

Belton, V. (1997): Implementing Multicriteria Decision Analysis. *Working Paper Series in the Theory, Method & Practice of Management Science, University of Strathclyde* 97/10.

Belton V, Ackermann F, Shepherd I. (1997): Integrated Support from Problem Structuring through to Alternative Evaluation Using COPE & VISA. *Journal of Multi-Criteria Decision Analysis* 6:115-30.

Bouyssou D, Perny P, Pirlot M, Tsoukias A, Vincke P. (1993): A Manifesto for the New MCDA Era. *Journal of Multi-Criteria Decision Analysis* 2:125-7.

Buede, DM. (1986): Structuring Value Attributes. *Interfaces* 16:52-62.

Eden, C. (1988): Cognitive Mapping: A Review. *European Journal of Operational Research* 36:1-13.

Eden C. (1989): Using Cognitive Mapping for Strategic Options Development & Analysis. In Rosenhead J, (ed.): *Rational Analysis for a Problematic World.* Wiley. Chichester, p 21-42.

Eden, C. and Ackermann, F. (1998): *Making Strategy: The Journey of Strategic Management.* London: Sage.Eden C, Jones S, Sims D. (1983): *Messing About in Problems.* Pergamon, Oxford.

Healey MC. (1984): Multiattribute Analysis and the Concept of Optimum Yield. *Can. J. Fish. Aquat. Sci.* 41:1393-406.

Keeney RL. (1992): *Value-Focused Thinking: A Path to Creative Decisionmaking.* Harvard University Press, Cambridge, MA.

Kelly, G. A. (1970): A Brief Introduction to Personal Construct Theory. In Bannister D (ed): *Perspectives in Personal Construct Theory.* Academic Press Inc.

Leschine TM. (1988): Policy Analysis and the Incorporation of Biological Objectives into Fisheries Management Decisions. In Wooster WS (ed.*): Lecture Notes on Coastal and Estuarine Studies.* Springer-Verlag. p 141-63.

Ormerod, R. J. (1997): The Role of OR in Shaping the Future: Smart Bits, Helpful Ways and Things That Matter. *Journal of the Operational Research Society* 48(11):1045-56.

Ostanello A. (1997): Complexity Issues and New Trends in Multiple Criteria Decision Aid. In Climaco J, (ed.): *Multicriteria Analysis: Proceedings of the XIth International Conference on MCDM,* Springer-Verlag.

Parsons LS. (1993): Management of Marine Fisheries in Canada. *Can. Bull. Fish. Aquat. Sci. 225.*

Phillips LD. (1990): Decision Analysis for Group Decision Support. In Eden C, Radford J, (eds): *Tackling Strategic Problems : The role of group decision support.* Sage, London.

Phillips LD and Phillips MC. (1993): Facilitated Work Groups: Theory & practice. *Journal of the Operational Research Society* 44(6):533-49.

Rosenhead, J. (1996): What's the Problem? An Introduction to Problem Structuring Methods. *Interfaces* 26(6):117-31.

Stephenson RL and Lane DE. (1995): Fisheries Management Science: A plea for conceptual change. *Can. J. Fish. Aquat. Sci.* 52:2051-6.

Stewart TJ and Scott L. (1995): A scenario-based framework for multicriteria decision analysis in water resources planning. *Water Resources Research* 31(11):2835-43.

Problem Structuring and Evaluation Integration: A Case Study in Transportation Planning

Sule Onsel[1], Fusun Ulengin[2], and Y. Ilker Topcu[3]

Istanbul Technical University, Management Faculty,
80680 Maçka, Istanbul TURKEY
[1] onsel@ayasofya.isl.itu.edu.tr
[2] ulengin@sariyer.cc.itu.edu.tr
[3] topcu@ayasofya.isl.itu.edu.tr

Abstract. This study focuses on the integration of problem structuring with the evaluation of alternatives. The multiple criteria analysis is generally based on the latter with a hidden assumption that the problem is well defined with explicitly known criteria and alternatives. However, especially the strategic planning problems are ill-defined and, thus, the first step of the analysis must be concerned with the structuring of the problem before any careful evaluation of the alternatives is conducted. In this paper, the water crossing infrastructure selection problem is selected as a case study. The cognitive map derived through knowledge acqusition from 19 experts is converted into a hierarchical structure with the goal at the top and the alternatives at the bottom. At the second stage of the analysis, an expert system approach is used to transform tha data of the multiattribute problem into a manageable decision matrix according to the rules. Finally this matrix will be ready to proceed by any appropriate multiattribute method.

Keywords. Multiattribute decision making, cognitive mapping, expert system, transportation planning

1. Introduction

The general tendency in decision making is to focus on the evaluation and choice of alternatives. The basic assumption is that the problem is well-defined, the set of criteria against which the alternatives are to be evaluated are specified beforehand, and the only step that remains is to find out an efficient evaluation procedure to help decisionmakers in their selection of the best alternative. However, in many real life situations, there are only broadly defined, ill-structured problems where even the objectives and subobjectives to be used during the evaluation are not known in advance. For such problems, the initial step of decision making must be the problem structuring which will then be integrated with the problem evaluation phase. Additionally, these types of problems necessitate the participation of a group of experts, each with a different perspective on the problem at hand. Therefore, the specification of the objectives and alternatives should be the result of a collective effort in order to construct a realistic and appropriate model of the problem.

This paper proposes the integration of the cognitive map and the rule-based expert systems in order to support the decisionmaker throughout the entire process of problem structuring. A step by step explanation of the iterative problem structuring and evaluation phases is followed by the application of the proposed framework to a case study concerning the selection of the appropriate alternative for the Bosphorus water crossing problem.

2. Integration Framework

2.1. Data Gathering

The initial step for such an integration is the knowledge acquisition from the experts using a Delphi type group decision making procedure. This step focuses on values and necessitates constraint-free thinking on the part of the experts.

The experts are first asked to pinpoint the values that should be guiding a decision-making process. The objectives will make explicit the values that one cares about in a decision context, and the means objectives will provide the means to the achievement of the fundamental objectives. The premise is that focusing on values when facing difficult problems will lead to the understanding and the use of these values to make meaningful decisions, to find better alternatives than those already identified, and to evaluate the desirability of the alternatives more carefully.

2.2. The Relationship Matrix

The second step of the proposed integration process is the identification of the relationships between pairs of objectives. For this purpose, all the objectives presented by the experts in the first step are combined in a list. The list is sent back to the experts in order for them to pair the items on the list according to their positive/negative or zero relationship with one another.

2.3. Cognitive Mapping

The third step is the aggregation of these pairwise comparison matrices. In the aggregated matrix, the sign of each cell is obtained through the majority rule. Based on this matrix, a cognitive map is drawn which represents the expert opinion on the origins of the problem.

The use of cause maps to explore the cognitive structures of individuals has become popular in recent years (Huff, 1990). However, the theoretical basis for cause mapping, which allows an interpretation of analysis, differ. Most methods follow a "cognitive mapping" method presented by Axelrod (1976). One clear exception is the approach of Eden (1988) developed to reflect the personal construct theory.

As is well known, in cases where the problem relates to a situation where it is described redominantly by qualitative notions, mapping is an excellent starting point. Additionally, if there are likely to be a number of important and different perspectives on a problem, mapping will be a good way of drawing these perspectives together and negotiating a new vision of the problem that will enable all interested parties to work as a group. However, a cognitive map as such may include generally between 50 and 100 ideas. That is why, the analysis of such a complex map generally necessitates a software tool such as COPE (1995), in order to draw the map appropriately, and to capture the contents and the associated structure. Therefore, once the cognitive map is drawn using the COPE software, the "domain" analysis is conducted to calculate the total number of in-arrows and out-arrows from each node. The nodes with the most complex immediate domain are considered to be those cognitively most central. The cognitive centrality of each node, computed as mentioned above, will in fact show the importance of that node within the cognitive map as a whole. This value can, thus, be used as the weight of that node.

Another important analysis may be based on "heads" and "tails". A "head" is a concept with no further consequence and can be accepted as a goal. It is a node having no out-going arrow. In fact, the ratio of the number of "heads" to the total number of nodes will be the sign of the complexity of the map. An idealised way of thinking about a topic tends to generate maps with a small number of "heads" (fewer number of goals, or, if possible, a single goal). Conversely, a map with a relatively large number of "heads" will indicate a concern for multiple and possibly conflicting goals. The analysis of "tails", on the other hand, will show the nodes at the lowest part of the map; i.e. those nodes which do not have incoming arrows. In fact, in a cognitive map based on the objectives gathered from the experts, those lowest level nodes will correspond to the lowest level subobjectives (Eden et al., 1992).

Whenever the map is complex, the corresponding problem can be broken down into a number of related and interacting parts using "cluster analysis". The intention here, is to attempt the formation of clusters where the nodes in each cluster are tightly linked to one another and the number of links with other clusters is minimised.

2.4. Constructing a Hierarchical Model and Creating Alternatives

In the problem structuring phase, the fifth step after cognitive mapping is to structure the map hierarchically. For this purpose, first of all, the existence of causal loops should be investigated. If such causal loops exist and if they are not due to a coding accident, it means that there is a feedback relationship, and the problem cannot be represented hierarchically. In this case, the problem should be analysed using system dynamics (Ackermann and Tait,1994). On the other hand, if there are no causal loops, and hence no feedback relationship, the cognitive map can be represented hierarchically. This paper proposes the iterative application of the "head" analysis for such a hierarchical model.

First of all, all the heads of the map are listed in order to specify the nodes that are at the top of the hierarchy. Then, erasing these nodes from the map, a second "head" analysis is conducted in order to specify the nodes at the second level of the hierarchy. This process goes on until the whole map is represented hierarchically.

As can be appreciated, the "problem structuring" framework proposed in this study aims to realise the process of building a value tree based on top-down (value-focused) thinking through building and elaborating a cognitive map (Belton et. al., 1997).

Once the hierarchical representation of the objectives is obtained, it is time to think about how to achieve these objectives. This step is helpful in creating the alternatives (Keeney, 1992). Alternatives themselves can also trigger thought process that generate new alternatives.

2.5. Expert System Architecture

At the final stage of problem structuring, the data of the hierarchical multiattribute problem will be transformed into a manageable decision matrix according to the rules. This will offer an "intelligent " system which, starting from the decision maker(s)' data, fills in a multiattribute decision matrix. This matrix constitutes the input of the multiattribute decision support system. In fact, the decision matrix obtained is ready to be processed by any appropriate multiattribute method.

In order to obtain this matrix, the rules will be used to aggregate some objectives into new objectives of a hierarchical level (Lévine et al, 1990). This will provide an explicit and readable means of making aggregation as opposed to numerical aggregation made by computation, as can be seen in Saaty (1986). In this proposed setting, the knowledge base plays the role of the aggregation function to obtain a multiattribute decision matrix. The ultimate choice, however, is made by the decision maker(s). In other words, the rules are used to aggregate the means objectives step by step in order to reach the fundamental objectives, which then constitute the decision matrix. For the quantitative criteria which are naturally ranked by a numerical scale, it is not difficult to obtain these aggregated values. However, for qualitative criteria, a knowledge based interface is proposed that replaces the calculation algorithms adapted to numerical data. An example of such an aggregation by rule is presented later in the case study, namely node 3 (Dependence on outside sources for construction cost) and node 29 (Suitability for the transportation policy). These two nodes are directly connected to node 26 (Suitability of the technology used in construction for the means of the country at present and in the foreseeable future). But while node 3 has a negative direct impact on node 26, node 29 has a positive direct impact as follows:

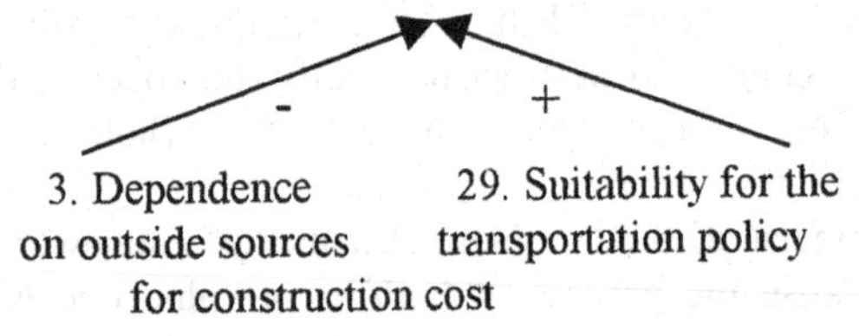

Fig. 1. An Example of Cause-effect Relationship

The above reasoning takes the form of rules and offers an expert system interface between the decision maker(s) files and the decision matrix. In fact the rule corresponding to the above given example will be:

IF "dependence on outside sources for construction cost" is high
AND "suitability for the transportation policy" is high
THEN "suitability of the technology used in construction to the means of the country at present and in the foreseeable future" is medium.

This aggregation continues until the top of the hierarchy is reached. As a result, a decision matrix having the alternatives at the vertical axis and the fundamental objectives at the horizontal axis is obtained. Each cell v_{ij} of this matrix will be computed through aggregation rules.

In order to obtain such a matrix using the aggregation rules mentioned above, it is necessary to prepare an information file for each alternative. Consequently, for each alternative i, a set of facts F_i summarising the information about the alternative is obtained. In fact, this information is expected to be related to the performance of each alternative with respect to the subobjectives at the lowest level of the objective hierarchy.

For instance, "the dependence on outside sources for construction cost = high", and "suitability to the transportation policy = high" may be set in the fact-base related to alternative i. If numerical values can be gathered, they will be used instead of these subjective values. Also subjective (qualitative) values can be expressed on a numerical scale. The development of rules also necessitates the definition of the criteria which will include an agreement on the scale and meaning of, let's say, "high", "medium", and "low".

The advantage of such a rule-based approach is that the system will allow easy updating by adding new alternatives just by describing them. The decision matrix will be automatically updated thanks to the rules that remain unchanged because the fact-base depends on the alternative and the set of rules depends on the criterion.

2.6. Evaluation

Once the final decision matrix is obtained, it is ready to be processed by any multiattribute method such as AHP (Saaty, 1986), PRIAM (Lévine & Pomerol, 1986), VISA (Belton et. al., 1997), Visual Interactive Method (Korhonen & Laakso, 1986), VIG (Korhonen & Wallenius, 1990; Korhonen et. al., 1992), VIMDA (Korhonen, 1988), UTA (Jacquet-Lagreze & Siskos, 1982), PRAGMA (Matarazzo, 1988), IMGP (Stewart, 1992), ELECTRE (Vincke, 1992), PROMETHEE (Brans et. al., 1986; Mareschal & Brans, 1988), Conflict Analysis Method (Huylenbroeck, 1995).

In fact, it is theoretically possible to use these types of methods without using the aggregation rules. However, in case the hierarchical map obtained is very complex (with excessive number of levels in the hierarchy and with complex direct relationships between the nodes from different levels) the direct use of the methods become impractical and the reduction to one decision matrix becomes unavoidable. The selection of the most suitable multiattribute method to evaluate the final decision matrix may itself necessitate a rule-based expert system and is beyond the scope of this study.

3. A Case Study in Transportation Planning

The application of the proposed framework presented in this case study is the choice of the most suitable alternative for the Bosphorus water crossing infrastructure. The traffic congestion on the currently existing two bridges (Boğaziçi and Fatih Bridges) calls for an urgent solution. The data of this case study are drawn from Ulengin & Topcu (1997). As explained in Ulengin & Topcu (1997), a questionnaire survey is conducted with 19 experts to reveal all the factors that might be relevant to the water crossing problem at hand. The purpose of this step is to develop a comprehensive list of their objectives. This first step is identical to that presented by Ulengin & Topcu (1997). However, in the present paper, the step corresponding to the reduction of the objectives, which is, in fact, an "artificial" restriction of the variables, is totally omitted. As a result, instead of the 11 objectives which constituted the final configuration of Ulengin & Topcu (1997), 49 objectives are obtained. The relationship assignments between the whole set of objectives are obtained from 19 experts and then aggregated using the majority rule.

The "loop" analysis conducted on the map did not detect any loops and hence, the problem is found to be suitable to a hierarchical representation. The head analysis conducted, as explained in section 2, resulted in the hierarchical form shown in Figure 2.

As can be seen, in this hierarchical map, a node in a higher level may have direct relationships with other nodes in any lower levels which shows the complexity of the resulting map. An example of the direct relationship structure between the nodes and the levels of the nodes can be seen in Figure 3.

The domain analysis conducted shows that node 9 (Contribution to the economy of the region and the country during operation) is the central issue of the map with 11 links followed by node 2 (Construction cost) with 8 links, node 4 (Cost of operation and maintenance) with 7 links, and node 21 (Possible damage on the inhabitants of the surrounding area caused by crossing) with 7 links.

Based on this hierarchical map, the experts are then asked to specify the alternatives that may satisfy these objectives in some way. The step of creating alternatives also differs from that of Ulengin & Topcu (1997) where the alternatives were dictated to the decision makers and were limited in number. As a result 9 alternatives were created: bridge for wheeled vehicles (A1), tunnel for wheeled vehicles (A2), bridge for railway vehicles (A3), tunnel for railway vehicles (A4), bridge for both wheeled and railway vehicles (A5), tunnel for both wheeled and railway vehicles (A6), improvement of the currently existing Boğaziçi and Fatih Bridges and especially of the connecting roads (A7), increasing the number and improving the condition of the boats currently run across the Bosphorus (A8), increasing the number and improving the condition of the sea buses currently run across the Bosphorus (A9).

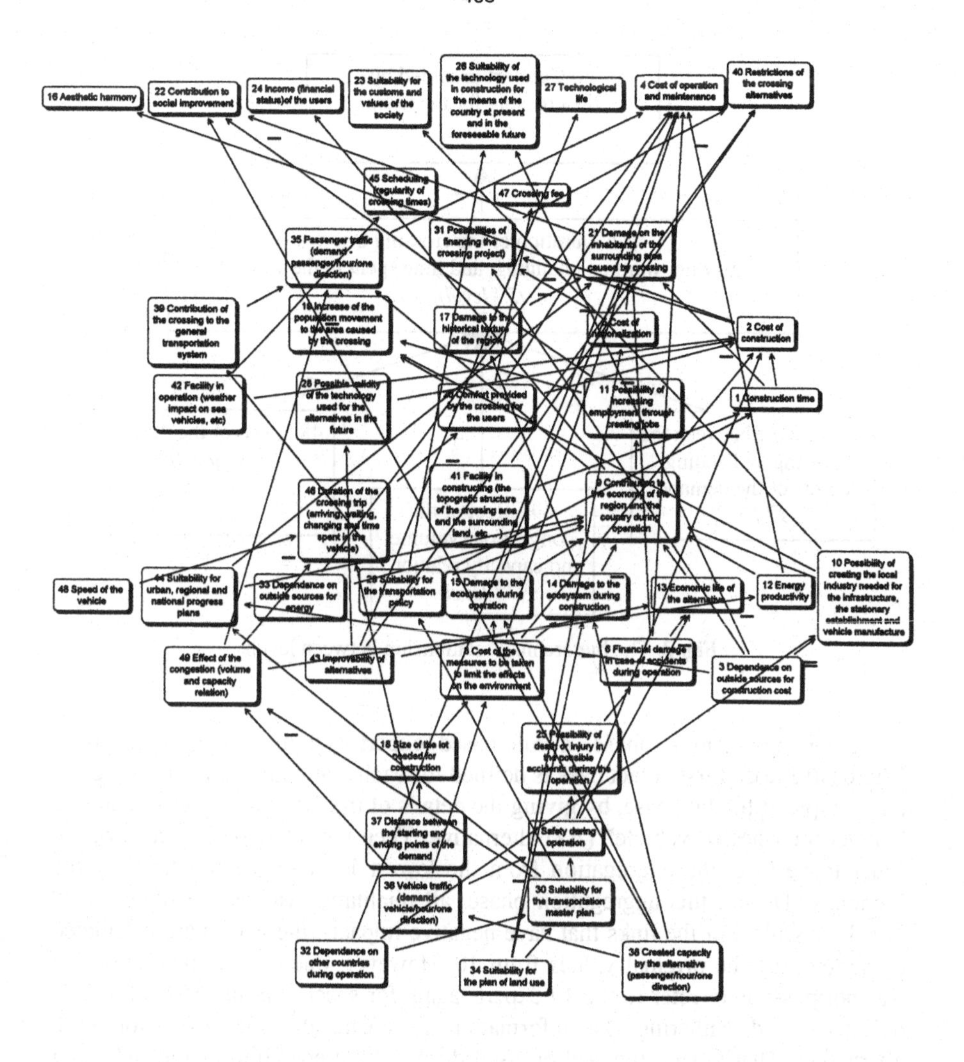

Fig. 2. The hierarchical form of the cognitive map

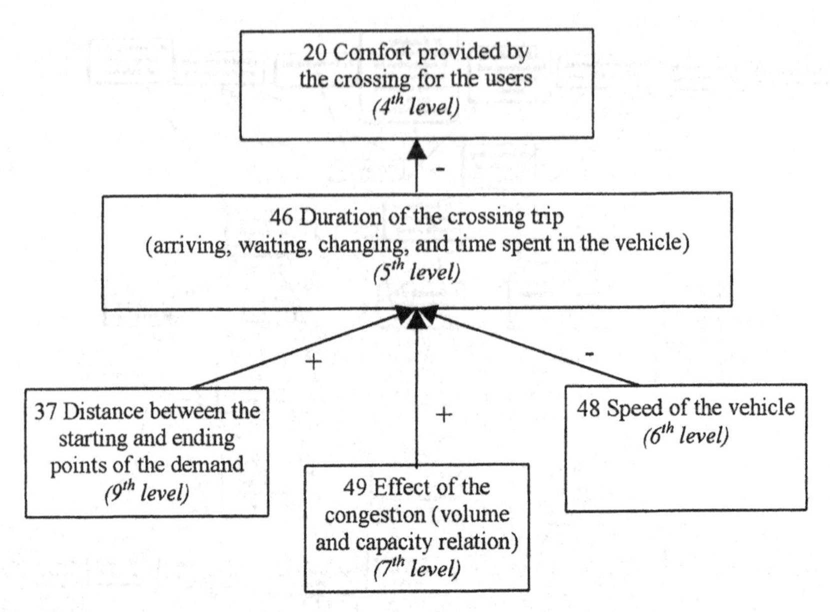

Fig. 3. The hierarchical relationship between levels

For the aggregation phase, a rule-based expert system is developed using Microsoft Excel. First, a fact base is defined for each alternative. Appendix I gives an example of the fact base, by giving the details of the fact base corresponding to "tunnel for wheeled vehicle" (A2) alternative. Then the rule-based models run in order to perform the aggregation from the lowest level to the top level of the hierarchy. During this aggregation phase, all qualitative values are transformed into 1-10 scale. For the links that have negative impacts, the values are calculated by subtracting the original values from 10. However, due to time restrictions, it was not possible to refer to the 19 experts again for specifying the scales for each objective, and gathering the information to create the fact base for each alternative. That's why, the scaling used during the preparation of the rules and the fact base corresponding to the alternatives were prepared by the authors of this paper for illustration purposes.

Appendix II gives an example of some of the rules based on the opinion of the authors for realizing the aggregation from the lowest level to the top of the hierarchy in order to find to value of the fundamental objective.

For example, the value of alternative i for criterion 26 (suitability of the technology used in construction to the means of the country at present and in the foreseeable future) is denoted as $v_{i,26}$ and is computed as

$$v_{i,26} = ((10-v_{i,3})* w_3 + v_{i,29} * w_{29}) / (w_3 + w_{29}) \quad (1)$$

where w_j indicates the cognitive centrality (weight) of criterion j.

This aggregation continues until the first level of the hierarchy where the fundamental objectives lie. As a result, the value of each alternative for each fundamental objective is calculated through aggregation rules and using these values as entries, the final decision matrix is constructed (Appendix III).

As explained in Section 2, once the decision matrix is obtained, any appropriate multiattribute method can be used to evaluate the alternatives. The selection of the most suitable method is, itself, a subject of detailed research. Again for illustration purposes, the authors used a linear additive value approach

$$V_i = \sum_j (v_{i,j} * w_j) \ / \sum_j w_j \quad j \in J_F \qquad (2)$$

where V_i is the overall score of alternative i and J_F is the set of fundamental objectives at the first level of hierarchy.

The overall score of the alternatives can be found in the last column of the decision matrix, and, as mentioned before, reflects only the subjective opinion of the authors.

4. Conclusion

As in the case of transportation planning, important strategic decisions incite controversy because they involve several and generally conflicting objectives which should be neatly specified before any alternatives are evaluated. The aim of this paper is to propose a framework for the integration of problem structuring and evaluation, and thus to present an approach to guide such type of decisions. For this purpose, an iterative process is suggested. The advantage of the proposed framework is the possibility of separating the reasoning behind the content of the criteria and the evaluation of the alternatives. In other words, it separates the filling of the decision matrix and the decision making itself by integrating cognitive maps and expert systems in a multicriteria decision process. The discussion does not address the best methods for choosing among alternatives; but, instead, focuses on clearly articulating expert values, using them as the basis for creating an improved set of alternatives, and finally, for developing a decision matrix which will, then, be used for the evaluation purposes.

The approach described in this paper attempts to open up the policy planning process considerably by involving representatives of all the interested parties. In a practical sense, this requires including representatives of all major groups in the identification of objectives and alternatives.

Once the problem is structured in this way, the selection of the appropriate evaluation procedure, which is generally the focus of the works in the field of multicriteria analyses, can be easily realised.

As a further suggestion, instead of aggregating the relationship assignment gathered from each expert according to a majority rule, the cognitive map of each expert can be drawn individually and their differences can be measured (Langfield-Smith and Wirth, 1992). The results can be sent back to the experts as a subsequent step of the Delphi analysis in order to let them agree on a final aggregated map. In fact, such a process can be more effective and less time consuming, if experts are invited to a workshop where the SODA (Strategic Options Development and Analysis) (Eden, 1988) is conducted. The SODA will allow the group members to refine the objectives and alternatives to be used during the evaluation. Unfortunately, it was impossible to have such a workshop in our case study due to the busy schedules of the experts.

Appendices

I. Criteria Values for A2

Node #	Node (criterion) name	Value
32	Dependence on other countries during operation	6
34	Suitability for the plan of land use	5
38	Capacity created by the alternative (passenger/hour/one direction)	2.56
42	Facility in operation (weather impact on sea vehicles, etc)	7
18	Size of the lot needed for construction	6
43	Improvability of alternatives	1
3	Dependence on outside sources for construction cost	8
33	Dependence on outside sources for energy	9
48	Speed of the vehicle	3
41	Facility in constructing (the topographic structure of the crossing area and the surrounding land, etc.)	1

II. Aggregation Rules for "Restrictions of the Crossing Alternatives"

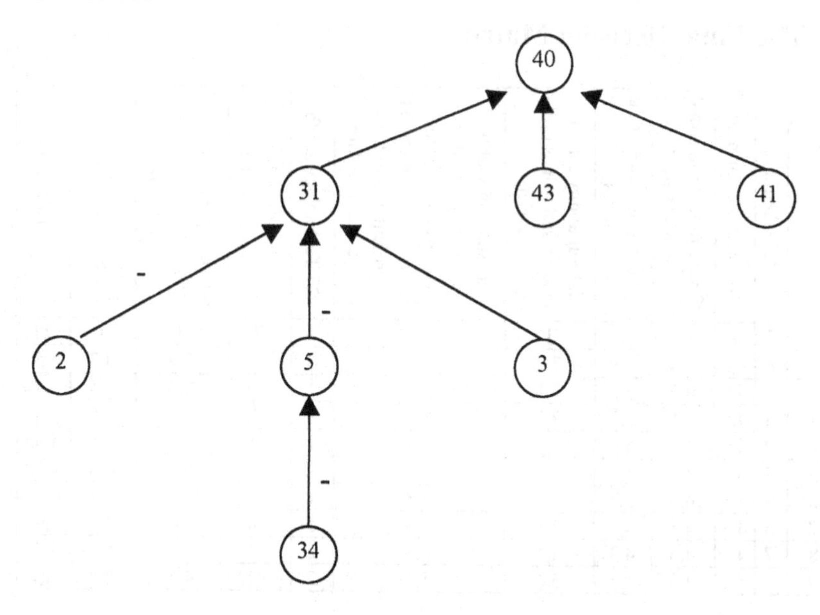

IF the scale value of alternative i for "suitability for the plan of land use" (criterion 34) is a
THEN the value of alternative i for "cost of nationalization" (criterion 5) is
$b = 10 - a$

IF value of alternative i for "cost of nationalization" (criterion 5) is b
AND value of alternative i for "cost of construction" (criterion 2) is c
AND value of alternative i for "dependence on outside sources for construction cost" (criterion 3) is d
THEN value of alternative i for "possibilities of financing the crossing project" (criterion 31) is
$e = (3 * (10 - b) + 8 * (10 - c) + 4 * d) / (3 + 8 + 4)$
'where 3, 8, and 4 are centrality (weight) of criteria 5, 2, and 3 respectively.

IF value of alternative i for "possibilities of financing the crossing project" (criterion 31) is e
AND value of alternative i for "improvability of alternatives" (criterion 43) is f
AND value of alternative i for "facility in constructing" (criterion 41) is g
THEN value of alternative i for "restrictions of the crossing alternatives" (criterion 40) is
$h = (4 * e + 4 * f + 2 * g) / (4 + 4 + 2)$
'where 4, 4, and 2 are centrality (weight) of criteria 31, 43, and 41 respectively.

III. The Final Decision Matrix

	45 Scheduling	16 Aesthetic harmony	22 Contribution to social improvement	24 Financial status of the users	23 Suitability for the customs and values of the society	26 Suitability of the technology used in construction for the means of the country at present and in the foreseeable future	27 Technological life	4 Cost of operation and maintenance	40 The upper bounds of the alternatives of the crossing	47 Fee (price of crossing)	**Overall score**
W	1	1	3	1	1	2	1	7	3	1	21
A1	6	4,6	4,4	4,9	7,0	3,6	3	5,4	3,9	4,7	4,55
A2	7	4,7	4,5	5,3	7,1	3,7	1	5,3	3,0	5,5	4,41
A3	9	4,8	5,1	5,5	7,1	3,7	3	5,3	4,2	5,6	4,87
A4	9	5,0	5,4	5,9	7,7	3,8	1	5,4	3,2	6,1	4,69
A5	7	5,2	5,3	5,4	7,1	3,7	2	5,6	3,1	4,8	4,52
A6	8	5,3	6,0	5,7	7,7	3,8	1	5,9	3,0	4,7	4,59
A7	6	4,8	4,8	4,6	4,7	5,9	5	5,4	5,0	4,9	4,94
A8	7	4,1	4,5	4,3	4,1	6,3	2	5,1	4,4	5,6	4,77
A9	5	5,1	4,9	5,5	5,6	5,7	2	5,2	4,4	7,1	4,7

Criterion 4 and 47 have negative impact on overall score

List of Figures and Tables

References

Ackermann, F., Tait, A.J., Eden, C., and Williams, T.M., *COPE-ing with System Dynamics-a story about soft and hard OR*, Working Paper 94/14, Management Science, Theory, Method and Practice Series, University of Strathclyde (1994).

Axelrod, R., *The Structure of Decision*, Princeton University Press, NJ (1976).

Belton, V., Ackermann, F. and Shepherd, I., Integrated support from problem structuring through to alternative evaluation using COPE and V.I.S.A.. *J. Multi-Crit. Decis. Anal.* **6**, 115-130 (1997).

Brans, J.P., Vincke, Ph. and Mareschal, B., How to select and how to rank projects: the PROMETHEE method. *Eur. J. Oper. Res.* **24**, 228-238 (1986).

Eden, C., Ackermann, F. and Cropper, S., The analysis of cause maps. *Journal of Management Studies* **29**, 3, 309-324 (1992)

Eden, C., Cognitive Mapping: A Review. *Eur. J. Oper. Res.* **36**, 1-13 (1988).

GRAPHICS COPE 2.0, Computer software. The University of Strathclyde and Banxia Software Ltd. (1995).

Huff, A. (Ed.), *Mapping Strategic Thought*, Wiley, New York (1995).

Huylenbroeck, G., The conflict analysis method bridging the gap between ELECTRE, PROMETHEE and ORESTE. *Eur. J. Oper. Res.* **82**, 490-502 (1995).

Jacquet-Lagreze, E., and Siskos, J., Assessing a set of additive utility functions for multicriteria decision making, the UTA method. *Eur. J. Oper. Res.* **10**, 151-164 (1982).

Keeney, R.L., *Value-Focused Thinking: A Path to Creative Decision Making*, Harvard University Press, Cambridge (1992).

Korhonen, P., A visual reference direction approach to solving discrete multicriteria problems. *Eur. J. Oper. Res.* **34**, 152-159 (1988).

Korhonen, P. and Laakso, J., A visual interactive method for solving the multiple criteria problem. *Eur. J. Oper. Res.* **24**, 277-287 (1986).

Korhonen, P. and Wallenius, J., Using qualitative data in multiple objective linear programming. *Eur. J. Oper. Res.* **48**, 81-87 (1990).

Korhonen, P., Wallenius, J. and Zionts, S., A computer graphics based decision support system for multiple objective linear programming. *Eur. J. Oper. Res.* **60**, 280-286 (1992).

Langfield-Smith, K. and Wirth, A., Measuring differences between cognitive maps. *Eur. J. Oper. Res.* **43**, 1135-1150 (1992).

Lévine, P. & Pomerol, M., PRIAM- An interactive program for choosing among multiple attribute alternatives. *Eur. J. Oper. Res.* **25**, 272-280 (1986).

Lévine, P., Pomerol, M. and Saneh, R., Rules integrate data in a multicriteria decision support system. *IEEE Transactions On Systems, Man and Cybernetics* **20**, 3, 678-686 (1990).

Mareschal, B. and Brans, J.P., Geometrical representation for MCDA. *Eur. J. Oper. Res.* **34**, 69-77 (1988).

Matarazzo, B., Preference ranking global frequencies in multicriterion analysis (PRAGMA). *Eur. J. Oper. Res.* **36**, 36-49 (1988).

Saaty, T.L., Axiomatic foundation of the Analytic Hierarchy Process. *Management Science* **32**, 841-855 (1986).

Stewart, T.J., A critical survey on the status of the multiple criteria decision making theory and practice. *Omega* **20**, 5-6, 569-586 (1992).

Ülengin, F. and Topcu, I., Cognitive Map-KBDSS integration in transportation planning. *J. Opl. Res. Soc.* **48**, 11, 1065-1076 (1997).

Vincke, P., *Multicriteria Decision-Aid*, Wiley, West Sussex (1992).

Management of Surgical Waiting Lists in Public Hospitals[1].

Rodríguez Uría, M.V.; Arenas Parra, M.; Bilbao Terol, A.
Dpto. Economía Cuantitativa. Universidad de Oviedo. Spain.
Cerdá Tena, E.
Dpto. de Análisis Económico. Universidad Complutense, Madrid.

Abstract.

Hospitals are considered public utility companies and are supposed to be non-profit institutions [12]; usually they function in a political and highly complex environment. Managerial authority is shared between doctors and administrators and each of the two groups aims to formulate its own individual policies and pursue objectives that may not coincide or may even be in direct opposition to one another [1]. This situation is certain to affect hospital performance and it must be taken into account when proposing any kind of results analysis. Improving the efficiency of Hospital performance, establishing quantitative target values to those objectives, some of them involving intangible benefits, is the main objective of any Hospital Administration [10]. The purpose of this study is to analyze through the M.C.D.M. approach [7], the inner coherency of the goals expressed by administrative Authorities. Also, applying Multicriteria Decision techniques we intend to design the real performance of surgical services at a local general hospital offering the decision centre a suitable methodology that allows us to analyze whether or not it is possible to improve the running of the services, taking into account all the real constraints, e.g. space, staff availability and financial support.

Keywords: Multiobjective Programming, Dynamic Optimization, Decision-Making, Hospital Management.

1 Introduction

Many real world problems may be modelled as multicriteria decision problems and a variety of modelling approaches to problems have been developed; in general health care literature, the usefulness of optimization models has been demonstrated for problems of different types in the area [5] but most of them are not applicable to hospital administration, because they have not accounted for

[1]This work has been developed with the inestimable help of Arias, E.; G. Clouté, P.; Gordo, J.L. and Ramírez, A. from the hospital where the work has been done and Pérez Gladish, B. PHD student

nonfinancial, intangible benefits such as improving quality of service or covering community requirements [8, 9].

To present the model in this paper it is necessary to look briefly at The Spanish National Health Service: The Spanish Constitution assures the right to healthcare of all Spanish people. The Spanish National Health Service covers 98% of the population - and is mainly run by the public sector. Sixty per cent of health care funding comes from general taxation, 29.4% from social insurance and 4.5% from other sources. Public health care providers are paid through regional health services by TSNHS (INSALUD, in spanish) with global budgets.

In order to have some autonomous management, at the end of each year, hospital managers must draw up a document called *Management Contract*, that expresses the objectives to be reached during the next year and also intends to establish quantitative target values to those objectives. This contract must be ratified by the Department of Health.

The Government aims to decentralise responsibility for management, as it searches for a way to improve efficiency and effectiveness and to control public health costs. Public hospitals will have more autonomy than at present from central/regional government, especially in the areas of operational management and human resource policy.

Several measures will be taken to control the increase in health care expenditure. Nevertheless, the percentage of GDP spent on health care will grow moderately reaching 6.86% of GDP by 1998. It is believed that a competitive environment will be created inside the public sector which will help optimise access, quality and cost of health care delivery [11].

Equity would be the basis and the principal goal of a National Health Service, but the only possibility of permanence as a Public Service is based on its efficiency. In order to preserve it, public sector management responsibilities will be further decentralised and increased and evaluation criteria will be introduced to measure the effectiveness and efficiency of providers.

There are many factors behind inefficient delivery of health care services and increasing costs, among them the inefficient allocation of resources or the inefficient use of existing facilities. Some of these factors are social and political in nature and they are hard to change. This means that hospital administration has become a very complex management process and it is essential to apply modern management techniques to assure the efficient utilization of medical facilities and resources.

This study attempts to apply a management science technique to improve the efficiency of Hospital Administration. An efficient delivery of health services through a more scientific management could be developed using Multiobjective Programming (MP) techniques [2, 3, 6].

We aim to elaborate a multicriteria model to design the performance of two surgical services at a General Hospital. Through the MP approach we intend to offer the Decision Center a suitable methodology that allows it to plan surgical scheduling over one year in order to reduce both waiting list numbers and waiting time for each process, taking into account all the real constraints, e.g. space, staff availability and financial support.

The general characteristics of the Hospital where the scheduling has been done are the following: Number of beds, 407; Operating rooms for scheduled processes, 8 ; Number of consulting rooms, 70; It has to serve a population of 305.000, that are the potential patients. In 1997, 10.107 surgical processes were carried out.

2 Formulation of the problem

In the introduction we mentioned Management Contract which must set out the objectives for each hospital. For 1998 the main political goal in all of them is to reduce the waiting time for surgical processes .

The processes with higher numbers in waiting lists, not just in our Hospital but in most of the hospitals in Madrid are: Cataracts, Hallux Valgus, Knee Operations and Osteoartrosis. In this paper we want to optimise the management of waiting lists for these four processes.

In our Hospital, according to the decision maker, the bottleneck is neither in the human resources nor in the number of beds available (at least initially), but in the operating rooms.

During the year the maximun waiting time may be nine months, but at the end of 1998 no one will be required to be on a waiting list more than six months. So the challenge for all hospitals involves getting waiting time down but mantaining costs within certain limits [4, 6].

To reach the previously mentioned objectives, hospitals are allowed to use several methods of operating scheduling:

1.- Within regular-operating hours.
2.- Overtime.
3.- Private hospital contracts.

We are working with data from two hospital services: *Oftalmology and Orthopedic surgery* and *Traumatology*. In the former we will study one process *Cataracts* (C) and for the second we will consider three processes: *Hallux Valgus* (H), *Knee Operations* (K) and *Osteoarthritis* (O).

2.1 Program Variables and data

In order to reflect the dynamic waiting list of these four processes, we will define the following variables:

Table 1: *Program Variables* (i=1,2,...,12):

	Program State Variables	Program	Decision	Variables
Process	Waiting List at 1st. of (i+1)th-month	Regular	Overtime	Private
C	CL_{i+1}	CR_i	CO_i	
H	HL_{i+1}	HR_i		HP_i
K	KL_{i+1}	KR_i		KP_i
O	OL_{i+1}	OR_i		

CR_i,HR_i, KR_i OR_i represent the number of each process scheduled within regular-operating hours; CO_i the number of Cataracts done on overtime and HP_i, and KP_i the number of Hallux Valgus and Knee operations done in Private hospitals respectively; all of them during the ith month, i=1, 2, ...12.

We denote by: CL_1=480; HL_1=199; KL_1=132; OL_1=128 *the initial state* of the waiting list at 1 January 1998.

The Hospital has some monthly estimations about patient flow for each process, during the present year; they are expresed in table 2, *expected admissions*, and in table 3 *expected exit* without surgical process:

Table 2: *Admission.*

Process/mths	1	2	3	4	5	6	7	8	9	10	11	12
C	84	85	82	94	78	104	125	42	78	98	94	86
H	28	28	22	22	34	45	31	12	20	24	12	33
K	21	22	18	15	30	18	15	12	24	18	21	13
O	10	22	15	14	30	24	5	5	17	34	14	21

Table 3: *Exit.*

Process/mth	1	2	3	4	5	6	7	8	9	10	11	12
C	8	13	16	16	20	37	53	12	19	20	17	7
H	4	8	13	6	10	31	22	3	5	12	2	19
K	3	5	4	3	10	14	4	0	7	9	1	5
O	5	2	9	7	9	7	7	5	7	13	2	5

We denote by CA_i, HA_i, KA_i and OA_i the number of expected admissions for each process and for each month.

CE_i, HE_i, KE_i and OE_i represent the number of expected exit without surgical process for each process and for each month.

To assure spatial requirements, we take into account the mean time spent in each process : $CT = 80$; $HT = 80$; $KT = 120$; $OT = 160$ expresed in minutes (for each process, this time includes 20 minutes needed to clean and prepare the operating room), and the global availability of Operating Rooms by month and service (in fact, for the processes that we schedule for each service: 80% of the total time assigned to Oftalmology and 45% of the total time assigned to Traumatology), also expresed in minutes:

Table 4: *Operating Rooms Time.*

Time/mths	1	2	3	4	5	6
$Oftal.(OQ_i)$	6000	6400	6240	5280	6400	6400
$Traumt.(TQ_i)$	4640	4800	4800	4160	4800	4800
Time/mths	7	8	9	10	11	12
$Oftal.(OQ_i)$	2640	3280	2640	7200	6400	4480
$Traumt.(TQ_i)$	3680	3360	3200	5440	4800	4160

There are also monthly upper limits established for operating scheduling in some processes:

Table 5: *Monthly Upper Limits.*

months	1	2	3	4	5	6	7	8	9	10	11	12
Overtime C.	0	0	68	40	64	72	0	0	44	52	48	24
Private H.	0	20	25	35	35	35	35	35	35	35	35	35
Private K.	0	0	8	21	21	20	20	20	20	10	10	0

Also we know the proccess *costs* of each way of scheduling, expressed in pesetas:

Table 6: *Costs.*

	Cataracts	Hallux Valgus	Knee operations	Osteoarthritis.
Regular	110852	125899	287973	853338
Overtime	123733			
Private		106605	141120	

We are going to denote by: CRC, HRC, KRC and ORC the cost of each process on regular time; COC the cost of cataracts on overtime; HPC and KPC are the cost for Hallux Valgus and Knee Operations in a private hospital.

2.1.1 Constraints

a) *State equations*, i=1, 2, ...12.

$$CL_(i+1) \quad = \quad CL_i + CA_i - CE_i - CR_i - CO_i$$

$$
\begin{aligned}
HL_(i+1) &= HL_i + HA_i - HE_i - HR_i - HP_i \\
KL_(i+1) &= KL_i + KA_i - KE_i - KR_i - KP_i \\
OL_(i+1) &= OL_i + OA_i - OE_i - OR_i
\end{aligned}
$$

b) *Operating Room by Services:*

These constraints only affect operation plans in regular hours:

b-1 Oftalmology: $80CR_i \leq OQ_i$; i=1,...,12

b-2. Traumatology: $80HR_i + 125KR_i + 160OR_i \leq TQ_i$ $i = 1, ..., 12$

See OQ_i and TQ_i data on table 4.

c) *Bound to the number of processes on private and over time scheduling:*

$CO_i \leq l_i$; $HP_i \leq m_i$; $KP_i \leq n_i$ i=1,...,12

Table 7:

	1	2	3	4	5	6	7	8	9	10	11	12
l_i	0	0	68	40	64	72	0	0	44	52	48	24
m_i	0	20	25	35	35	35	35	35	35	35	35	35
n_i	0	0	8	21	21	20	20	20	20	10	10	0

d) *Waiting list time upper limit: no more than nine months:* With the following equations we reflect that along the year the maximun time for patients to be in waiting list should be nine months.

$$
\begin{aligned}
&\textstyle\sum_{i=1}^{k}(CR_i + CO_i) \geq a_k \\
&\textstyle\sum_{i=1}^{k}(HR_i + HP_i) \geq b_k \ \sum_{i=1}^{k}(KR_i + KP_i) \geq c_k \\
&\textstyle\sum_{i=1}^{k} OR_i \geq K_k
\end{aligned}
$$

where parameters are defined in **Table 8:**

months	1	2	3	4	5	6	7	8	9	10	11	12
a_k	11	26	50	116	153	224	309	398	480	556	628	694
b_k	9	24	62	92	103	130	153	165	199	223	243	252
c_k	4	23	35	49	59	82	100	119	132	150	167	181
K_k	3	17	34	38	42	61	94	107	128	133	153	159

e) *No more than six months waiting at the end of 1998:*

$$CL_{13} \leq 395 \ ; \ HL_{13} \leq 69 \ ; \ KL_{13} \leq 77 \ ; \ OL_{13} \leq 57.$$

f) *All the variables would be nonnegative integers.*

2.1.2 Objectives functions

The priority objective in this problem is, as we have said, to minimize waiting list time at the end of 1998, then:

$$\min \quad f_1 = 80CL_{13} + 80HL_{13} + 120KL_{13} + 160OL_{13}$$

The second objective is to minimize operational costs:

$$\begin{aligned}
\min f_2 \;=\; & 110852(\sum_{i=1}^{12} CRi) + 125899(\sum_{i=1}^{12} HRi) + \\
& +287973\left(\sum_{i=1}^{12} KRi\right) + 853338\left(\sum_{i=1}^{12} ORi\right) + \\
& +123733(\sum_{i=1}^{12} COi) + 106605(\sum_{i=1}^{12} HPi) + 141120\left(\sum_{i=1}^{12} KPi\right)
\end{aligned}$$

Once all the equations have been developed we may model the problem as a bi-objective one which was developed through the weighted approach. We use weights reflecting Decision Maker's opinion: 0.8 for the first one and 0.2 for the second.

In order to solve the problem of homogeneity of objectives it is convenient to normalize the weights before introduce them in the joint function. The normalization procedure to be used will consist on dividing every weighting coefficient by the difference between anti-ideal and ideal points.

2.1.3 Results

The problem was computing using HYPERLINDO and integer variables. We have found the optimal solution of the integer program with the Integer Programing Optimality Tolerance (IPTOL) equal 0.0045. That means that the objective value of the final solution returned by the program will be within 0.45% of the true optimal objective value.

The problem dimension hinders to make explicit the complete solution: we will specify the objectives values of significative variables in **tables 9 and 10:**

Table 9:

Process/moths	1	2	3	4	5	6
Cataracts Regular	75	80	78	66	80	80
Cataracts Overtime			68	40	64	72
Hallux Valgus Regular	10	4	3	0	0	0
Hallux Valgus Private		20	25	35	35	35
Knee Op. Regular	4	20	38	0	0	0
Knee Op. Private			8	21	21	20
Osteoarthritis Regular	21	13	0	26	30	30

Process/moths	7	8	9	10	11	12
Cataracts Regular	33	41	33	90	80	56
Cataracts Overtime			44	52	48	24
Hallux Valgus Regular	0	0	0	0	0	0
Hallux Valgus Private	35	35	35	33	35	35
Knee Op. Regular	20	28	0	0	0	34
Knee Op. Private	20	20	20	10	10	
Osteoarthritis Regular	8	0	20	34	30	0

Table 10:

Monthly Waiting List	2	3	4	5	6	7
Cataracts	481	473	393	365	279	194
Hallux Valgus	213	209	190	171	160	139
Knee Opeations	146	143	111	102	101	85
Osteoarthritis.	112	119	125	106	97	84

Monthly Waiting List	8	9	10	11	12	13
Cataracts	233	222	204	140	89	88
Hallux Valgus	113	87	67	46	21	0
Knee Opeations	56	20	17	16	26	0
Osteoarthritis.	74	74	64	51	22	49

An upper bound of activity has been reached. And the most restrictive constrain was also verified, i.e., no one patient will be more than six months on waiting list at the end of 1998.

3 Conclusions

An important property in our model is the fact that can be easily made adaptive, in the sense that in every month in the year, where we have new information about current waiting list or updated forecasting for admision/exit of patients, it is possible to adapt the model in such a way that it incorporates the new information in substitution of the old one and we can obtain updated values after optimization, from that month to the end of the year.

The obtained results have been introduced and commented on with the decision maker. Even though we have yet to do a deeper study, the decisor maker's opinion is very positive (he had previously doubted whether he would be able to comply with all the conditions laid down by the Health Ministry, particulary those related to six months maximun waiting time after the 30th of June).

We asked him specifically about whether there had been a problem of bed availability in the results coming out of Traumatology surgical processes (which require hospitalization). He replied that, in general, there had been no problems of this kind , any difficulties could, in principle, be dealt with easily.

References

[1] ANDERSEN CONSULTING (1993): "The Future of European Health Care". A study by Andersen Consulting in Co-operation with Burson-Marsteller.

[2] ARENAS, M.; BILBAO, A.; RODRIGUEZ, M.V. (1992): "Aplicación de la programación por objetivos a la toma de decisiones de un centro de salud". *VI Reunión Anual ASEPELT - ESPAÑA*. Granada.

[3] ARENAS, M.; LAFUENTE, E.; RODRÍGUEZ URÍA, M.V. (1997). "Goal Programming model for evaluating an Hospital service performance". *Advances in Multiple Objective and Goal Programming.* Ed. Caballero/ Ruiz/ Steuer. Springer

[4] BERTSEKAS, D.(1995):"Dynamic Programming and Optimal Control".Vol. 1. *Athena Scientific.*

[5] BITRAN, G.R.; VALOR-SABATIER, J. (1987): "Some Mathematical Programming Based Measures of Efficiency in Health Care Institutions". *Advances in Mathematical Programming and Financial Planning.* Vol 1: 61-84.

[6] CHAE, Y.; SUVER, J.; CHOU, D. (1985): "Goal programming as a capital investment tool for teaching hospitals". *HCM review*, Winter, 27-35.

[7] CHARNES,A; COOPER,W.W.(1977) "Goal Programming multiple objetive optimization", part 1. *European Journal of Operational Research 1*, 39-54.

[8] DONABEDIAN, A (1984). *La calidad de la atención médica: Definición y mètodos de evaluación. Ed.Cientifica la Prensa Médica.* Mexicana S.A.; Mèxico D. C.

[9] FRANZ, L.; BAKER, H.; LEONG, G.K.; RAKES, T.R. (1989): "A Mathematical Model for Scheduling and Staffing Multiclinic Health Regions". *Elsevier Science Publishers* B.V. 277-287.

[10] LEE, S. (1973): "An Aggregative Resource Allocation Model for Hospital Administration". *Socio-Economic Planning Science.* Vol. 7, 381-395.

[11] ORTÙN RUBIO, V; (1990) *La economia en sanidad y medicina: Instrumentos y limitaciones.* Ed. Euge, Barcelona.

[12] WATCH, R.F.; WHITFORD, D.T. (1976): "A Goal Programming Model for Capital Investment Analysis in Nonprofit Hospitals". *Financial Management 5*: 37-56.

[13] WHITTLE, P.(1982): *Optimization over time.* Vol 1 and 2. J.Wiley.

On Designing Health Care Plans and Systems from the Multiple Criteria Decision Making (MCDM) Perspective

Charles H. Smith and H. Roland Weistroffer
School of Business, Virginia Commonwealth University, Richmond, Virginia, USA

Abstract. Many studies of medical decision making focus on the cost effectiveness of taking action A for disease X. These focused evaluations do not consider that the recommended actions must compete in aggregate for available resources. Relatively fewer studies have considered the aggregate decision making required for the rational design of health care. The operations research perspective and, in particular, the MCDM perspective can aid our understanding of the design of health care plans from the insurer's and payer's perspectives. This paper considers issues in defining health benefits plans and suggests how MCDM methodology can assist in structuring the appropriate decision context. Issues to be addressed in designing health plans include global budgets, new technology, system objectives, the hierarchy of decision makers in medical care, and data requirements. While MCDM methods may be useful in health policy making, considering these complex applications can also provide a focus for further conceptual and methodological developments in MCDM.

Keywords. Health care planning, multiple criteria, cost effectiveness

1 Introduction

How should society decide which uses of health care resources are affordable? How should a health care payer decide which benefits to offer to policy holders or members? Finite health care resources are currently allocated in a non-systematic manner. Should insurers and other health care payers apply rational decision making principles to select the prevention, screening, diagnostic, and treatment procedures that they cover when these decisions can literally mean life and death? The strategic issues of system design are much more complex and much more political than the tactical issues of comparing the relative worth of two treatments for a specific condition. It is a basic premise, if not emphasized, of managed care in the United States that the patient and the physician cannot be permitted to make all health resource allocation decisions alone.

Consider a few recent, real examples of resource rationing in health care.

1. In 1995 a health insurer denied a last chance bone marrow transplant to a dying child in Richmond, VA. It was estimated that the cost was $250,000 and the chance of success was 14%.

2. Some national governments choose not to cover adult dental health care in benefits packages.

3. When the AIDS drugs called protease inhibitors were introduced in the US, special programs that were state-administered, but received two-thirds of their funds from the federal government, faced a dilemma. These programs supported a relatively small group of patients, 70,000 nationwide, who were without private insurance but who did not qualify for the Medicaid program for the poor. The cost of the new treatment was $12,000 per year per patient. In the state of Washington the total cost to taxpayers rose from $600,000 to $6 million in a single year. States responded in a variety of ways to ration care. One approach was to bar new patients and cap the number that could receive the protease drugs. Some states refused to offer the new drugs through their programs. Some states offered the new drugs, but eliminated other medications, such as antibiotics, that were previously covered.

4. Managed care providers have sometimes refused to cover the $25,000 cost of implantable morphine pumps to ease the pain of terminal cancer patients. Some doctors argue the pumps are greatly overused, and critically needed in only about 5% of these cases.

5. In May 1998 a study was published that demonstrated people with average cholesterol levels could reduce their risk of heart attack by 37% if they took the drug lovastatin at a cost of $1400 per year.

As operations researchers, we should know the dangers of making decisions based on anecdotes, and limited information. However, reflecting on the examples above suggests the following observations.

(1) Rationing health care is unavoidable, but it can take several forms. Decisions about health benefits are literally decisions about who will live and who will die.

(2) In health care there are many conflicting objectives that are valued differently by the different participants in the system: patient, provider, payer, insurer, society (government).

(3) Technological change in health care brings promise of system improvement, but it also requires new, complex decisions to be made. Policy makers want change that results in the same or increased benefits, but at lower cost.

(4) Decision making about benefits packages is complicated by issues of personal responsibility and lifestyle, and by issues of equity and access.

(5) Inconsistency in medical practice is an important reality resulting in widely different uses of resources.

2 Cost-Effectiveness and Cost-Benefit Analysis

Very little research has been reported on applying multiple objective optimization to health care policy; in fact, there seems to be very little application of cost-effectiveness analysis, which could perhaps be viewed as one of the simplest multiple criteria approaches. Sloan and Conover (1996) state: "Cost-benefit/ cost-effectiveness analyses are rarely used in benefits coverage decisions....The actual use of cost-effectiveness/cost-benefit analysis appears to fall far short of its potential...." Likewise, Luce (1998) states in a recent article: "... a common theme emerges, which is that cost-effectiveness is not routinely used for decision making regarding technology coverage, formulary inclusion, or reimbursement, including pharmaceuticals among the technologies considered. There is little or no evidence that cost-effectiveness analysis is used for establishing practice guidelines; thus, it probably is not."

A number of concerns have been raised about the conduct and use of cost-effectiveness studies. Since evaluations are often at least partially funded by parties with a vested interest in the outcome, especially pharmaceutical companies, concerns about bias and use in promotional claims have been raised. In evaluating effectiveness, one must distinguish clearly between efficacy in clinical trials and effectiveness under conditions of actual use with its issues of noncompliance, etc. Furthermore, cost-effectiveness depends on the perspective of the decision maker. Society's perspective can differ from the payer's perspective. Sloan and Conover (1996) observe that "even if such methods were in greater use, a critical question is whether such analyses performed by private payers should adopt a social or a payer perspective. For example, an employer may get no benefit from a heart transplant performed on a worker who will soon retire, because all the longevity benefits will occur after the worker leaves the firm."

Cost-effectiveness criteria have been cited in guidelines used for approval of coverage decisions in a few instances, such as Ontario and Australia. By its nature, cost-effectiveness analysis guides decision making at the margin, evaluating the effect of incremental changes to the status quo. It is not an ideal tool for the global design of a health benefits package.

3 Other Attempts at Rational Design of Health Care Systems

Health care plans can define the types of coverage they provide in several ways. The benefits can be described based on the process to be followed, the treatments included, treatments excluded, and blends of these features and others, such as the possibilities for experimental treatments. Buchanan et al. (1991) performed a simulation of expenditures under structural changes to plans, such as deductibles.

The original version of the Oregon Medicaid Plan aimed for universal access by expanding the state Medicaid program, a program for the poor, funded by state and federal funds. It later implemented several health insurance reforms, but the feature that attracted attention was the definition of specific condition/treatment pairs that would be covered or not covered. This plan was able to fund an expanded number of Medicaid recipients by limiting the services covered. While the only explicit objective was cost, health actions were ranked and about six hundred chosen for funding. No longer covered are such procedures as tonsillectomies and treatment of viral sore throats, hay fever, and the common cold. The initial plan was not approved by the federal government, but a revised plan that had to abandon some of the more explicit rationing features went into effect about five years ago (Chase 1994, Gold 1997, Rutledge 1997).

Sainfort and Booske (1993) recognized that designing health plans effectively requires considering attributes of the entire portfolio of benefits, not just characteristics of individual items. They described the use of portfolio theory to set priorities in designing health plans. The authors reported results that showed the Oregon method, a heuristic priority setting approach, was far better than traditional non-systematic methods in improving the performance of attributes. Yet the Oregon method did not produce portfolios of benefits as good as Sainfort and Booske's method. They concluded that while setting health care priorities is complex, medical decision making researchers need to explore it. They recognized that designing health plans is inherently multiattribute and requires a complex attribute structure.

4 Multicriteria Integer Program Approach for Health System Design

The following highly simplified example illustrates the possible development of a multicriteria integer programming model for designing a health benefits plan.

Suppose there are only five possible medical procedures to consider for a health insurance plan. Let

$x_i = 0$ if medical procedure *i* is not covered by the plan, and
$x_i = 1$ if medical procedure *i* is covered, $i = 1,\ldots,5$.

Objective one is to minimize cost and objective two is to maximize the quality-adjusted life years (QALY) of the plan's participants. The model might include the following objective functions:

$$\text{Min } Z_1 = 3\,x_1 + 4\,x_2 + 2\,x_3 + 1\,x_4 + 5\,x_5$$
$$\text{Max } Z_2 = 2\,x_1 + 2\,x_2 + 2\,x_3 + .5\,x_4 + 3\,x_5$$

One constraint in our simple example is the budget,

$$3\,x_1 + 4\,x_2 + 2\,x_3 + 1\,x_4 + 5\,x_5 \leq 9.$$

Also, dependencies can be modeled. For example, the constraint

$$x_1 + x_3 \leq 1$$

insures that either procedure 1 or procedure 3 can be included, but not both.

While the Oregon plan was not modeled as an optimization problem, they considered several hundred medical procedures, rather than just five as in the example above. This simple example does illustrate the general approach, however, and also the need to model dependencies. One way that dependencies can enter the problem is illustrated by observing that the costs and benefits of treating disease X depend on whether a vaccination for X is funded in the plan. It may sometimes be necessary to include non-linear terms to permit the costs and benefits of a service to depend on inclusion or exclusion of another service.

Due to the multitude of health care services that can be provided and the relative lack of satisfactory data, it would typically not be possible to rely exclusively on solutions to the conceptually satisfactory model described above. Yet it can be useful to perceive the decision context in this global manner, and the model can provide a tool for exploring the problem and thereby gaining insight.

After a health plan has been designed, a more prescriptive use of multiattribute techniques can follow. For example, a multiattribute decision analysis can be performed for a payer, e.g., a self-insured employer, who needs to evaluate two or more competing plans previously designed.

5 Similar Models in Other Application Areas

It should be observed that models of the type illustrated by the example in the previous section, i.e., large 0-1 multicriteria integer programs, represent an area that needs more development of user-friendly solution support tools. To provide additional motivation for the development of these tools, some problems in other applied areas with similar characteristics as the health care design problem are described below.

The health plan design optimization model presented above is similar to some project selection situations that have been treated in the literature. For example, Santhanam and Kyparisis (1995) considered project interdependencies in their work. The project selection problems described in the literature are typically small relative to the health plan design problem described above.

A recent application of 0-1 programming in health care is described by Jacobson et al. (1998). In that study the decision by a purchasing organization as to which vaccine combinations to obtain to meet childhood immunization schedules was examined. Their only objective was cost, but the problem could be expanded to include other objectives, such as minimizing the number of patient visits, maximizing the number of different manufacturers used, etc.

6 Potential Objectives in Designing Health Plans

Even without an explicit model and solution, the MCDM perspective can assist in bringing some structure to the complex problem of health plan design by explicitly identifying an objectives hierarchy. Here we merely list examples of the conflicting objectives that might be considered in designing health care plans:

- guarantee access for all to medically justified treatment
- minimize costs
- maximize comfort for terminal cases
- retain sufficient numbers of appropriately skilled health care personnel
- maximize profit
- maximize value-added health care
- minimize inappropriate use of health care resources
- promote wellness lifestyles by individuals
- obtain appropriate medical outcomes
- prevent unnecessary deaths
- improve quality of life
- obtain maximum market share
- maintain workforce stability
- maximize customer satisfaction

Readers from nations with national health care systems may especially notice the remarkable mix of business objectives and medical objectives in the list above.

7 Additional Critical Issues

Before concluding this brief look at health care plans from the MCDM perspective, the following issues also provide significant material for thought.

Who should make which decisions?

There are decisions that a health care plan should leave to the patient; other decisions should be left to the doctor; still others should be described in the plan's policy descriptions, etc. It is an extremely challenging exercise to define a health system that results in all decisions being considered by the most appropriate decision maker.

How long must a benefit commitment last?

If a health plan is constructed by a rational process, it should be changed to reflect new realities, including availability of new data and new technology. On the other hand, participants in the system should be entitled to dependability in the services provided to them.

How does a plan cope with rapid turnover of participating patients?

Under competition and managed care in the United States, many people have the opportunity to select annually from a set of health care plans made available by their employer. The potential for sizable changes in the number and characteristics of the plan's patient population increases the difficulties of designing an optimal plan.

Would it be ethical for health maintenance organizations to offer services that vary with the availability of resources at the time a situation presents itself? For example, could a plan decide to offer a particular service for the last two months of its fiscal year when it discovers it has sufficient funds? Or must a service be offered 100% of the time or not at all?

8 Conclusions

Many decision makers in health care are reluctant to employ cost-effectiveness and cost-benefit analysis. Concerning cost-benefit analysis, Pauly noted (1996), "As long as policymakers believe that the main challenge to health policy is to find those new programs that will enable them to lower costs without reducing benefits at all, they will not be eager to measure benefits in monetary terms; they will not even need to measure benefits in monetary terms." He saw the possibility of increased use of such analysis, however, in the current US system, and stated a "stimulus to cost-benefit analysis is the emergence of private insurance coverage in a managed care setting. Such managed care plans, by necessity, must decide which programs are worth the cost.... Especially for programs or products such as drugs that sometimes do some good but at a high cost, the plan must ask whether the benefit from the program can be collected in the form of higher premiums."

While even cost-effectiveness analysis may be little used in health care policy making, we have indicated the need to move beyond this approach that is essentially incremental in nature. An approach that captures the entire breadth of a system's effects is needed. There are many difficulties in applying a comprehensive multiple criteria optimization model to the design of health care plans. These include inadequate data, changing technology, and the political discomfort that such analysis creates for policy makers. Such a global view of the system, however, can be a tool for developing understanding, leading to a different perception of the decision context and thus focusing discussions in a new way.

Regardless of the health care system, intelligent resource allocation decision making is needed, whether made by private insurers, health maintenance organizations, or government institutions. Health care payers must explore and gain understanding of the coverage that can be supported with a given budget, balancing the system view with knowledge that physicians treat individuals, not average patients with average responses to treatment.

References

Buchanan, J.L., Keeler, E.B., Rolph, J.E., Holmer, M.R. (1991): Simulating Health Expenditures Under Alternative Insurance Plans. *Management Science* 37(9), 1067-1090

Chase, M. (1994): Oregon's New Health Rationing Means More Care for Some but Less for Others. *The Wall Street Journal*, January 28, 1994

Gold, M. (1997): Markets and Public Programs: Insights from Oregon and Tennessee. *Journal of Health Politics, Policy and Law* 22(2), 633

Jacobson, S. H., Sewell, E. C., Deuson, R., Weniger, B.G. (1998): An Integer Programming Model for Vaccine Procurement and Delivery in the National Childhood Immunization Program: A Pilot Study, unpublished manuscript

Luce, B. R. (1998): Pharmacoeconomics and Managed Care: Methodologic and Policy Issues. *Medical Decision Making* 18(2,Supplement) S4

Pauly, M. V. (1996): Valuing Health Care Benefits in Money Terms. In: Sloan, F. A. (Ed.) *Valuing Health Care: Costs, Benefits, and Effectiveness of Pharmaceuticals and Other Medical Technologies*, Cambridge University Press, Cambridge, UK, 99-124

Rutledge, K. M. (1997): The Oregon Health Plan: Lessons Learned. *Healthcare Financial Management* 51(4), 48

Sainfort, F., Booske, B.C. (1993): Portfolio Theory for Priority Setting in Designing Health Plans. Presentation at the 15th Annual Meeting of the Society for Medical Decision Making, Research Triangle Park, North Carolina

Santhanam, R., Kyparisis, J. (1995): A Multiple Criteria Decision Model for Information System Project Selection. *Computers Ops. Res.* 22(8), 807-818

Sloan, F. A., Conover, C.J. (1996): The Use of Cost-Effectiveness / Cost-Benefit Analysis in Actual Decision Making: Current Status and Prospects. In: Sloan, F. A. (Ed.) *Valuing Health Care: Costs, Benefits, and Effectiveness of Pharmaceuticals and Other Medical Technologies*, Cambridge University Press, Cambridge, UK, 207-232

Case Studies on the Application of Adaptive Risk Analysis to USDA's Resource Conservation Programs

Mark A. Tumeo [1], David A. Mauriello [2], Ali M. Sadeghi [3], and Ronald Meekhof [4]

ABSTRACT

As part of its role of ensuring that major regulations proposed by the US Department of Agriculture (USDA) are based on sound scientific and economic analysis, the Office of Risk Assessment and Cost-Benefit Analysis (ORACBA) is undertaking a series of case studies to apply a multiple criteria decision making (MCDM) risk analysis technique to USDA's Resource Conservation Programs. The MCDM tool being developed and tested is aimed at providing a clear understanding of the environmental and human health benefits and associated level of uncertainty of various management practices that are implemented under the Resource Conservation Programs. The case studies examine two different programs: 1) the Environmental Quality Incentives Program (EQIP), and 2) the Conservation Reserve Program (CRP). The EQIP case study will focus on specific watersheds and determine the impact of a range of manure management practices on multiple environmental and human health objectives and will provide the basis for a comparison of benefits, including those related to the reduction or prevention of risk to the costs associated with a set of potential alternative management strategies. The CRP case study will evaluate the effects of various kinds of managed disturbances to grasslands enrolled in the CRP with the objectives of reducing nutrient runoff and enhancing wildlife values. Both projects were in their initial stages at the time this paper was prepared and therefore only the progress to-date and the underlying theory being applied to the case studies are discussed.

[1] AAAS/USDA Risk Assessment Fellow, Cleveland State University, 1899 E. 22nd St., Cleveland, OH 44114

[2] US Environmental Protection Agency, OPPT/RAD/SSB 7403, 401 M St., SLO, Washington, DC 20460

[3] Agricultural Research Service, US Dept. of Agriculture, 103000 Baltimore Blvd., Beltsville, MD 20705

[4] Food Safety and Inspection Service, USDA, 1400 Independence Avenue, SW, USDA, Washington, DC 20250

1. Multiple Criteria Decision Making in the Regulatory Context

The process of regulatory decision making is a unique study in the application of multiple criteria decision making. Decision making in the Federal government involves not only the balancing of the normally complex factors of science, technology, economics, social and cultural concerns, but also must reflect statutory requirements, Administration policy, Legislative intent, competing political interests, and public perception. In this milieu, it is not surprising that the regulatory decision making process is often neither easily explained to, nor understood by, the outside observer. However, ever more increasingly, the public and the Congress want agency decision making to be transparent. This idea has been reinforced by the courts in numerous decisions in which agency decisions were vacated because of a lack of a clear and concise documentation of the decision-making criteria and process.

As a result of this pressure, there has been considerable movement in the past few years towards the application of risk and cost-benefit analyses in decision-making techniques. While regulatory decisions have traditionally applied concepts of risk and cost-benefit to decision making, the methodology and documentation of this application has often been lacking. In the current trend, risk and cost-benefit analyses are conducted to provide structured information useful for regulatory decision making. While neither is "science," both make use of applied science and are subsets of decision analysis. The purpose of conducting a risk analysis is to make clear the nature of the risks the program will address, alternative ways of reducing those risks, the reasoning that justifies selecting the alternative described in the proposed rule, and a comparison of the likely costs and the benefits of reducing the risk.

2. Risk Analysis and Regulatory Decision-Making

The concept of using some sort of formalized risk analysis to assess the performance of proposed or ongoing activities of the Federal government developed in the late 1960s. As the public became more aware of environmental degradation, there was a demand that government become more transparent in the way in which decisions were made that could have an adverse impact on the environment and human health. The National Environmental Policy Act (NEPA) was passed mainly in response to this public pressure and created the requirement that an Environmental Impact Statement (EIS) be prepared for "*any major Federal action ... that significantly affects the quality of the human environment*" (§102(c)(3)). While not explicitly requiring a risk analysis, the law does require an examination of the tradeoffs between short-term benefits and long-term commitments of resources, as well as a thorough examination of alternatives.

As the process of preparing documentation to support decisions which affected the environment developed through the 1970s, it became apparent that better methods were needed to evaluate the uncertainties and trade-offs faced by government decision-makers. Consequently, the concept of incorporating risk analysis into government decision processes in order to improve the quality of decisions gained attention. On February 17, 1981, President Reagan issued Executive Order 12291 which required that all major regulations be accompanied by an analysis of the risks being regulated and the costs and benefits of the regulation. A "major regulation" was defined as one that impacts the environment and has an annual economic impact of at least $100 million. This requirement was reiterated in 1993 when President Clinton issued Executive Order 12866 entitled "Regulatory Planning and Review." The application of risk and cost-benefit analyses was codified in the *Federal Crop Insurance Reform and Department of Agriculture Reorganization Act* of 1994 (PL 103-354) which requires USDA to conduct a thorough analysis that makes clear the nature of the risk, alternative ways of reducing it, the reasoning that justifies the proposed rule, and a comparison of the likely costs and the benefits of reducing the risk. At the same time, Congress established within USDA the Office of Risk Assessment and Cost-Benefit Analysis (ORACBA) to provide guidance and technical assistance, coordinate risk analysis work across the Department, and certify that statutory requirements are met.

ORACBA pulled together a team of risk assessors and conservation program experts from the Environmental Protection Agency and from within USDA's Animal and Plant Health Inspection Service, Agricultural Research Service, Economic Research Service, and ORACBA to discuss methodologies available for conducting risk assessments of this scope and scale. The team reached consensus that the EPA "Framework for Ecological Risk Assessment" (USEPA, 1992) and the associated draft "Proposed Guidelines for Ecological Risk Assessment" (USEPA, 1996) provided a reasonable basis on which to start develop the assessments. However, USDA has found that these documents, in places, are not completely adequate for the scope, scale, and purpose of risk assessment within the Department, especially with respect to conservation programs. As a result, modifications have been made with regard to the development of assessment endpoints, the analysis of ecological effects, and the focus of the risk characterization sections of the EPA approach. Overall, these modifications have been developed into an adaptive risk analysis approach (Meekhof, et al., 1998).

3. Focus of the Paper

This paper focuses on the use of a "risk-based" approach to structure information about environmental hazards and their consequences, identify sound programs to prevent or reduce those hazards, and determine whether program objectives are being achieved in a cost effective manner. A risk analysis approach is one methodology to assist regulatory decision making in the resource conservation programs within the USDA. The paper reviews the approach being developed within USDA, provides an overview of its application in two resource conservation programs (the Conservation Reserve Program and the Environmental Quality Incentives Program), and presents the underlying models and techniques to be applied in the risk assessment portion of the process.

4. An Adaptive Risk Analysis Approach

An "adaptive" risk analysis approach provides a basis for ongoing improvement in program management by providing feedback on program performance in reducing environmental hazards and provides a framework for the program development and evaluation process by identifying the technical and management linkages between objectives, program alternatives, assessment endpoints, monitoring, and program performance. The principal components of an adaptive risk analysis process are:

1) Program identification;
2) Risk assessment;
3) Program analysis;
4) Implementation; and
5) Monitoring and evaluation.

Each component, while listed in a specific order, is actually part of an iterative process; and depending on the regulatory program involved, various steps may be omitted or highlighted in different ways. This is the essence of the "adaptiveness" of the methodology. Figure 1 graphically presents the overall adaptive risk analysis process and shows some of the interactions of the various steps.

various steps may be omitted or highlighted in different ways. This is the essence of the "adaptiveness" of the methodology. Figure 1 graphically presents the overall adaptive risk analysis process and shows some of the interactions of the various steps.

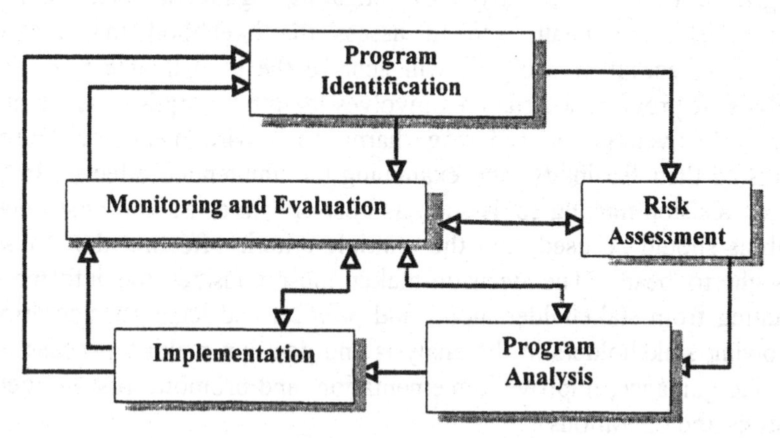

FIGURE 1: Adaptive Risk Analysis Process

4.1 Risk Assessment: Risk assessment is the process of using scientifically supported relationships to delineate the impacts of specific actions, events or hazards on measurable, quantitative endpoints. Risk assessment provides a consistent method to a) identify a hazard, b) evaluate the extent to which a group of people or a resource would be exposed to the hazard, and c) determine the magnitude of harm to the environment and human health and safety that would result from the exposure. Risk assessment uses quantitative and qualitative methods and reasonably obtainable and sound scientific, technical, economic and other information. The risk assessment should identify the consequences of the hazards for ecological relationships and the corresponding assessment endpoints; clearly delineate areas of uncertainty, and be structured in a way that provides the types of information needed to evaluate alternatives. An important task of the risk assessment is to identify assessment endpoints that are ecologically significant relative to program component can help identify the questions to be addressed and the types of information needed and will guide the selection of endpoints.

4.2 **Program Analysis**: The purpose of this component is to examine program options or alternatives for mitigating negative impacts identified in the risk assessment, and to assess the likelihood that a given regulation, action or program will achieve the overall objectives. An analysis of program alternatives involves identifying options, evaluating their effectiveness in reducing harm to environmental resources, assessing their feasibility, and examining the unintended effects. In this stage, decision-making tools such as cost-effectiveness and cost-benefit analysis should be used, and the multiple criteria affecting the decision brought to bear. The decision-maker must consider the information resulting from stakeholder input, and political and legal considerations. Involving stakeholders in the analysis and decision making process may provide guidance, improve implementation, and promote trust in agency analysis and operations.

The problem identification and assessments conducted in previous stages help identify and narrow the range of reasonable alternatives. The information resulting from the program identification and risk assessment stages should help the risk manager identify which alternative is the most suitable for achieving program objectives and the associated level of uncertainty. Given the statutory requirements, the complexity of USDA resource conservation programs, decentralized decision making, and the need for flexibility, the identification of the alternatives may require an iterative process. Broad policy goals, program alternatives and implementation issues may be modified based on the results of the analysis, and may require additional assessment of risk to allow a clearly supported and well documented decision to be made. In addition, for large scale national programs which address multiple assessment endpoints, achieve multiple objectives, and operate in multiple ecosystems like those administered by USDA, the program analysis phase can require extensive negotiation and compromise among risk managers, risk assessors, stakeholders, program managers, field staff, and others.

4.3 Implementation: The implementation component involves taking the necessary steps to put the selected option, action or program into effect. Implementation of flexible, voluntary programs require a high level of coordination at the national, state, and local levels. Analyses of the types of problems which may arise at various steps in the implementation process and the resource requirements may be required if the program is new or is being significantly modified.

4.4 Monitoring and Evaluation: The successful implementation of programs over time requires the establishment of institutional procedures for monitoring and evaluation. Such feedback must be integrated into the decision making framework to provide the necessary information for determining the effectiveness of program implementation, to respond to changing political and economic concerns, and to ensure that the agency remains responsive to the public. Monitoring provides the means to evaluate the management options that have been implemented and to determine the level of success in achieving program objectives. Unfortunately, monitoring often receives the least attention from the decision-maker, and rarely are adequate resources allocated to conduct the type of monitoring necessary to gage overall success of the program. In many instances, this is the result of both Congressional failure to appropriate the funds for monitoring activities, and the fact that decision makers seldom have the time to review past decisions in the face of ongoing or impending decision making processes.

Monitoring and evaluation efforts should be tightly coupled with program objectives at each level at which the program is implemented. This implies that monitoring and evaluation endpoints be selected which are most effective in measuring program benefits. An additional requirement is that the measurement endpoints identified must be scalable from local to national level in order to provide for the aggregation of information collected. Thus, the establishment of an experimental design for a monitoring and evaluation purposes, involving the selection of indicators and thresholds, should occur concurrently with the conceptual phase of Program Identification, at the same time as management objectives, assessment endpoints, and implementation options are selected.

5. Example Applications of an Adaptive Risk Analysis Approach

As stated in the section above on Risk Analysis and Regulatory Decision Making, the Federal Crop Insurance Reform and Department of Agriculture Reorganization Act of 1994 set new standards for regulatory analysis of major rules proposed by USDA. Included in this law is the requirement that a risk assessment and cost-benefit analysis accompany all proposed major regulations. Two of the first major USDA programs to come under this law were the new Environmental Quality Incentives Program (EQIP) and the re-vamped Conservation Reserve Program (CRP).

In February of 1997, nation-wide, multiple stressor ecological risk assessments for EQIP and CRP were completed by USDA's Natural Resources Conservation Service (NRCS) and Farm Service Agency (FSA), respectively. The objectives of the assessments included: 1) identification of those agricultural activities and practices that place natural resource values at risk; 2) characterization of the mechanisms that result in risk; 3) characterization of the magnitude and extent of the environmental risk; and, where possible, 4) recommendations to risk managers and decision makers on how the results of the assessment can be used in program development and implementation.

The EQIP and CRP risk assessments were the first assessments conducted under the risk assessment requirement of the 1994 Act. The process being used was found to be unduly prescriptive, restrictive, or inadequate for the scope, scale, and purpose of the assessments. This was due, in part, to the varied physical and chemical nature of the multiple stressors under evaluation in USDA programs, and to the difference between the management goals for the USDA conservation programs and those of EPA, whose ecological risk assessment guidance was being used. It was decided that additional information should be assembled and a more detailed assessment and cost-benefit analysis of program options should be completed to examine the decisions that were made in implementation of the programs with regard to mitigation practices supported and approval of acre enrollment or project applications. The two case studies discussed below are part of NRCS's and FSA's ongoing activities in administering and implementing the EQIP and CRP programs.

5.1 Case Study 1: *Manure Management through the Environmental Quality Incentives Program (EQIP)*

5.1.1 ***Program Description***: The Environmental Quality Incentive Program was created in the Federal Agriculture Improvement and Reform Act of 1996 (commonly called "the 1996 Farm Bill"). EQIP provides technical, educational, and financial assistance to eligible farmers and ranchers to address soil, water, and related natural resource concerns on their lands in an environmentally beneficial and cost-effective manner. The program provides assistance to farmers and ranchers in complying with Federal, State, and tribal environmental laws, and encourages environmental enhancement. The purposes of the program are achieved through the implementation of a conservation plan, which includes structural, vegetative, and land management practices on eligible land. Five- to ten-year contracts are made with eligible producers. Cost-share payments may be made to implement one or more eligible structural or vegetative practice, such as animal waste management facilities, terraces, filter strips, tree planting, and permanent wildlife habitat. Incentive payments can be made to implement one or more land management practice, such as nutrient management, pest management, and grazing land management. Fifty percent of the funding available for the program will be targeted at natural resource concerns relating to livestock production. The program is carried-out primarily in priority areas that may be watersheds, regions, or multi-state areas, and for significant statewide natural resource concerns that are outside of geographic priority areas. The President's budget includes $300 million for this program in Fiscal Year 1999.

5.1.2 ***Background and Objectives of Case Study***: Animal agriculture in the United States has evolved into large production facilities that are geographically concentrated. In some cases, multiple animal industries have located in the same geographic region. Increased size and concentration of animal operations has increased the concentration of manure and wastes, and the emission of hydrogen sulfide, the most prominent of the compounds attributed to odor. Concentrated animal production sites are a concern because of the potential for nutrient and bacterial contamination of water resources, transmission of disease to humans, and nuisance problems, such as odors, affecting neighboring communities. EQIP funds targeted to environmental problems associated with livestock are aimed at mitigating many of these problems.

The case study will serve two purposes: 1) to examine the nature of the risks the EQIP will address, alternative ways of reducing those risks, and a comparison of the likely costs and the benefits of reducing the risk; and 2) to demonstrate the use of risk analysis to assist NRCS in meeting the statutory mandate for risk assessment and cost-benefit analysis set forth in the *Federal Crop Insurance Reform and Department of Agriculture Reorganization Act of 1994* (PL 103-354).

5.1.3 ***Case Study Progress To-Date***: The overall plan for the case study has been approved and work formally began in May.

5.1.4 *Steps in the Adaptive Risk Analysis Process*

5.1.4.1 *Program Identification*: NRCS, working with ORACBA, has developed the focus of this case study to address questions relevant to current and future conservation management activities under EQIP. In general, the case study is intended to serve as an example of a process that NRCS intends to apply to future EQIP regulatory decision processes.

The case study will involve examining four watersheds that have been identified through a locally-led process as EQIP priority areas. These watersheds will be selected based on the primary resource concern, the type and extent of livestock production in the watershed, and the availability of monitoring data in the watershed. All data will be aggregated such that individual field-level data are indistinguishable and confidential information thereby protected.

5.1.4.2 *Risk Assessment*: The risk assessment portion of the analysis will answer the following questions through the use of the techniques and monitoring endpoints identified below.

Issue 1: ***The type and magnitude of the risks to water, soil, human health, and law compliance associated with manure from livestock production.*** This analysis will be drawn from the original EQIP Ecological Risk Assessment (ERA) completed in February 1997, and serves as the basis for selecting the specific monitoring endpoints discussed below.

Issue 2: ***The magnitude of reduction in risks to water, soil, and human health from the implementation of EQIP as currently formulated on key monitoring variables.*** The study has been focused to address animal feeding operations (AFOs), an issue that is currently relevant to EQIP activities and which was referenced in the original ERA. NRCS recently modified its approach to nutrient management by changing from a nitrogen-based standard to a phosphorous-based standard. In addition, USDA and EPA are reviewing nutrient loading strategies for AFOs. Therefore, the proposed case study will use seasonal concentrations of nitrogen, phosphorus and pathogens in the selected watersheds as the endpoints against which to measure EQIP environmental risk reduction impacts. NRCS models will be modified to conduct this assessment using existing data from four watersheds.

Issue 3: ***The magnitude of reduction in risk to farmers (economic, liability) from the implementation of EQIP as currently formulated***. This evaluation will include an analysis of the likelihood of the AFO operators in the study-watersheds being subjected to liability for violating existing water quality standards for nitrogen, phosphorus and microbial concentrations (where standards exist). These liabilities, along with the economic impacts of the EQIP program will be analyzed as part of the cost-benefit analysis.

Issue 4: ***A comparison of the critical risk points as related to the criteria used by the State in which the subject watershed resides***. Each state selects priority areas for EQIP programs based on their own development of State criteria. The case study will examine the State criteria that resulted in the study-watershed being selected as a priority area and compare it to the environmental risks identified in the original ERA.

Issue 5: ***Areas in which changes in EQIP program implementation might be considered***. This aspect of the case study will include two types of analysis: a) a modeling study in the four case study-watersheds of the reduction in environmental risk (nitrogen, phosphorus and pathogen concentrations) as a result of moving to phosphorus uptake rate-based land application of manure; and b) a comparison of the concentration of livestock activity and changes in that activity since 1969 with the spending patterns of the predecessors to the EQIP program.

5.2 Case Study 2: *Enhancing Wildlife Values on Grasslands in the Conservation Reserve Program (CRP)*:

5.2.1 ***<u>Program Description</u>***: The Conservation Reserve Program (CRP) is the Federal government's single largest environmental improvement program, with over $1.7 billion in payments to farmers in fiscal year 1998. The CRP is designed to safeguard topsoil from erosion, increase wildlife habitat, and protect ground and surface water by reducing water runoff and sedimentation. The CRP is a voluntary partnership between individuals and government. Agricultural land owners remove highly erodible or other environmentally sensitive land from production and implement conservation practices on that land, usually for a minimum of ten years. In return, USDA pays participants an annual rent and half the cost of implementing the conservation practice. The vast majority (82%) of lands enrolled under the CRP are converted by the landowners to grasslands.

5.2.2 ***<u>Background and Objectives of Case Study</u>***: Grasslands are recognized by many as the most imperiled ecosystem worldwide. Native North American grasslands that once extended from Canada into Mexico and from the foothills of the Rocky Mountains to western Indiana and Wisconsin have dramatically declined in area. The largest percentage of grassland habitat occurs in the Central Grasslands, which covers the entire central portion of the United States. The Central Grasslands include the three commonly recognized grassland types - eastern tallgrass, central mixed-grass, and western shortgrass - as well as Sonoran/Chihuahuan desert grasslands. Grassland bird populations have shown steeper, more consistent, and more geographically widespread declines than any other behavioral or ecological guild of North American's Breeding Bird Survey (BBS) data from 1966.

The status of wildlife on CRP enrolled grasslands has been the subject of research undertaken by the USDA and other state and Federal wildlife management organizations. While there have been undeniable benefits resulting from the CRP program, there are other factors which have prevented the CRP program from achieving the full potential of increased wildlife benefit. These factors include weed control and haying on CRP lands. Others concerns are the possibility that the quality of wildlife habitat on unmanaged CRP grasslands will deteriorate over time due to the increased density of vegetation and litter. It has been suggested that controlled grazing, haying, light disking, and prescribed burning could be incorporated into a management plan for CRP lands to maintain the diversity and vigor of CRP grasslands and ensure the continuance of a high level of wildlife benefits. An increased focus on the role of CRP in wildlife habitat enhancement has motivated the Farm Service Agency (FSA) to undertake a collaboration with ORACBA on a case study to use the adaptive risk analysis protocol to examine the issue of managing CRP enrolled grasslands so as to maximize and maintain wildlife habitat values throughout the length of CRP contracts.

5.2.3 ***Case Study Progress To-Date***: This case study is still in its initial stages of development. To date, the program identification step has been essentially completed. Once approval for the plan is gained, the next step in the process will be to clearly define the risk assessment outline and focus. This will also mean revisiting the program identification step to ensure that resources are appropriately allocated to the project and that the risk assessment is adequately addressing the issues identified.

5.2.4 *Steps in the Adaptive Risk Analysis Process*

5.2.4.1 *Program Identification*: The primary goal of the case study is to evaluate the effects of various kinds of managed disturbances to CRP enrolled grasslands upon the objectives of reducing nutrient runoff and enhancing wildlife values. While lying outside the direct scope of the case study, an implicit objective is the quantification of wildlife benefits of various CRP practices so that they may be incorporated in a logical fashion into the determination of the Environmental Benefits Index (EBI), which is used to identify parcels of land most suitable for enrollement into the CRP program. While the case study can evaluate the suitability of quantitative measures of habitat quality, a successful adaptive management framework must link such measures to the expected environmental benefits to be achieved by the CRP program.

There are still some issues to be completed in this step, which will greatly effect, and be affected by, the risk assessment work. Major issues of the case study still to be answered involve the selection of particular grassland habitats for study and a description of the vegetation characteristics which will be used for the study comparisons. In addition, the scope of wildlife populations to be included must be determined. Most of the current and past research on CRP lands has focused on habitat suitability for birds. The feasibility of incorporating insects and small mammals into the study is also being considered.

5.2.4.2 Risk Assessment: Risk assessment scenarios for the effects of managed disturbances require the development of conceptual models linking the stressors of grazing, mowing, burning, etc., to potential ecological hazards and appropriate measurement and assessment endpoints. Risk scenarios are drawn from the proposed CRP management practices and will examine various potential ecological hazards by examining impacts on endpoints relating to water quality and habitat suitability for wildlife.

In order to conduct the risk assessment, several areas have been identified where information or data are needed. For example, data are needed which demonstrate the effects of alternative CRP grassland management practices, particularly light disking and haying, on grassland species, especially for mixed-grass and tallgrass prairies. Specific to mixed-grass and shortgrass prairie is the need to understand the effects of grasshopper control on avian communities. In tallgrass prairie and desert grasslands, there is a need to evaluate the effectiveness of habitat restoration activities. Issues surrounding the use of prescribed burning include the effects of block size for burning, frequency of burn, timing of burn, and the importance of providing refuge for wildlife displaced by burning. Fire and grazing represent a continuum in which grazing tends to assume greater importance as one moves west from tallgrass into mixed-grass and shortgrass, and southwest to the wintering grounds, while fire assumes greater importance as one moves east toward tallgrass prairie.

The primary issue is the degree to which current research and data collection efforts can provide sufficient information to characterize the temporal and spatial components of habitat quality. For example, can statistical models alone predict changes in wildlife habitat quality over time? To what extent can ecosystem simulation models be used to predict changes in habitat quality using different disturbance scenarios? Other major issues to be addressed are the selection and parameterization of grassland models, the modifications required to incorporate habitat suitability endpoints, the treatment of uncertainty, and the need to aggregate results in order to scale projections.

6. References

Commission on Risk Assessment and Risk Management. (1997): *Risk Assessment and Risk Management in Regulatory Decision-Making*. Environmental Protection Agency, Washington, D.C.

Meekhof, R, J Kuzma, D Mauriello, T Osborn, M Powell, C Rice, and S Shafer (1998): "Adaptive Risk Analysis for Resource Conservation Programs." *Proceedings of the Eighth Engineering Foundation Conference on Risk Based Decision Making in Water Resources VIII*, October 12-17, 1997, Santa Barbara, California.

U.S. Environmental Protection Agency (1992): *Framework for Ecological Risk Assessment*, EPA/630/R-92/001. Risk Assessment Forum, Office of Research and Development, Washington, D.C.

U.S. Environmental Protection Agency (1996): "Draft Proposed Guidelines for Ecological Risk Assessment", Federal Register Vol. 61, No. 175, Monday, September 9, 1996, pp. 47552-47631.

Fuzzy Multi-objective Reconstruction Plan for Post-earthquake Road-network by Genetic Algorithm

Gwo-Hshiung Tzeng [1] Yuh-Wen Chen [1] and Chien-Yuan Lin [2]

[1] Energy and Environmental Research Group, Institute of Traffic and Transportation, and Institute of Information Management, National Chiao Tung University, 4F-114, Sec. 1, Chung Hsiao W. Rd., Taipei 100, Taiwan.

[2] Graduate Institute of Building and Planning, National Taiwan University, Taipei 106, Taiwan.

Abstract- According to the seismic experience of Japan and America, earthquakes have often caused damage to the road-networks, which are important for maintaining the quality of life and the daily transit after a disaster. Taiwan and Japan are both located in the Pacific earthquake region, which is very active and unstable. The Taiwanese people will suffer seriously after a large-scale earthquake because the population and road-network are both highly concentrated nowadays. If the necessary reconstruction strategies to cope with quakes are not available, mass travelers can't be efficiently conducted via the post-earthquake road-network. Thus, the convenience of transit after earthquake would be seriously hampered. To aid the reconstruction decision for post-earthquake road-networks, we intend to establish a fuzzy multi-objective model, which is an integration of work scheduling and task assignment for many work-troops. Multi-objective optimization is applied because of the following reasons: first, we do want to minimize the travel-time of travelers during reconstruction; secondly, we intend to minimize total time needed for reconstruction; furthermore, we also expect that each available and homogeneous work-troop on duty will share almost the same work-load during reconstruction. Since the aspiration level of the aforementioned goals are vague, a fuzzy multi-objective approach is used. The algorithm of this combinatorial optimization problem based on a two-step genetic algorithm is then developed and employed to reduce the computation complexity of such a problem. Study results show that a satisfying solution of this problem can be efficiently derived by thirty generations of our modified genetic algorithm-- this solution not only instructs the reconstruction order for each damage point in road-network, but also assign the appropriate reconstruction work to relevant work-troops. Thus, for reasons of computational efficiency and practical applicability in this study, we do strongly suggest this research can't only be a basis for seismic simulation but can also be the reference of pre-quake exercises for relevant authorities.

Keywords: Fuzzy multi-objective optimization, network, earthquake, genetic algorithm (GA).

1 Introduction

Taiwan. Japan and America are all located in the earthquake region, which is very active and unstable. For the sake of transportation effectiveness and safety of the post-earthquake road-network in Taiwan, a well considered reconstruction plan for the road network has become a must.

To design an appropriate reconstruction plan for the seismic road-network, we should firstly take the damage characteristics of seismic road-network into consideration. Many damage points will be generated and scattered throughout the post-earthquake road-network– this real situation of Northridge earthquake in America is shown in Fig. 1.

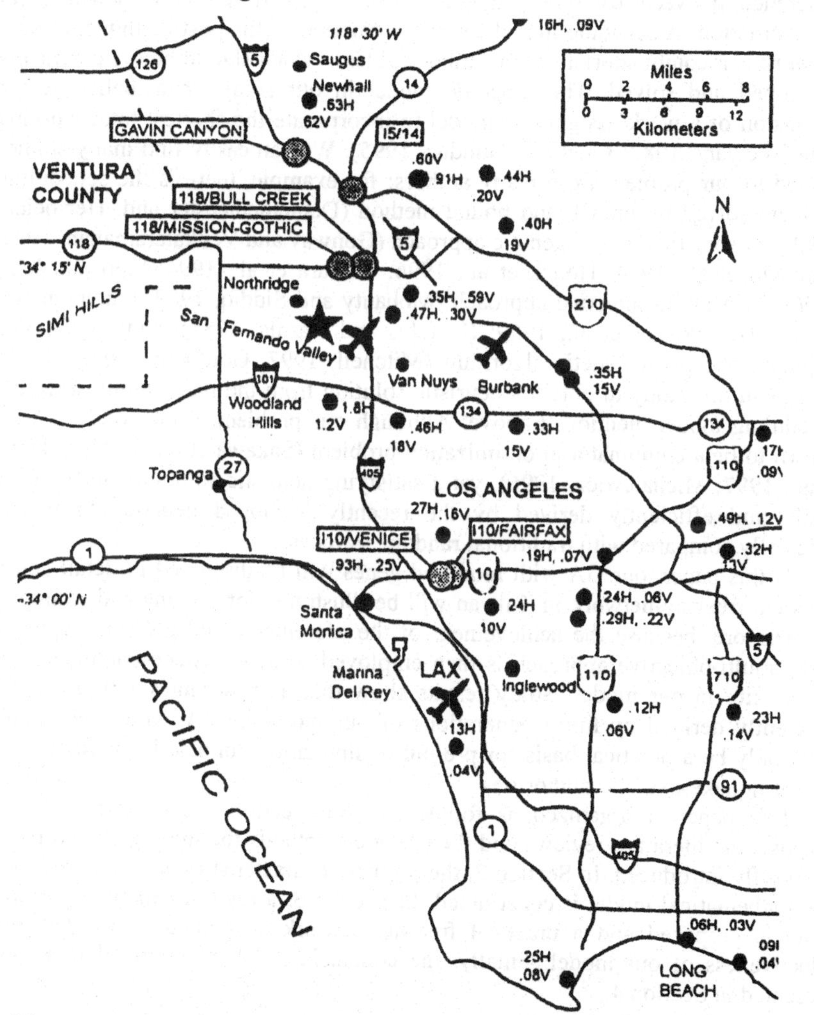

Fig. 1 A Real Situation of Post-earthquake Road-network in Northridge

Fig. 1 is copied from the special report of the 1994 Northridge earthquake (Finn et al., 1995), where the black spot denotes the scale of the earthquake measured by local quake-meters, and the gray spot shows the slump area of the road-network.

Thus, the reconstruction task for each damage point should be reasonably assigned to each homogeneous work-troop on duty so as to maximize the reconstruction effectiveness of a post-earthquake road-network. The reconstruction effectiveness can be maximized when the following goals are achieved: (a) minimizing the total reconstruction time, (b) maximizing the convenience for travelers during reconstruction; (c) minimizing the work-load difference between each homogeneous work-troop which is responsible for reconstruction. A schedule model for reconstruction of the post-earthquake road-network is a combinatorial optimization problem, which should be systematically developed and solved. The schedule model in our study, in actuality, is the expansion of a work assignment model to incorporate the shortest total working time (Weglarz, 1980; Linet and Gunduz, 1995). We can easily find many studies related to our problem from many authors; for example, there is the scheduling problem solved by branch and bound method (Demeulemeester and. Herroelen, 1992; Stinson, 1979), by a genetic approach (Gonway and Venkataramanan, 1994; Ishibuchi et al., 1994; Houck et al., 1996; Tamura et al., 1994; Sato and Ichii 1996), by a multi-objective approach (Mohanty and Siddiq, 1989; Patterson and Roth, 1976; Prisker et al., 1969; Kurtulus and Narula, 1985), and so on. The original concept of genetic algorithm (Mitchell, 1997; Goldberg, 1989) is also applied in this study to obtain a heuristic solution from the fuzzy multi-objective scheduling model mentioned above. Although our problem in this study is well known to be a combinatorial optimization problem (Sakawa et al., 1994; Sakawa et al., 1997; Michalewicz, 1996), yet a satisfying heuristic solution can be more easily and efficiently derived by the recently-developed genetic algorithm– especially compared with traditional random searches.

In this paper, our GA with modified genes will be discussed in detail and a practical freeway-network in Taiwan will be illustrated for a numerical example. Furthermore, because the achievement of the aforementioned goals is vague, a fuzzy multi-objective approach is then employed so as to reduce computational complexity in our model. Study results show that a satisfying solution can be efficiently derived by thirty generations of our modified GA. Thus, this study can't only be a practical basis for pre-quake simulation but also be available for the earthquake-exercise authorities.

This paper is organized as follows: in this section, the background, the purpose, the literature reviews and the previous methods for solving our problem are briefly introduced. In Section 2, the problem characteristics are described and the mathematical model is constructed. In Section 3, a revised genetic algorithm (GA) is proposed and a practical freeway-network is applied to validate the effectiveness of our model. Finally, the conclusions and recommendations are presented in Section 4.

2 Problem Description and Model Construction

According to the seismic experience of Japan and America, the quake disaster varied with time can be simply divided into three periods: the first period can be defined as a chaos period, which begins as the earthquake suddenly occurs. At the same time, the life-lines: e.g. the computer-networks, electric power lines, telephone lines, road-networks, etc. will be seriously destroyed. This chaos period ends as the instant rescue activities are proposed. The second period can be defined as a rescue period, which allows fire-extinguishing, emergency rescue for lives, etc. according to damage surveys. The third period can be defined as a restoration period, which begins after the necessary rescue for lives are adequately provided, and a long-term reconstruction plan should also be drawn out and executed in this period. This study is trying to establish a simulation model so as to efficiently design a schedule for road-network reconstruction during the restoration period. Moreover, our model may be installed on a portable computer to be a radioactive command center– this will prevent the command difficulty through a line communication (e.g., telephones) after a large-scale earthquake.

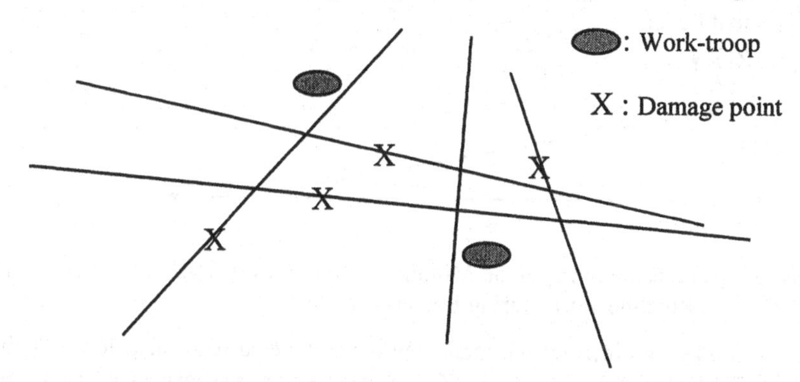

Fig. 2 An Abstracted Situation of Post-earthquake Road-network

After a large-scale earthquake occurs, there might be many damage points scattered throughout a road-network (see Fig. 1)– this abstracted situation is shown in Fig. 2 to clarify our formulation concept. In the mean time, suppose that there are also a few homogeneous work-troops available for reconstructing damage points– this condition is also shown in Fig. 2. In order to reconstruct each damage point by each work-troop which is eventually constrained in resources, the guide of reconstruction for each work-troop must be arranged appropriately so as to save total reconstruction time, travel-time for travelers and balance work-load between homogeneous work-troops during reconstruction. The guide of reconstruction should not only instruct each work-troop how many damage points should be tackled but also in what order these damage points are to be reconstructed. Furthermore, the work-troops for reconstruction only can arrive at any damage point by a well-reconstructed or non-decimated route.

The travel-time in a post-quake road-network will be sharply increased and unacceptable when compared with a normal traffic condition, but this post-quake travel-time can be slowly reduced if appropriate reconstruction activities are executed. Thus, we use the first objective (Z_1) to reflect the convenience of traveling by minimizing total travel-time of all travelers in the road-network during reconstruction. This objective is measured by summing up the product of the traffic volume by travel-time in each link during reconstruction. To construct fuzzy membership functions of our objectives, the relevant optimistic and pessimistic values in Equations (1)-(3) are all consulted from experienced engineers in the transportation-safety department of the Institute of Transportation (IOT) in Taiwan. The achievement level of Z_1 can be expressed as a membership function of $\mu_H(Z_1)$, which is graphically shown in Fig. 3, where H indicates that Z_1 belongs to the membership function of hours (H: hours).

$$\begin{cases} \mu_H(Z_1) = 0, & \text{if } Z_1 \geq 100000 \text{ hours} \\ \mu_H(Z_1) = (100000\text{-}Z_1)/100000, & \text{if } 100000 \text{ hours} \geq Z_1 > 0 \text{ hours} \\ \mu_H(Z_1) = 0, & \text{if } Z_1 \leq 0 \text{ hours} \end{cases} \quad (1)$$

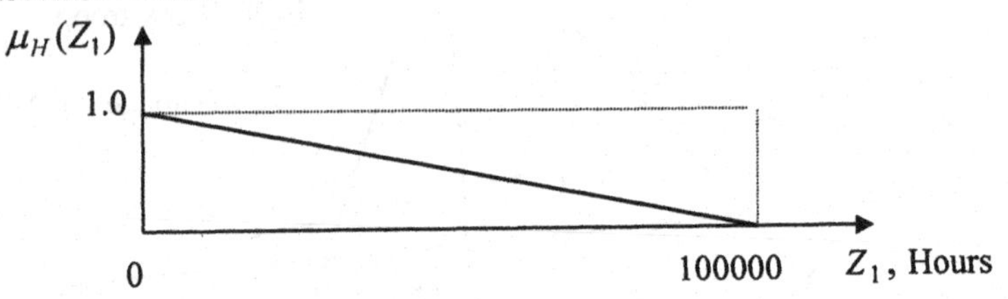

Fig. 3 Achievement Level of Minimizing the Travel-time for Travelers in a Road-network During Reconstruction

Secondly, minimizing the total time for completed reconstruction is defined as Z_2. The achievement level of Z_2 can also be similarly expressed as a piece-wise membership function of $\mu_D(Z_2)$, which is shown in Equation (2) and as a graph in Fig. 4, where D indicates that Z_2 belongs to the membership function of days (D: days).

Thirdly, the final fuzzy objective (Z_3) is designed to minimize the work-load difference between any two work-troops. This work-load difference is measured by the longest time spent by a work-troop on heavy duty minus the shortest time spent by another work-troop on light duty. The work-load difference between the longest working time and the shortest working time within any two work-troops should be as small as possible. The achievement level of Z_3 can also be similarly expressed as a piece-wise membership function of $\mu_D(Z_3)$, which is shown in Equation (3) and as a graph in Fig. 5, where D indicates that Z_3 belongs to the membership function of days (D: days).

$$\begin{cases} \mu_D(Z_2) = 1, & \text{if } Z_2 \le 60 \text{ days} \\ \mu_D(Z_2) = (80 - Z_2)/20, & \text{if } 70 \text{ days } \ge Z_2 > 60 \text{ days} \\ \mu_D(Z_2) = (90 - Z_2)/40, & \text{if } 90 \text{ days } \ge Z_2 > 70 \text{ days} \\ \mu_D(Z_2) = 0, & \text{if } Z_2 \ge 70 \text{ days} \end{cases} \quad (2)$$

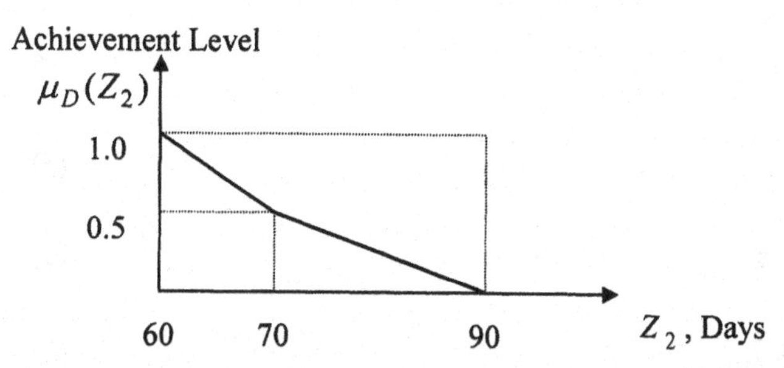

Fig. 4 Achievement Level of Minimizing the Total Time Needed for Reconstruction

$$\begin{cases} \mu_D(Z_3) = 1, & \text{if } Z_3 \le 10 \text{ days} \\ \mu_D(Z_3) = (60 - Z_3)/60, & \text{if } 60 \text{ days } \ge Z_3 > 10 \text{ days} \\ \mu_D(Z_3) = 0, & \text{if } Z_3 > 60 \text{ days} \end{cases} \quad (3)$$

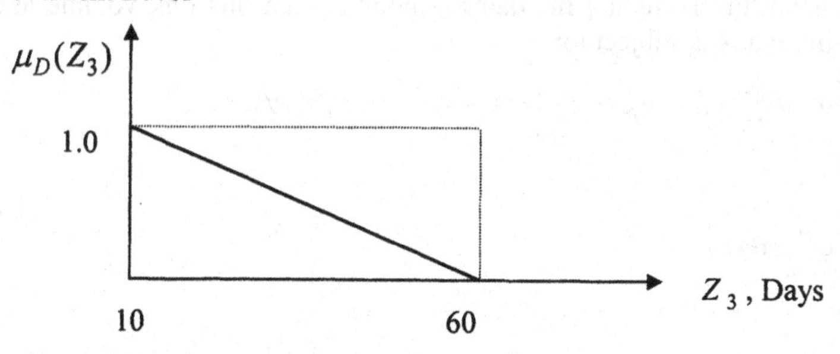

Fig. 5 Achievement Level of Minimizing the Work-load Difference between any Two Work-troops

Finally, we assume that a work-troop is continuously reconstructing an assigned damage point till this point is well-reconstructed; therefore, our multi-objective model in this study can be shown in Equations (4) - (13):

$$\left.\begin{aligned} Min\ \ Z_1 &= \sum_{i\in o}\sum_{j=1}^{n}\sum_{k=1}^{h}\sum_{l=1}^{nl} Q_k^l(x_{ijk})\times t_k^l(x_{ijk}) \\ Min\ \ Z_2 &= \ Max_j\,\{\,RT_j\,\} \\ Min\ \ Z_3 &= \ Max_j\,\{\,RT_j\,\} - Min_j\,\{\,RT_j\,\} \end{aligned}\right\} \tag{4}$$

s.t

$$\text{if}\ \ x_{ijk}\neq 0,\ \text{then}\ \sum_{j=1}^{n} x_{ijk}=1\ ,\ \ \forall\, i,\ \ \forall\, k \tag{5}$$

$$\sum_{k=1}^{h}\ \sum_{i\in o} x_{ijk} \geq WT_i^j\ \ ,\ \ \forall j \tag{6}$$

$$\sum_{k\in u} x_{ijk} = WT_i^j\ \ ,\forall\, i,\forall\, j \tag{7}$$

$$\sum_{k=1}^{h}\sum_{i\in o}\sum_{j=1}^{n} x_{ijk} = \sum_{i\in o}\sum_{j=1}^{n} WT_i^j \tag{8}$$

$$\sum_{k=1}^{h}\sum_{i\in o}\sum_{j=1}^{n} y_{ijk} = \sum_{\substack{i,s\in o\\ i\neq s}}\ \sum_{j=1}^{n} TT_{is}^j, \tag{9}$$

$o = \{i|$ the damage point i is assigned to the work-troop $j\}$
$u = \{k|$ the damage point i is continuously reconstructed by the work-troop j within the time-interval $k\}$
$x_{ijk}\in\{0,1\};\ \ y_{ijk} = 1 - x_{ijk},\ \ \forall\, i,\ \ \forall j,\ \ \forall\, k$

if l is a damage link including the damage point i, then this link volume at any given time-interval k is subject to:

$$Q_k^l(x_{ijk}) = \begin{cases} 0,\ \ if \sum_{\tau=1}^{k}(x_{ij\tau}+y_{ij\tau}) \leq \sum_{\substack{i,s\in o\\ i\neq s}} WT_i^j + TT_{is}^j, & \forall i,\forall j,\forall k, \\ Q_k^{l*},\, otherwise & \end{cases} \tag{10}$$

but if l is a well link, then the link volume is subject to:

$$Q_k^l(x_{ijk}) = Q_k^{l*},\quad \forall k \tag{11}$$

and the varying travel-time of a well link l in each time-interval k is subject to:

$$t_k^l(x_{ijk}) = \beta_l\, L_l\, Q_k^l\,(x_{ijk}),\forall l,\quad \forall k \tag{12}$$

and for each work-troop j, the total working time is subject to:

$$RT_j = \sum_{i\in o}\sum_{k=1}^{h}(x_{ijk}+y_{ijk}),\quad \forall j \tag{13}$$

where:

i : damage points, $i = 1,2,\ldots,m$;

j : available work-troops for reconstruction, $j = 1,2,3,\ldots n$;

k : time-interval, $k = 1,2,\ldots,h$;

l : all physical links in a post-earthquake road-network, $l = 1,2,\ldots,nl$;

x_{ijk} : the decision variable, if damage point i is reconstructed by work-troop j at time interval k, then its value is 1, otherwise is 0;

y_{ijk} : the dummy variable, if $x_{ijk} = 1$, this value equals 0, otherwise is 1;

WT_i^j : the time needed for work-troop j to completely reconstruct damage point i;

TT_{io}^j : the travel time for work-troop j traveling from a well-reconstructed point i to another damage point o;

$t_k^l(x_{ijk})$: the travel-time function of link l at time-interval k, this function is related to the reconstruction state of any damage point i;

$Q_k^l(x_{ijk})$: the traffic-volume function of link l at time-interval k, this function is also related to the reconstruction state of any damage point i;

L_l : the length of link l;

Q_k^{l*} : the traffic volume of a well link l at time interval k, this volume is recomputed by asymmetric assignment when a damage point is completely reconstructed;

β_l : the translation coefficient between traveling-speed and traffic-volume of a well link l, where $L_l\, Q_k^{l*} / t_k^l = 1/\beta_l$, β_l is set to be 0.0001 for all physical links in this study;

RT_j : the total reconstruction time for work-troop j;

Eventually, the aforementioned model is integrated by the work scheduling and work assignment for reconstruction. The meanings of the equations are explained as follows:

Equation (5): any damage point is exactly reconstructed by only one work-troop;

Equation (6): any work-troop should at least reconstruct one damage point;

Equation (7): any damage point will be continuously reconstructed till this point is recovered;

Equations (8)-(9): total damage points should all be tackled during reconstruction;

Equations (10)-(11): a test rule established so as to take the recovered links into the first objective's (Z_1) consideration;

Equation (12): a function established based on the traffic-flow theory in order to compute the travel-time of all travelers for the first objective (Z_1);

Equation (13): a function established to compute the reconstruction time spent by each work-troop;

Equations (5)-(9) are modified structures of traditional transportation problems in operations research field (OR), but Equations (10)-(12) are established so as to dynamically catch the traffic variation in this study by traffic-flow theory. Furthermore, according to the concepts of λ transformation for fuzzy multi-objective approach (Martison, 1993; Sakawa et al., 1994, 1997), Equations (4)-(13) can be efficiently transformed in Equation (14):

$$\begin{aligned} &\underset{x_{ijk}}{Max}\ \lambda \\ &\text{s.t} \\ &\quad \lambda \le \mu_H(Z_1) \\ &\quad \lambda \le \mu_D(Z_2) \\ &\quad \lambda \le \mu_D(Z_3) \\ &\quad \text{and Equations (5)-(13)} \end{aligned} \tag{14}$$

After inputting the available data (e.g., traffic-volume, travel-time, locations of work-troops and damage points, etc.) to Equation (14), which will output a plan so as to instruct each work-troop for its own reconstruction task. This model can be solved by many approaches, which we discussed in the introduction. Since GA has been recently applied to many problems related to our study and reveals its potential power in such combinatorial optimization problems, we try to develop a revised GA with a lattice-gene in this study. In the next section, we will use a numerical example to illustrate how our GA works.

3.3 Numerical Example and Resolution

In this section, we use a practical freeway-network in Taiwan to illustrate how we collect required data for Equations (4)-(13) and how our GA works so as to validate the effectiveness of our model.

3.3.1 Data Input

Since the freeway-network of Taiwan had never been decimated by a large-scale earthquake like that in Japan. Thus, we only do our best to imagine and assume the most possible situation of a post-earthquake freeway-network in Taiwan. We firstly collect historical earthquake data from 1897 to 1996 from Central Weather Bureau (CWB). Secondly, we statistically analyze these data on a Taiwan map to check where big quakes are possible to occur– this assumed damage points are shown in Fig. 6– the number from "1" to "15" represents the damage points by our assumption. Thirdly, we survey available work-troops in Taiwan and locate their sites in Fig. 6– the symbol from "A" to "F" represents the location of each work-troop. Of course, all work-troops are simply assumed to be homogeneous in this study. Fourthly, the reconstruction time for a single damage point is estimated about 12~24 work-days by consulting civil engineers– not including the travel time form a work-troop to arrive at a damage point. The travel time between any two nodes are assumed from 1 to 24 hours.

Finally, we also survey the traffic volume between required nodes in our network– this data is shown in *Reference Table*. Moreover, to reflect the traffic variation during reconstruction, the traffic volume in each link is varied with two different states during reconstruction periods. These states are: (a) before-earthquake state; and (b) reconstruction-completed state. The traffic volume in each state is reasonably assigned to each link by equal-travel-time assumption, which means the traffic volume of a reconstruction-completed link will continuously increase until the travel time of this well-reconstructed link is equal to that of the nearby substituted routes (see Fig. 6).

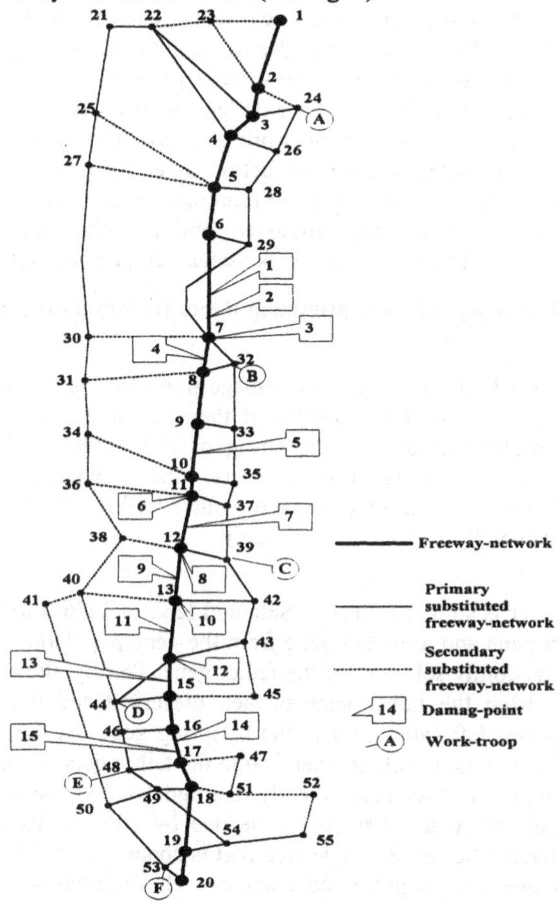

Fig. 6 An Assumed Situation of Post-earthquake Freeway-network

3.3.2 Model Resolution by Two-step Genetic Algorithm

To solve our mathematical model in Equation (14), a two-step approach is necessary. The first step is to decide the reconstruction order for each damage point, the second step then uses the results from the first step to assign the reconstruction task to each work-troop. Although there are many approaches developed and applied to solving our problem, we also use GA for three reasons:

(a) the concept of GA is easy to be understood for readers who lack a strong background in mathematics; (b) the GA is full of transferability– it means that the other problems which are similar to ours can be solved with almost the same steps for resolution; (c) our model is a NP-hard problem, and GA is proved to be effective in solving such a combinatorial optimization problem (Sakawa et al., 1994, 1997). To resolve the scheduling and work assignment for reconstruction, two different GAs are then developed– we will discuss these details in Section 3.3.2.1 and 3.3.2.2. The traditional and basic notion of gene type in GA is to translate decimal values into binary-digit value: such as decimal 2 is equivalent to binary 0010 (Goldberg, 1989). But considering the characteristics of our model, we use a lattice-gene to describe the work assigned to each work-troop and a string-gene used to express the reconstruction order for damage points. In any case, the computational spirit of genetic reproduction here is similar to that of Goldberg's, but the gene type, crossover, and mutation are re-defined for reconstruction order (scheduling) and work assignment, respectively as follows:

3.3.2.1 GA of String-gene for Optimizing Reconstruction Order

(a) Gene type

The reconstruction order for each damage point can be defined as a string-gene of Arabic numerals. For example, if there are 8 damage points:1,2,...,8 needed to be reconstructed, then the string of "13245687" indicates that the damage point 1 will be reconstructed first, the damage point 3 will be reconstructed secondly,...,and the damage point 7 will be reconstructed finally– this encoding way will be applied in this study.

(b) Crossover

Our crossover is similar to that of Sato and Ichii in 1996. First, we randomly choose a father gene and a mother gene from the gene population. Secondly, two cut-points are randomly selected for the father gene. Thirdly, the content between two cut-points from the father gene is then preserved and transmitted to the beginning location (left-hand side) of the offspring gene. At the same time, the mother gene's elements, which are identical to the father gene's preserved content are deleted from the mother gene. Finally, the undeleted elements of the mother gene are transmitted to the offspring gene one by one and from left to right. Furthermore, the mother gene's undeleted part is connected to the tail (right-hand side) of the father gene's preserved content. This aforementioned process is illustrated as follows:

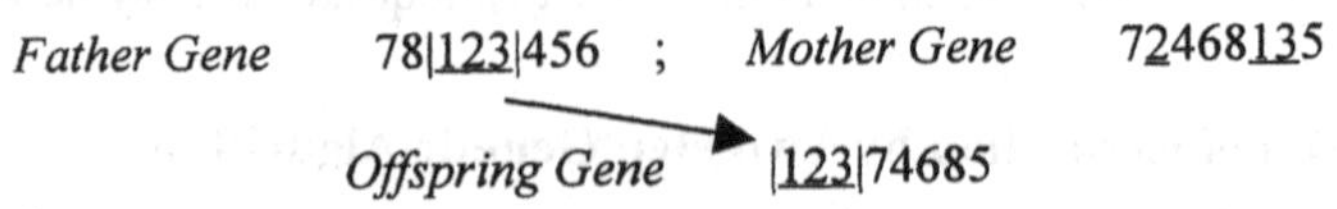

| : cut-point,

Note: 1, 2 and 3 are the same elements between parent genes. The sequence remained in the mother gene are 7, 4, 6, 8 and 5.

(c) Mutation

The mutation proceeds as follows: first, we randomly choose a gene from the gene population. Secondly, two cut-points are randomly selected from this gene and the content between cut-points is then preserved. Finally, the Arabic numerals preserved are inversely transmitted to the offspring gene one by one. This mutation can be illustrated as follows:

Parent Gene 78|123|456

↓

Offspring Gene 78|321|456

| : cut-point

3.3.2.2 GA of Lattice-gene for Optimizing Work Assignment

We also develop a lattice-gene in this study for work assignment, this GA is shown as follows:

(a) Gene type

In view of Fig. 6, there are 15 damage points (e.g., 1,2,...,15) needed to be reconstructed by 6 work-troops (e.g., A, B,...,F), then a reconstruction plan can be expressed by a lattice-gene as follows:

	Damage pt. 1	Damage pt. 2	...	Damage pt. 15
Work-troop A	1	1	...	0
Work-troop B	0	0	...	1
...	...	...	...	...
Work-troop F	0	0	...	0

The damage points in the first row are arranged by the scheduling result from GA for reconstruction order above (see Section 3.3.2.1). The emergency of reconstruction for damage points is listed in descending order from left to right– we assume the damage point 1 should be tackled first, then the damage point 2,..., finally the damage point 15 in the case above. Moreover, the value of “1” indicates a damage point is reconstructed by a specified work-troop; by contrast, the value of “0” indicates no task to be assigned. To match the assumption that each damage point is reconstructed by only one work-troop at the same time, there is at most one “1” in each column. For example, if work-troop A is assigned to tackle damage point 1 and 2, and damage point 1 should be reconstructed first, and damage point 2 secondly. And the other work-troops (e.g., B, C,...,F) would not be assigned to reconstruct damage point 1 and 2 again.

(b) Crossover

The two-point crossover for the lattice-gene is defined as follows: first, we randomly choose a father gene and a mother gene from the gene population. Secondly, two cut-points are randomly selected from the father gene as well as the mother gene. Finally, two offspring genes are formed when the father gene is partially replaced with the content, which is preserved between two cut-points of the mother gene; and when the mother gene is partially replaced with the content, which came from the cut part of the father gene. This crossover is shown as follows:

Father's Content between Two Cut-points

Father Gene

	Damage pt. 1	Damage pt. 2	...	Damage pt. 15
Work-troop A	0	1	...	0
Work-troop B	1	0	...	1
...	...	...	...	...
Work-troop F	0	0	...	0

Mother's Content between Two Cut-points

Mother Gene

	Damage pt. 1	Damage pt. 2	...	Damage pt. 15
Work-troop A	0	0	...	0
Work-troop B	1	0	...	0
...	...	...	...	...
Work-troop F	0	0	...	1

Crossover

Offspring Gene 1 (Mother Gene is Partially Replaced by Father's Cut Content)

	Damage pt. 1	Damage pt. 2	...	Damage pt. 15
Work-troop A	0	1	...	0
Work-troop B	1	0	...	0
...	...	...	...	...
Work-troop F	0	0	...	1

Offspring Gene2 (Father Gene is Partially Replaced by Mother's Cut Content)

	Damage pt. 1	Damage pt. 2	...	Damage pt. 15
Work-troop A	0	1	...	0
Work-troop B	1	0	...	1
...	...	...	...	...
Work-troop F	0	0	...	0

(c) Mutation

The mutation is defined as exchanging contents between rows and columns. Firstly, we randomly select two rows and two columns to mutate for a randomly selected gene. Considering the row mutation, the contents of two selected rows are then exchanged each other. For the column mutation, the contents of two selected columns are also exchanged each other. Two types of mutation in this study are shown as follows:

***Mutation* 1**

Parent Gene

	Damage pt. 1	Damage pt. 2	...	Damage pt. 15
Work-troop A	0	1	...	0
Work-troop B	0	0	...	1
...	...	...	...	...
Work-troop F	1	0	...	0

Row Mutation

	Damage pt. 1	Damage pt. 2	...	Damage pt. 15
Work-troop A	1	0	...	0
Work-troop B	0	0	...	1
...	...	...	...	...
Work-troop F	0	1	...	0

***Mutation* 2**

Parent Gene

	Damage pt. 1	Damage pt. 2	...	Damage pt. 15
Work-troop A	1	0	...	0
Work-troop B	0	1	...	1
...	...	...	...	...
Work-troop F	0	0	...	0

Column Mutation

	Damage pt. 1	Damage pt. 2	...	Damage pt. 15
Work-troop A	0	0	...	1
Work-troop B	1	1	...	0
...	...	...	...	...
Work-troop F	0	0	...	0

To sum up, the GA in Section 3.3.3.2 is developed and designed to exchange the task assignment between work-troops.

3.3.2.3 Optimizing the Reconstruction Order and Work Assignment

The aforementioned GA for scheduling and work assignment in this study will derive satisfying solutions after 30 generations by our experiment. A generation is defined as a process to make gene population undergo rank-selection once, crossover once and mutation once. Since there are two different GA modules developed in this study, this special optimization process is illustrated as follows:

***Step* 1**
Input the required data of Equations (4)-(13): for example, the data of traffic volume of each link, the travel time of each link, the travel time between required nodes, locations of damage points, locations of available work-troops, the reconstruction time of each damage point, etc.

***Step* 2**
Encode the reconstruction order by method in Section 3.3.2.1, then use the relevant crossover, mutation of GA to extend offspring genes– the initial population size of string-genes is set to 20 genes.

***Step* 3**
The reconstruction order is transmitted into the lattice-genes in Section 3.3.2.2 by the initial 20 string-genes from step 2. The population of lattice-genes is then expanded by relevant crossover and mutation in Section 3.3.2.2.

***Step* 4**
The achievement level of each lattice-gene will be evaluated by λ value in Equation (14), then the best 20 genes will be kept by rank-selection after *Step* 4 is terminated. The achievement level of each gene is defined as the smaller value between $\mu_H(Z_1)$, $\mu_D(Z_2)$ and $\mu_D(Z_3)$ of this gene by a fuzzy multi-objective approach.

***Step* 5**
The reconstruction order found in the final population of lattice-genes will be extracted to *Step* 2 to form a new population of string-genes, and the steps from 2 to 5 will be iterated until 30 iterations are reached.

3.3.3 Computational Results and Discussion

After 20 runs– each run includes aforementioned five steps, a heuristic solution of our model can be easily derived. In view of Fig. 7, the work-troop A should tackle the damage point 1 first, then the damage point 2; the work-troop B should reconstruct the damage point 3 first, then the damage point 4, finally the damage point 5. The reconstruction plans of the other work-troops can be recognized in a similar way. The total reconstruction time is 56 days with an aspiration level of 1.0; the cumulative travel time is 29,537 hours with an achievement level of 0.70 and the significant work-load difference between work-troops is 28 days with a achievement level of 1.0. Thus, this implies the global achievement level of three objectives is 0.70. Furthermore, our GA did show its power when we analyze the global achievement value (λ) between generations– this is shown in Fig. 8. This reconstruction schedule in Fig. 7 is practically available for relevant authorities for pre-quake exercises so as to reduce the negative impact of a transportation network.

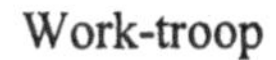

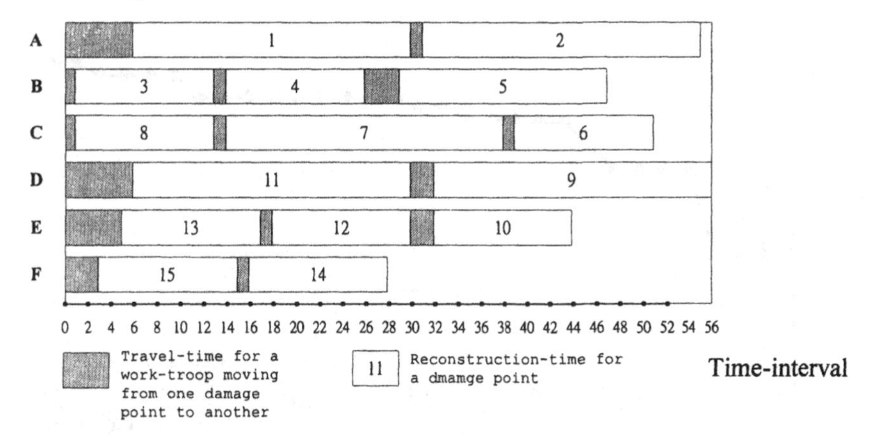

Fig. 7 A Heuristic Reconstruction Schedule for Each Work-troop

Thus, our model is reasonably constructed and practically validated by a freeway-network so far. Both the aspiration level of Z_2 and Z_3 are 1.0– this indicates that the acceptable threshold subjectively set up by us may be too optimistic for these two objectives. Our model did show its practicability and operational utility from Equations (4)-(13) and Fig. 7, 8. Since this study is developed by mathematical programming, with more precise data for our model, a reconstruction schedule can be scientifically made instead of messy decisions after a large-scale earthquake. Furthermore, since this effective reconstruction schedule is powerful and efficient, it can't only be a basis for pre-earthquake exercises of work-troops, but can also be potentially installed on a portable computer for real-time reaction of restoring damage road-networks by radioactive communication.

4 Conclusions and Recommendations

, This study explores and formulates the complex part of the restoration process for a post-disaster transportation network, and transforming traditional disaster-response concepts into an operable simulation model. Our efforts will help generate more objective, scientific and logical reinforcements for a transportation network against quakes. The earlier researchers, such as Tamura et al. (1994), Sato and Ichii (1996) only tried to find the optimal reconstruction sequence (priority) of a post-earthquake road-network without considering many conflicting objectives and available work-troops. Thus, we extend their researches to a general form so as to implement an efficient reconstruction plan. In view of this study, there are many advantages: firstly, it can shorten the total time needed for reconstruction, minimize the inconvenience of travelers during reconstruction and balance the work-load between work-troops. Since many economic activities

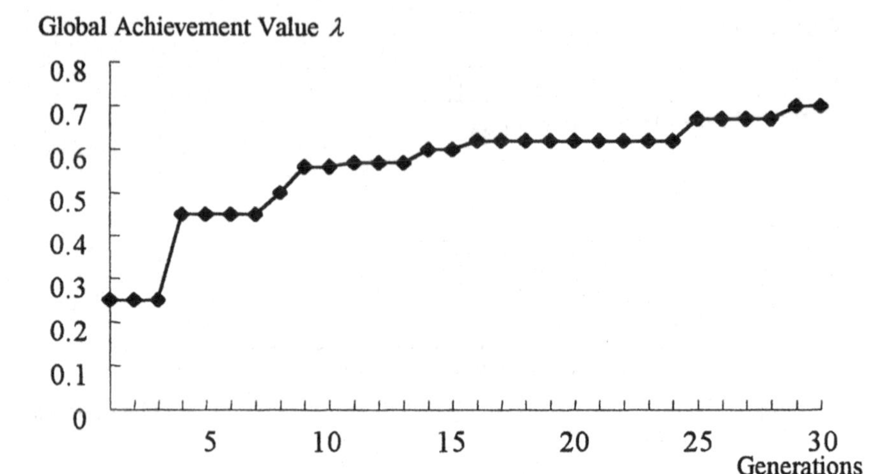

Fig. 8 The GA Evolution between Generations

are supported by road-network, the post-quake economic-productivity can be recovered as soon as possible by a well-considered reconstruction plan. Secondly, since the restoration process can be formulated by an optimization model, we can efficiently utilize the scarce resources for reconstruction and prevent resource wasting by optimizing techniques. Thirdly, an efficient reconstruction schedule can systematically command the work-troops instead of messy decisions; therefore, this will promote the reputation of post-earthquake reaction for the government. For these reasons, this study tries to establish a fuzzy multi-objective model for reconstruction scheduling and work assignment of a post-earthquake road-network. Because our model is essentially a combinatorial optimization problem with NP-hard complexity; thus, our modified GA with string-gene and lattice-gene is then developed and applied to show its power in this study. This formulation concept and primary results of this study are available for relevant authorities to predetermine and stimulate an effective reconstruction plan against quakes.

With more precise data and extension for our model in the near future, this study can be regarded as a basis of disaster decision support system (DDSS) for the restoration simulation in many disaster situations. Furthermore, the process of programming done in this study has being constructed on computerized software, which is now being developed with some improvements, such as considering the post-quake traffic assignment during reconstruction, combined with fuzzy time window and fuzzy working time, integrated with geographic information system (GIS), etc. We will continuously devote our efforts to improve this study so as to greater reflect real situations.

References

Demeulemeester, E. and Herroelen, W. (1992), "A branch-and bound procedure for the multiple resource-constrained project scheduling problem," Management Science, Vol. 38, No. 12, pp. 1903-1818.

Finn, W. D. L., Ventura C. E. and Schuster N. D. (1995), "Ground motions during the 1994 Northridge earthquake," Journal of Canada Civil Engineering, Vol. 22, pp. 300-315.

Goldberg, D. (1989), Genetic algorithms in search, optimization and machine learning, Addison Wesley Publishing Co., Massachusetts.

Gonway, D. G. and Venkataramanan, M. A. (1994), "Genetic search and the dynamic facility layout problem," Computers and Operations Research, Vol. 21, No. 8, pp. 955-960.

Houck, C.R., Joines, J. A., and Kay, M. G. (1996), "Comparison of genetic algorithms, random restart and two-opt switching for solving large location-allocation problems," Computers and Operational Research, Vol.23, No.6, pp. 587-596.

Ishibuchi, H., Yamamoto, T., and Tanaka, H. (1994), "Genetic algorithms and neighborhood search algorithms for fuzzy flow-shop scheduling problems," Fuzzy Sets and Systems, Vol. 67, No. 3, pp. 81-100.

Kurtulus, I.S., and Narula, S. C. (1985), "Multi-project scheduling analysis of project performance," IIE Transactions, vol. 24, No. 1, pp. 58-66.

Linet, O. and Gunduz, U. (1995), "A survey on resource-constrained project scheduling problem," IIE Transactions, Vol. 29, No. 5, pp. 574-586.

Martison, F. K. (1993), "Fuzzy vs. Minmax Weighted Multiobjective Linear Programming: Illustrative Comparison," Decision Sciences, Vol. 24, No. 4, pp. 809-824.

Michalewicz, Z. (1996), Genetic algorithms + Data Structures = Evolution Programs, Springer-Verlag Press, Berlin.

Mitchell, M. (1997), An Introduction to Genetic algorithms, MIT Press.

Mohanty, R. P. and Siddiq, M. K. (1989), "Multiple project-multiple resource constrained scheduling: a multiobjective analysis," Engineering Costs and Production Economics, Vol. 18, No. 1, pp. 83-92.

Patterson, J. H. and Roth, G. W. (1976), "Scheduling a project under multiple resource constraints: A zero-one programming approach," AIIE Transaction, Vol.8, No.4, pp. 449-455.

Prisker, A. A. B., Watters, L. J. and Wolfe, P. M. (1969), "Multiproject scheduling with limited resources: A zero-one programming approach," Management Science, Vol. 16, No. 1, pp. 93-108.

Sakawa, M., Kato, K., Sunada, H. and Shibano, T. (1994), "Fuzzy Programming for Multiobjective 0-1 Programming Problems through Revised Genetic Algorithms," European Journal of Operational Research, Vol. 97, No. 2, pp. 149-158.

Sakawa, M., Inuiquchi, M., Sunada, H. and Sawada, K. (1994), "Fuzzy Multiobjective Combinatorial Optimization through Revised Genetic Algorithm," Journal of Japan Society for Fuzzy Theory and Systems, Vol. 6, No. 1, pp. 177-185.

Stinson, J. P. (1979), "Multiple resource-constraint scheduling using branch and bound," AIIE Transaction, Vol. 10, No. 3, pp. 252-259.

Tamura, T., Sugimoto, H. and Kamimae, T. (1994), "Application of genetic algorithms to determining priority of urban road improvement," Japan Society of Civil Engineers, No. 482/IV-22, pp. 37-46

Sato, T., and Ichii, K. (1996), "Optimization of post-earthquake restoration of lifeline networks using genetic algorithms," Japan Society of Civil Engineers, No. 537/I-25, pp. 245-256.

Weglarz, J. (1980), "Control in resource allocation systems," Foundation of Control Engineering, Vol. 5, No. 3, pp. 159-180.

Reference Table

The Daily Traffic Volume between Nodes in Fig. 6

Node	1	2	3	4	5	6	7	8	9	10	11	12	13	14	15	16	17	18	19	20
1		1,844	4,502	2,006	1,397	1,315	367	555	89	403	298	168	159	115	178	57	50	117	244	184
2			4,542	2,039	1,442	1,336	403	765	105	429	317	184	176	135	185	64	56	133	229	138
3				6,840	4,712	5,033	1,037	1,998	275	1,150	792	506	448	676	500	172	129	325	677	371
4					4,163	5,296	1,029	2,063	308	1,214	895	1,338	433	670	1,039	188	186	424	631	535
5						3,212	993	1,429	234	1,018	571	853	310	413	687	226	158	444	386	332
6							577	1,270	125	595	469	513	469	356	384	137	102	253	419	230
7								1,083	239	872	385	642	398	607	768	141	229	330	355	417
8									176	763	599	531	430	475	702	253	195	486	361	487
9										564	445	406	454	608	627	207	158	410	691	334
10											912	1,356	798	1,157	1,766	476	339	888	1,486	1,123
11												780	616	602	909	320	255	511	801	737
12													1,050	1,601	1,739	550	421	1,105	1,944	1,215
13														1,067	1,608	561	327	719	1,347	909
14															5,228	1,918	1,425	3,607	4,085	3,086
15																1,649	1,157	3,173	5,892	2,583
16																	512	1,305	2,286	1,649
17																		3,143	5,410	3,877
18																			14,577	9,671
19																				12,334
20																				

Fair Allocations using Multicriteria Power Indices

R. C. van den Honert[1]

[1] Department of Statistical Sciences, University of Cape Town, Rondebosch 7701, South Africa

Abstract. In an allocation problem to determine a fair distribution of some divisible benefit or cost amongst members of a group, there are often a number of input criteria which may be used to make the allocation. We propose the weighted multicriteria power index, which combines traditional voting power in committees with the concept of criteria weighting, to yield a fair allocation. The multicriteria power index can be refined to allow for a moderation or amplification of the voting power of the members of the group, which in the limit would lead to parity or priority (respectively) of the members' power. Two well-documented power indices (the Shapley-Shubik and Banzhaf-Coleman power indices) are extended to operate under multiple criteria, and are used in an illustrative example.

Key Words. Salience, multicriteria power index, allocation of a divisible commodity

1 Introduction

One of the strengths of discrete multicriteria decision analysis methodology as it has developed is the requirement to structure the problem in a hierarchical manner, i.e. to allow the decision maker to think about the objectives, the attributes or criteria, and finally the range of alternatives under consideration. In a group decision making problem (and assuming a consensus-seeking group) this approach lets all the players see which aspects are going to impact on the final decision, and how.

But exactly how each group member will react to each alternative, and hence to contributing to an overall group decision, depends to a large extent on the *salience* (or relative prominence) afforded to each criterion by that group member. This naturally implies some sort of *power gradient* across the group members under each criterion, meaning that the group members will have differing powers to influence the overall decision. Indeed this has long been recognised, and has been explicitly included in a number of multicriteria decision making (MCDM) methodologies and software (for example, the AHP and the associated software, Expert Choice).

Actual determination of each group member's salience or power to influence the overall decision is probably done implicitly in many group decision processes anyway. However we advocate that paying attention to this will focus the search for creative alternatives, and will be more likely to arrive at a consensus-based decision.

In this paper we will examine the concept of a player's power as a multidimensional (or multicriteria) variable.

2 Measurement of Power - The Power Index

The analysis of power has received much attention in the literature in recent years, both from a methodological viewpoint and in applied problems (see, for example, Gambarelli (1990), Holler and Li (1995), Holubiec and Mercik (1994), Roth (1988), Turnovec (1996) and Widgrén (1994,1995)). In the current paper we will borrow from co-operative n-person game theory (see, for example, Owen (1982)) the concept of a *power index* for the members of the decision making group. A power index effectively assigns to each group member a proportion of the total power (where total power is normalised to, say, 1), based on (1) the members' weight allocations according to some criterion, and (2) some "rule" which will determine the members' contribution to forming successful decision coalitions with other group members.

In this context we can define an n-person consensus-seeking group as an (n+1)-tuple

$$(w_1\ w_2\ \ldots.\ w_n \mid \gamma)$$

where $w_1\ w_2\ \ldots.\ w_n$ are the weight allocations of the n group members under some criterion (normalised, so that $\Sigma w_i = 1$ and $w_i \geq 0\ \forall i$), and $0 \leq \gamma \leq 1$ is some *quota*, which if exceeded by some coalition of the group members' weights, implies "sufficient consensus" having been reached amongst group members.

Now let Γ be the set of all such n-person groups, i.e.
$\Gamma = \{(w_1\ w_2\ \ldots.\ w_n \mid \gamma) \in \mathbb{R}_{n+1} \mid \Sigma w_i = 1,\ w_i \geq 0,\ 0 \leq \gamma \leq 1\}$ and let $\mathscr{E}$ be the unit simplex, i.e. $\mathscr{E} = \{e \in \mathbb{R}_n \mid \Sigma e_i = 1,\ e_i \geq 0\ \forall i\}$. Then a *power index* is a vector-valued function

$$\wp : \Gamma \rightarrow \mathscr{E}$$

So let $\wp_i(W \mid \gamma)$ stand for the power index of group member *i* in an n-person group where the weight allocation is given by the set $W = (w_1\ w_2\ \ldots.\ w_n)$. This is a reasonable expectation of the share of decisional power vested in each group member, given by each group member's ability to contribute to forming successful decision coalitions.

Allingham (1975) introduced an axiomatic characterisation of power indices, these being a minimal intuitively acceptable set of properties which should be satisfied by any reasonable power index $\wp$. These axioms are as follows:

Axiom 1: (*Dummy*)
If a group member i cannot contribute to forming any successful decision coalitions then i has no decisional power. (By "successful decision coalitions" we mean that without member i we do not have sufficient consensus in the group, but with member i there is sufficient consensus).

Axiom 2: (*Anonymity*)
The decisional power of group member i remains the same, regardless of the ordering of the group members.

Axiom 3: (*Symmetry*)
If two group members i and j are symmetric i.e. their benefit to any decision coalition is the same), then i and j have equal decisional power.

Axiom 4: (*Local Monotonicity*)
A group member i with weight allocation greater than that of some other group member j cannot have less decisional power than group member j.

The most well-known power indices are the Shapley-Shubik power index (Shapley and Shubik, 1954) and the Banzhaf-Coleman power index (Banzhaf (1965) and Coleman (1971)).

The Shapley-Shubik power index for group member j is computed as

$$P_j^{SS}(W|\gamma) = \sum_{r=1}^{n} \frac{(r-1)!(n-r)!}{n!} \cdot [[C_{jr}(W|\gamma)]]$$

where $[[C_{jr}(W|\gamma)]]$ is the number of coalitions of size r in which group member j is *critical* (in the sense that by joining this coalition group member j ensures that the coalition has imparted "sufficient" consensus on the group, which was not the case without this member).

The Banzhaf-Coleman power index is calculated as

$$P_j^{BC}(W|\gamma) = \frac{[[C_j(W|\gamma)]]}{\sum_{k=1}^{n} [[C_k(W|\gamma)]]}$$

where $[[C_j(W|\gamma)]]$ is the number of coalitions of *any* size in which group member j is critical.

These two power indices can be shown to satisfy the four axioms (see, for example, Turnovec, 1998), and can thus be considered "good" measures of power in a decision making group. [There are other power indices which are *incomplete* in the sense that they violate one or more of Allingham's axioms. An example of this is the Holler-Packel power index (Holler (1978) and Holler and Packel (1983)), which violates the axiom of local monotonicity (see Turnovec, 1998)].

2. The Multicriteria Power Index

2.1 The Multicriteria Nature of Decisional Power

The following example illustrates the fact that decisional power can often be deemed to be multicriteria in nature. Consider the case of a game reserve traversed by a river. Upstream of the game reserve the landscape is rural/agricultural. Abstractors of water from the river are several: large scale farming activity in the vicinity use water for irrigation purposes, there is a nearby forestry industry which extracts water for use in their sawmills, and rural dwellers (who have resided in the area for generations) use water for subsistence farming. Since rainfall in the area is very seasonal and unpredictable the flow in the river is too. To ensure a more constant supply of water to all abstractors the Department of Water Affairs some years ago had a dam constructed in the river upstream from the game reserve. To capitalise on this, a large electricity utility successfully negotiated to install hydroelectric power generators in the dam. However the dam had the effect of altering the natural flow of the river downstream of the wall, which in turn upset the natural ecological systems and habitats downstream (i.e. in the game reserve). A decision making group was convened, whose objective was to plan a water release/discharge policy or strategy for the dam. All interested and affected parties were invited to join the group: the abstractors (farmers, forestry industry, rural dwellers); the electricity utility; ecologists, wildlife experts and environmental groups; and the Department of Water Affairs, in their capacity as "problem owners".

Clearly the different group members derived decisional power or salience from different sources (or criteria). For example, the farming community added to the regional economy through their agricultural produce, and were large-scale employers of the local indigenous population. The objective of the ecologists and other environmental groups was to ensure sustainable natural habitats, which in turn would lead to sustainable viability of the game reserve. This would result in continued (or increased) ecotourism to the area, thus also adding to the regional economy. This group also provided an unquantifiable "feel good" factor by ensuring survival of natural species for the enjoyment of future generations. The Department of Water Affairs, as "problem owners", had eventual absolute control over the water release/discharge patterns: their decisional power derived from this overall control.

In cases like this it is necessary to define a *multicriteria power index* to represent an overall measure of the decisional power vested in each group member across all relevant criteria. Thus for criteria $i=1,2,\ldots,m$ we define the multicriteria power index of group member $j=1,2,\ldots,n$ as

$$\Pi_j(\bar{W}|\bar{\gamma}) = \sum_{i=1}^{m} c_i P_{ij}(w_i|\gamma_i) \tag{1}$$

where c_i is the normalised weight allocated to criterion i (with $\sum_{i=1}^{m} c_i = 1$ and $c_i \geq 0$), and $P_{ij}(w_i|\gamma_i)$ is the (unicriterion) power index of group member j under criterion i. Note that $\Pi_j(\bar{W}|\bar{\gamma})$ will in general need to be normalised.

2.2 Properties of Multicriteria Power Indices

In this section we examine the properties of the multicriteria power index.

Since $\Pi_j(\bar{W}|\bar{\gamma})$ is merely a weighted average of the $P_{ij}(w_i|\gamma_i)$ it follows that $\Pi_j(\bar{W}|\bar{\gamma})$ is also a power index (after normalisation). We show now that providing $P_{ij}(w_i|\gamma_i)$ satisfies the Allingham axioms, then $\Pi_j(\bar{W}|\bar{\gamma})$ will satisfy multicriteria equivalents of these axioms. Note that proofs of the following lemmas depend critically on the arguments of Turnovec (1998).

Lemma 1: (*Dummy*) If group member j does not benefit any decision coalition for any criterion i by joining such coalition, then group member j has no multicriteria power, i.e. $\Pi_j(\bar{W}|\bar{\gamma}) = 0$.

Proof: Since $P_{ij}(w_i|\gamma_i) = 0 \;\; \forall i=1,\ldots,m$ (Turnovec, 1998) we have that

$$\Pi_j(\bar{W}|\bar{\gamma}) = \sum_{i=1}^{m} c_i \, . \, 0 = 0$$

Lemma 2: (*Anonymity*) If a decision making group is permuted identically for each criterion $i=1,\ldots,m$, then the power of group member k remains the same regardless of his position in the group, i.e. $\Pi_k(\bar{W}|\bar{\gamma}) = \Pi_{\hat{k}}(\bar{W}|\bar{\gamma})$ where $\hat{k}$ is the position of group member k under the permutation.

Proof: By definition

$$\begin{aligned}\Pi_k(\bar{W}|\bar{\gamma}) &= \sum_{i=1}^{m} c_i P_{ik}(w_i|\gamma_i) \\ &= \sum_{i=1}^{m} c_i P_{i\hat{k}}(w_i|\gamma_i) \qquad \text{(Turnovec, 1998)} \\ &= \Pi_{\hat{k}}(\bar{W}|\bar{\gamma})\end{aligned}$$

Lemma 3: **(*Symmetry*)** If two group members f and g are symmetric insofar as their benefit to any decision coalition is the same for each criterion i, then f and g have equal multicriteria power, i.e. $\Pi_f(\bar{W}|\bar{\gamma}) = \Pi_g(\bar{W}|\bar{\gamma})$.

Proof: By definition

$$\Pi_f(\bar{W}|\bar{\gamma}) = \sum_{i=1}^{m} c_i P_{if}(w_i|\gamma_i)$$

and

$$\Pi_g(\bar{W}|\bar{\gamma}) = \sum_{i=1}^{m} c_i P_{ig}(w_i|\gamma_i)$$

Since $P_{if}(w_i|\gamma_i) = P_{ig}(w_i|\gamma_i)$ (Turnovec, 1998) we have that

$$\Pi_f(\bar{W}|\bar{\gamma}) = \Pi_g(\bar{W}|\bar{\gamma})$$

Lemma 4: **(*Local Monotonicity*)** If a member f of a decision making group has a weight allocation equal or greater than that of some other group member g for each criterion, then member f cannot be less powerful than group member g, i.e. $\Pi_f(\bar{W}|\bar{\gamma}) \geq \Pi_g(\bar{W}|\bar{\gamma})$.

Proof: By definition $P_{if}(w_i|\gamma_i) \geq P_{ig}(w_i|\gamma_i) \quad \forall i=1,\ldots,m$ (Turnovec, 1998)

Now
$$\Pi_f(\bar{W}|\bar{\gamma}) = \sum_{i=1}^{m} c_i P_{if}(w_i|\gamma_i) \geq \sum_{i=1}^{m} c_i P_{ig}(w_i|\gamma_i) = \Pi_g(\bar{W}|\bar{\gamma})$$

Furthermore $\Pi_j(\bar{W}|\bar{\gamma})$ is simply an additive value function, providing that the $P_{ij}(w_i|\gamma_i)$ are additively independent and lie on an interval scale of preferences, which can be ensured if the criteria are carefully chosen. Additive value functions are widely used in multicriteria decision making because they are transparent and easy to understand.

These properties ensure that $\Pi_j(\bar{W}|\bar{\gamma})$ is an appealing measure of multicriteria power to both game theorists and MCDM users.

The multicriteria power index $\Pi_j(\bar{W}|\bar{\gamma})$ can be considered the *proportional* allocation of decisional power: allocations are made proportional to weighted power indices. However a *progressive* allocation of power could be made, i.e. one in which the allocation is a moderation or an amplification of the proportional allocation. This approach is easily included in the present model by including a non-negative amplification factor or power r, i.e.

$$\Pi_j(\bar{W}|\bar{\gamma}) = \sum_{i=1}^{m} [c_i P_{ij}(w_i|\gamma_i)]^r$$

If $r>1$ we have an *amplification* of the allocation (if $r \to \infty$ we arrive at a *priority* allocation, where one group member is awarded the entire allocation). If $0 \leq r < 1$ we have a *moderation* of the allocation (if $r=0$ we have *parity*, where all group members are awarded an equal allocation).

3. Application to Fair Allocations

In this section we apply the multicriteria power index to the fair allocation of some divisible commodity. Obvious examples of this are the allocation of a worker's time to a set of projects, or the allocation of funds amongst a set of research projects or departments. The use of multicriteria decision making tools in setting fair allocations is not new: recent work includes that by Lootsma, Ramanathan and Schuijt (1998). The method of fair allocation using multicriteria power indices is a normative one, to be used to determine a "best" or fairest outcome.

The example we concentrate on is one from the public domain in South Africa. Since the general election in South Africa in April 1994 the country has undergone a great deal of change. For one, there are nine provinces currently, as opposed to only four previously, implying that any allocation of funds to the provincial structures from central government has to be done in a completely different way to previously. Real thought has had to be given to identifying the fundamental criteria on which the new allocation will be based.

In accordance with the country's new Constitution adopted in October 1996, a Financial and Fiscal Commission has been established, who have advised on the *Division of Revenue Bill*, soon to be debated in Parliament. This Bill will provide for

> "... the determination of each province's equitable share of [the] revenue [raised nationally]
>
> .
> .
> .
>
> and must take account of
>
> - the need to ensure that provinces are able to provide basic services and perform the functions allocated to them;
> - the fiscal capacity and efficiency of the provinces;
> - developmental needs of the provinces;
> - economic disparities within and among the provinces;
>
> .
> .
> .

(Quoted from the Constitution of the Republic of South Africa, Act 108 of 1996).

The measurements/criteria used as surrogates for the above issues mentioned in the Constitution will naturally differ from one particular allocation problem to another, and will need to be carefully sought out and weighted. For example "developmental needs" of a province would typically relate to the levels of infrastructure in the province, which might be measured differently if the allocation is one relating to health and welfare compared to an allocation relating to the environment and/or tourism.

The Constitution does not indicate exactly how the allocation should be made, but nonetheless there is the acceptance that public sector allocation problems of this sort are multicriteria in nature. In the empirical work that follows we will use the idea that provinces derive power from their relative performance under the measures or criteria used to satisfy the issues stated in the Constitution.

Consider the allocation from the national fiscus to the capital expenditure and current account of the nine provinces. In late 1997 the preliminary results of the first fully inclusive census in South Africa (which took place in October 1996) appeared, and these results showed that Gauteng province had a significantly larger population than was previously thought (see Table 1). Provincial spokesmen representing Gauteng insisted that the province be given a greater share of the upcoming budget to the provinces, thereby explicitly stating that *population size* was, in their opinion, the most important factor in the allocation of funds. This was a clear power-play to gain a competitive edge, since Gauteng would derive a greater power from their large population. Other provinces, on the other hand, could identify and propose suitable criteria under which they were powerful, and which would have improved their relative proportion of the budget allocation. For example, the Northern Cape (a rural, semi-arid, sparsely-populated and thus underdeveloped area) would be detrimentally affected under the population criterion, but would benefit from criteria which focused on its low level of development and large geographical size (for example, the cost of developing new infrastructure (roads, electricity, water supply etc.) would be related to some extent to distances within the province, i.e. to geographical size). There are enormous economic disparities between the provinces (as measured by the GDP). Provinces with lowest power derived from the GDP criterion might argue for a greater share of the budget as they are less able to generate their own revenues. For details of population size (1996 census), geographic area and GDP by province, see Table 1.

It is not the objective here to attempt to model the actual budget awarded to the nine provinces, since in practice the financial authorities in central government would consider a large number of factors or criteria when setting up such a budget (and we have only considered three here). We merely wish to illustrate how a multicriteria power index could be used in a real allocation problem; this could easily be extended to include any number of criteria.

Table 1. Populations, geographic area, GDP and actual 1998 budget of the provinces

	Population (millions)	Area ($x10^3km^2$)	GDP ($Rx10^9$)	Actual Budget (March '98)
KwaZulu Natal	7.672	92.10	57.007	15.508
Gauteng	7.171	17.01	144.359	11.701
Eastern Cape	5.865	169.58	29.049	14.073
Northern Province	4.128	123.91	14.158	10.424
Western Cape	4.118	129.37	53.874	7.965
North West	3.043	116.32	21.252	6.837
Mpumalanga	2.646	79.49	31.175	5.213
Free State	2.470	129.48	23.688	5.073
Northern Cape	0.746	361.83	8.000	1.964

Firstly we calculate the Shapley-Shubik and Banzhaf-Coleman power indices for each province under each of the three criteria (assuming $\gamma=0.67$ (i.e. a two-thirds majority, common in much government policy/strategy decision making) to imply "sufficient" consensus). This is shown in Table 2.

Table 2. Shapley-Shubik and Banzhaf-Coleman power indices for the criteria population, geographic area and GDP for the provinces

	Shapley-Shubik			**Banzhaf-Coleman**		
	Popn	Area	GDP	Popn	Area	GDP
KwaZulu Natal	.2202	.0738	.1171	.2120	.0838	.1575
Gauteng	.2004	.0079	.5067	.1902	.0051	.3543
Eastern Cape	.1639	.1270	.0663	.1522	.1421	.0787
North Province	.1000	.0933	.0218	.1087	.1015	.0341
Western Cape	.0980	.0968	.1119	.1060	.1066	.1496
North West	.0774	.0933	.0496	.0815	.1015	.0577
Mpumalanga	.0655	.0667	.0580	.0679	.0736	.0840
Free State	.0575	.0968	.0571	.0625	.1066	.0656
Northern Cape	.0171	.3448	.0115	.0190	.2792	.0184

Note that in general the Banzhaf-Coleman power index proves to be less extreme when allocating power across group members.

A decision making group with representatives from each province as well as from the Ministry of Finance would be required to determine weights for each criterion. The choice of weights for each criterion might be evoked using procedures common in MCDM practice (e.g. pairwise comparisons, swing weighting approach etc.). This will naturally lead to debate and even conflict between provincial representatives, as there is no widely accepted method for determining a single set of weights (e.g. Northern Cape would prefer a high weight on area, while KwaZulu Natal would prefer a high weight on population). In the case of consensus not being reached the Ministry of Finance could make the final decision on criterion weights, based on argument put forward by provincial representatives. Using values for the weights as $c_{popn}=0.85$, $c_{area}=0.075$ and $c_{GDP}=0.075$, power index values as in Table 2, and formula (1) we arrive at the multicriteria budget allocations as in Table 3.

Table 3. Multicriteria budget allocations, and actual allocation of total 1998 budget to the provinces ($Rx10^9$)

	Shapley-Shubik allocation	Banzhaf-Coleman allocation	Actual allocation
KwaZulu Natal	13.372	12.894	15.508
Gauteng	10.738	10.870	11.701
Eastern Cape	11.303	10.800	14.073
Northern Province	8.540	8.906	10.424
Western Cape	8.117	8.360	7.965
North West	7.440	7.621	6.837
Mpumalanga	6.782	6.814	5.213
Free State	6.551	6.776	5.073
Northern Cape	5.915	5.718	1.964

Finally, Table 4 shows the allocations after the introduction of a progression factor of $r=1.55$. In almost all instances it is seen that the multicriteria power index allocation is very close to the actual allocation, despite only making use of three criteria.

Table 4. Multicriteria budget allocations, and actual allocation of total 1998 budget to the provinces (Rx10^9), with progression factor r=1.55

	Shapley-Shubik allocation	Banzhaf-Coleman allocation	Actual allocation
KwaZulu Natal	16.379	15.538	15.508
Gauteng	11.659	11.925	11.701
Eastern Cape	12.623	11.806	14.073
Northern Province	8.174	8.756	10.424
Western Cape	7.555	7.938	7.965
North West	6.602	6.877	6.837
Mpumalanga	5.719	5.781	5.213
Free State	5.420	5.732	5.073
Northern Cape	4.626	4.405	1.964

4. Conclusions

The multicriteria power index proposed and developed here is a simple extension of the traditional power index used to measure voting power of members in committees. Decisional power can often be deemed to be multicriteria in nature. As such the multicriteria power index may prove useful in arriving at a fair allocation of a divisible commodity.

Acknowledgements

The financial assistance of the Centre for Science Development of the Human Sciences Research Council (South Africa) towards this research is hereby acknowledged. Opinions expressed in this work, or conclusions arrived at, are those of the author and are not to be attributed to the Centre for Science Development.

References

Allingham, M.G. (1975). "Economic Power and Values of Games", *Zeitschrift für Nationalökonomie*, 35, 293-299.

Banzhaf, J.F. (1965). "Weighted Voting Doesn't Work: A Mathematical Analysis", *Rutgers Law Review*, 19, 317-343.

Coleman, J.S. (1971). "Control of Collectivities and the Power of the Collectivity to Act". In B. Liberman (ed.) *Social Choice*, New York, 277-287.

Gambarelli, G. (1990). " A New Approach for Evaluating the Shapley Value", *Optimization*, 21, 445-452.

Holler, M.J. (1978). "A Priori Party Power and Government Formation", *Munich Social Science Review*, 4, 25-41.

Holler, M.J. and Packel, E.W. (1983). "Power, Luck and the Right Index", *Journal of Economics*, 43, 21-29.

Holler, M.J. and Li, X. (1995). "From Public Good Index to Public Value: An Axiomatic Approach and Generalization", *Control and Cybernetics*, 24, 257-270.

Holubiec, J.W. and Mercik, J.W. (1994). *Inside Voting Procedures*, Accedo Verlag, Munich.

Lootsma, F.A., Ramanathan, R. and Schuijt, H. (1998). "Fairness and Equity via Concepts of Multi-Criteria Decision Analysis. In T.J. Stewart and R.C. van den Honert (eds.) *Trends in Multicriteria Decision Making*, Springer, Heidelberg, 210-222.

Owen, G. (1982). *Game Theory*, Academic Press, New York.

Roth, A.E. (ed.) (1988). *The Shapley Value: Essays in Honour of Lloyd S. Shapley*, Cambridge University Press, Cambridge.

Shapley, L.S. and Shubik, M. (1954). "A Method for Evaluating the Distribution of Power in a Committee System", *American Political Science Review*, 48, 787-792.

Turnovec, F. (1996). "Weights and Votes in European Union: Extension and Institutional Reform", *Prague Economic Papers*, 2, 161-174.

Turnovec, F. (1998). "Monotonicity of Power Indices". In T.J. Stewart and R.C. van den Honert (eds.) *Trends in Multicriteria Decision Making*, Springer, Heidelberg, 199-214.

Widgrén, M. (1994). "Voting Power in the EC and the Consequences of Two Different Enlargements", *European Economic Review*, 38, 1153-1170.

Widgrén, M. (1995). "Probabilistic Voting Power in the EU Council: the Cases of Trade Policy and Social Regulation", *Scandinavian Journal of Economics*, 97, 345-356.

The College Selection Process from a Multi-Criteria Decision Analysis Perspective

H. Roland Weistroffer
Charles H. Smith

School of Business, Virginia Commonwealth University,
Richmond, VA 23284-4000, USA. E-mail: hrweistr@vcu.edu

Abstract. Given the large number of colleges and universities in the USA alone, and the diversity among these institutions, the problem faced by an American high school senior to make the "right" choice is not trivial. Decision criteria may include the geographic location of the institution, its perceived quality and prestige, the perceived feasibility of attending that school (including the likelihood of getting accepted, and the expected financial burden of attending), the expected quality of life at the institution, etc. Typically, the decision-making procedure consists of at least three stages: (1) selecting a manageable number of colleges for further investigation; (2) selecting a subset for formal application, after information on the institutions has been obtained; and (3) making the final decision after receiving acceptances as well as information on financial support from some of the institutions to which applications were submitted. The suitability of multi-criteria methodology in any of these stage and the general obstacles faced are investigated.

Keywords: College selection, multiple criteria.

1 Introduction

There are probably close to 2000 colleges and universities in the USA alone that offer four-year bachelor's degrees. The problem faced by an American high school senior to select the "right" college is thus not always a simple one to solve. Though hundreds of thousands, if not millions of people are confronted with this decision situation every year, the operations research literature seems to have disregarded this problem so far. A few scholarly papers deal with the reverse problem, the college admission problem, taking the perspective of the college or university rather than the perspective of the student. Edwards and Bader (1988) developed an expert system to help the admissions tutor of a British university decide which students to accept. Molinero and Qing (1990) also describe a decision support system for admissions planning for a British university.

Typically, the decision-making process for selecting a college consists of multiple stages. In the first stage, a selection of a manageable number of colleges is made for further investigation. Possible decision criteria may include the geographic location of the institution, its perceived quality and prestige, the perceived feasibility of attending that school (including the likelihood of getting accepted, and the expected financial burden of attending), the expected quality of life at the institution, etc. In the second stage, after acquiring literature from the previously selected institutions and after possibly visiting some of them, a subset is selected for formal application. Additional criteria (not contemplated in the first stage) may include the cost of the application and the ease or the arduousness of the application process. The third stage is making the final decision after receiving acceptances as well as information on financial support from some of the institutions to which applications were submitted. This process is summarized in Figure 1 below.

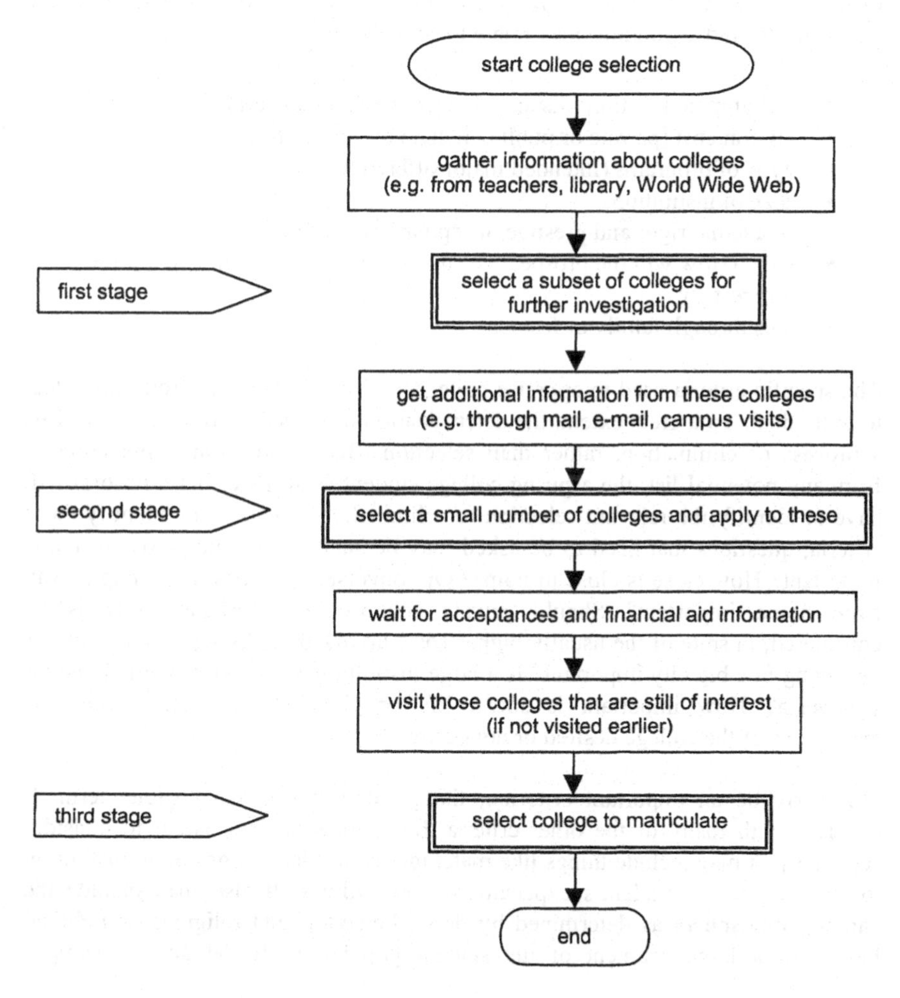

Fig. 1. Overview of College Selection Process

In the rest of the paper, the three decision making stages are discussed in more detail as essentially separate decision problems. The use of multi-criteria methodology to support the decision making process in each of these sub-problems is investigated.

2 Selecting a Manageable Number of Colleges for Further Investigation

As was stated above, the sheer number of colleges and universities in the USA alone preclude an in-depth investigation of each and every institution. Thus the first decision to be made is to determine a manageable subset of colleges that seem to fit the requirements and aspirations of the prospective college student. These requirements and aspirations may relate to such things as

- geographic location (distance from home, climate, etc);
- tuition/cost (private or public, in-state or out-of-state);
- type of programs (intended major offered?);
- size of institution;
- academic rigor and prestige or reputation of school;
- "fit" (How will the prospective student fit in, with regards to abilities, interests?);
- ties through family or friends.

The specific criteria used to appraise colleges will of course vary from individual to individual. To a great extent, the determination of the subset of colleges will be a process of elimination, rather than selection. But before eliminating colleges from any potential list, the aspiring college student (and his or her advisors) will have to identify the relevant selection (or elimination) criteria. To identify these criteria, questions that need to be asked may be such as: Is living close to home important? How close is close to home? Or conversely, is being far enough away from home important? Should private schools or out-of-state schools be considered, in spite of the usually higher costs involved? Is living in a big city, or not living in a big city important? Is a large institution preferred to a small one, or vice-versa? How important is social life or climate? Are certain religious affiliations of the college desired or not desired?

"Fit" may be an important criterion, though it is a somewhat vague term. It overlaps with many of the other criteria, but is nevertheless useful to consider separately. It may include things like matching the academic rigor of an institution to the prospective student's expectations and abilities. It also may include the campus atmosphere as determined by dress, behavior, and religious or political beliefs of a large segment of the student population. If the general campus atmosphere conflicts with the character and interests of the student, he or she may spend a miserable four years, if not dropping out.

Identifying the criteria that are important to the prospective student (and to his or her advisors) is one step. A further step is to find those schools that perform well on these criteria, and/or eliminate from consideration those that do not. Some criteria attributes are easy to evaluate, such as distance from home, size of the institution, size of city, etc., but others are not so easy to assess. To make a reasonable preliminary selection of colleges based on the identified criteria, a certain amount of data gathering must be performed. Such data may come from teachers or guidance counselors at the student's high school, from family and friends, but also from college guides available in libraries, by browsing the Internet, from "college nights" organized at some high schools, and from unsolicited mailings from some colleges.

The decision procedure in this stage usually consists of collecting data, sifting through this wealth of available information, and eliminating from further consideration those schools that clearly do not meet the identified selection criteria of the prospective student. Traditional MCDM software unfortunately will not provide much help here. What might be helpful would be on-line college guides that could be searched, based on a large range of criteria.

A number of existing Web pages do address this problem, allowing prospective college students to specify certain selection criteria and providing a list of colleges that meet these criteria. For example, a site by CollegeView[1] allows searching on fields of study, location by state, size of student body, coed or single sex, ethnic mix of student body, and religious affiliation of institution. Unfortunately these criteria may not be the ones that the prospective student finds most important. Several other sites allow searching on similar criteria. Only one site found by the authors, a site by College Board Online[2], allows prospective students to specify criteria such as average SAT scores or average GPAs of admitted students, criteria that may indicate how selective a college is. Unfortunately this site did not seem to have a very complete list of colleges: institutions like Harvard and MIT did not show up on any search.

Of-course, even if a Web site were available that included all colleges and that allowed searching on a wider variety of criteria, this would not render further human decision making needless in compiling the list of institutions to be considered further. For example, it is hard to imagine how the criterion of "fit" could be accurately quantified in such a system. Also, some schools may be placed

[1] http://www.collegeview.com
[2] http://cbweb1.collegeboard.org

on the list because of "legacy": It may be a family tradition to attend a particular college or university.

Further, selections or eliminations of colleges based on cost may need to take possible financial aid into consideration. Most colleges, private and public, claim that they will meet all established financial needs of the applicant (and family) through various forms of financial aid. Problems with these claims are that the prospective student's family often sees their financial need differently than the colleges do, and are often unable or unwilling to meet the financial obligation as established by a college. Nevertheless, a college should not be eliminated from the list because of high tuition, if there is a reasonable prospect of substantial financial aid. The exact amount of financial aid will normally not be made available until a student has been accepted at that college.

The final outcome of this decision stage should preferably be a list of maybe ten to thirty institutions. It is usually desirable to have a certain amount of diversity represented by this list, as it is by necessity based on incomplete information and compiled at a time when the prospective student's requirements and aspirations, as represented by the identified criteria, may still be evolving.

3 Selecting a Small Number of Colleges for Formal Application

Most colleges in the USA charge a non-refundable application fee of between $30 and $70. Also, most colleges require students to complete long application forms, including one or more essays, as well as recommendation letters from teachers and guidance counselors at their high school. Though some essays may be recycled, many colleges set specific themes for the essays that preclude using the same essay for all applications without modifications. The *Common Application* form, accepted by some colleges, is meant to make it easier for students to apply to multiple institutions. However, many colleges that do accept this form also require students to fill out an institution specific *supplement.* The arduousness and cost of the application process keeps most students from applying to a large number of institutions. Thus the list of candidate institutions established in stage one of the college selection process must be reduced further.

The student's selection criteria for further reducing the number of candidate institutions are basically the same as in stage one, though, as mentioned earlier, these criteria may still be evolving. The cost and arduousness of applying to a specific college may enter as an additional criterion, however. More data gathering must take place at this stage, to reduce the set of candidate colleges to include only those that the prospective student is willing to attend, if admitted. Materials may be requested directly from the institutions of interest; they may be obtained from their Web pages; or preferably, information may be gathered from personal campus visits. Often, students may decide that they do not like a school after visiting it, even if until then it appeared to be ideal. And conversely, a school that only seemed marginally acceptable previously, may be viewed more favorably after a campus visit.

The decision problem at hand, after gathering sufficient information, is to reduce the set of maybe 10 to 30 institutions to a set of, perhaps 3 to 5. Some traditional MCDM software could be used to rank the institutions of the larger set, thus allowing the selection of the top subset. However, an important consideration in selecting the reduced list is diversity in the resultant subset. Rather than selecting the, say, five highest ranked institutions from the initial, larger list, the subset should include at least one institution that the prospective student feels is a definite *acceptance*, based on the selectiveness of the school and the student's credentials. It may also include one or two institutions that the prospective student feels is a *reach*, i.e. an institution that he or she would like to attend, but is not very likely to get accepted to. Further, it should include institutions that the student feels are a best "fit" in terms of his/her qualifications and the colleges' demands with respect to academic performance. The decision problem thus is a *portfolio selection* problem, rather than a traditional multi-criteria problem.

Though we are not aware of previous work directly on the problem of finding a suitable subset of colleges, the problem is similar to subset selection models in other fields. For example, some authors have explored the selection of research and development projects. Both single and multiple criteria versions of this problem have been investigated. Some procedures merely select the highest ranked projects, but others deal with the interdependence of the selected projects. The collective properties of the entire selected subset are critical in the current problem. Karabakal et al. (1994) consider a single criterion collection of replacement decisions with economic interdependence among the assets being replaced. Gackstatter (1997) developed a multiple criteria decision support system for research and development project planning. Some of his criteria are characteristic of the selected subset and not just descriptive of individual project ratings. Earlier, Stewart (1991) also proposed a multi-criteria decision support system for R & D project selection.

4 Selecting a College to Matriculate

Application deadlines for US colleges are typically in December for enrollment the following fall. Notifications of acceptance (together with notifications of financial aid) are typically mailed around April 1. The prospective student then, typically, has to around May 1 to make a commitment to one of the institutions that accepted him or her. The decision problem here varies, depending on how many acceptances were received. If no college accepted the applicant, then new alternatives must be investigated, such as attending a local community college, and then, maybe, trying to transfer to a four-year institution the following year. If only one college accepted the applicant, there may not be much of a decision problem at all, except perhaps to decide whether the financial burden can be met, having received information on financial support. If multiple acceptances have been received, a typical MCDM problem arises.

The decision criteria are likely to have shifted somewhat for stage three. Assuming that all institutions that the prospective student applied to are considered acceptable, some of the criteria that were used in determining the set of candidate colleges may no longer be as significant. The criteria that may still be of concern to the student (or his or her family) are possibly the financial burden that the family is expected to assume (this can now be assessed more precisely, given the scholarship information), the academic quality of the program into which the student has been accepted, the perceived prestige of the institution, the expected quality of life at the college, and the convenience or inconvenience of visiting home due to distance or travel connections. Of course, this may vary from individual to individual, and a reassessment of criteria is appropriate at this stage. For example, the aforementioned "legacy" criterion may figure more prominently now than it did in the earlier stages. In stages one and two, it would have been easy to just add a "legacy" institution to the list compiled on other criteria. Now that only one institution is to be selected, this criterion may be treated like the other criteria.

Personal visits to all institutions still under consideration are essential at this stage, if these visits were not done earlier. Discussions with friends or relatives attending the colleges of interest, if available, also are useful for information gathering.

Conventional MCDM software is most appropriate at this stage. For example, Expert Choice[3], implementing the analytic hierarchy process (Saaty, 1990) may be used to weight the criteria and also to evaluate the colleges with respect to these criteria, to achieve an overall ranking. Figure 2 shows an example hierarchy.

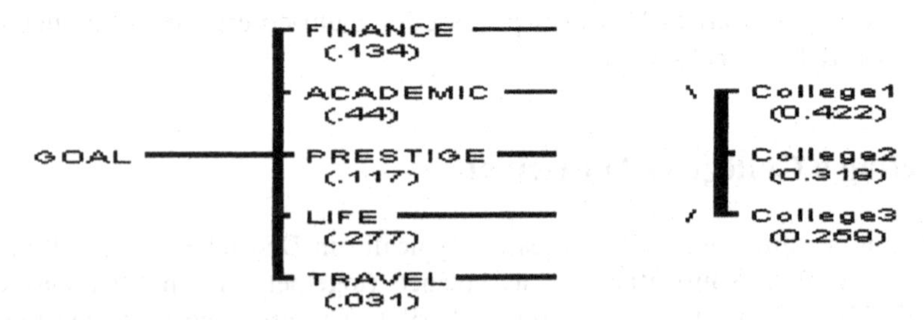

Fig. 2: Example Hierarchy for Stage 3

[3] Expert Choice, Inc., 4922 Ellsworth Ave., Pittsburgh, PA, 15213, USA

Figure 3 shows an example of how the alternative colleges from Figure 2 may perform on the decision criteria, as well as the relative weights of the decision criteria.

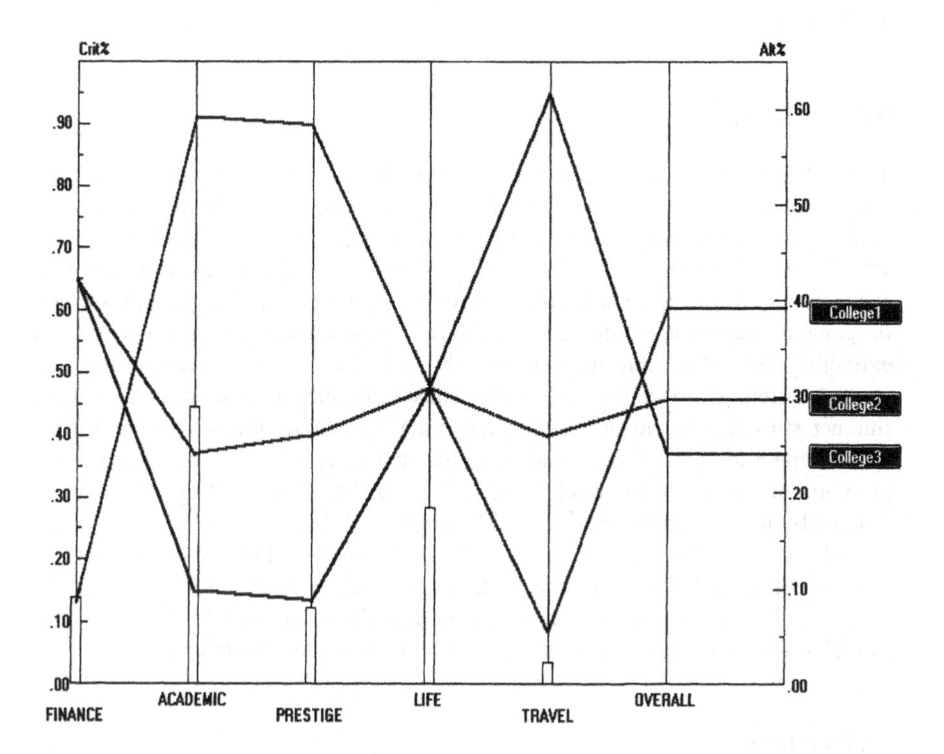

Fig. 3: Performance of Colleges Relative to Criteria

5 Early Decision Applications

The above described three stages represent the most common decision scenarios. One variation is where students in stage two apply to one institution for early decision admission. These applications typically are submitted in November, and students are notified in early December as to acceptance, deferral, or rejection. In most cases, an early decision application is viewed by the college as a commitment to attend, if accepted. A deferral means that the student's application is held over for the regular admission process, in which case the commitment on the part of the student is no longer binding. In the case that a student gets deferred or rejected, he or she can then follow the normal stage two and stage three decision process, as described above (though less time is available, since regular applications often are due in December.)

If a student chooses to follow the early decision route, a single, most preferred college must be determined early. This is different though, from the normal step three, as the student does not have the knowledge that his or her most preferred selection will admit him or her. Also, information on financial aid may not be available at this time.

6. Conclusion

The college selection decision is likely one of the most important choices made by a high school student. Despite the fact that this decision is made annually by perhaps millions of people, the existing decision aids are very limited. The three stages, or sub-problems, of the college selection problem present different requirements for decision support. Clearly the students and their parents are not, in general, experienced decision analysts. Simple interfaces are needed for extracting the critical information from the decision-makers. Currently, the first stage decision, identifying a preliminary subset of colleges, is supported by some Internet sites that aid in performing the initial screening, though vast room for improvement remains. The second stage has neither specific DSSs to support the problem, nor are there any easily available general software packages that can be utilized by inexperienced end-users. Whereas the third stage does fall in the realm of traditional MCDM software, no specific DSSs for the college selection problem seem to be available as yet. This lack of generally available assistance for the decisions at the second and third stages, and the crudeness of the assistance available for the first stage, present both an opportunity and a challenge.

References

Edwards, J.S. and J.L. Bader (1988), "Expert Systems for University Admissions", *Journal of the Operational Research Society*, **39** (1), 33-40.

Gackstatter, S. (1997), "A DSS for Planning the R & D Project Portfolio," unpublished manuscript, Universität Hohenheim.

Karabakal, N., J.R. Lohmann, and J.C. Bean (1994), "Parallel Replacement Under Capital Rationing Constraints," *Management Science*, **40**(3), 305-319.

Molinero, C.M. and M. Qing (1990), "Decision Support Systems for University Undergraduate Admissions," *Journal of the Operational Research Society*, **41**(3), 219-228.

Saaty, T.L. (1990), "The Analytic Hierarchy Process – Planning, Priority Setting, Resource Allocation. RWS Publications, Pittsburgh, Pennsylvania.

Stewart, T.J. (1991), "A Multi-Criteria Decision Support System for R & D Project Selection," *Journal of the Operational Research Society*, **42**(1), 17-26.

Index of Authors

Lecture Notes in Economics and Mathematical Systems

For information about Vols. 1–295
please contact your bookseller or Springer-Verlag

Vol. 300: U. Derigs, Programming in Networks and Graphs. XI, 315 pages. 1988.

Vol. 301: J. Kacprzyk, M. Roubens (Eds.), Non-Conventional Preference Relations in Decision Making. VII, 155 pages. 1988.

Vol. 302: H.A. Eiselt, G. Pederzoli (Eds.), Advances in Optimization and Control. Proceedings, 1986. VIII, 372 pages. 1988.

Vol. 303: F.X. Diebold, Empirical Modeling of Exchange Rate Dynamics. VII, 143 pages. 1988.

Vol. 304: A. Kurzhanski, K. Neumann, D. Pallaschke (Eds.), Optimization, Parallel Processing and Applications. Proceedings, 1987. VI, 292 pages. 1988.

Vol. 305: G.-J.C.Th. van Schijndel, Dynamic Firm and Investor Behaviour under Progressive Personal Taxation. X, 215 pages.1988.

Vol. 306: Ch. Klein, A Static Microeconomic Model of Pure Competition. VIII, 139 pages. 1988.

Vol. 307: T.K. Dijkstra (Ed.), On Model Uncertainty and its Statistical Implications. VII, 138 pages. 1988.

Vol. 308: J.R. Daduna, A. Wren (Eds.), Computer-Aided Transit Scheduling. VIII, 339 pages. 1988.

Vol. 309: G. Ricci, K. Velupillai (Eds.), Growth Cycles and Multisectoral Economics: The Goodwin Tradition. III, 126 pages. 1988.

Vol. 310: J. Kacprzyk, M. Fedrizzi (Eds.), Combining Fuzzy Imprecision with Probabilistic Uncertainty in Decision Making. IX, 399 pages. 1988.

Vol. 311: R. Färe, Fundamentals of Production Theory. IX, 163 pages. 1988.

Vol. 312: J. Krishnakumar, Estimation of Simultaneous Equation Models with Error Components Structure. X, 357 pages. 1988.

Vol. 313: W. Jammernegg, Sequential Binary Investment Decisions. VI, 156 pages. 1988.

Vol. 314: R. Tietz, W. Albers, R. Selten (Eds.), Bounded Rational Behavior in Experimental Games and Markets. VI, 368 pages. 1988.

Vol. 315: I. Orishimo, G.J.D. Hewings, P. Nijkamp (Eds), Information Technology: Social and Spatial Perspectives. Proceedings 1986. VI, 268 pages. 1988.

Vol. 316: R.L. Basmann, D.J. Slottje, K. Hayes, J.D. Johnson, D.J. Molina, The Generalized Fechner-Thurstone Direct Utility Function and Some of its Uses. VIII, 159 pages. 1988.

Vol. 317: L. Bianco, A. La Bella (Eds.), Freight Transport Planning and Logistics. Proceedings, 1987. X, 568 pages. 1988.

Vol. 318: T. Doup, Simplicial Algorithms on the Simplotope. VIII, 262 pages. 1988.

Vol. 319: D.T. Luc, Theory of Vector Optimization. VIII, 173 pages. 1989.

Vol. 320: D. van der Wijst, Financial Structure in Small Business. VII, 181 pages. 1989.

Vol. 321: M. Di Matteo, R.M. Goodwin, A. Vercelli (Eds.), Technological and Social Factors in Long Term Fluctuations. Proceedings. IX, 442 pages. 1989.

Vol. 322: T. Kollintzas (Ed.), The Rational Expectations Equilibrium Inventory Model. XI, 269 pages. 1989.

Vol. 323: M.B.M. de Koster, Capacity Oriented Analysis and Design of Production Systems. XII, 245 pages. 1989.

Vol. 324: I.M. Bomze, B.M. Pötscher, Game Theoretical Foundations of Evolutionary Stability. VI, 145 pages. 1989.

Vol. 325: P. Ferri, E. Greenberg, The Labor Market and Business Cycle Theories. X, 183 pages. 1989.

Vol. 326: Ch. Sauer, Alternative Theories of Output, Unemployment, and Inflation in Germany: 1960–1985. XIII, 206 pages. 1989.

Vol. 327: M. Tawada, Production Structure and International Trade. V, 132 pages. 1989.

Vol. 328: W. Güth, B. Kalkofen, Unique Solutions for Strategic Games. VII, 200 pages. 1989.

Vol. 329: G. Tillmann, Equity, Incentives, and Taxation. VI, 132 pages. 1989.

Vol. 330: P.M. Kort, Optimal Dynamic Investment Policies of a Value Maximizing Firm. VII, 185 pages. 1989.

Vol. 331: A. Lewandowski, A.P. Wierzbicki (Eds.), Aspiration Based Decision Support Systems. X, 400 pages. 1989.

Vol. 332: T.R. Gulledge, Jr., L.A. Litteral (Eds.), Cost Analysis Applications of Economics and Operations Research. Proceedings. VII, 422 pages. 1989.

Vol. 333: N. Dellaert, Production to Order. VII, 158 pages. 1989.

Vol. 334: H.-W. Lorenz, Nonlinear Dynamical Economics and Chaotic Motion. XI, 248 pages. 1989.

Vol. 335: A.G. Lockett, G. Islei (Eds.), Improving Decision Making in Organisations. Proceedings. IX, 606 pages. 1989.

Vol. 336: T. Puu, Nonlinear Economic Dynamics. VII, 119 pages. 1989.

Vol. 337: A. Lewandowski, I. Stanchev (Eds.), Methodology and Software for Interactive Decision Support. VIII, 309 pages. 1989.

Vol. 338: J.K. Ho, R.P. Sundarraj, DECOMP: An Implementation of Dantzig-Wolfe Decomposition for Linear Programming. VI, 206 pages.

Vol. 339: J. Terceiro Lomba, Estimation of Dynamic Econometric Models with Errors in Variables. VIII, 116 pages. 1990.

Vol. 340: T. Vasko, R. Ayres, L. Fontvieille (Eds.), Life Cycles and Long Waves. XIV, 293 pages. 1990.

Vol. 341: G.R. Uhlich, Descriptive Theories of Bargaining. IX, 165 pages. 1990.

Vol. 342: K. Okuguchi, F. Szidarovszky, The Theory of Oligopoly with Multi-Product Firms. V, 167 pages. 1990.

Vol. 343: C. Chiarella, The Elements of a Nonlinear Theory of Economic Dynamics. IX, 149 pages. 1990.

Vol. 344: K. Neumann, Stochastic Project Networks. XI, 237 pages. 1990.

Vol. 345: A. Cambini, E. Castagnoli, L. Martein, P Mazzoleni, S. Schaible (Eds.), Generalized Convexity and Fractional Programming with Economic Applications. Proceedings, 1988. VII, 361 pages. 1990.

Vol. 346: R. von Randow (Ed.), Integer Programming and Related Areas. A Classified Bibliography 1984–1987. XIII, 514 pages. 1990.

Vol. 347: D. Ríos Insua, Sensitivity Analysis in Multi-objective Decision Making. XI, 193 pages. 1990.

Vol. 348: H. Störmer, Binary Functions and their Applications. VIII, 151 pages. 1990.

Vol. 349: G.A. Pfann, Dynamic Modelling of Stochastic Demand for Manufacturing Employment. VI, 158 pages. 1990.

Vol. 350: W.-B. Zhang, Economic Dynamics. X, 232 pages. 1990.

Vol. 351: A. Lewandowski, V. Volkovich (Eds.), Multiobjective Problems of Mathematical Programming. Proceedings, 1988. VII, 315 pages. 1991.

Vol. 352: O. van Hilten, Optimal Firm Behaviour in the Context of Technological Progress and a Business Cycle. XII, 229 pages. 1991.

Vol. 353: G. Ricci (Ed.), Decision Processes in Economics. Proceedings, 1989. III, 209 pages 1991.

Vol. 354: M. Ivaldi, A Structural Analysis of Expectation Formation. XII, 230 pages. 1991.

Vol. 355: M. Salomon. Deterministic Lotsizing Models for Production Planning. VII, 158 pages. 1991.

Vol. 356: P. Korhonen, A. Lewandowski, J . Wallenius (Eds.), Multiple Criteria Decision Support. Proceedings, 1989. XII, 393 pages. 1991.

Vol. 357: P. Zörnig, Degeneracy Graphs and Simplex Cycling. XV, 194 pages. 1991.

Vol. 358: P. Knottnerus, Linear Models with Correlated Disturbances. VIII, 196 pages. 1991.

Vol. 359: E. de Jong, Exchange Rate Determination and Optimal Economic Policy Under Various Exchange Rate Regimes. VII, 270 pages. 1991.

Vol. 360: P. Stalder, Regime Translations, Spillovers and Buffer Stocks. VI, 193 pages . 1991.

Vol. 361: C. F. Daganzo, Logistics Systems Analysis. X, 321 pages. 1991.

Vol. 362: F. Gehrels, Essays in Macroeconomics of an Open Economy. VII, 183 pages. 1991.

Vol. 363: C. Puppe, Distorted Probabilities and Choice under Risk. VIII, 100 pages . 1991

Vol. 364: B. Horvath, Are Policy Variables Exogenous? XII, 162 pages. 1991.

Vol. 365: G. A. Heuer, U. Leopold-Wildburger. Balanced Silverman Games on General Discrete Sets. V, 140 pages. 1991.

Vol. 366: J. Gruber (Ed.), Econometric Decision Models. Proceedings, 1989. VIII, 636 pages. 1991.

Vol. 367: M. Grauer, D. B. Pressmar (Eds.), Parallel Computing and Mathematical Optimization. Proceedings. V, 208 pages. 1991.

Vol. 368: M. Fedrizzi, J. Kacprzyk, M. Roubens (Eds.), Interactive Fuzzy Optimization. VII, 216 pages. 1991.

Vol. 369: R. Koblo, The Visible Hand. VIII, 131 pages.1991.

Vol. 370: M. J. Beckmann, M. N. Gopalan, R. Subramanian (Eds.), Stochastic Processes and their Applications. Proceedings, 1990. XLI, 292 pages. 1991.

Vol. 371: A. Schmutzler, Flexibility and Adjustment to Information in Sequential Decision Problems. VIII, 198 pages. 1991.

Vol. 372: J. Esteban, The Social Viability of Money. X, 202 pages. 1991.

Vol. 373: A. Billot, Economic Theory of Fuzzy Equilibria. XIII, 164 pages. 1992.

Vol. 374: G. Pflug, U. Dieter (Eds.), Simulation and Optimization. Proceedings, 1990. X, 162 pages. 1992.

Vol. 375: S.-J. Chen, Ch.-L. Hwang, Fuzzy Multiple Attribute Decision Making. XII, 536 pages. 1992.

Vol. 376: K.-H. Jöckel, G. Rothe, W. Sendler (Eds.), Bootstrapping and Related Techniques. Proceedings, 1990. VIII, 247 pages. 1992.

Vol. 377: A. Villar, Operator Theorems with Applications to Distributive Problems and Equilibrium Models. XVI, 160 pages. 1992.

Vol. 378: W. Krabs, J. Zowe (Eds.), Modern Methods of Optimization. Proceedings, 1990. VIII, 348 pages. 1992.

Vol. 379: K. Marti (Ed.), Stochastic Optimization. Proceedings, 1990. VII, 182 pages. 1992.

Vol. 380: J. Odelstad, Invariance and Structural Dependence. XII, 245 pages. 1992.

Vol. 381: C. Giannini, Topics in Structural VAR Econometrics. XI, 131 pages. 1992.

Vol. 382: W. Oettli, D. Pallaschke (Eds.), Advances in Optimization. Proceedings, 1991. X, 527 pages. 1992.

Vol. 383: J. Vartiainen, Capital Accumulation in a Corporatist Economy. VII, 177 pages. 1992.

Vol. 384: A. Martina, Lectures on the Economic Theory of Taxation. XII, 313 pages. 1992.

Vol. 385: J. Gardeazabal, M. Regúlez, The Monetary Model of Exchange Rates and Cointegration. X, 194 pages. 1992.

Vol. 386: M. Desrochers, J.-M. Rousseau (Eds.), Computer-Aided Transit Scheduling. Proceedings, 1990. XIII, 432 pages. 1992.

Vol. 387: W. Gaertner, M. Klemisch-Ahlert, Social Choice and Bargaining Perspectives on Distributive Justice. VIII, 131 pages. 1992.

Vol. 388: D. Bartmann, M. J. Beckmann, Inventory Control. XV, 252 pages. 1992.

Vol. 389: B. Dutta, D. Mookherjee, T. Parthasarathy, T. Raghavan, D. Ray, S. Tijs (Eds.), Game Theory and Economic Applications. Proceedings, 1990. IX, 454 pages. 1992.

Vol. 390: G. Sorger, Minimum Impatience Theorem for Recursive Economic Models. X, 162 pages. 1992.

Vol. 391: C. Keser, Experimental Duopoly Markets with Demand Inertia. X, 150 pages. 1992.

Vol. 392: K. Frauendorfer, Stochastic Two-Stage Programming. VIII, 228 pages. 1992.

Vol. 393: B. Lucke, Price Stabilization on World Agricultural Markets. XI, 274 pages. 1992.

Vol. 394: Y.-J. Lai, C.-L. Hwang, Fuzzy Mathematical Programming. XIII, 301 pages. 1992.

Vol. 395: G. Haag, U. Mueller, K. G. Troitzsch (Eds.), Economic Evolution and Demographic Change. XVI, 409 pages. 1992.

Vol. 396: R. V. V. Vidal (Ed.), Applied Simulated Annealing. VIII, 358 pages. 1992.

Vol. 397: J. Wessels, A. P. Wierzbicki (Eds.), User-Oriented Methodology and Techniques of Decision Analysis and Support. Proceedings, 1991. XII, 295 pages. 1993.

Vol. 398: J.-P. Urbain, Exogeneity in Error Correction Models. XI, 189 pages. 1993.

Vol. 399: F. Gori, L. Geronazzo, M. Galeotti (Eds.), Nonlinear Dynamics in Economics and Social Sciences. Proceedings, 1991. VIII, 367 pages. 1993.

Vol. 400: H. Tanizaki, Nonlinear Filters. XII, 203 pages. 1993.

Vol. 401: K. Mosler, M. Scarsini, Stochastic Orders and Applications. V, 379 pages. 1993.

Vol. 402: A. van den Elzen, Adjustment Processes for Exchange Economies and Noncooperative Games. VII, 146 pages. 1993.

Vol. 403: G. Brennscheidt, Predictive Behavior. VI, 227 pages. 1993.

Vol. 404: Y.-J. Lai, Ch.-L. Hwang, Fuzzy Multiple Objective Decision Making. XIV, 475 pages. 1994.

Vol. 405: S. Komlósi, T. Rapcsák, S. Schaible (Eds.), Generalized Convexity. Proceedings, 1992. VIII, 404 pages. 1994.

Vol. 406: N. M. Hung, N. V. Quyen, Dynamic Timing Decisions Under Uncertainty. X, 194 pages. 1994.

Vol. 407: M. Ooms, Empirical Vector Autoregressive Modeling. XIII, 380 pages. 1994.

Vol. 408: K. Haase, Lotsizing and Scheduling for Production Planning. VIII, 118 pages. 1994.

Vol. 409: A. Sprecher, Resource-Constrained Project Scheduling. XII, 142 pages. 1994.

Vol. 410: R. Winkelmann, Count Data Models. XI, 213 pages. 1994.

Vol. 411: S. Dauzère-Péres, J.-B. Lasserre, An Integrated Approach in Production Planning and Scheduling. XVI, 137 pages. 1994.

Vol. 412: B. Kuon, Two-Person Bargaining Experiments with Incomplete Information. IX, 293 pages. 1994.

Vol. 413: R. Fiorito (Ed.), Inventory, Business Cycles and Monetary Transmission. VI, 287 pages. 1994.

Vol. 414: Y. Crama, A. Oerlemans, F. Spieksma, Production Planning in Automated Manufacturing. X, 210 pages. 1994.

Vol. 415: P. C. Nicola, Imperfect General Equilibrium. XI, 167 pages. 1994.

Vol. 416: H. S. J. Cesar, Control and Game Models of the Greenhouse Effect. XI, 225 pages. 1994.

Vol. 417: B. Ran, D. E. Boyce, Dynamic Urban Transportation Network Models. XV, 391 pages. 1994.

Vol. 418: P. Bogetoft, Non-Cooperative Planning Theory. XI, 309 pages. 1994.

Vol. 419: T. Maruyama, W. Takahashi (Eds.), Nonlinear and Convex Analysis in Economic Theory. VIII, 306 pages. 1995.

Vol. 420: M. Peeters, Time-To-Build. Interrelated Investment and Labour Demand Modelling. With Applications to Six OECD Countries. IX, 204 pages. 1995.

Vol. 421: C. Dang, Triangulations and Simplicial Methods. IX, 196 pages. 1995.

Vol. 422: D. S. Bridges, G. B. Mehta, Representations of Preference Orderings. X, 165 pages. 1995.

Vol. 423: K. Marti, P. Kall (Eds.), Stochastic Programming. Numerical Techniques and Engineering Applications. VIII, 351 pages. 1995.

Vol. 424: G. A. Heuer, U. Leopold-Wildburger, Silverman's Game. X, 283 pages. 1995.

Vol. 425: J. Kohlas, P.-A. Monney, A Mathematical Theory of Hints. XIII, 419 pages, 1995.

Vol. 426: B. Finkenstädt, Nonlinear Dynamics in Economics. IX, 156 pages. 1995.

Vol. 427: F. W. van Tongeren, Microsimulation Modelling of the Corporate Firm. XVII, 275 pages. 1995.

Vol. 428: A. A. Powell, Ch. W. Murphy, Inside a Modern Macroeconometric Model. XVIII, 424 pages. 1995.

Vol. 429: R. Durier, C. Michelot, Recent Developments in Optimization. VIII, 356 pages. 1995.

Vol. 430: J. R. Daduna, I. Branco, J. M. Pinto Paixão (Eds.), Computer-Aided Transit Scheduling. XIV, 374 pages. 1995.

Vol. 431: A. Aulin, Causal and Stochastic Elements in Business Cycles. XI, 116 pages. 1996.

Vol. 432: M. Tamiz (Ed.), Multi-Objective Programming and Goal Programming. VI, 359 pages. 1996.

Vol. 433: J. Menon, Exchange Rates and Prices. XIV, 313 pages. 1996.

Vol. 434: M. W. J. Blok, Dynamic Models of the Firm. VII, 193 pages. 1996.

Vol. 435: L. Chen, Interest Rate Dynamics, Derivatives Pricing, and Risk Management. XII, 149 pages. 1996.

Vol. 436: M. Klemisch-Ahlert, Bargaining in Economic and Ethical Environments. IX, 155 pages. 1996.

Vol. 437: C. Jordan, Batching and Scheduling. IX, 178 pages. 1996.

Vol. 438: A. Villar, General Equilibrium with Increasing Returns. XIII, 164 pages. 1996.

Vol. 439: M. Zenner, Learning to Become Rational. VII, 201 pages. 1996.

Vol. 440: W. Ryll, Litigation and Settlement in a Game with Incomplete Information. VIII, 174 pages. 1996.

Vol. 441: H. Dawid, Adaptive Learning by Genetic Algorithms. IX, 166 pages.1996.

Vol. 442: L. Corchón, Theories of Imperfectly Competitive Markets. XIII, 163 pages. 1996.

Vol. 443: G. Lang, On Overlapping Generations Models with Productive Capital. X, 98 pages. 1996.

Vol. 444: S. Jørgensen, G. Zaccour (Eds.), Dynamic Competitive Analysis in Marketing. X, 285 pages. 1996.

Vol. 445: A. H. Christer, S. Osaki, L. C. Thomas (Eds.), Stochastic Modelling in Innovative Manufactoring. X, 361 pages. 1997.

Vol. 446: G. Dhaene, Encompassing. X, 160 pages. 1997.

Vol. 447: A. Artale, Rings in Auctions. X, 172 pages. 1997.

Vol. 448: G. Fandel, T. Gal (Eds.), Multiple Criteria Decision Making. XII, 678 pages. 1997.

Vol. 449: F. Fang, M. Sanglier (Eds.), Complexity and Self-Organization in Social and Economic Systems. IX, 317 pages, 1997.

Vol. 450: P. M. Pardalos, D.W. Hearn, W.W. Hager, (Eds.), Network Optimization. VIII, 485 pages, 1997.

Vol. 451: M. Salge, Rational Bubbles. Theoretical Basis, Economic Relevance, and Empirical Evidence with a Special Emphasis on the German Stock Market.IX, 265 pages. 1997.

Vol. 452: P. Gritzmann, R. Horst, E. Sachs, R. Tichatschke (Eds.), Recent Advances in Optimization. VIII, 379 pages. 1997.

Vol. 453: A. S. Tangian, J. Gruber (Eds.), Constructing Scalar-Valued Objective Functions. VIII, 298 pages. 1997.

Vol. 454: H.-M. Krolzig, Markov-Switching Vector Autoregressions. XIV, 358 pages. 1997.

Vol. 455: R. Caballero, F. Ruiz, R. E. Steuer (Eds.), Advances in Multiple Objective and Goal Programming. VIII, 391 pages. 1997.

Vol. 456: R. Conte, R. Hegselmann, P. Terna (Eds.), Simulating Social Phenomena. VIII, 536 pages. 1997.

Vol. 457: C. Hsu, Volume and the Nonlinear Dynamics of Stock Returns. VIII, 133 pages. 1998.

Vol. 458: K. Marti, P. Kall (Eds.), Stochastic Programming Methods and Technical Applications. X, 437 pages. 1998.

Vol. 459: H. K. Ryu, D. J. Slottje, Measuring Trends in U.S. Income Inequality. XI, 195 pages. 1998.

Vol. 460: B. Fleischmann, J. A. E. E. van Nunen, M. G. Speranza, P. Stähly, Advances in Distribution Logistic. XI, 535 pages. 1998.

Vol. 461: U. Schmidt, Axiomatic Utility Theory under Risk. XV, 201 pages. 1998.

Vol. 462: L. von Auer, Dynamic Preferences, Choice Mechanisms, and Welfare. XII, 226 pages. 1998.

Vol. 463: G. Abraham-Frois (Ed.), Non-Linear Dynamics and Endogenous Cycles. VI, 204 pages. 1998.

Vol. 464: A. Aulin, The Impact of Science on Economic Growth and its Cycles. IX, 204 pages. 1998.

Vol. 465: T. J. Stewart, R. C. van den Honert (Eds.), Trends in Multicriteria Decision Making. X, 448 pages. 1998.

Vol. 466: A. Sadrieh, The Alternating Double Auction Market. VII, 350 pages. 1998.

Vol. 467: H. Hennig-Schmidt, Bargaining in a Video Experiment. Determinants of Boundedly Rational Behavior. XII, 221 pages. 1999.

Vol. 468: A. Ziegler, A Game Theory Analysis of Options. XIV, 145 pages. 1999.

Vol. 469: M. P. Vogel, Environmental Kuznets Curves. XIII, 197 pages. 1999.

Vol. 470: M. Ammann, Pricing Derivative Credit Risk. XII, 228 pages. 1999.

Vol. 471: N. H. M. Wilson (Ed.), Computer-Aided Transit Scheduling. XI, 444 pages. 1999.

Vol. 472: J.-R. Tyran, Money Illusion and Strategic Complementarity as Causes of Monetary Non-Neutrality. X, 228 pages. 1999.

Vol. 473: S. Helber, Performance Analysis of Flow Lines with Non-Linear Flow of Material. IX, 280 pages. 1999.

Vol. 474: U. Schwalbe, The Core of Economies with Asymmetric Information. IX, 141 pages. 1999.

Vol. 475: L. Kaas, Dynamic Macroeconomics with Imperfect Competition. XI, 155 pages. 1999.

Vol. 476: R. Demel, Fiscal Policy, Public Debt and the Term Structure of Interest Rates. X, 279 pages. 1999.

Vol. 477: M. Théra, R. Tichatschke (Eds.), Ill-posed Variational Problems and Regularization Techniques. VIII, 274 pages. 1999.

Vol. 478: S. Hartmann, Project Scheduling under Limited Resources. XII, 221 pages. 1999.

Vol. 479: L. v. Thadden, Money, Inflation, and Capital Formation. IX, 192 pages. 1999.

Vol. 480: M. Grazia Speranza, P. Stähly (Eds.), New Trends in Distribution Logistics. X, 336 pages. 1999.

Vol. 481: V. H. Nguyen, J. J. Strodiot, P. Tossings (Eds.). Optimation. IX, 498 pages. 2000.

Vol. 482: W. B. Zhang, A Theory of International Trade. XI, 192 pages. 2000.

Vol. 483: M. Königstein, Equity, Efficiency and Evolutionary Stability in Bargaining Games with Joint Production. XII, 197 pages. 2000.

Vol. 484: D. D. Gatti, M. Gallegati, A. Kirman, Interaction and Market Structure. VI, 298 pages. 2000.

Vol. 485: A. Garnaev, Search Games and Other Applications of Game Theory. VIII, 145 pages. 2000.

Vol. 486: M. Neugart, Nonlinear Labor Market Dynamics. X, 175 pages. 2000.

Vol. 487: Y. Y. Haimes, R. E. Steuer (Eds.), Research and Practice in Multiple Criteria Decision Making. XVII, 553 pages. 2000.